FRONTIERS OF POLYMER RESEARCH

FRONTIERS OF POLYMER RESEARCH

Edited by
Paras N. Prasad
State University of New York at Buffalo
Buffalo, New York

and
Jai K. Nigam
Shriram Institute for Industrial Research
Delhi, India

SPRINGER SCIENCE+BUSINESS MEDIA, LLC

Library of Congress Cataloging-in-Publication Data

Frontiers of polymer research / edited by Paras N. Prasad and Jai K.
Nigam.
        p.   cm.
    "Proceedings of ... [the First] International Conference on
Frontiers of Polymer Research, held January 20-25, 1991, in New
Delhi, India"--T.p. verso.
    Includes bibliographical references and index.
    ISBN 978-0-306-44096-0   ISBN 978-1-4615-3856-1 (eBook)
    DOI 10.1007/978-1-4615-3856-1
    1. Polymers--Congresses. I. Prasad, Paras N. II. Nigam, J. K.
(Jai Krishna), 1945-   . III. International Conference on Frontiers
of Polymer Research (1st : 1991 : New Delhi, India)
    TP1081.F76   1992
    668.9--dc20   91-40013
                                                         CIP

Proceedings of an international conference on Frontiers of Polymer Research,
held January 20–25, 1991, in New Delhi, India

ISBN 978-0-306-44096-0

© 1991 Springer Science+Business Media New York
Originally published by Plenum Press, New York in 1991

PREFACE

This book represents the proceedings of the First International
Conference on Frontiers of Polymer Research held in New Delhi, India during
January 20-25, 1991. Polymers have usually been perceived as substances to
be used in insulations, coatings, fabrics, and structural materials.
Defying this classical view, polymers are emerging as a new class of
materials with potential applications in many new technologies. They also
offer challenging opportunities for fundamental research. Recognizing a
tremendous growth in world wide interest in polymer research and
technology, a truly global "1st International Conference on Frontiers of
Polymer Research" was organized by P. N. Prasad (SUNY at Buffalo), F. E.
Karasz (University of Massachusetts) and J. K. Nigam (Shriram Institute for
Industrial Research, India). The 225 participants represented 25 countries
and a wide variety of academic, industrial and government groups. The
conference was inaugurated by the Prime Minister of India, Mr. Chandra
Shekhar and had a high level media coverage.

The focus of the conference was on three frontier areas of polymer
research: (i) Polymers for photonics, where nonlinear optical properties
of polymers show great promise, (ii) Polymers for electronics, where new
conduction mechanisms and photophysics have generated considerable
enthusiasm and (iii) High performance polymers as new advanced polymers
have exhibited exceptionally high mechanical strength coupled with light
weight.

The conference emphasized the cross-fertilization of these areas,
which traditionally have been covered in separate scientific meetings. A
unique feature of the conference was the session on Emerging New
Technologies, in which selected high-level industrial representatives from
many countries were invited to present their views on the applications of
polymeric materials in new future technologies.

This book presents a compendium of invited, review, contributed and
poster papers presented at the conference. The approach presented in the
book is multidisciplinary, containing a blend of chemistry, physics,
materials processing and technology applications. It is our belief that
this book will bring into focus new directions of research and development
in the field of polymeric materials as well as emphasize important
interfacing of polymeric materials with Emerging New Technologies. Review
papers will assist in the cross-fertilization of different areas, where as
the research papers will serve as useful reference materials summarizing
the current status and projecting future directions.

We wish to acknowledge various agencies which provided support for
the conference. The international organizations were: Air Force Office of
Scientific Research (USA), National Science Foundation (USA), International
Center for Theoretical Physics, Italy (UNESCO), European Office of

Aerospace Research & Development (UK), Army Research Office (USA), Foster-Miller, Inc. (USA), Hoechst-Celanese (USA), Allied-Signal (USA), Unitika, Ltd. (Japan), Lucky, Ltd. (Korea), Loctite (Ireland), I. C. I. (UK), and B. P. America. The national (Indian) organizations were: Department of Electronics (Government of India), Indian Petrochemicals Corporation, Ltd., Shriram Rayons (Kota), Uniplas India, Ltd., Shriram Fertilizers & Chemicals (Kota), LML Fibers, Ltd., National Organic Chemical Industries, Ltd., Polychem Ltd., Pearl Polymers, Ltd., Shivathene Linopack, Ltd., Synthetics & Chemicals, Ltd., Nulab Equipment Company Pvt., Ltd., Spectronic Instruments (I) Pvt., Ltd., Techno Industruments, Nucon Engineers/AIMIL, Nuchem Plastics, Ltd., Gujarat Binil Chemicals, Ltd., and Automatic Electric, Ltd.

Finally, we all want to dedicate this book to the memory of Dr. Donald R. Ulrich of Air Force Office of Scientific Research whose sudden tragic death saddened all of us. Dr. Ulrich's guidance as a member of the committee for the conference was very pivotal. We have lost a valuable colleague who was a strong supporter of polymer research and emphasized strong international collaboration.

Paras N. Prasad
Buffalo, USA

J. K. Nigam
Delhi, India

CONTENTS

_____

*From the Emerging New Technologies Session
+From the Review Session

POLYMERS FOR ELECTRONICS

HIGH PERFORMANCE POLYMERS

INTRODUCTION

Paras N. Prasad
Photonics Research Laboratory
Department of Chemistry
State University of New York at Buffalo
Buffalo, NY 14214

EMERGENCE OF POLYMERS AS MULTIFUNCTIONAL MATERIALS

Polymers are emerging as an important class of materials which offer challenging opportunities for both fundamental research and new technological applications. An important advantage of polymers lies in their structural flexibility both at the molecular and bulk levels. Thus, one can use molecular engineering to tailor the structural features and functionality by chemical modification of the polymer backbone or by building side chains. By chemical modification and physical processing one can also control the conformation of a given polymer. These approaches provide exciting opportunities to impart multifunctional characteristics to a polymer in that the same polymeric material can perform more than one function. In addition, polymers in the bulk form can be fabricated into various structures suitable for specialized applications. They can be used as fibers or films or channeled waveguides for exploring new nonlinear optical processes.

There has been tremendous growth of interest in the field of polymers both at universities and in industries. Polymer science also offers great opportunities for fundamental research. The photophysics and chemistry of polymers have shown intriguing prospects. New conduction mechanisms involving solitons, polarons and bipolarons have been proposed. Ultrafast photoinduced conformation deformations have been reported for one dimensional conjugated polymers. Conjugated polymers have also revealed the largest nonresonant nonlinear optical effects. The materials science of polymer blends, composites, and the structural property of polymers have shown phenomenal new developments. Polymers have emerged from traditional usage as passive structural materials to add new dimensions such as conducting polymers and photonic polymers. This realization has brought a resurgence of interest to the field of polymers from the academic community worldwide. At this stage of heightened enthusiasm for polymer research and emerging novel applications, we consider bringing together the frontier areas of polymer research highly timely.

SCOPE OF THE BOOK

We focus on the three areas which we consider are currently at the forefront of polymer research. They are described below.

Frontiers of Polymer Research, Edited by P.N. Prasad and
J.K. Nigam, Plenum Press, New York, 1991

(i) Polymers for Photonics: Photonics is the newly emerging field in which photons instead of electrons will be used to acquire, transmit, process, and store information. Photonics will be used in broad band optical signal processing and optical computing, telecommunications and image analysis. The advantages of photonics are a tremendous gain in the speed of transmission and increase of the bandwidth, compatibility with fiber optics link, and non-suceptibility to electrical and magnetic interference. Polymers have emerged as a dominant class of photonic materials because they exhibit very interesting photophysics and the largest non-resonant nonlinear optical effects with fastest (~10^{-14} sec) response times derived from π-electrons. The nonlinear optical effects provide methods to produce functions of optical switching, optical frequency conversion, optical logic and optical memory operations. The nonlinear optical devices will use integrated optical circuits involving optical waveguides. The topics included in the proceeding cover chemical synthesis of advanced polymers for photonics, linear and nonlinear optical properties studied by using ultrafast laser pulses, photophysics of polymers and biopolymers, polymers for telecommunication, polymer waveguides for integrated optics, optical switching and optical bistability in polymeric materials.

(ii) Polymers for Electronics: Polymers are being increasingly recognized for their applications in electronics as active components and not merely as passive insulation. New conduction mechanisms involving conformational deformations such as solitons, polarons and bipolarons have been proposed for conjugated linear polymers. Recent advances have shown that in polyacetylenes, the electrical conductivity on weight basis can achieve a value higher than that provided by copper. Important battery applications of polymers have been demonstrated. Photoconductivity of polymers finds important applications in xerography. Polymers also show very interesting piezoelectric and ferroelectric behavior. The topics in the proceedings cover conducting polymers, dielectric polymers, piezoelectric and ferroelectric polymers, photoconductivity and photovoltaic effects in polymers.

(iii) High performance polymers: New polymers with exceptionally high mechanical strength and environmental stability have been produced. A very promising avenue to improve on the performance has been through composites and blends. The topics covered are physics and chemistry of high performance polymers, composites and blends, structural and spectroscopic characterization, mechanical properties, elasticity, thermal behavior, structural relaxation, and radiation damage.

COMMONALITY

The three areas covered in the proceedings have traditionally not been considered together. There is a certain commonality which again emphasizes the mutlifunctional nature of polymers. A class of linear conjugated polymers have shown that they are excellent nonlinear optical materials; in doped states they can be good electronic conductors; and they also have large mechanical strength because of the rigid linear chain structure. One example is poly(p-phenylenevinylene). In uniaxially stretched polymers, the third-order nonlinear optical properties, the electronic conductivity, and the mechanical strength (elastic constants) all have the largest component of their respective tensors along the draw direction (i.e., the direction in which the polymer chains preferentially orient). In addition, the mechanical strength, the damage threshold, and the environmental stability are very important considerations for photonics and electronics. This book is unique in its approach that it brings together scientists and engineers from these three areas to cross-fertilize their ideas and expertise.

2

INTERFACING WITH EMERGING NEW TECHNOLOGIES

The development of new technology is crucially dependent on the availability of novel advanced materials which would fulfill the technical requirements. Therefore, interfacing of advanced materials with new technologies is of great significance. The conference focused on this important topic by bringing high level industrial representatives to present their industrial perspectives. This goal was accomplished through the session on "Emerging New Technologies", in which selected high-level industrial representatives from many countries were invited to present their views on the applications of polymeric materials in new future technologies. Their contributions are also included in this proceeding.

AN OVERVIEW OF THE DEVELOPMENT OF
PLASTIC OPTICAL FIBERS

Fumio Ide [*] and Akira Hasegawa [**]

* Mitsubishi Rayon Co.Ltd.
    Executive managing director,Tokyo, Japan

** Mitsubishi Rayon Co.Ltd.
    Deputy general manager of central research labo,Hiroshima,Japan

Introduction

Plastic optical fibers are already in wide use in a large number of fields and both the range of applications and the market are expected to continue to expand steadily.

Mitsubishi Rayon introduced Super ESKA plastic optical fiber with an attenuation of approximately 300 dB/Km to the market in Japan in 1976 and also the lower transmission loss type of ESKA EXTRA in 1983 as well as the several improved new grades.

This paper gives an overview of the field including the advantages of plastic optical fibers, the structure and materials for plastic optical fibers, the progress that has been made in lowering the transmission loss of the fibers, the improvement of the heat resistant plastic optical fibers, the development of multi fiber endoscope and several other items which have recently been developed as well as the scope of the development of plastic optical fiber in the future.

Structure of POF and its distinctive feature

Plastic optical fibers (so called POF) have a plastic core and cladding, and all the commercially available ones are of the multi-mode step-index type. The graded-index type is not on the market yet.

The diameter of POF is roughly 3 to 10 times larger than that of sillica ones,but, even though they are larger,they are much more flexible and much tougher, which is one of the advantages of plastic. Also, the cladding layer is far thinner than the core diameter. This is another feature which distinguishes plastic from silica fibers.

Frontiers of Polymer Research, Edited by P.N. Prasad and
J.K. Nigam, Plenum Press, New York, 1991

POF, which can be used to transmit optical signals along plastic fibers, are in many ways superior to their counterpart—glass.

They are easier to handle because of their good ductility and light weight, easier to splice together and to light sources because of their large core diameter and high numerical aperture (NA), and they are cheap.

One of the main properties distinguish POF from silica fiber is that amorphous polymers sustain much higher strains before suffering permanent damage in contrast to the brittle behaviour of silica. These properties permit fibers of large diameter to be bent to tight radii of curvature without damage. Presently available fibers constructed from PMMA are available in and around 1mm diameter which can sustain bend radii of less than 10mm.

Table—1 shows the comparison of the properties of POF with that of silica optical fibers.

Table 1. Plastic OF vs. Silica OF.

	POF	Silica OF
Trans. loss (dB/㎞)	125(650nm)	0. 2~(1300nm)
Bandwidth	~10MHz・㎞	10000MH・㎞
N. A	0. 3 ~0. 6	0. 1~0. 25
Available wavelength	Visible	Visible~ Infrared
Mechanical properties	Better Flexibility	Brittle
Fiber diameter (Bulk Fiber)	100~3000μ	100~500μ
Processability	Easier handling	Special tooling
Heat resist (dry)	<85℃	150 ℃
Chemical resist.	poor	good
Specific gravity	1. 2 ~	2. 4 ~
Cost (Bulk, System)	Cheap	Expensive

Plastic fibers are classified into two main categories depending on the core material (Table 2). Fibers with a PMMA core are the most readily available in Japan and have about 90% share of the market for plastic fibers. But recently, other materials have been developed to improve the heat resistance. They include polycarbonate (PC) and the thermosetting polymers, such as polysiloxane, both of which are on the market.

Table 2. Types of Plastic Optical Fibers.

Core		Cladding	Main Application	Maker (in Japan)
Thermo-plastic	PMMA	Fluoro-polymer	Data-trans.,Sensor Lightguide Sign	Mitsubishi Rayon Asahi Chemical Toray
	PSt	PMMA	Display	
	PC	F-polymer (Polyolefine)	Heat resist. Data-trans., Sensor	Mitsubishi Rayon Fujitsu, Teijin Idemitsu, Asahi
Thermo-setting	Polysiloxane	F-polymer	Heat resist. sensor	Sumitomo Electric
	Crosslinked structure	F-polymer	Heat resist. sensor	Hitachi

The core materials for POF are usually amorphous homopolymers, because crystalline polymers, graft copolymers, blend polymers and random copolymers tend to give the fluctuations in density and refractive index distribution, which cause the increase in the transmission loss. At present, there have been used methylmethacrylate polymer (PMMA), polystyrene (PS), polycarbonate (PC) and polysiloxane for the core of the commercially available POF.

The most important characteristics in the choice of a polymer material for POF is that of attenuation in the visible or near infrared region for light transmission. Several important properties of the typical optical plastics including PMMA, PSt and PC are shown in Table 3. The plastic of choice up till now has been PMMA which shows an exceptionally high transparency in the visible region.

Table 3. Properties of Optical Plastics.

	Refractive Index N_D	Abbe No. V_D	Trans-mittance (%)	Working temp. (°C)	Specific gravity (g/cc)	Thermal expansion (m/m °C)$\times 10^{-6}$
P S t	1.591	30.9	90	70	1.06	80
P C	1.586	30.3	89	120	1.20	70
C R – 3 9	1.498	57.8	92	70	1.32	117
P M M A	1.492	57.2	92	70~100	1.19	63
T P X *	1.466	56.4	—	180	0.87	117

* : poly-4 methl-pentene-1

7

Lowering the transmission loss

The most important challenge to the forcing up POF at the application in long distances is how to bring close to the attenuation of POF to its loss limit.

Figure 1 shows the transmission loss spectra of PMMA core current type and that of improved version. Mitsubishi Rayon succeeded to introduce this improved version in the market whose transmission loss is not exceeding 125 dB/Km at 650nm wavelength.

Taking into consideration with the developing stage of industrial manufactur— ing, the attenuation obtained has reached to the well enough level, comparing with the loss limit of 106 dB/Km calculated by Dr. Kaino in the experimental stage. [1] The progress that has been made in lowering the transmission loss of a PMMA core type of fiber, which has a nice balance between performance and cost and the most widely used, is refered to in this chapter.

Transmission loss is one of the most important measures of the performance of optical fibers and Table 4 shows the factors which cause it. The inherent absorption and scattering arise basically from the molecular structure of the core polymer. They are much more serious in plastic fiber than in silica fiber.

In material with C–H bonds, the higher harmonics of the vibration of this bond have effects that are felt all the way up to the visible region.

Transmission loss of PMMA core POF is composed of inherent absorptions due to the carbon–hydrogen (C–H) vibrations and the electronic transition, the int— rinsic Rayleigh scattering due to imperfections in the waveguide structure such as core diameter fluctuations as well as the particulates and other impurities from the environment.

Table 4. Factors Causing Transmission Loss.

Inherent factors	Absorption	● Higher harmonics of C-H vibration
		○ Electronic transition
	Scattering	○ Rayleigh scattering
External factors	Absorption	○ Transition metals
		○ Organic contaminants
		○ OH group
	Scattering	● Dust and Micro void
		○ Disturbance at core/clad. interface
		○ Fluctuation of core diameter
		○ Microbending
		○ Birefringence due to orientation

Absorption loss

One of the intrinsic losses in absorption is that due to the vibrational structure of the fundamental bond stretching frequencies. These are absorptions due to C–H, C–O and other molecular bonds found in the polymers.

The fundamental absorptions of the vibration spectra are found in the region 1.8 to 3μm but overtones of these frequencies occur in the visible regions.

The major absorption peaks in this region are due to overtones of the fundamental CH vibration spectra. These transmission frequencies may be shifted by replacing the hydrogen by other atoms such as deuterium which lowers the oscillation frequency and causes it to shift further into the infrared.

Schleinitz already has reported ductile plastic optical fibers with improved visible and near infrared transmission using poly (methylmethacrylate–d8) as a core material. [2]

The replacement of hydrogen in the PMMA by deuterium brings about the reduction of C–H vibrational absorption in the infrared wavelength region and its overtones in the near infrared to visible region. The reduction of the absorption loss due to C–H vibration brings about a reduction of the total absorption of POF's especially in the visible to near infrared region.

Figure 2 shows the spectral attenuation of deuterated PMMA POF, compared the absorption of PMMA with partially–deuterated PMMA–d5 and more fully deuterated PMMA–d8. The PMMA–d8 shows a much lower absorption loss in the wavelength range from 600 to 800 nm.

Dr. Kaino (NTT) showed the loss limits of PMMA and PMMA–d8 core POF. (Table–5) The lowest attenuation of 55 dB/Km at 568 nm as well as 126 dB/Km at 650 nm wavelength using PMMA as a core, and 20 dB/Km at 680 nm wavelength using PMMA–d8 as a core are obtained. From the mentioned above, the loss limit of PMMA core POF is estimated to be 106 dB/Km at 650nm and that of PMMA–d8 core POF is to be 10 dB/Km at 680 nm.

These low loss POF's is expected to permit the use of cheaper light sources such as a display grade GaP light–emitting diode (LED) or GaAℓAs LED.

Table 5. Loss Factors and Loss Limits of PMMA and PMMa-d8 Core POF.

Core materials	PMMA			P(MMA–d8)		
Wave length (nm) \ loss factor	516	568	650	680	780	850
Total Loss	57	55	126	20	25	50
Absorption	11	17	96	1.6	9.7	36
Rayleigh scattering	26	18	10	7.5	4.3	3.1
Structural imperfection	20			11		
loss limit	37	35	106	9.1	14	39

$$(dB \cdot km^{-1})$$

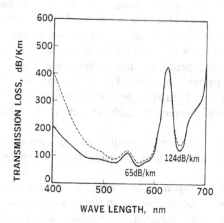

WAVE LENGTH, nm

Figure 1. Transmission Lost Spectra of PMMA Current Type and Improved Version.

--- Current Type

— Improved Version

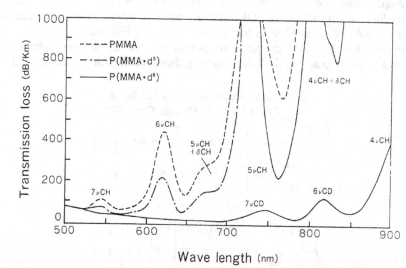

Wave length (nm)

Figure 2. Spectral Attenuation of Deuterated PMMA POF.

They are suitable to apply to short distance optical data links of about 160m in length at a rate up to 10 MHz using PMMA or PS as a core and 400 m using PMMA–d5 as a core. [1]

However, there are several problems with the deuterated PMMA core POF. One of the problems is the water absorption in a polymer which increases the attenuation loss by OH vibrational absorption. [3]

Another problem is the monomer cost of deuterated MMA which makes POFs so expensive and they are not yet commercially available.

Scattering loss

The transmission loss caused by intrinsic scattering is mainly composed of Rayleigh scattering which is indicated as the sum of isotropic fluctuations in density and that of anisotropic fluctuations. Table 6 shows the isotropic fluctuations in density for PSt and PMMA. [3] In case of PMMA, the isotropic fluctuations and that of anisotropic gives almost the same value.

On the contrary, anisotropic fluctuations of PSt gives much larger than that of isotropic fluctuations. In other words, PMMA can be the polymer whose scattering loss is the lowest caused by its isotropic fluctuation structure.

Table 6. Isotropic Fluctuations in Density for PSt and PMMA.

Polymer	T_d	T_d^{iso} (Exp.)	T_d^{iso} (cal.)
PSt	55	20.6	20.6
PMMA	13	12.6	11.6

The formula for Rayleigh scattering is as following.[4]

$$T_d^{iso} = \frac{8}{3} \pi^3 \frac{KT}{\lambda} \beta \left[\frac{(n_b^2 - 1)\,(n_b^1 + 2)}{3} \right]$$

where
- T : the absolute temperature
- λ : wavelength
- n_p : average refractive index
- K : Boltzmanns constant
- β : the isothermal compressibility

Structual consideration in lowering the transmission loss

As already mentioned above, transmission frequencies may be shifted by replacing the hydrogen by deuterium which considerably lowers the absorption. Other molecules which have higher molecular mass change the spectra radically and replacement of hydrogen by fluorine or chlorine would shift the absorption regions out into the middle infrared.

Table 7 shows the transmission loss limit of polyfluoroalkylmethacrylate. [5]

Table 7. Transmission Loss Limit of Polyfluoroalkylmethacrylate*.

Wave length (nm)	α_V (dB/km)	α_R (dB/km)	Loss limit (dB/km)
516	6.2	13.8	20.0
568	9.5	9.5	19.0
650	52.7	5.5	58.2

*) :1, 1, 1, 2, 3, 3-hexafluorobutylmethacylate

α_V :Vibration absorption

α_R :Rayleigh scattering

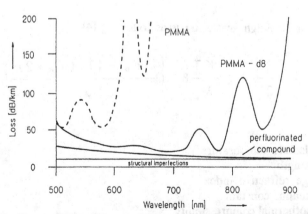

Figure 3. Comparison of Spectral Loss of PMMA, Deuterated-PMMA and Perfluorinated Compunds for Fibers.

Workers at Hoechst [6] have examined a number of perfluorinated aliphatic compounds which have high transparencies out to beyond 1000 nm. The calculated loss spectra of this fluorinated model compound is compared to deuterated PMMA in Figure 3. At the present time, almost all plastics are considerably inferior to PMMA in their optical transparency since this has been developed in a pure state over a number of years. However, the example shown here of fluorinated compounds show that, in the future, if a considerable effort is put into the research of pure materials and methods of fabrication that do not introduce additional losses, the promise for producing tranparent optical fibers of very low attenuation in the near infrared region hold considerable promise.

Du pont has succeeded to develop new type of the amorphous fluoropolymer Teflon AF which has many unique properties in the following ways; a true amorphous fluoropolymer, very low refractive index around 1.30, higher coefficient of friction, excellent mechanical and physical properties at end use temperature up to 300 C, excellent light transmission from UV through a good portion of IR, lowest dielectric constant and limited solubility. [9]

In spite of its highly expensive cost and narrow acceptance angle, Teflon AF is very interesting material for POF.

Table 8 shows the characteristics for Teflon AF [9].

Table 8. Comparison of Performance Characteristics for TEFLON AF and TEFLON PFA.

	TEFLON AF	TEFLON PFA
Morphology	Amorphous	Semicrystalline
Optical clarity	Clear:>95%	Translucent to opaque
Upper use temperature	285°C	260°C
Thermal stability	360°C	360°C
Thermal expansion(linear)	45 ppm/°C	150 ppm/°C
Water absorption	No	No
Weatherability	Outstabding	Outstabding
Flame resistant LOI	95%	95%
Tensile modulus	950~2150 MPa	271~338 MPa
Creep resistance	Good	Low
Solubility	Selected Solvents	No
Resistance to chemical attack	Excellent	Excellent
Surface free energy	Low	Low
Refractive index	1.29~1.31	1.34~1.35
Dielectric constant	1.89~1.93	2.1

Extrinsic factor

External factors also affect the absorption and scattering, which are subject to the influence of the method of polymerization and the way of spinning. In particular the elimination of contaminants, such as dust, is the most important way to improve the method of manufacture. [7]

The bulk method gives a lower transmission loss caused by the absence of the polymerization medium or dispersant.

Figure 4 shows the influence of the spinning method on the transmission loss.

The non–direct method, which means that the fiber is spun from pellets of polymer made by bulk polymerization, and the direct method in which means spinning and the bulk polymerization is carried out in a closed process. In the direct method, there is no chance for dust to enter, and the lower transmission loss can be obtained. By improving the process, the loss has reached to 124dB/Km at 650nm, 65dB/Km at 570nm and 200dB/Km at 400nm in commercially available PMMA fiber.

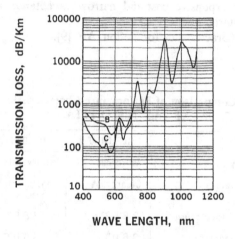

B : Non-direct spinning

C : Direct spinning

Figure 4. Influence of Spinning Method on Transmission Loss.

Figure 5 gives the influence of the orientation on birefringence for PMMA core POF. Various degrees of birefringence can be obtained by changing the drawing rate. The transmission loss of POF increases in proportion to the fiber orientation. It is also important to consider the effect of drawing conditions on the properties of POF.

Cladding material also affects the transmission loss.

When a light–emitting diode is used as a light source for data transmission, the optical properties of the cladding layer are also important. Table 9 shows the influence of cladding material on transmission loss under the condition of N.A as 0.10 and 0.65. In case of using 0.65 NA light source, some of the light rays are reflected more often at the core/cladding interface, and so they are more strongly affected by the scattering at the cladding layer.

Table 9. Influence of Cladding Material on Transmission Loss.

Optical Properties of Claddings			Transmission Loss		(dB/km)
			$\lambda=650$ nm (spectrometer)		$\lambda=664$ nm ※ (red LED)
Appearance		N_D	N.A.	(light source)	
			0.1	0.65	0.65
1	Slightly hazy	1.405	133	199	250
2	Transparent	1.417	124	136	185
	$\triangle = 1-2$		$\triangle = 9$	$\triangle = 63$	$\triangle = 65$

※　Toshiba TOTU 170A

Figure 6 gives the historial progress in reducing the transmission loss for PMMA core POF.

In 1965, Du pont in the USA developed CROFON the first fiber with a PMMA core. And for the last 10 years and more, Mitsubishi Rayon have been active in trying to reduce the loss. At present, the loss has nearly reached to the theoretical limit.

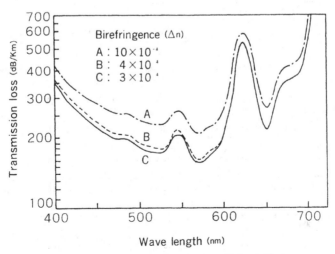

Figure 5. Birefringence Due to Orientation.

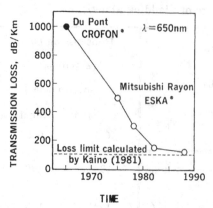

TIME

Figure 6. Progress in Reducing Transmission Loss (POF with PMMA Core).

Development of heat resistant POF

To improve the heat resistance is also extremely important target for POF. Mitsubishi Rayon has taken the lead in developing the high heat resistant POF and succeeded in manufacturing of ESKA–DH (115 C type) as well as ESKA–FH(125 C type) in the first.

Since then many candidates have vigorously tried to develop the heat resistant POF. Table 10 gives the update of the heat resistant POF developed including the newly developed ESKA–PH.

Table 10. Update of Heat Resistant POF.

	Core	Clad	Transmission loss(dB/m)	Heat Shrinkage	Heat resist(°C)	Accept. angle
MRC FH	P C	Fluoro polymer	650	0.2(125°C)	125	
PH	P C	Teflon-AF	800~1000	0% (135°C)	135	128°
Fujitsu	P C	TPX	800	—	125	78°
Teijin Chemical	P C	Fluoro-PC	870~950	0.33% (130°C)	120	62°
Idemitsu Petro chemi.	P C	Fluoro-PC TPX	560	0.80% (135°C)	120	78°
Asahi Chemical	P C	F-vinyliden polymer	700	—	120	
Bayer	P C	Urethan acrylate	1200	—	125	
Sumitomo Electric	Polysiloxane	Teflon FEP	240		150	64°
Hitachi Wire	Crosslinked PMMA	Teflon FEP	1500		170	81°
Hitachi & Shinetsu.		Teflon FEP	800		180	

One of the principle of giving the heat resistance to POF is how to release the influence of molecular orientation caused by fiber drawing.

The heat stability of a PMMA core POF is attributable to decrease its influence of drawing effect as shown in Figure 7. [8]

Another way of giving high heat resistance to POF is to use a high Tg material for the core. Giving a structure of crosslinking is also the possible way of high heat POF.

For the last 10 or 15 years, we have been trying to improve the heat resistance of plastic optical fibers to expand the field of applications and Figure 8 shows the progress we have made. ESKA PH is the heat resistant fiber that we finished developing last summer.

The core of ESKA PH is made of PC and the cladding is made of Teflon AF co-developed with Du Pont. Teflon AF is a new heat-resistant amorphous fluoropolymer that Du Pont announced to the world in July 1990. [9]

ESKA PH has the loss of 1dB/m at 770nm which gives us a long enough transmission distance for automobiles. And at 135 C in long-term heat

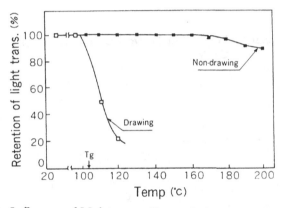

Figure 7. Influence of Moisture on Transmission Loss of PMMA-d8.

durability is hardly any loss after 2000 hours where there is no shrinkage of the fiber at all (Figure 9). The fiber has a numerical aperture of 0.9. This is so large that the fiber can be wound or bent into a much tighter configuration, so this makes ESKA PH much more suitable for the confined spaces of automobiles and other equipment (Figure 10). Moreover, there are no influence of water on the transmission loss at the windows of 660nm and 770nm, high-power light-emitting diodes with an output at 770nm can effectivly be used (Figure 11).

ESKA PH not only has a much higher heat resistance than previous fibers, it is also easier to handle and has excellent resistance to humidity. We believe that this fiber satisfies the strict requirements of flexibility, high performance, and low cost for automobile use.

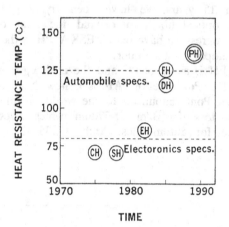

**Figure 8. Progress in Thermal Durability of ESKA®
Cables.**

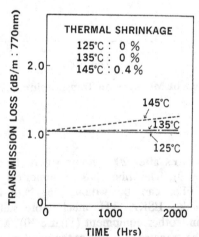

Figure 9. Long-term Heat Durability of PH 4001.

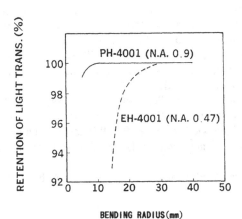

Figure 10. Bending Radius vs. Retention of Light Trans.
of PH4001 and EH 4001.

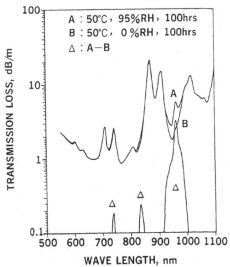

Figure 11. Influence of Water on Transmission Loss of PH-4001.

Bandwidth and related study

Estimation

     With the recent advances in laser diodes operating in the visible region, it is becoming increasingly important to have detailed information on the transmission bandwidth of POF's so that they can be used effectively as data links. A few studies on this subject have been done, but the launching conditions have not yet been adequately examined. The measurement of the −3dB transmission bandwidth under different launching conditions was undertaken, and a large dependence was observed.

The transmission bandwidth of an POF strongly depends on the launching conditions, and the characteristics of the detector must be taken into account in the measurements. A bandwidth of over 190 MHz can be achieved at a length of 100m, for a small−NA launch with a laser diode operating in the visible region. [10]

Fig 12 gives the −3dB bandwidth measured at launch NA being 0.10 and 0.65 respectively.

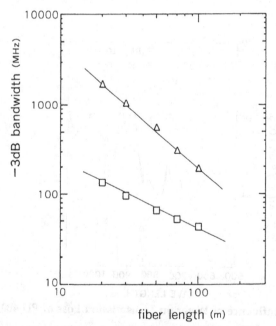

Figure 12. -3dB Bandwidth Measured Under Different Launching Conditions.

△ : Launch NA=0.10
□ : Launch NA=0.65

The gradient—index POF

One of the most promising try to obtain the higher data rate POF will be developing the graded index type POF about which Dr.Otuka, Dr.Koike have studied.

A light focusing plastic rod could be successfully fabricated by two different processes, two—step polymerization and photo copolymerization. By the latter process the rod can be readily heat drawn into a light—focusing

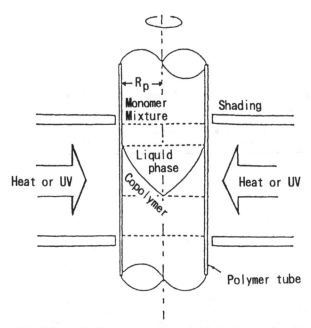

Figure 13. Schematic Representation of Fabricating a Gradient-Index Distribution in a Rod.

plastic fiber. The multimode fiber is suitable for optical telecomunications due to the wide bandwidth capability [11]

The attenuation level as a 142.5 dB/Km at 650 nm was reported under ex—perimental stage.

Figure 13 shows the schematic representation of fablicating a gradient—index distribution in a rod.[12]

Development of Multi–Fiber type POF

A multi–fiber type of POF is used for image guides, such as medical endoscopes, industrial bore scopes, automobile viewscopes and so on.

The fiber has a rectangular cross section 0.3✕0.6mm in size. As many as 1500 core fibers, each with a diameter of 10μm are closely packed into a regular array (Figure 14). This type of multi–fiber is very flexible and very difficult to break in contrast to glass fibers. Morever, the resolution is as high as 40 line pairs/mm, which is the same level as that of glass fibers.

To make this, all the cores are spun continuously at the same time together with the cladding of fluoropolymer. This ultra–fine spinning technology enables these image guides to be mass–produced at a lower cost. The better performance and low cost of this type of multi–fiber will open up new fields of application.

Table 11. Comparison of Properties of Multi-fiber.

| | Quartz multi fiber | Multi compornent glass | POF | |
			Mono filament	1500hole multi fiber
unit fiber diameter	5~12 μ	12~30 μ	250 μ~	10~20 μ
resolution	high	high	low	high
aceptance angle	△	○	○	○
transmission loss	○	○	△	△
ductility	△	○	○	○
processability	○	△	△	○
	be broken		excellent flexibility	
applications	industrial imagescope (inspection furnace inside)	medical endoscope (digestive organ urination) Common image scope	common imagescope (inspection sewer inside)	blood vessel endoscope image scope high resolution sensor

In the field of medical endoscope, it is seriously dangerous to be broken inside the blood vessel. This point is the strongest advantage of plastic multifibers against quartz multi–fibers or multi–component glass fibers. The extremely improved new type is now being developed. [13]

Table 11 gives the comparison of the properties of plastic multi–fibers with quartz and multi component glass fibers.

Applications of ESKA products

Table 12 lists the performance and main applications of ESKA family of plastic optical fibers. ESKA fiber spectral attenuations are shown in Figure 15. Table 13 shows to what uses plastic optical fibers are put in the various fields of application. [14]

Fiber optics networks are not affected by electromagnetic interference, even if they are laid parallel to power lines or very close to electric motors. So such networks using both silica and plastic fibers have been installed on all the new bullet trains of Japan Railways. These have been in daily service since October, 1985. This demonstrates the high reliability of plastic optical fibers.

Table 12. ESKA Family.

	Eska	Super Eska	Extra*	D	F	P	Multi -Fiber
Core	PMMA	PMMA	PMMA	PMMA	Eng. plastic	Eng. plastic	PMMA
Cladding	F-polymer	F-polymer	F-polymer	F-polymer	F-polymer	F-polymer	F-polymer
Trans. Loss (dB/Km)	<500	<400	125	200	600	1000	500
N.A. Number	0.5	0.5	0.47	0.54	0.75	0.90	0.5
N.A. Angle	60	60	56	65	97	128	60
Heat resist. (°C)	70	70	85	115	125	135	70
Main applications	Display Inf. panel	Light guide Sensor	Data- trans. Sensor	Heat resist. Data-trans.	Heat resist. Data-trans.	Heat resist. Data-trans.	Image guide

※　Band width : 63.6MHz/100m (Pulse Ar+ laser)

In the automobile applications, many types of POF products have been used. In their role as light guides, they are widely used as light monitors and for panel illumination. In the field of data transmission, they are used as point—point data links in luxurious cars. And in their role as sensors, their use is expected to expand a great deal in the near future. Of course, the heat resistant type of fiber can be used in all these applications.

Cross section

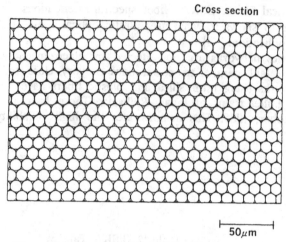

50μm

Figure 14. Photograph of Multi-fiber Type POF.

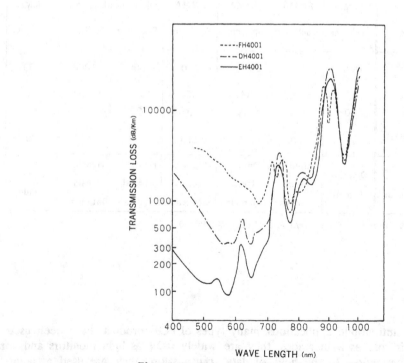

WAVE LENGTH (nm)

Figure 15. Eska Fiber Spectral Attenuation.

The reasons for the extensive application of plastic type fiber in the automobile industry can be found in the following factors:

* Core diameter is large enough to provide low connection loss, easy end termination and workability

* Numerical Aperture is large enough to mitigate the adverse effects of axial displacement and end gaps

* Connectors and related components do not require a high degree of precision

* Resilient against bending stresses

* Light—weight

* Low cost (compared to glass fiber)

In the field of sensors, many types of POF products have been commercialized, including fiber type photo electric switches and the array type sensor—heads arranged for various shapes as desired, and are becoming very popular as products ultilizing the merits of plastics.

Actual application examples in sensor field are as follows: Scanning head with a resolution of 254 points per inch for optical scanner to capture drawings and documents. Circular defect detector used at a bottling plant.

Sensor—head for quantitative detection of liquid level etc. Sensor—head for detection of paper position and light intensity of light source in printers, copying machines and etc.

In the field of data links, multiplexers and LAN, plastic optical fibers are used for short—distance data transmission in relatively low frequency band such as in transportation including automobiles and trains, factory automation including numerical control machines, robots and the systems to connect these machines and robots with each other, office machines and audio—sets.

Plastic optical fibers are already in wide use in a large number of fields, and the range of applications and the market are both expected to continue to expand steadily.

Table 13. Applications of ESKA Products.

Application Fields		Product Concept			
		Light guide illumination	Photo switch sensor	Data link LAN	Image guide
Transportation	Automobile		R & D		R & D
	Train				
	Ship				
Sign	In-door				
	Out-door				
	Traffic Sign				
Equipment	Industrial Machine				R & D
	Office Machine				
	Audio/Appliance				
Communication	Factory				
	Office				
	Home			R & D	

☐ widely in use   ☐ growing   R & D under development

Reference

[1] T. Kaino, et al: Review of the E.C.L.,vol 32, No.3 (1984)

[2] H. M. Schleiniz: Int. Wire & Cable Symp. 25th, 352 (1977)

[3] T. Kaino: Appl. Phys. Letter 48 (12), 24 March (1986)

[4] P. Debye, M. Buech: J.Chem. Phys vol 18, 1423 (1950)

[5] T. Kaino: Japanese J. of Polymer Sci. and Tech. vol. 47 No. 12 (1990)

[6] W. Groh, et al: FOC/LAN 1988, Atlanta, Sept, 1988

[7] F. Ide, H.Terada: " Optical Fiber & Optical Material," edited by High Polymer Society Japan, (1988), P76 78

[8] Y. Ohtsuka: Optoronics " the present situation of the material for POF" P67, Feb. (1985)

[9] P. R. Resnick : presented at 1989 MRS Fall Meeting, Boston, NA, 1989

[10] S. Takahashi (Mitsubishi Rayon, Tokyo reserch labo.): in contribution to Electronics Letters

[11] Y. Koike, Y.Ohtsuka: Applied Optics, vol. 21, No.6 1057 (15 March 1982)

[12] Y. Koike et al: Symp. of Japan Soci. of Synthetic Fiber. 13–39, (1989)

[13] F. Suzuki: The Japan Society of Appl.Physics and Related Societics (The 38th spring Meeting, 1991), in press.

[14] F. Ide, T. Yamamoto: presented at the 1989 MRS Fall meeting, Boston, MA.

MULTIFUNCTIONAL POLYMERS - AN INDUSTRIAL PERSPECTIVE

John P. Riggs, Harris A. Goldberg and James B. Stamatoff

Hoechst Celanese Corporation
Advanced Technology Group
Summit, NJ 07090

ABSTRACT

This paper overviews, from an industrial viewpoint, the developing unusually broad and many-faceted utility of organic polymeric-based materials, including both emerging technology and product developments, in structural and functional polymers. Included is a positioning of the role and significance of polymers, in the general context of materials development, as an increasingly important enabling technology, with specific identification of materials types, emerging markets and growth rates.

A more detailed discussion focuses on functional organic polymers - and in particular on recent efforts to design more than a single functionality into a molecular structure for various applications. This considers combinations of functions such as nonlinear optical response, piezo- and pyro-electric effects, photoconductivity and electrical conductivity. Issues in molecular design, synthesis, materials characterization, figures of merit and macroscopic materials engineering are considered, along with examples of specific applications.

INTRODUCTION

Range of the Influence and Impact of Polymers

Polymeric Materials - i.,e., plastics, in particular those often referred to as commodity thermoplastics - continue to have a broad impact on current industrial technology and the economy, with uses familiar to all of us - clothing, furniture and furnishings, packaging and containers of all sorts and construction and building applications. Polymers are also, as shown in Figure 1, playing a critical role in new advanced materials developments with cross-cutting impact on the growth of a range of new technologies and emerging industries - these include high performance structural polymers (including composite materials) with applications dominant in automotive/transportation and aircraft/aerospace industries, polymeric materials for electrical applications, polymeric-based gas and liquid membrane separations uses, biopolymers, and - the newest area of impact - polymer materials for optical applications, with a range of uses developing in both passive and active devices and components in telecommunications, data communications, information storage and other areas. It is this increasing scope of impact that makes the area of polymeric materials science and engineering so fascinating, so challenging and so rich with opportunity. The emerging area of multifunctional polymers that is the principal focus and theme of this discussion is a particularly powerful example of still further, future opportunities.

POLYMERIC MATERIALS . . .

- **Broad Impact on Current Industrial Technology and Economy**

- **Critical Role in New, Advanced Materials Developments with Cross- Cutting Impact on Growth of New Technologies and Emerging Industries**

High Performance Structural Polymers
- High Strength/High Modulus
- High Temperature
- Matrices for Composites
- Reinforcing Fibers

Polymer Materials for Electrical Applications
- Lithographic Materials
- Packaging Materials
- Encapsulant Materials
- Conductive Polymers

Gas and Liquid Membrane Separation
- Ultra Microfiltration
- Dialysis
- Reverse Osmosis
- Gas Separations

Biopolymers
- Artificial Organs/Prostheses
- Controlled Release
- Targeted Delivery

Polymer Materials for Optical Applications
- Nonlinear Optics
- Polymer Optical Fibers
- Optical Storage
- Photoactive Materials
- Holography

Figure 1. Critical Role of Polymers in New Technologies and Emeging Industries

In Figure 2, the overall role of these advanced polymeric based materials is compared to metals and other inorganics in the total structural and functional arena, showing that over 50% of the high performance materials application area is based on polymers. Total dollar volume associated with these higher performance structural materials is projected to grow from $15-20B in 1988 to $45-50B in 2000, and for the functional materials to grow from $10-15B to $30-35B. In total, almost $40B would - or could - be polymeric-based, with the highest growth areas being in polymer based materials, particularly in the functional area.

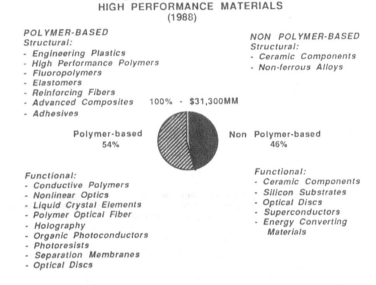

POLYMER-BASED VS. OTHER
HIGH PERFORMANCE MATERIALS
(1988)

POLYMER-BASED
Structural:
- Engineering Plastics
- High Performance Polymers
- Fluoropolymers
- Elastomers
- Reinforcing Fibers
- Advanced Composites
- Adhesives

NON POLYMER-BASED
Structural:
- Ceramic Components
- Non-ferrous Alloys

100% - $31,300MM

Polymer-based
54%

Non Polymer-based
46%

Functional:
- Conductive Polymers
- Nonlinear Optics
- Liquid Crystal Elements
- Polymer Optical Fiber
- Holography
- Organic Photoconductors
- Photoresists
- Separation Membranes
- Optical Discs

Functional:
- Ceramic Components
- Silicon Substrates
- Optical Discs
- Superconductors
- Energy Converting
  Materials

Figure 2. The Impact of Polymers as High-Performance Materials. Source: HCC; Various Market Forecasts

Opportunity Positioning - Structural and Functional Polymers

In Figure 3 and 4, a comparison is made of structural and functional advanced materials with respect to relative stage of technology and business development for various specific materials types. It is clear from this that most structural advanced materials are in the take-off to mid-life stage of development (more towards the latter), while the functional materials cover a broader range of development stages and with more early stage opportunities for portfolio renewal than in the structural area. Note that multifunctional polymers are in an early concept stage. It is also important to recognize that the functional material opportunity is not so much one of materials per se, but more one of the ability to develop devices, components or other value-added products that make innovative use of unique materials characteristics.

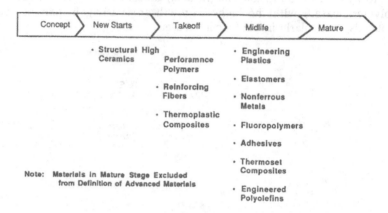

**Most Structural Advanced Materials Are in the Takeoff
to Midlife Stage of Development**

**Structural Advanced Materials
By Stage of Development**

| Concept | New Starts | Takeoff | Midlife | Mature |

- Structural High Ceramics
  - Performance Polymers
  - Reinforcing Fibers
  - Thermoplastic Composites
- Engineering Plastics
- Elastomers
- Nonferrous Metals
- Fluoropolymers
- Adhesives
- Thermoset Composites
- Engineered Polyolefins

Note: Materials in Mature Stage Excluded from Definition of Advanced Materials

Figure 3. Structural Materials Position in Development Cycle

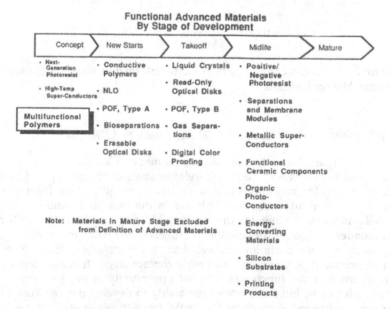

**Functional Advanced Materials Cover a Broad Range of Development
Stages, but with More Early Stage Opportunities for Portfolio Renewal
Than in the Structural Area**

**Functional Advanced Materials
By Stage of Development**

| Concept | New Starts | Takeoff | Midlife | Mature |

- Next-Generation Photoresist
- High-Temp Super-Conductors
- Multifunctional Polymers

- Conductive Polymers
- NLO
- POF, Type A
- Bioseparations
- Erasable Optical Disks

- Liquid Crystals
- Read-Only Optical Disks
- POF, Type B
- Gas Separations
- Digital Color Proofing

- Positive/ Negative Photoresist
- Separations and Membrane Modules
- Metallic Super-Conductors
- Functional Ceramic Components
- Organic Photo-Conductors
- Energy-Converting Materials
- Silicon Substrates
- Printing Products

Note: Materials in Mature Stage Excluded from Definition of Advanced Materials

Figure 4. Functional Materials Position in Development Cycle

Functional Materials Markets Are Also Large and Growing Rapidly.

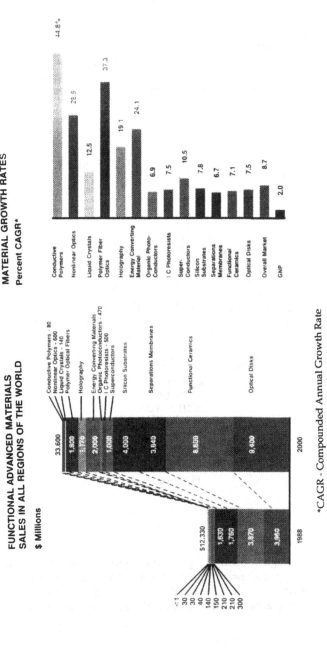

*CAGR - Compounded Annual Growth Rate

Source: HCC; various market forecast

Figure 5. Functional Materials: Market and Growth Rates

31

Some specific estimates are given in Figure 5 for functional material products growth rates for the period 1988 to 2000. This provides further appreciation for the total scope and impact of the area and for the truly explosive growth rates projected in some segments. Again, the importance of organic, polymeric materials systems is evident.

Before closing this introductory discussion, it would be a serious omission not to note and emphasize the four elements of materials science and engineering and their interactions - as shown schematically in Figure 6 (and eloquently developed in a recent U.S. National Research Council Study)*. The control of the interactions among the four areas of synthesis/

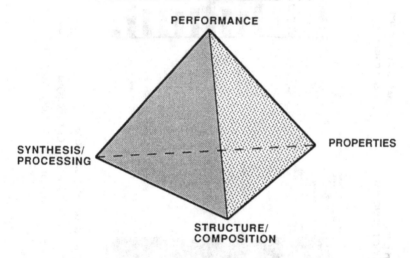

THE FOUR ELEMENTS OF MATERIALS SCIENCE AND ENGINEERING . . .

PERFORMANCE

SYNTHESIS/
PROCESSING

PROPERTIES

STRUCTURE/
COMPOSITION

. . . Particularly Relevant to Polymers

Figure 6

processing, structure/composition, properties and performance is key to successful technology and product development based on materials - both structural and functional - and is particularly relevant to polymers in general and to the emerging area of multifunctional polymers discussed in more depth below.

---

*   Materials Science and Engineering for the 1990s, National Academy Press, Washington, D.C., 1989.

MULTIFUNCTIONAL POLYMERS (MACROMOLECULES)

Multifunctional Material Science Organization

The development of new multifunctional materials and demonstrating device concepts using these materials is the objective of an international program funded in part by the U. S. Government. The material science program has developed in three steps as indicated in the figure below. We will describe developments for each of the three steps and give one example of a multifunctional device concept.

• **Material Science Developed in Three Steps:**

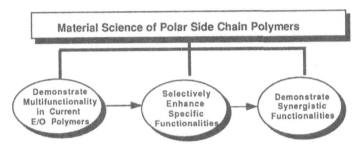

• **Device Application based on new material properties:**

--- **Light Modulated Deformable Mirror**

--- **Photorefractive Holographic Elements**

Figure 7. Program to Develop Multifunctional Polymeric Materials.

The program is a collaborative effort between Hoechst Celanese, Prof. A. Windle, Prof. I. Ward, Prof. D. Haarer, and GEC of England at the Marconi Research Labs. The areas of research are give in Figure 8.

Demonstration of Multifunctionality in Current E/O Polymers

Numerous studies have been performed which showed the advantages of organic molecules in non-linear optics [1]. Through the use of donor and acceptor molecules, polarizable organic units such as stilbenes, biphenyls, or azomethines; may be made polar and electronically noncentric. These molecules possess large second order nonlinearities which may be used to create exceptional electro-optical materials. In addition, the non-linear polarization is intramolecular in origin and is intrinsically very fast (on the order of femtoseconds) which leads to materials with equally rapid response. More importantly from an electro-optical point of view, is the fact that the polarization is predominantly electronic with virtually no nuclear motion. The leads to low dispersion in the polarizability of these materials from optical frequencies to D.C. which in turn results in a low dielectric constant. Low ε results in the ability to create very high speed electro-optical devices [2].

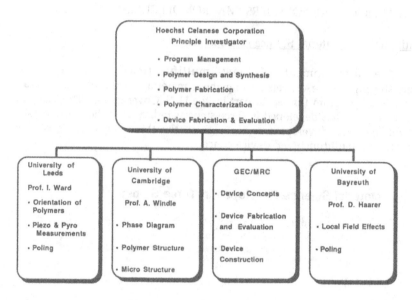

Figure 8. Project Organization.

For a material to possess second order nonlinearity and therefore electro-optic activity, the material must be noncentric. The organic molecules described above have by the very design which imparts molecular noncentrosymmetry, a large ground state dipole moment. As a result many of these molecules crystallize in a centric or nearly centric habit leading to a crystalline material with little or no activity. By attaching the same molecules to polymers, crystalline formation may be avoided. Appropriately designed polymers contain high concentrations of the NLO chromophore and are glassy. By heating these polymers to a temperature above the glass transition temperature, the dipolar NLO chromophores may be poled in a strong electric field imparting a net noncentric alignment to the ensemble of active molecules. By cooling the polymer below the glass transition temperature, the alignment may be maintain in a stable glass. The formalism which describes this process and connects the molecular nonlinearities with the material nonlinearities is given in Figure 9.

An example of one class of NLO polymers is given in Figure 10. This class was developed at Hoechst Celanese and has the NLO chromophore attached to a flexible polymer backbone via an alkyl side chain of variable length [3]. The polymer is a random copolymer with an inactive comonomer unit so that the chromophore content can be adjusted.

Most electro-optical materials have the proper symmetry to display both piezoelectric and pyroelectric activity. As a baseline study, the pyroelectric activity of a side chain NLO type polymer was determined. For these studies, 4,4' oxy-nitrostilbene was used in place of the 4,4' amino-nitrostilbene shown in Figure 10. This structure is shown below in Figure 11.

Fundamentals

Molecular

$$p_i(E_j, E_k, E_l, \ldots) = \alpha_{ij} E_j + \beta_{ijk} E_j E_k + \gamma_{ijkl} E_j E_k E_l + \ldots$$

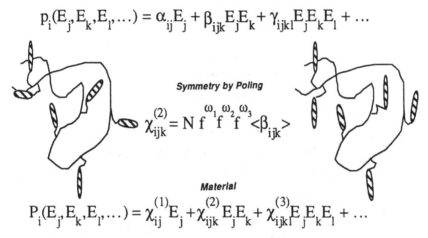

Symmetry by Poling

$$\chi_{ijk}^{(2)} = N f^{\omega_1} f^{\omega_2} f^{\omega_3} \langle \beta_{ijk} \rangle$$

Material

$$P_i(E_j, E_k, E_l, \ldots) = \chi_{ij}^{(1)} E_j + \chi_{ijk}^{(2)} E_j E_k + \chi_{ijkl}^{(3)} E_j E_k E_l + \ldots$$

Figure 9.

Formalism relating Molecular Nonlinearities to
Material Nonlinearities. p_i and P_i are the molecular and material
polarization respectively. α and $\chi^{(1)}$ are the molecular and
material coefficients of linear polarizability. β and $\chi^{(2)}$ are the
non-linear second order coefficients of polarizability.

f indicates local field factors and < > indicates orientational
averaging of the β tensor over all chromophore orientations.

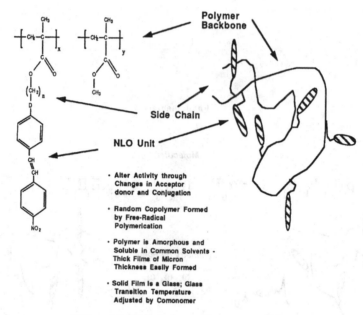

Figure 10. NLO Polymer Structure Design.

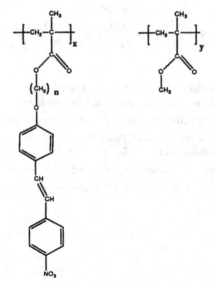

Figure 11.   E/O Polymer Used for Multifunctional Property Studies

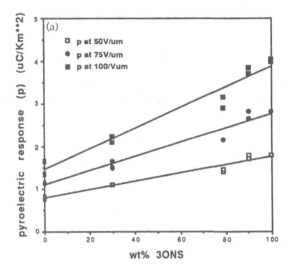

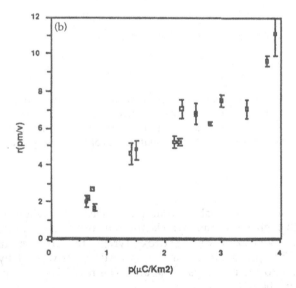

Figure 12. Pyroelectric activity of poled 4,4' oxynitrostilbene methacrylate copolymers as a function of weight % of the chromophore (a) and the electro-optical coefficient as a function of the pyroelectric coefficient (b). Poling voltages are indicated in the first graph.

The pyroelectric and electro-optic coefficients were measured as previously described [4]. Measurements were made for a range of compositions and the pyroelectric activities were correlated with both the weight % of the chromophore and the electro-optical coefficient. These results are given in Figure 12.

Designed Polymers to Selectively Enhance Specific Functionalities

Using the copolymer aspect of these polymers, it is straightforward to design new polymers with new enhanced functionalities. The basic approach is shown in Figure 13.

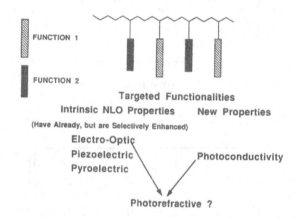

Figure 13. A Multifunctional Polymer Concept: Old functionalities may be selectively enhanced and new functionalities may be added by copolymerizing specifically designed monomers. Several functionalities may interact giving rise to an entirely new material property (e.g. photorefraction).

As an example of the above concept, we selected vinylidene dicyanide as a unit with high piezo and pyroelectric but low electro-optical activity. This was copolymerized with hydroxystyrene to form an amorphous copolymer. This precursor polymer was further derivatized by grafting oxynitrostilbene to the hydroxystyrene unit. The result is a new polymer as shown in Figure 14.

Results of measurements for this polymer are shown in Table 1. Polymers containing only the oxynitrostilbene unit show low pyroelectric activity but substantial electro-optical activity. Poly vinylidene dicyanide-vinylacetate polymers show a high pyroelectric (and piezoelectric) activity but very low electro-optical activity. The copolymers which contain both types of units show substantial activity for both properties.

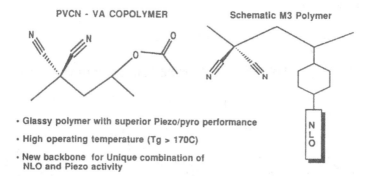

PVCN - VA COPOLYMER

Schematic M3 Polymer

- Glassy polymer with superior Piezo/pyro performance

- High operating temperature (Tg > 170C)

- New backbone for Unique combination of NLO and Piezo activity

N
L
O

Figure 14. Vinylidene Dicyanide based Multifunctional Polymers

Table 1.   Selective Enhancement of Pyroelectric and Linear Electro-optic Properties.

POLYMER	p (µC/mK)	r(pm/V)	E(V/µm)
MO3ONS-MMA(50-50)	3	10	100
MO6ONS-MMA(50-50)		11	100
PVCN-VA	7	1	100
PVCN-MO6ONS	5	10	90

Demonstration of Synergistic Functionalities

Using the same methods, organic photoconductors may be attached to a glassy polymer so that photoconductive piezoelectric polymers or photoconductive electro-optic polymers may be designed. The latter polymer should also have the ability to create large internal electric fields upon illumination with light. This internal field is due to the conduction of charge (electrons) throughout the polymers leaving holes behind. Such charge segregation should result in light induced changes in the index of refraction of the polymer or photorefraction. This is in exact anaolgy to what happens in inorganic crystalline photorefractors [5]. Examples of both types of polymers are given in Figure 15.

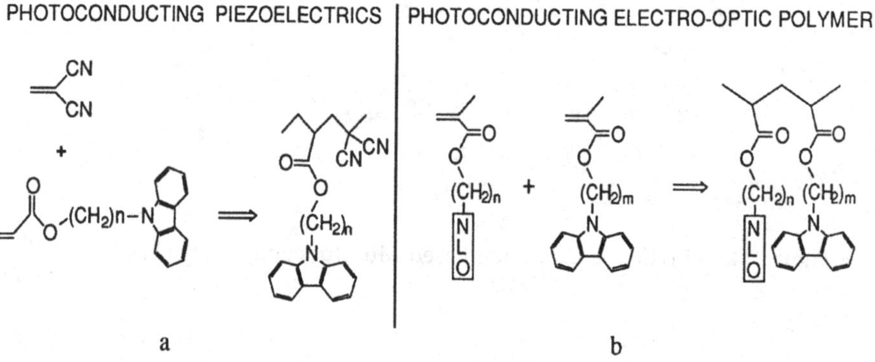

Figure 15. New Multifunctional Polymer Designs

Two examples of photorefracting electro-optic polymers have recently been reported in the literature [6,7]. Both of these were formed by adding the photoconducting functionalities as small molecules in an electro-optic polymer. This is typical of what is done in organic photoconductors [8]. It is also the approach taken in the early work in which one wanted to incorporate electro-optic functionality into polymers; i.e. guest-host systems were prepared [9]. It is expected that a copolymer route to photorefractive polymers will provide significant advantages in control over microstructure, intermolecular interactions, optical properties, and processability.

Applications of Multifunctional Polymers

For many devices, multifunctional materials permit unique device designs. One example is the light modulated deformable mirror. A schematic of this device is shown in Figure 16. Light impinges on a photoconductor allowing to applied voltage to drop across the piezoelectric layer. The resulting deformation of the piezoelectric layer alters the curvature of the mirror resulting in a spatial modulation of reflected light. The construction

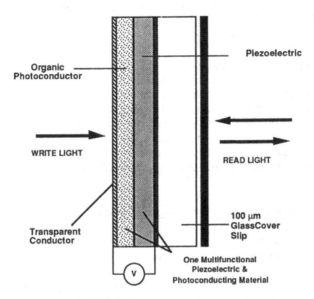

Figure 16. LMDM using organic functional materials.

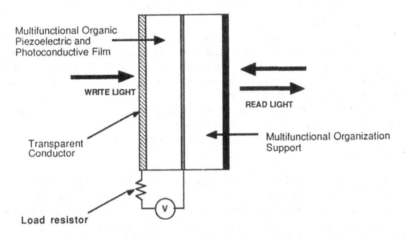

Figure 17. LMDM Utilizing Multifunctional Polymer

and performance of this device has been described by Worboy's et al [10]. This device utilizes two separate functional materials: an organic piezoelectric and an organic photoconductor. In addition, it utilizes glass as a mechanical support.

This device could benefit from multifunctional materials by combining functions into a single polymer. This would significantly simplify construction. In addition, if the material could also be used as the supporting structure, one would not have to worry about thermal expansion mismatch and the resultant mechanical strains caused by temperature variations. The multifunctional materials concept of a light deformable mirror is depicted in Figure 17.

The load resistor is added in Figure 17 in the external circuit and is needed so that the effective voltage across the multifunctional material will depend on the current (which depends on the light intensity).

The applications of photorefractive polymers are numerous. They include all the applications currently being considered for inorganic photorefractors such as active holographic elements for beam steering, phase conjugate mirrors, and optical computing. Our copolymer approach should provide distinct advantages in the design and ultimate optimization of the photorefractive performance for particular applications. In addition, polymers will be lighter and more easily processable into large area films than inorganics, thus opening up new potential applications.

Multifunctional Polymers: Conclusions and Trends

Although research in this area is still very early on, there are certain conclusions and trends that seem apparent:

* Organic, polymeric materials offer a significant opportunity for the tailored design of (multi-) functional materials with characteristics critically enabling to a range of emerging technologies and applications.
* In contrast to structural polymers, the search for new molecules with desired properties will dominate.
* Morphology and orientation control through novel processing are major needs and opportunities - there will develop a combination of synthetic and processing approaches to give materials of extremely high structural uniformity and, consequently, greatly enhanced effects in various electronic and optical applications.
* There will be continued development of improved processes to respond to a need for materials of exceptional purity.
* Computer-aided molecular design and process simulation will have a dominate role.

In addition, it is important to recognize that high functional activity coefficients in the polymer, although a necessary primary focus of early synthesis work and materials development, are far from meeting the total materials requirements. The total materials performance requirement is, indeed, a formidable challenge and encompasses areas such as the following:

* High functional activity coefficients
* Low optical loss
* Low absorption at the operating wavelength
* Long term stability of the poled state
* Controlled electrical conductivity
* Relatively high glass transition temperature
* Solubility ~10-20% in suitable solvents
* Good mechanical stability in thick films (~20μ)
* Good adhesion on surfaces
* Thermal stability at high temperature (~200°C) for ~5 hours, and long-term stability at elevated (50-85°C) use temperatures.

Acknowledgements

The authors acknowledge the support of an SDIO/AFOSR contract for the development of these materials. We are especially indebted to the guidance and support provided to this program by the late Dr. Don Ulrich at AFOSR.

The authors also acknowledge contributions to this effort by A. East, R. Johnson, C. Shu, I. Kalnin, R. Carney, F. Battito, G. Kim, L. Charbonneau, and A. Buckley.

The program is a collaborative effort and the discussions with Prof. I. Ward, Prof. A. Windle, Prof. D. Haarer, and GEC Marconi Labs have proved to be very valuable.

References

[1] Non-linear Optical Effects in Organic Polymers, ed by Messier, Kajzar, Prasad, and Ulrich, Kluwer Academic Publishers, Dordrecht (1989).

[2] D. Haas, H.T. Man , C.C. Teng, K.P. Chiang, H.N. Yoon, T.K. Findakly High Frequency Analog Fiber Optic Systems, SPIE 1371 ed by Paul Dierak, p 56 (1990).

[3] J. Stamatoff, R. DeMartino, D. Haas, G. Khanarian, H.T. Man, R. Norwood, H.N. Yoon, Critical Requirements for Non-Linear Optical Polymeric Materials in Active Optical Devices: The Present State and Prospects for the Future. Die Angewandte Makromolekulare Chemie 183 (1990) 151-166 (321).

[4] H.A. Goldberg, A.J. East, I.L. Kalnin, R.E. Johnson, H.T. Man, R.A. Keosian, and D. Karim; Mat. Res. Soc. Symp. Proc. Vol 175: Multifunctional Materials; ed by Buckley, Gallagher-Daggit, Karasz, and Ulrich; p113 (1990).

[5] Photorefractive Materials and Their Applications I: Fundamental Phenomena; ed. by P. Gunter and J.-P. Huignard; Springer Verlag; Berlin (1988).

[6] J. S. Schildkraut; Appl. Phys. Lett. 58, #4, p 340 (1991).

[7] S. Ducharme, J.C. Scott, R.J. Twieg, and W.E. Moerner; Phys. Rev. Lett. 66, #14, p 1846 (1991).

[8] W. Wiedmann, Chemiker Zeitung, 106, 1982.No. 7-8, pp. 275-278 (Part I); No. 9, pp. 313-326 (Part II).

[9] J. B. Stamatoff, A. Buckley, G. Calundann, E.W. Choe, R.N. DeMartino, G. Khanarian, T.M. Leslie, G.V. Nelson, D. Stuetz, C.C. Teng, H.N. Yoon, Proc. SPIE, 682, 85 (1986) and K.D. Singer, M.G. Kuzyk, J.E. Sohr; J. Opt Soc. Am B 4 p 968, (1987).

[10] M.R. Worboys, M.S. Griffith, and N.A. Davies; <u>Mat. Res. Soc. Symp. Proc. Vol 175: Multifunctional Materials;</u> ed by Buckley, Gallagher-Daggit, Karasz, and Ulrich; p135 (1990).

POLYMERS FOR PHOTONICS

Paras N. Prasad

Photonics Research Laboratory
Department of Chemistry
State University of New York at Buffalo
Buffalo, NY 14214

ABSTRACT

Photonics has been labeled by many as the technology for the 21st century. Polymers have emerged as an important class of materials for applications in photonics. In this review, a brief background is presented on photonics and nonlinear optical processes, the latter providing many of the operational functions for the photonics technology. Nonlinear optical processes in polymeric materials are discussed along with the needed structural requirements. Technologically relevant issues and the current status of the field are summarized. This review finally concludes with a discussion of the potential areas of opportunities for polymer scientists and engineers.

NEW TECHNOLOGIES AND ADVANCED POLYMERIC MATERIALS

We are living in an era of technological revolution. Our daily life has grown highly accustomed to advanced technology - from high definition TV to microcomputers. As we march towards the next century, our expectations from new technologies are growing at an unprecedented rate. We place a very stringent demand on new technologies. Some of the important features we require from a new technology suitable for the next generation are listed in Figure 1. First, operating speed of the devices is very important; second the technology should involve light-weight components so that it can also easily interface with space based systems. Other important requirements are compactness, ability to integrate into a system network, and device components which have exceptional durability

Frontiers of Polymer Research, Edited by P.N. Prasad and
J.K. Nigam, Plenum Press, New York, 1991

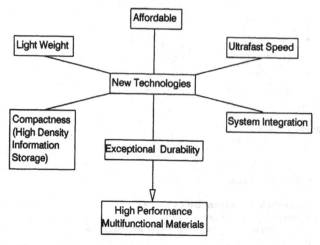

Figure 1. Desirable Features of New Technologies.

(life cycle). The success of these new technologies is crucially dependent
on the availability of advanced new materials which are highly efficient,
durable in their performance and simultaneously perform more than one
function. These are the high performance multifunctional materials.

Polymeric materials are molecular hierarchical systems in which the
structure and functionalities can be controlled from the atomic level to
the bulk level. This tremendous structural flexibility provides one with a
true opportunity to introduce multifunctionality in a polymeric material.
In essence, one can attempt to mimic nature. Like the structures of
proteins, DNA and other biological system, one can control the primary,
secondary, and tertiary structures by a sequential chemical synthesis
approach. The schematics of a multifunctional polymer are shown in Figure
2. One can vary the main chain structure by making alternating or block
co-polymers; one can graft side chains of various functionalities. Also,
by introducing flexible chain segments in the main chain or in the side
chain, the polymer can be made soluble. By using a combination of these
modifications one can introduce different functionalities such as optical
response, electronic response, mechanical strength and improved
processibility. Further improvements can be brought by making composites
or blends of polymers. Use can also be made of induced orientational
alignment to produce or enhance a desired functional response.

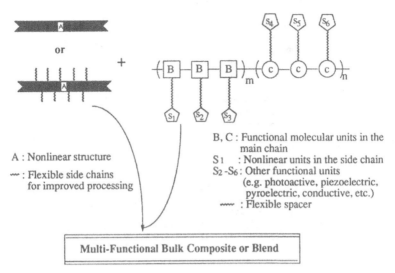

A : Nonlinear structure

~ : Flexible side chains
for improved processing

B, C : Functional molecular units in the
main chain
S_1 : Nonlinear units in the side chain
S_2 -S_6 : Other functional units
(e.g. photoactive, piezoelectric,
pyroelectric, conductive, etc.)
~~~ : Flexible spacer

Multi-Functional Bulk Composite or Blend

Figure 2.   Schematic representation of a multifunctional polymer.

PHOTONICS TECHNOLOGY

We are living in an age where speed and efficiency are very important
to us. Light-wave technology offers promise in these regards where photons
instead of electrons are used to acquire, process, store and transmit
information. In analogy with electronics, we call this technology
photonics. The most important advantage of photonics over electronics is
the gain in speed; this results from the simple fact that a photon travels
much faster than an electron. Also, one can use information storage more
compactly. The components are lightweight because plastics or glasses
instead of metals can be involved. An example is a fiber optics telephone
line much lighter than traditional lines using copper wires. Another
advantage is that there is no electrical or magnetic interference. The
developing photonics technology is compatible with the existing fiber
optics communication systems.

There are a whole range of applications of photonics which are listed
in Table I. Optical telecommunication is already partially achieved; the
most exciting prospect is optical processing of information and optical
computing. The functions needed are frequency conversion for image
analysis as well as for high density optical data storage, light modulation

TABLE I. Applications of Photonics

and optical switching for an optical transistor type operation. Just like
electronic circuits, one can describe a photonic circuit shown in Figure 3.
In this example, photons are conducted through channels. These channels
can be fibers or channel waveguides embedded in a film strip. Light can be
switched from one channel to another at a certain junction point using
appropriate nonlinear optical effect.

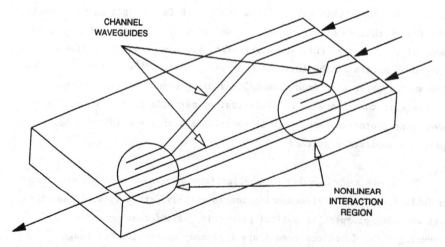

Figure 3. Schematics of a photonic circuit.

The needed functions of frequency conversion, light modulation, optical switching and logic gate are provided by the nonlinear optical effects. The most spectacular demonstration of a nonlinear optical effect is frequency doubling which is frequently used in laser laboratories to generate a 532 nm green beam from the 1064 nm near-IR lasing output of a Nd:Yag solid state laser. Other types of nonlinear processes, shown in Figure 4 as an optical output from a device at different optical input, are power limiter action to limit the output power, switching of optical output level from a low value to a high value and optical bistability in which there are two output levels within a given range of input. This latter device can be used for optical memory storage in computing.

NONLINEAR OPTICAL PROCESSES IN POLYMERIC MATERIALS

Light is nothing but a rapidly oscillating electric and magnetic field. Nonlinear optical processes occur when a medium is subjected to an intense electric field E such as that associated with a strong pulse of light. The field polarizes the medium. At the molecular level, the change of electronic distribution creates an induced dipole moment which can be expressed as a power series [1]:

$$\mu_{ind} = \alpha \cdot E + \beta : EE + \gamma : EEE + \ldots \tag{1}$$

The linear term involving polarizability α is generally considered and it describes the linear response such as ordinary refraction and absorption. The higher hyperpolarizability terms β and γ describe the molecular nonlinear optical responses.

An analogous power expansion is used for the induced polarization P at the bulk level[1,2]:

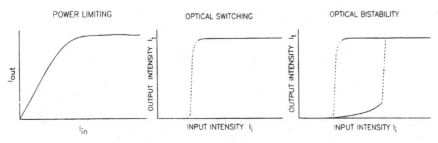

Figure 4. Nonlinear optical input-output relation.

$$P = \chi^{(1)} \cdot E + \chi^{(2)} : EE + \chi^{(3)} \vdots EEE + \ldots = \chi_{eff} \cdot E \qquad (2)$$

In equation (2), $\chi^{(1)}$ is the linear susceptibility which is generally adequate to describe the optical response in the case of a weak optical field (ordinary light). The terms $\chi^{(2)}$ and $\chi^{(3)}$ are the second and third-order nonlinear optical susceptibilities which describe the nonlinear optical response of the medium. Since the electric field E is oscillating at a frequency ω (in the case of an optical pulse, this frequency is the corresponding optical frequency), a more precise notation for the nonlinear susceptibilities carries the frequency specification. An example is $\chi^{(2)}(-2\omega;\omega,\omega)$, the second order susceptibility for second harmonic generation, where two input fundamental waves of frequency ω interact in the medium to generate the second harmonic wave at frequency 2ω. The generalized notation for the nonlinear susceptibility, for example the third order susceptibility, in the frequency representation is $\chi^{(3)}(-\omega_4;\omega_1,\omega_2,\omega_3)$ for the process in which three input waves of frequencies ω_1, ω_2 and ω_3 interact in the medium to generate an output wave at frequency ω_4. The bulk susceptibilities can be related to corresponding microscopic coefficients α, β and γ of equation (1) if one uses the weak intermolecular coupling limit of an oriented gas model. Under this model, the bulk susceptibilities $\chi^{(n)}$ are derived from the corresponding microscopic coefficients by using simple orientationally-averaged site sums with appropriate local field correction factors which relate the applied field to the local field at a molecular site. Under this approximation[1]

$$\chi^{(2)}(-\omega_3;\ \omega_1,\omega_2) = F(\omega_1)F(\omega_2)F(\omega_3)\sum_n <\beta^n(\theta,\phi)> \qquad (3)$$

$$\chi^{(3)}(-\omega_4;\ \omega_1,\omega_2,\omega_3) = F(\omega_1)F(\omega_2)F(\omega_3)F(\omega_4)\sum_n <\gamma^n(\theta,\phi)> \qquad (4)$$

In the above equation, β^n and γ^n represent the microscopic coefficients at site n which are averaged over molecular orientations θ and ϕ and summed over all sites n. The terms $F(\omega_i)$ are the local field corrections for a wave of frequency ω_i. Generally, one utilizes the Lorentz approximation for the local field, in which case[1,2]

$$F(\omega_i) = \frac{n_o^2(\omega_i)+2}{3} . \qquad (5)$$

In this equation, $n_o(\omega_i)$ is the linear refractive index of the medium at frequency ω_i.

From equation (3), it is clear that even for molecular systems with $\beta \neq 0$, the bulk second order nonlinearity, determined by the second order nonlinear susceptibility $\chi^{(2)}$, will be absent if the bulk structure is centrosymmetric or amorphous in which case $\Sigma_n < \beta^n(\theta, \phi) = 0$. Therefore, for a polymeric system to give rise to second-order effect the conditions are that (i) $\beta \neq 0$ and (ii) the bulk structure is non-centrosymmetric.

Since γ is a fourth rank tensor, its average does not vanish even in a centrosymmetric structure. Therefore, an isotropic medium such as an amorphous polymer will exhibit third-order nonlinear optical response. However, a system may still show large differences in the $\chi^{(3)}$ value if it is oriented. An example is a conjugated polymeric structure which has the largest component of the γ tensor along the polymer chain[3]. If only this component contributes, then the largest value of $\chi^{(3)}$ will correspond to a bulk in which all the polymeric chains are oriented in the same direction. The largest component of $\chi^{(3)}$ in this case will be along the orientation direction. In contrast, the $\chi^{(3)}$ value in a truly amorphous phase of the same polymer will be reduced by a factor of five. These orientation effects have been observed in stretch-oriented polymers.[1]

The manifestations of nonlinear optical effects can conveniently be examined in the picture of the dielectric theory. For a linear dielectric, application of an electric field polarizes the medium to produce a polarization P that is linearly proportional to the applied field, the proportionality constant $\chi^{(1)}$ being the linear susceptibility. If the electric field is due to an optical field, then the response is also described by a refractive index n. Thus at optical frequencies, $n^2(\omega) = 1 + 4\pi\chi^{(1)}(\omega)$. For a plane wave propagation[1]

$$E = E_\omega(z,t) = E_0 \cos(kz - \omega t) \text{ or } 1/2 [E_0 e^{i(\omega t - kz)} + cc] \qquad (6)$$

The refractive index n, the wave vector $k = n\omega/c$, and the phase velocity $v = c/n$ are all independent of the field strength E.

For a nonlinear dielectric one has the polarization given by equation (2). χ_{eff}, the effective susceptibility tensor, now depends on the field strength E. Therefore, n, k, and v are all dependent on E. Two important manifestations of optical nonlinearities are harmonic generation and refractive index modulation by electric and optical fields. Their origin can conveniently be explained by considering a plane wave propagation through the nonlinear medium using equations (2 and 6). The polarization is then given by[1]

$$P = \chi^{(1)}E_o\cos \alpha + \chi^{(2)}E_o^2\cos^2 \alpha + \chi^{(3)}E_o^3\cos^3 \alpha =$$

$$\chi^{(1)}E_o\cos \alpha + 1/2 \; \chi^{(2)}E_o^2(\cos 2\alpha + 1) + \tag{7}$$

$$\chi^{(3)}E_o^3[3/4 \cos \alpha + 1/4 \cos 3\alpha]$$

where $\alpha = (kz - \omega t)$. Equation 7 shows that due to nonlinear optical effects, higher frequency components (2α and 3α terms) are generated that describe higher harmonic, examples being second harmonic generation due to $\chi^{(2)}$ and third-harmonic generation due to $\chi^{(3)}$. In addition, $\chi^{(3)}$ leads to a term with cos α that describes the intensity dependence of refractive index. The dependence of refractive index on the electric field actually consists of two terms: (i) one derived from $\chi^{(2)}$ that is linearly dependent on E and describes the electrooptic effect, also known as Pockels effect, in which the application of an electric field modulates the refractive index; (ii) one derived from $\chi^{(3)}$ that is quadratically dependent on E and hence linearly dependent on I; it describes the optical Kerr effect. It is the latter that provides a mechanism for light control by light because an intense beam can be used to change the refractive index of the medium and influence either its own propagation or propagation of another beam of different or same frequency.

The intensity dependence of the refractive index can also be used in a degenerate four-wave mixing process. With this process in the backward wave geometry, the nonlinear medium acts as a phase-conjugate mirror that reverses the direction of an incoming carrier wave when two counter-propagating waves of the same frequency are applied as shown in Figure 5. This process of phase conjugation corrects for any phase distortion of the carrier wave by reversing the phase in the outgoing beam. Since the process involves four optical waves of the same frequency, it is called degenerate four-wave mixing. In contrast, reflection from an ordinary mirror (also shown in Figure 5) does not reverse the path of the carrier wave and hence does not correct the phase distortion. This phase conjugation is of great significance in relation to real-time holography

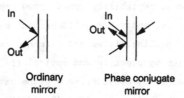

Ordinary   Phase conjugate
mirror   mirror

Figure 5. Comparison of reflections from an ordinary mirror and from a phase conjugate mirror.

because using phase reversal one can correct for phase aberrations, introduced by various optical elements and/or environmental distortions, to reconstruct a high quality image.

To take advantage of the flexibility offered by a molecular material, one needs to be able to project what chemical structures will contribute to optical nonlinearity. This is one area where our theoretical understanding still needs to be developed. However, by using existing theoretical models, we can make some structural projections. A useful model for second-order effect has been the two-level model. In this case one treats a molecule as having only two levels, a ground state, g, and an excited state, i. In such a case and far from resonance, β is given as[1]

$$\beta(-2\omega;\omega,\omega) = \left(\frac{3e^2}{2\hbar m}\right)\frac{\omega_{ig}^2}{(\omega_{ig}^2 - \omega^2)(\omega_{ig}^2 - 4\omega^2)}f\Delta\mu \qquad (8)$$

In the above equation, f is the oscillator strength of the transition $g \to i$ and $\Delta\mu$ is the difference of dipole moment between the excited state and the ground state. Therefore, a molecular structure that possesses a low-lying excited state with a large oscillator strength and a partly ionic character (to give large $\Delta\mu$) will possess large β. A suitable structure for this purpose is[1]

In the above structure, a conjugated unit (benzene ring in the above example) separates an electron-donor group (D) such as NH_2 and an electron acceptor group (A) such as NO_2. The lowest lying excited state in such structures involves a charge transfer from group D to group A, which gives rise to a large change of dipole moment. Some examples of this type of chromophores used for the 2nd-order effect are shown in Table II along with their respective β values.

Although a polymeric structure is not required, it is desirable both for mechanical strength and for ease of fabrication of device structures. For this reason, polymers with nonlinear active groups built into the side chain, as shown in Figure 6, are becoming very attractive materials.[4] In this example the nonlinear side group conforms to the molecular design in which a biphenyl π-electron structure separates an electron donor, oxy group, from an electron acceptor, nitro group.

TABLE II
β values of some chromophores

structure	$\beta \times 10^{30}$ (esu)
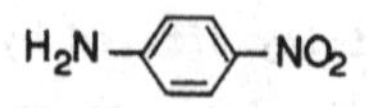 H_2N—⟨benzene⟩—NO_2	34.5 *
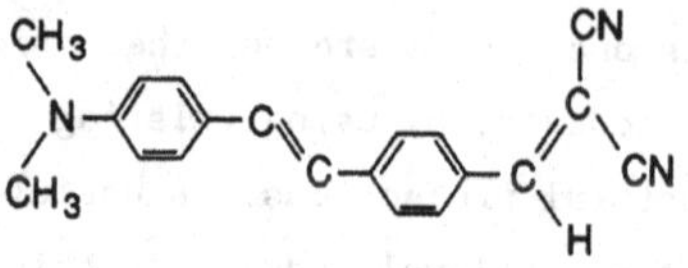 (CH₃)₂N—⟨benzene⟩—C=C—⟨benzene⟩—C=C(CN)₂ with H	323*
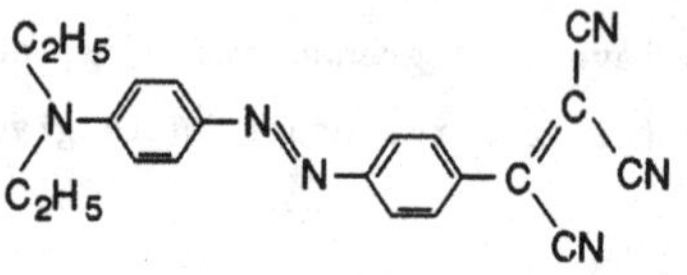 (C₂H₅)₂N—⟨benzene⟩—N=N—⟨benzene⟩—C=C(CN)₂ with CN	390*

* J. L. Oudar and D. S. Chemla, <u>J. Chem. Phys.</u>
66:2664 (1977).

** H. E. Katz, C. W. Dirk, M. L. Schilling, K. D.
Singer, and J. E. Sohn, <u>in</u>: "Nonlinear Optical
Properties of Polymers," A. J. Heeger, J.
Orenstein, and D. Ulrich, eds., MRS, Pittsburgh,
(1988), p 127.

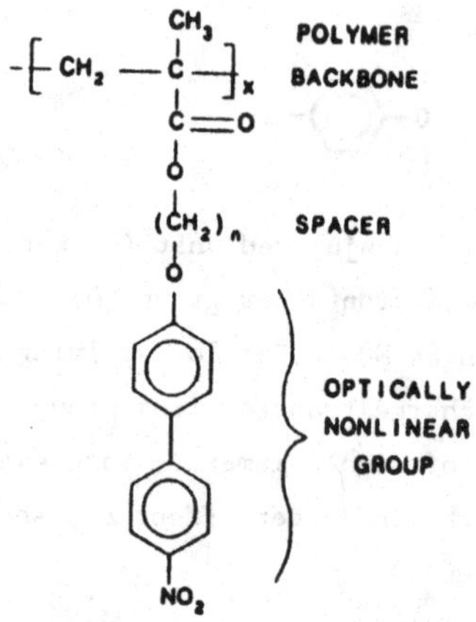

Figure 6.   Optically nonlinear side-chain polymers.

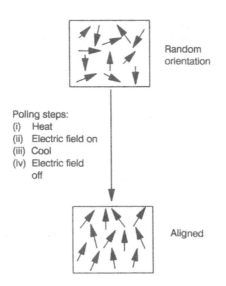

Figure 7. Electric field poling process.

Since $\chi^{(2)}$ is a third-rank tensor, it is nonvanishing only for noncentrosymmetric media. Therefore, to produce a net second-order nonlinear optical effect in a bulk substance, the medium must have a noncentrosymmetric ordering of dipoles, created either by spontaneous ordering or by electric field poling of the medium.[1,4,6] Electric field poling is a four-step process as shown in Figure 7. In the first step, the polymer, which has a random orientation of dipoles, is heated to a temperature at which molecular motions are significant. An electric field-typically 100 kV/cm is applied to align the dipoles along the field direction. The field is kept on for some time, with the material remaining at a high temperature. The material is then cooled to lock the dipole orientation, and finally, the electric field is turned off. The resultant $\chi^{(2)}$ in an electrically poled system is maximum only when the largest component of the microscopic nonlinearity, β tensor is along the direction of the molecular dipole. One important concern is reorientation of the ordered dipoles in the poled sample, which can be induced by various relaxation processes.[1] Therefore, current research is focusing on minimizing these dipolar relaxations by various methods.

For third-order nonlinear optical effects, the structural requirements are different. Extended π-conjugation is found to enhance the nonlinearity. Consequently, conjugated polymers have emerged as an

TABLE III

The $\chi^{(3)}$ values of some molecular and polymeric materials measured in the thin film form[1].

STRUCTURE	WAVELENGTH	METHOD	$\chi^{(3)}$ (esu)
POLY-P-PHENYLENEVINYLENE	602 nm	DFWM	-4×10^{-10}
POLYDIACETYLENE R—CH₂—O—S—⟨⟩—CH₃ PTS	2.62 μ	THG	1.6×10^{-10} parallel to chain
POLY-4-BCMU	585 nm	DFWM	4×10^{-10} red form
PBT	585 nm	DFWM	-10^{-11}
POLYTHIOPHENE	602 nm	DFWM	-10^{-9}
POLYACETELYNE	1.06 μ	THG	4×10^{-10}

important class of third-order nonlinear material. Although order in polymeric structures does influence the third-order nonlinear behavior, it is not required. Therefore, even an amorphous polymer can be used for applications of third-order nonlinear optical effects. Some examples of the conjugated polymers with their respective $\chi^{(3)}$ values are shown in Table III.

TECHNOLOGICALLY RELEVANT ISSUES AND CURRENT STATUS

Technologically relevant issues differ somewhat for different applications, although some material issues from technology perspectives are common. For example, need for reliable performance under large fluctuations in ambient conditions as well as for higher optical damage thresholds and high optical throughput of the device are essential to all applications.

For second order processes, high efficiency requires a large $\chi^{(2)}$ value, the second-order nonlinear bulk susceptibility. For second harmonic generation two additional requirements that must be met for efficient frequency conversion are phase-matching and broad band transparency. Phase-matching simply ensures that the fundamental input wave and the generated second harmonic wave travel through the medium with the same phase velocity so that power builds up in the second-harmonic generation, as the interaction length is increased.

For electro-optic applications such as electro-optic modulation, optical beam steering, and spatial light modulation, an electric field is used to affect the light signal. In addition to the need for a large $\chi^{(2)}$ value, which will reduce the needed voltage for modulation, it is also imperative for electro-optic devices that the material has a low dc dielectric constant. A low dielectric constant helps in two ways: (i) it enhances the field across the sample for a given applied voltage, and (ii) it enables a faster time response because of a reduced RC time constant.

Most third-order nonlinear optical devices are based on the intensity dependence of the refractive index which provides a mechanism for all optical processing and control. In order for a device to operate at low optical power, the corresponding nonlinear susceptibility $\chi^{(3)}$ must again be large. For broad band response and femtosecond operation, non-resonant purely electronic nonlinearity is needed. Since the major attractive feature of all optical processing is the tremendous gain in speed, the fastest nonlinear optical response is provided by the nonresonant nonlinearity. The nonresonant nonlinearity is also desirable because no optical loss from absorption and subsequent thermal effects are present. An especially complicating thermal effect is thermal nonlinearity (thermally-induced refractive index) which can build up because of its extremely slow relaxation.

For frequency doubling (second harmonic generation), crystalline materials are currently more suitable. The reason is that the natural birefringence of a crystalline medium can be used for phase-matching.[1,2] The concept of quasi-phase matching has recently been applied to polymeric materials by using periodic poling.[4] However, the observed second harmonic conversion efficiency is not as yet comparable to that achieved in the crystalline materials containing small molecules. Therefore, the usefulness of polymeric materials for efficient frequency doubling has not been demonstrated.

For electro-optic modulation, electrically poled polymers have emerged as a promising class of materials because of the ease with which a polymer material can be fabricated into device structures such as fibers, channel waveguides, etc. Also, polymeric structures are thermally and environmentally more stable and they do not have the high vapor pressure problems associated with small organic molecules. Polymers with nonlinearly active groups in the side-chain are very promising.[1] Another advantage offered by the polymeric structures is their low dielectric constant and resultant high bandwidth and speed. Potential problems encountered have been the dipolar relaxation of the poled alignment which will lead to device failure in time. Another concern for waveguide applications is the fabrication of a structure with extremely low optical loss.

For third-order applications such as optical switching, optical bistability, optical gates, power limiting, etc., currently achievable nonresonant $\chi^{(3)}$ values ($< 10^{-9}$) are not sufficient for low pulse energy operations. Conjugated polymeric structures have shown the largest non-resonant $\chi^{(3)}$, but still not large enough for applications.[1,3] These conjugated polymeric structures are often insoluble, and, therefore, not processible into useful device structures. Furthermore, they are optically lossy and, because of the reduced bandgap due to large conjugation effect, their region of transparency is limited.

OPPORTUNITIES FOR POLYMER SCIENTISTS AND ENGINEERS

The field of photonics offers exciting opportunities for polymer scientists and engineers. It is a multidisciplinary field which will need input from chemists, physicists, material scientists and engineers to find solutions for technologically relevant issues addressed in the previous section. Some of the areas of opportunities are briefly revealed in this section.

Increase of nonlinearity, particularly $\chi^{(3)}$

For all optical applications, the biggest hurdle to overcome is the low value of currently achievable nonresonant $\chi^{(3)}$. For molecular and polymeric materials, enhancement of the $\chi^{(3)}$ value, however, requires a better understanding of the relation between molecular structure and microscopic nonlinearity so that one is able to identify structural units with enhanced nonlinear optical response. Two important approaches to

58

understand the needed structure-nonlinear optical property relationship are
(a) theoretical modeling and (b) synthesis and nonlinear optical
measurements of sequentially built and systematically derivatized
structures.[6]

Chemical Processing To Reduce Dipolar Relaxation in Solid Polymers

In the case of poled polymers for second-order nonlinear processes, a
major issue is the long-term stability of the induced metastable dipole
aligned state. Dipole-dipole interaction is unfavorable in this
configuration, and dipolar relaxation occurs to produce
randomization of dipoles in the polymeric phase. Chemical processing can
be used to minimize this dipolar relaxation and overcome the loss of
alignment. Ye et al.[5] showed that by using an appropriate structure
capable of hydrogen bonding, one can considerably increase the stability of
poled structures. They prepared the following copolymer in which the
composition n was varied between 0.15 and 0.48:

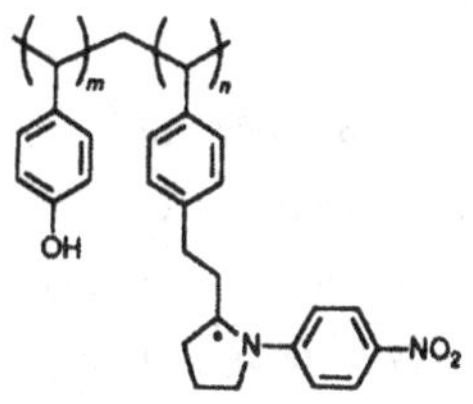

The hydrogen bonding between the OH group and the electron-rich amine
unit produces a weak cross-linked network that contributes to the enhanced
stability of the poled structure.

Recently, Robello et al.[7] and Eich et al.[8] used an approach where
photoinduced cross-links in multiacrylic systems or diepoxide-diamine
condensation reactions lead to considerable increase in the stability of
poling. Also, Hubbard et al.[9] recently used an approach in which a high-β
guest is dispersed in an optically transparent host matrix that can
simultaneously be poled and chemically cross-linked. In this work, a
two-component optically transparent thermosetting epoxy, IPO-TEK 301-2, was
employed as the cross-linkable host matrix.

Materials Processing

An important issue concerning the conjugated polymeric structure as
third-order nonlinear materials is their processibility. The conjugated

linear polymeric structures tend to be insoluble, and, therefore, cannot readily be processed into device structures. The lack of processibility may render a material totally useless for practical application even if it may have a large $\chi^{(3)}$ value.

New chemical approaches for processing of nonlinear materials can play an important role. Two specific examples presented here are: (a) Soluble precursor route and (b) Chemical derivatization for improving solubility. In the soluble precursor route, a suitable precursor is synthesized which can be cast into a device structure (i.e., film) by using solution processing. Then it can be converted into the final nonlinear structure upon subsequent treatment (such as heat treatment). This approach has been used for poly-p-phenylenevinylene (PPV) as shown below:[10]

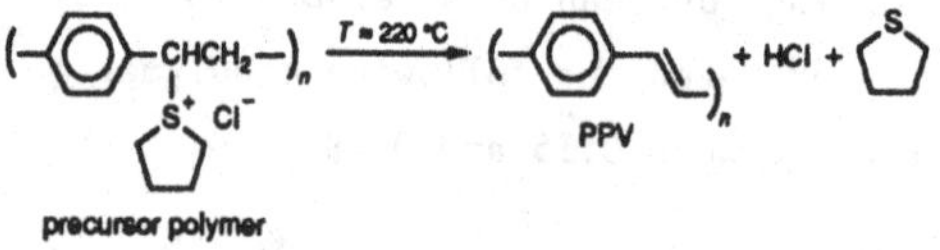

precursor polymer

In the chemical derivatization approach one introduces a long pendant alkyl or alkoxy group to increase solubility. Polythiophene itself is insoluble but poly(3-dodecylthiophene), is soluble in common organic solvents. We have also successfully used poly(3-dodecylthiophene) to form mono and multilayer Langmuir-Blodgett films.[11]

Optical Quality

Optical quality of materials is of prime concern for integrated optics applications which will involve waveguide configurations. Most conjugated polymeric structures are optically lossy. There is a need for approaches which will provide a better control of structural homogeneities so that optical losses can be minimized. Another approach is through the use of composite structures where both the optical quality and $\chi^{(3)}$ can be optimized by a judicious choice of the two components. The best optical quality medium is provided by inorganic glasses such as silica. However, they by themselves have very low $\chi^{(3)}$. A composite structure such as that of silica and a conjugated polymer may be a suitable choice. Chemical processing of an oxide glass using the sol-gel chemistry provides a suitable approach to make such composite structures. Using the sol-gel method, a composite of silica glass and poly(p-phenylene vinylene) has been prepared[12] in which the composition can be varied up to 50%. The procedure

60

involves molecular mixing of the silica sol-gel precursor and the polymer precursor in a solvent in which both are soluble. During the gelation, a film is cast. Subsequent heat treatment converts the precursor polymer to the conjugated poly(p-phenylene vinylene) polymeric structure. The optical quality of the film was found to be significantly improved and high enough to use them as optical waveguides at 1.06μ.

CONCLUSIONS

To conclude this article, it is hoped that the discussion of relevant issues and opportunities for polymer scientists and engineers presented here will sufficiently stimulate the interest of the polymer science community. Their active participation is vital for building our understanding of optical nonlinearities in polymeric systems as well as for the development of useful nonlinear optical polymers.

ACKNOWLEDGEMENTS

The research work at Photonics Research Laboratory is supported in part by the Air Force Office of Scientific Research, Directorate of Chemical and Atmospheric Sciences and Polymer Branch, Air Force Wright Laboratory, through contract number F49620-90-C-0021 and in part by the Office of Innovative Science and Technology-Defense Initiative Organization/Air Force Office of Scientific Research through contract number F49620-87-C-0097.

REFERENCES

1.	P. N. Prasad and D. J. Williams, "Introduction to Nonlinear Optical Effects in Molecules and Polymers," Wiley, New York (1991).

2.	Y. R. Shen, "The Principles of Nonlinear Optics," Wiley, New York (1984).

3.	"Nonlinear Optical and Electroactive Polymers," P. N. Prasad and D. R. Ulrich, eds., Plenum, New York (1988).

4.	G. Khanarian, D. Haas, R. Keosian, D. Karim, and P. Landi, CLEO abstract, paper THBI (1989).

5.	C. Ye, N. Minami, T. J. Marks, J. Yang, and G. K. Wong, Macromolecules 21:2899 (1988).

6.	P. N. Prasad and B. A. Reinhardt, Chem. Mater. 2:660 (1990).

7.	D. R. Robello, M. Scozzafova, C. S. Willand, and A.Ullman, ACS Symposium on New Materials for Nonlinear Optics, Boston (April 1990).

8. M. Eich, B. Peck, D. Y. Yoon, C. G. Wilson, and G. C. Bjorklund, _J. Appl. Phys._ 66:3241 (1989).

9. M. A. Hubbard, T. J. Marks, J. Yang, and G. K. Wong, _Chem. Mater._ 1:167 (1989).

10. D. R. Gagnon, J. D. Capistran, F. E. Karasz, R. W. Lenz, and S. Antoun, _Polymer_ 28:567 (1987).

11. P. Logsdon, J. Pfleger, and P. N. Prasad, _Synth. Met._ 26:369 (1988).

12. C. J. Wung, Y. Pang, P. N. Prasad, and F. E. Karasz, _Polymer_ 32:605 (1991).

NONLINEAR OPTICAL DEVICES:
RELATIVE STATUS OF POLYMERIC MATERIALS

George I. Stegeman

Center for Research in Electro-Optics and Lasers
University of Central Florida
Orlando, FL 32826

INTRODUCTION

Activity in nonlinear optics of polymer materials has been escalating for the last five years, driven by potential applications the most promising of which are:

1. Efficient second harmonic generation of GaAs-based semiconductor lasers for data storage and xerography;
2. All-optical signal processing for information manipulation in the all-optical domain;
3. Optical limiting for sensor protection.

In each case it appears that nonlinear organic materials have a great deal to offer and will be a serious contendor for the material of choice 1), because of their unique physics that leads to large nonlinearities and 2), because of the inherent flexibility offered by "molecular engineering".

The starting point for any discussion of nonlinear optics is the nonlinear polarization P^{NL} induced in a material by the presence of one or more high intensity optical fields (1). The standard expression for this polarization is

$$P_i^{NL}(\omega) = \varepsilon_0 [\chi_{ijk}^{(2)}(-\omega; \omega_1, \omega_2) E_j(\omega_1) E_k(\omega_2)$$
$$+ \chi_{ijkl}^{(3)}(-\omega; \omega_1, \omega_2, \omega_3) E_j(\omega_1) E_k(\omega_2) E_l(\omega_3)]$$

where the $\chi^{(p)}$ are material parameters, called the nonlinear susceptibilities. This polarization source term radiates a signal which is the desired product of the nonlinear interaction. The resulting signal frequency is given by $\omega = \omega_1 \pm \omega_2 \pm \omega_3$ and it dictates the application. The terms proportional to $\chi^{(2)}$ give rise to phenomena such as second harmonic generation, parametric oscillators etc. The third order terms can lead to a large variety of effects including third harmonic generation, coherent anti-Stokes Raman scattering, intensity-dependent refractive index [$\propto \chi^{(3)}_{1111}(-\omega; \omega; -\omega; \omega) |E_1(\omega)|^2 E_1(\omega)$] etc. The goal is to implement such interactions as efficiently as possible. Clearly large values for the susceptibilities are very important, in spectral regions where the loss is as small as possible.

Frontiers of Polymer Research, Edited by P.N. Prasad and
J.K. Nigam, Plenum Press, New York, 1991

SECOND ORDER NONLINEAR MATERIALS

A compact, long-lived source of a few milliwatts power in the blue spectral region has recently become an urgent need for optical data storage and xerography. Second harmonic generation of semiconductor lasers and diodes operating in the near infrared can be doubled efficiently to provide such sources. There are currently in excess of 30 research programs worldwide whose goal it is to produce efficient doubling into the blue region of the spectrum. Materials which are non-centrosymmetric on a macroscopic scale have non-zero first order coefficients, $d^{(2)}_{ijk}(-\omega_1-\omega_2;\omega_1,\omega_2)$. There are two ways to achieve this goal with organic materials. The obvious way is to crystallize molecules with large first order hyperpolarizabilities into solids which are non-centrosymmetric.[2-5] This approach has produced some very highly nonlinear materials to date, although the trend is that highly nonlinear molecules do not crystallize into non-centrosymmetric crystal classes. Another approach, unique to organics, is to induce partial alignment of highly nonlinear molecules in a polymer host.[6,7] Here, the degree of induced order determines the net nonlinearity which in turn depends on the properties of the host material and the magnitude of the DC electric field applied. Nonlinearities as large as 50 pm/V have been induced in this way.

The principal reason which makes organic materials attractive for doublers etc. is the unique physics of these materials which allows charge transfer nonlinearities as well as linear optical properties to be engineered. This leads to large nonlinear coefficients $d^{(2)}_{ijk}$ when compared to other standard doubling materials, as shown in Table I.[3-5,8-13] However, one of the current problems with using these materials to double GaAs lasers is that the wavelength of the transmission edge typically increases with increasing nonlinear activity. This is indeed a serious problem and is the current focus of much materials research.

Table I.   Nonlinear coefficients d_{ij} (in Voigt notation) for some representative inorganic and organic materials.

MATERIAL	d_{ii} pm/V	d_{ij} pm/V	Transparency nm
LiNbO$_3$ [11]	41	5.8	400-2500
β-BaB$_2$O$_4$[12]	2.0	0.3	190-3500
KTiOPO$_4$[13]	4	7	350-4500
MMONS [2]	184	71	530-1600
MNA [3]	165	25	480-2000
NPP [4]	30	84	480-1800
DMNP [5]	30	90	450-
DAN [8]	5	50	485-2270
DCV/MNA[9]	19	6	
DCANP [10]	8	2	400->2000

MNONS   3methyl-4-methoxy-4-nitrostilbene
NPP   N-(4-Nitrophenyl)-(L)-prolinol
DAN   4-(N,N-Dimethylamino)-3-Acetamidonitrobenzene
DCANP   2-docosylamino-5-nitropyridine (Langmuir-Blodgett film, Y-herringbone structure)
MNA   2-Methyl-4-nitroaniline
DCV/PMMA dicyanovinyl azo dye in copolymer (poled polymer)
DMNP   3,5dimethyl-1-(4-nitrophenyl) pyrazole

Optimizing the efficiency of second harmonic doublers with sub-watt input powers requires using waveguide geometries, that is the light is trapped into structures which maintain beam cross-sections of square microns for long distances. Another important advantage of organic materials is that they offer a great deal of flexibility in making such waveguides. There are the standard approaches of ion-exchange, ion milling, etching etc. In addition, organic materials have been crystallized directly into waveguides, for example as the core of a single mode fiber.[5]

There are other fabrication techniques unique to organic materials which have proven very useful. For poling, molecules with large nonlinear activities are initially "loaded" into a polymer matrix with a completely random orientation. During the poling process a preferential orientation is imparted onto the optically anisotropic molecules by a DC electric field, at a temperature at which the host softens and the molecules are free to partially reorient. For most molecules the molecular polarizability is a maximum along the direction of the permanent dipole moment and hence alignment leads to an increased refractive index along the poling direction, and a decreased index orthogonal to the poling direction. Therefore waveguides are formed for one field polarization (parallel to the DC field) via the electrode patterns.[14]

With organic materials, illumination with light can also lead to trans-cis photoisomerization,[15] breaking of bonds[16] etc. in the molecules, all of which lead to local changes in the refractive index. Typically, the refractive index and the nonlinearity are reduced by illumination at the appropriate frequency. The non-illuminated regions form the waveguides which are nonlinear.

Next to the magnitude of the nonlinear coefficient, the most critical parameter for efficient doubling is the phase-matching term $\Delta\phi \propto L[\beta(2\omega) - 2\beta(\omega)]$ which needs to be approximately zero. Here β is the guided wave wavevector (which is proportional to the waveguide core refractive index) and L is the effective interaction distance. For organic materials, the following approaches have been used for phase-matching ($\Delta\phi \rightarrow 0$):

1.   Birefringence Phase-Matching
      This well-established technique requires birefringent materials in which the refractive indices are different for different field polarizations. When orthogonal polarizations are used for the input and harmonic guided wave fields, the usual dispersion in index with increasing frequency in the visible can be canceled out by this refractive index difference so that the effective indices of the input and harmonic fields can be equal. This approach has been applied to doublers based on organic materials, none of which to date has involved optimized geometries.[17]

2.   Quasi-Phase-Matching
      The refractive index, or the nonlinearity are modulated along the propagation direction with period Λ so that $\Delta\phi = L[k_0 n_{eff}(2\omega) - k_0 n_{eff}(\omega) \pm 2\pi/\Lambda] \approx 0$ [4]. Therefore by adjusting the periodicity, phase-matching can be induced. So far only a few preliminary experiments with both nonlinearity and index modulation have been reported.[6,7,18]

3.   Cerenkov
      The second harmonic is generated at some angle θ to the channel (or fiber) axis such that $2k_0 n(2\omega)\cos\theta = 2k_0 n_{eff}(\omega)$, that is the SHG signal leaves the waveguide into the bounding medium. Here $n_{eff}(\omega) = \beta(\omega)/k_0(\omega)$ is the guided wave effective index at the fundamental frequency. Therefore phase-matching to a guided wave signal does not occur in the usual sense. The signal is phase-matched to a radiation field which reduces the

effectiveness of the interaction because the signal quickly leaves the waveguide region. This approach has been successfully demonstrated in a single crystal core fiber which led to a cylindrically symmetric second harmonic output.[5]

Although reports of doubling with organics are few in number and all very recent, the results are noteworthy even at this early stage. A comparison with state-of-the-art doublers in other material systems is given in Table II. Quasi-phase-matching has been implemented both by photobleaching and periodic poling.[6,7,18]

Table II. Conversion efficiencies, extrapolated to a 1 cmlong device for various, material and phase-matching conditions in channel waveguides, $\eta = P(2\omega)/P^2(\omega)$.

Reference	Waveguide	Phase Matching Technique	η %W^{-1}
Sohler[20]	LiNbO$_3$ (1)	Birefringence	45
Taniuchi[21]	LiNbO$_3$	Cerenkov	42
Uemiya[5]	DMNP	Cerenkov	40
Ishigame[22]	LiNbO$_3$	Quasi-Phase Matching	170
Cao[18]	Poled Polymer	Quasi-Phase	30

(1) Resonator at both ω and 2ω

Cerenkov second harmonic radiation has been obtained in a single mode fiber containing single crystal DMNP, resulting in an output with cylindrical symmetry.[5] Finally, we note that, although not very efficient, single harmonic generation has been obtained using Langmuir-Blodgett films.[19]

The best is yet to come and one can expect conversion efficiencies in the 1000's%/W in the future!

THIRD ORDER NONLINEAR MATERIALS

Although third order nonlinearities (characterized by an induced polarization proportional to the product of three optical fields) can have many possible applications, the most interesting ones are based on a refractive index which varies with the local intensity of light inside the material. This is commonly written as $n = n_0 + n_2I$ where n_0 is the low power (power-independent) refractive index, I is the local intensity and n_2 is the nonlinear coefficient in inverse intensity units, for example cm^2/W. A power-dependent refractive index leads to a power-dependent wavevector and a power-dependent phase shift after some propagation distance L. This in turn means that wavevector-matching interactions and interference effects can all be controlled by changing intensity. Most applications stem from these two concepts.

To date very few all-optical devices based on nonlinear organic materials have been reported. An important reason is that the search for appropriate materials is still in its infancy. Furthermore, the information required for designing prototype devices is just not yet widely available. It is known that the key to device operation is the maximum nonlinear phase shift which can be optically produced in a material , $\Delta\phi^{NL} = n_2Ik_0L_{eff}$

where L_{eff} is the effective interaction length.[23] A fundamental limitation to L_{eff} is the material absorption which has both a linear α_0 and a nonlinear contribution γI where γ is the two photon coefficient, usually in cm/GW. As the intensity is increased, a maximum in index change, Δn_{sat}, is always reached due to saturation of the nonlinear mechanism or due to material damage. This leads to a maximum nonlinear phase shift of $2\pi W = 2\pi\Delta n_{sat}/\alpha_0\lambda$. The minimum value of W required varies from device to device with W > 1 being typical. However, the value of W is known for only a handful of materials, none of them optimum. On the other hand, if the maximum phase shift is limited by two photon absorption, we can define another figure of merit $T = 2\lambda\gamma/n_2$ with T < 1 being required for device purposes.[24] Typically n_2 (or $\chi^{(3)}$) has been measured at only a few wavelengths near or on the electronic resonance of the material. And rarely is the absorption coefficient, linear or nonlinear also reported. In fact, summarized in Table III are the very few cases for which the needed parameters are known.[25-27]

Table III. Waveguide compatible organic materials with ultrafast nonlinearities and their all-optical switching figures of merit. An incident intensity of 1 GW/cm² was assumed.

Material	n_2 cm²/W	α cm^{-1}	W	T	λ Microns
PTS [26]	-3×10^{-11}	0.8	350	0.4	1.06
PTS [26]	-10^{-12}	0.8	13	23	1.06
poly 4BCMU[27]	-10^{-13}	0.2	3.5	127	1.06
	5×10^{-14}	1.7	0.3	< 0.7	1.3
DANS [25]	2×10^{-13}	< 1	> 1.4	≅ 1	1.06

Despite these limitations, a number of device related experiments have been performed on the best materials available at that time. These include absorption bleaching, optical bistability, nonlinear grating distributed feedback and coupling, optical limiting and all-optical switching.[25,28-34]

Bistability, a key device for optical computing, has now been observed in a number of organic materials, dating back to the early work in liquid crystals.[28-30] The recent work has dealt with materials chosen specifically for their large electronic nonlinearities, although frequently unwanted thermal effects were present, and in some cases dominant. The cleanest results were performed with a femtosecond laser and show a logic gate based on resonator bistability operating with a near-resonant nonlinearity.[31]

A nonlinear directional coupler for all-optical switching has been made out of poly-4BCMU. Switching due to thermal effects was observed, as well as due to two photon absorption.[33] In retrospect for this case the two photon absorption coefficient was too large, resulting in a value of T > 10. Subsequent material studies of poly-4BCMU at 1.32 microns indicate that switching due to electronic nonlinearities should be possible at that wavelength in this material.[27]

Optical limiting for protecting sensors against strong optical spikes has also been demonstrated with silicon-napth-phthalocyanine.[34] The device relied on an absorption coefficient which was proportional to the intensity I, mimicking that of two photon absorption. However, in this case there was one photon absorption to an excited state which was connected by another very efficient one photon process to a higher excited state. The cascading of these two processes exhibits a response similar to two photon absorption.

For the useful application of third order organic nonlinearities, the major hurdle is to identify appropriate materials. In order to choose materials, it is necessary to perform complete characterization of the linear and nonlinear optical properties as a function of wavelength. It is in this area that progress needs to be made.

SUMMARY

Nonlinear organic materials have shown real promise for applications to nonlinear optics over the last few years. New materials have been synthesized, characterized and in some cases already used in prototype devices. For second order nonlinearities the future is very promising indeed since efficient doublers appear to have found a niche in data storage for which their properties are near-ideal. In the case of third order nonlinearities it is still too early to predict where the ultimate applications will come first.

ACKNOWLEDGMENT

This research was sponsored by NSF under (ECS-8911960).

REFERENCES

1. see for example: Y.R. Shen Principles of Nonlinear Optics, (Wiley Interscience, New York, 1984).
2. I. Ledoux, D. Josse, P. Vidakovic and J. Zyss, Opt. Engin., 25, 202 (1986).
3. J.D. Bierlein, L.K. Cheng, Y. Wang and W. Tam, Appl. Phys. Lett., 56, 423 (1990).
4. B.F. Levine, C.G. Bethea, C.D. Thurmond, R.T. Lunch and J.L. Bernstein, J. Appl. Phys., 50, 2523 (1979).
5. T. Uemiya, U. Uenishi, Y. Shimizu, S. Okamoto, K. Chikuma, T. Tohma and S. Umegaki, SPIE Proceedings, 1148, 207 (1989).
6. G.L.J.A. Rikken, C.J.E. Seppen, S. Nijhuis and E.W. Meijer, Appl. Phys. Lett., 58, 435, (1991).
7. G. Khanarian, R.A. Norwood, D. Haas, B. Feuer and D. Karim, Appl. Phys. Lett., 57, 977 (1990).
8. P. Kerkoc, Ch. Brossard, H. Arend and P. Gunter, Appl. Phys. Lett., 54, 487 (1989).
9. J.E. Sohn, K.D. Singer, M.G. Kuzyk, W.R. Holland, H.E. Katz, C.W. Dirk, M.L. Schilling and R.B. Comizzoli, Proceedings of NATO Advanced Research Workshops on "Nonlinear Optical Effects in Organic Polymers," Series E: Applied Sciences, (Kluwer Acad. Pub., London, 1989), 291-7.
10. G. Decher, B. Tieke, Ch. Brossard and P. Gunter, J. Chem. Soc., Chem., Commun., 1988, 933 (1988).
11. G.D. Boyd, R.C. Miller, K. Nassau, W.L. Bond and A. Savage, Appl. Phys. Lett., 5, 234 (1964).
12. C.-T. Chen and G.-Z. Liu, Ann. Rev. Mater. Sci., 16, 203 (1986).
13. F.C. Zumsteg, J.D. Bierlein and T.E. Gier, J. Appl. Phys., 43, 4980 (1976).
14. J.I. Thackara, G.F. Lipscomb, M.A. Stiller, A.J. Ticknor and R. Lytel, Appl. Phys. Lett., 52, 1031 (1988).

15. Y. Shi, W.H. Steier, L. Yu, M. Chen and L.R. Dalton, Appl. Phys. Lett., **58**, 1131 (1991).

16. K.B. Rochford, R. Zanoni, Q. Gong and G.I. Stegeman, Appl. Phys. Lett., **55**, 1161 (1989).

17. for example, O. Sugihara, T. Toda, T. Ogura, T. Kinoshita and K. Sasaki, Opt. Lett., **16**, 702 (1991).

18. X.F. Cao, L.P. Yu and L.R. Dalton, Technical Digest of the 1990 Optical Society of America annual meeting, paper ThE3, pp 165 (1990).

19. Ch. Brossard, M. Flosheimer, M. Küpfer and P. Günter, Opt. Comm., in press

20. W. Sohler and H. Suche, in <u>Integrated Optics III</u>, L.D. Hutcheson and D.G. Hall, eds, Proc. SPIE, **408**, 163 (1983).

21. T. Tanuichi and K. Yamamoto, Oyo Buturi, **56**, 1637 (1987).

22. Y. Ishigame, T. Suhara and N. Nishihara, Opt. Lett., **16**, 375, (1991) .

23. G.I. Stegeman and E.M. Wright, J. Optical and Quant. Electron., **22**, 95 (1990).

24. V. Mizrahi, K.W. DeLong, G.I. Stegeman, M.A. Saifi and M.J. Andrejco, Opt. Lett., **14**, 1140 (1989).

25. M.B. Marques, G. Assanto, G.I. Stegeman, G.R. Mohlmann, E.W.P Erdhuisen and W.H.G. Horsthuis, Appl. Phys. Lett., **58**, 2613 (1991).

26. D.M. Krol and M. Thakur, Appl. Phys. Lett., **56**, 1406 (1990); S.T. Ho, M. Thakur and A. Laporta, IQEC Digest, paper **QTUB5**, 40 (1990).

27. K.B. Rochford, R. Zanoni, G.I. Stegeman, W. Krug, E. Miao and M.W. Beranek, Appl. Phys. Lett., **58**, 13 (1991).

28. W. Blau, Opt. Commun., **64**, 85 (1987).

29. J.W. Wu, J.R. Helfin, R.A. Norwood, K.Y. Wong, O. Zamani-Kamiri and A.F. Garito, J. Opt. Soc. Am. B, 1989, **4**, 707 (1989).

30. T.G. Harvey, W. Ji, A.K. Kar, B.S. Wherrett, D. Bloor and P. Norman, Technical Digest of the 1990 CLEO meeting, paper **CTUH67**, page 146

31. V.S. Williams, Z.Z. Ho, N. Peyghamberian, W.M. Gibbons, R.P. Grasso, M.K. O'Brien, P.J. Shannon and S.T. Sun, Appl. Phys. Lett., **57**, 2399 (1990).

32. R. Burzynski, P. Banhu, P. Prasad, R. Zanoni and G.I. Stegeman, Appl. Phys. Lett., **53**, 2011 (1988).

33. P.D. Townsend, J.L. Jackel, G.L. Baker, J.A. Shelbourne III, S. Etemad, Appl. Phys. Lett., **55**, 1829 (1989).

34. D.R. Coulter, V.M. Miskowski, J.W. Perry, T.H. Wei, E.W. Van Stryland and D.J. Hagan, Proceedings of SPIE Meeting on Materials for Optical Switches, Isolators and Limiters, SPIE **1105**, 42 (1989).

RECENT DEVELOPMENT OF POLYMERS FOR NONLINEAR OPTICS

T. Kurihara and T. Kaino

NTT Opto-electronics Laboratories
Nippon Telegraph and Telephone Corporation
Tokai, Ibaraki, 319-11 Japan

1. INTRODUCTION

Highly efficient third-order nonlinear optical materials are required for picosecond optical swiching devices. Organic materials are of especially great interest because of their potentially fast response times and their high damage thresholds[1]. Among them, most studies to date are on π-conjugated polymers such as polydiacetylenes and polyacetylene, which have been reported to show larger third-order nonlinear susceptibilities; $\chi^{(3)}$ in the order of 10^{-10}-10^{-9} esu[2-4]. However, these π-conjugated polymers are usually in crystalline states and they are difficult to process because they are rarely fusible and are insoluble in almost all solvents. Therefore, amorphous or low crystallinity π-conjugated materials with good processability are expected to be an excellent material for device use. Thus, $\chi^{(3)}$ has recently been investigated in various amorphous and low crystallinity polymers such as poly(arylene vinylene)s (PAVs)[5-7] and nBCMU-polydiacetylenes[8].

Furthermore, enhancement of the magnitude of $\chi^{(3)}$ is required for future practical application using laser diodes. Several attempts to enhance $\chi^{(3)}$ have been proposed, which include an absorption band sharpening[9], multiblock conjugated copolymerization[10] based on organic superlattices, aromatic substitution directly bound to the PDA main chain[11] and donor-acceptor substitution to π-conjugated liner chains[12].

This paper shows our recent developments of PAV as amorphous $\chi^{(3)}$ material. It also proposes a new concept for molecular design of highly efficient $\chi^{(3)}$ materials. The first part of this paper concerns the relation between $\chi^{(3)}$ and π-conjugation length in PAV. The $\chi^{(3)}$ increment with π-conjugation length has been discussed by numerous theoretical works in one dimensional π-conjugated systems. This matter have been investigated through experimental studies on the conversion dependence of $\chi^{(3)}$ spectra for poly(2,5-dimethoxy-p-phenylene)vinylene (MO-PPV)[13]. The second part of this paper concerns novel symmetrical π-conjugated molecules having a large $\chi^{(3)}$.

Frontiers of Polymer Research, Edited by P.N. Prasad and
J.K. Nigam, Plenum Press, New York, 1991

We show the $\chi^{(3)}$ spectra of these materials which are newly synthesized on the basis of a concept for transition moment control between excited states. The mechanism of $\chi^{(3)}$ enhancement for these materials is discussed[14]. Finally, some requirements for the molecular design of a novel polymer material with a large $\chi^{(3)}$ is suggested.

2. EXPERIMENTAL

2.1 Materials

2.1.1 Poly(2,5-dimethoxy-p-phenylene)vinylene (MO-PPV)

We used a non-ionic methoxy-pendant precursor polymer which is chemically more stable than a conventional sulfonium salt precursor polymer and is soluble in many organic solvents[15]. A conventional spin-coating process was applied to fabricate the precursor films. Colorless transparent films were obtained on schott D263 glasses from a 0.1 wt% of tetrahydrofuran solution. These precursor films were converted by heating with acid catalysts.

2.1.2 Terephthal-bis-(4-N,N-diethylaminoaniline) [SBA] and 2,5-dichloro-terephthal-bis-(4-N,N-diethylaminoaniline) [SBAC]

SBA : X=H   SBAC : X=Cl

SBA and SBAC are easily synthesized by condensation reactions between terephthalaldehyde or 2,5-dichloroterephthalaldehyde and excess N,N-diethyl-p-phenylene diamine at 80 °C in tetrahydrofuran, using benzenesulfonic acid as a catalyst. These materials are purified by recrystallization from chloroform or dichloromethane[14]. The film deposition of the SBA and SBAC were carried out at 2.6 X 10^{-6} Torr vacuum pressure.

The most important consideration during the fabrication of the SBA and SBAC films was to prevent thermally degraded products from depositing onto substrates when the temperature was about 100 °C lower than the deposition temperature.

2.2 THG Measurements

The output signal of the third harmonic generation (THG) using the Maker-fringe method was measured. Measurements were performed between 1.475

and 2.1 µm of fundamental wavelength, using the differential-frequency generation of a Q-switched Nd-YAG laser and a tunable dye laser. The pulse duration was 5.5 ns, the pulse repetition rate 10 Hz. The peak power density was 100 MW/cm^2. The samples on a schott D263 glass substrates were mounted on a goniometer and rotated around an axis perpendicular to the laser beam. The laser beam was polarized parallel to the rotation axis. Fused silica was used as a standard. All measurements were performed in air using a 50 mm focusing lens to eliminate the influence of the air[16].

2.3 Evaluation of $\chi^{(3)}$

The experimental $\chi^{(3)}$ value of the sample is calculated using equation(1). In this equation, the $\chi^{(3)}$ value is obtained from the ratio of the THG intensities of the sample and a reference material with a known $\chi^{(3)}$ such as fused silica.

$$\chi_{exp}^{(3)} = \frac{2}{\pi}\frac{1}{\ell}\left(\frac{I_{3\omega}}{I_{3\omega,S}}\right)^{1/2}\chi_s^{(3)}\ell_{c,s}\tag{1}$$

where I is the thickness of the MO-PPV film, which is much smaller than the coherence length. The coherence length of the MO-PPV film is about 1.5 µm, which is estimated from the equation $l_c=\lambda/6(n_{3\omega}-n_\omega)$ where fundamental wavelength λ is 1.9 µm and refractive indices $n_{3\omega}$ and n_ω are 1.98 and 1.77, respectively. $\chi^{(3)}{}_s$ and $l_{c,s}$ are third-order susceptibility and coherence length of the standard fused silica glass, respectively. For the calculation, a $\chi^{(3)}{}_s$ of 2.79 X 10^{-14} esu at 1.907 µm of fundamental wavelength is used as reported by Meredith et al.[17] for a fundamental wavelength of 1.475-2.1 µm. The used l_c values are obtained from the equation $l_c=\lambda/6(n_{3\omega}-n_\omega)$ for each fundamental wavelength. $I_{3\omega,s}$ is a peak THG intensity in the fringe patterns of the silica. $I_{3\omega}$ is the THG intensity from the MO-PPV film with correction for interference effect between film and substrate, which is obtained from the following simple equation[18]:

$$I_{3\omega} = (I_{max} + I_{min})/2 - I_{sub}/2\tag{2}$$

where I_{max} and I_{min} are obtained from envelopes of superimposed THG intensity pattern, and I_{sub} is peak THG intensity of the glass substrate.

The experimental $\chi^{(3)}$ values should be corrected for internal attenuation effects. We use an expression from Kanetake's study[4] for corrected $\chi^{(3)}$ in terms of experimental $\chi^{(3)}$ given by equation (2), as follows:

$$\chi_{corr}^{(3)} = \chi_{exp}^{(3)}\left|\frac{\tilde{n}_{3\omega} + n_{3\omega,S}}{n_{3\omega,S} + 1}\right|\left(\frac{(\Delta k\ell)^2 + (\alpha\ell/2)^2}{1 + \exp(-\alpha\ell) - 2\exp(-\alpha\ell/2)\cos(\Delta k\ell)}\right)^{1/2}\tag{3}$$

where α is the absorption constant of the film, κ is the wave vector mismatch between the fundamental and third-harmonic wave in the film, $n_{3\omega,s}$ is the refractive index of the glass substrate. Complex refractive indices $\tilde{n}(=n+i\kappa)$ were

73

determined from the absorption spectra of the same films by use of the procedure reported by Nilsson[19].

3. RESULTS AND DISCUSSIONS

3.1 MO-PPV

3.1.1 Conversion from precursor polymer to MO-PPV

The π-conjugation lengths in PAVs are variable by controlling temperature and heating time. The thermal conversion process of MO-PPV precursor polymer is shown in scheme 1. Conversion proceeds from a non-conjugated precursor polymer to a full-converted polymer via partially converted precursor copolymers. These precursor copolymers have conjugated and non-conjugated sequences for various conversion levels.

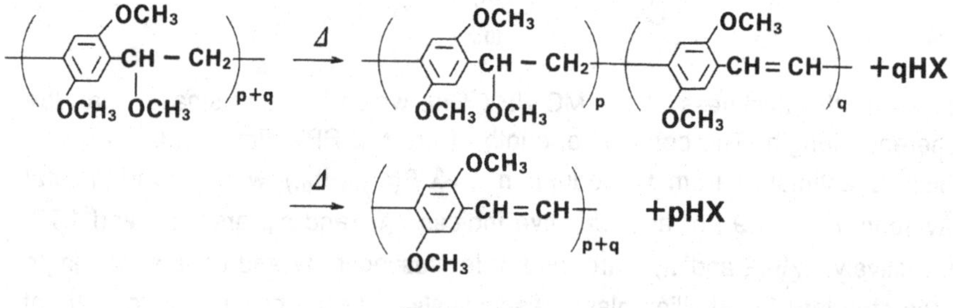

Scheme 1

The preparation conditions of MO-PPV films for various conversion levels are shown in Table 1. The extent of conversion is given by $q/(p+q)$, where p and q are fractions of non-conjugated and conjugated units, respectively. The extent of conversion was determined from an IR analysis; the intensity ratio of aliphatic C-O-C stretching at 1090 cm^{-1} to that of aromatic C-O-C stretching band at 1025 cm^{-1} was used.

Although the extent of conversion (%) shows stoichiometric change in the C-C bond structure, it is not always in accord with the extent of π-conjugation length, as described in the following sections.

Table I. The Preparation of MO-PPV by the Thermal Conversion of Precursors

Sample No.	conversion condition	conversion (%)	thickness(Å)
A	no conversion	0.0	342
B	25°C X 90 mim	56.7	302
C	(60°C X 120 min) + (80°C X 120 min)	67.7	310
D	90°C X 390 min	88.0	385
E	110°C X 160 min	83.1	347
F	(80°C X 30 min) + (120°C X 30 min) + (200°C X 60 min)	96.9	492

3.1.2 UV-VIS Spectra

Figure 1 shows optical absorption spectra of MO-PPV films for various conversion levels. Sample numbers in the figure correspond to those of Table 1. Increase of absorption intensities and red-shift of absorption maxima are observed with the increase in extent of conversion. The F film is approximately in a full-converted state. A specific absorption at about 340 nm in spectrum F is presumed to be a cis-structure of a vinylene unit.

Generally, at the initial conversion level for PAV, there are not cis-vinylenes but short trans sequences. This is because the activation energy of cis-vinylene formation is considered to be higher than that of trans-vinylene formation. At the final conversion level, the development of long rigid π-conjugated sequences from short trans sequences is completed, and even the highly bent non-conjugated parts between the rigid π-conjugated sequences are converted to cis-vinylene form by supplying sufficient thermal energy to cover its activation energy. Therefore, the appearance of cis structure can be used as an indiex for the existence of long rigid π-conjugated sequences.

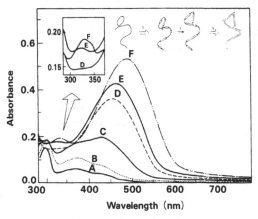

Fig. 1 Absorption spectra of MO-PPV films for various conversion levels.

As shown in Table 1, the B, C and D films were heated under the glass transition temperature (Tg) of the precursor polymer, which is about 80-100°C. Many short trans sequences are formed with low thermal energy in these films. In E and F films heated over Tg, short π-conjugated sequences are connected one after another to form long rigid π-conjugated sequences, and the cis-vinylenes were formed.

It should be noted that the relation between conversion levels and absorption characteristics is not invariant but dependent on the preparation condition. The E film, though in a slightly lower converted state than the D film, shows stronger absorption intensity than the D film. The cis structure, which appears due to the formation of rigid π-conjugated sequences, is not observed in the D film but observed in the E film. That is, in spite of its lower conversion level, the formation of the long rigid π-conjugated sequences in the E film was probably accelerated by preparation conditions, i.e., a short heating time at high temperature. This fact suggests that the formation of rigid π-conjugated sequences depends extensively on conversion conditions.

3.1.3 $\chi^{(3)}$ Spectra

Figure 2 shows the conversion dependence of the $\chi^{(3)}_{corr}$ spectra superimposed on that of the absorption spectra. Maximum $\chi^{(3)}$ value, which may be due to three-photon resonance, reached 1.6×10^{-10} esu[13]. This value is comparable to those of other amorphous π-conjugated polymers, such as nBCMU-polydiacetylene. Figure 2 clearly indicates that $\chi^{(3)}$ increases corresponding to the enlargement of the absorption intensity. The increase of $\chi^{(3)}$ from the D film to the F film is particularly significant. When comparing D film with E film, the E film with longer rigid π-conjugated sequences, shows larger $\chi^{(3)}$ than the D film with a higher extent of conversion The interrelation of D and E films in $\chi^{(3)}$ spectra resembles that in the absorption spectra, but the difference is more significant in the $\chi^{(3)}$ spectra than in the absorption spectra. The $\chi^{(3)}$ is more steeply enhanced by the formation of rigid π-conjugated sequences than absorption intensity. Thus one can say that the effective length of rigid π-conjugated sequences takes an important role in $\chi^{(3)}$ intensity.

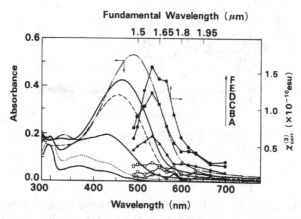

Fig. 2   Conversion dependence of the $\chi^{(3)}_{corr}$ spectra superimposed on that of the absorption spectra.

A detailed comparison between absorption spectra and $\chi^{(3)}$ spectra reveals the specific origin of resonant $\chi^{(3)}$ in MO-PPV films. As conversion proceeds, absorption peak shifts gradually to longer wavelength. On the other hand, $\chi^{(3)}$ peaks are fixed over all conversion levels. The wavelengths of $\chi^{(3)}$ peaks are 30 nm longer than the absorption peak wavelength of full-converted film. This indicates that there will be a dominant resonant level for $\chi^{(3)}$ at the position with a 30 nm longer wavelength than absorption peak. Also, considering the steep enhancement of $\chi^{(3)}$ associated with the formation of rigid π-conjugated sequences, we suggest that the $\chi^{(3)}$ spectrum peak and the absorption peak are ascribed to the different sequences as to the rigid π-conjugation length. The resonant level of $\chi^{(3)}$ is probably related to the existence of long rigid π-conjugated sequences[13].

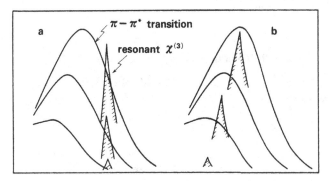

Fig. 3   Schematic representation of appearance of $\chi^{(3)}$ spectra

We have to take into account the distribution of the π-conjugation length concerning interaction between $\chi^{(3)}$ spectra change and absorption spectra change. If we assume that particular distribution of long rigid π-conjugated sequences is only effective for a large resonant $\chi^{(3)}$, $\chi^{(3)}$ spectra will not appear as shown in Fig.3(b) but will appear as shown in Fig.3(a). The actual observation shown in Fig.2 is consistent with Fig.3(a). Details on the validity of this interpretation will be discussed elswhere[20]. Measurements of the $\chi^{(3)}$ spectra of oriented MO-PPV films should support these interpretation; the drastic increase in components of the long rigid π-conjugated sequences through a stretching process will contribute to an enhancement of $\chi^{(3)}$.

3.2   SBA and SBAC

3.2.1 Molecular design for $\chi^{(3)}$ materials

Table 2 shows the molecular design for $\chi^{(3)}$ materials. Organic $\chi^{(3)}$ materials can be classified into two categories. One is π-conjugated polymers, such as poly(phenylene vinylene). The other is short π-conjugated molecules, such as dietylamino-nitrostilbene.

In the π-conjugated polymers, the construction of rigid π-conjugated sequences plays an important role in $\chi^{(3)}$ enhancement, as described in the preceding section. Theoretically, it is important to enhance the transition from the Bu excited state to the Ag excited state[21]. On the other hand, the $\chi^{(3)}$ of the short π-conjugated molecules depends on dipole moment change between the ground state(G) and the Bu excited state[22] due to intramolecular charge-transfer effect from donor to acceptor[23,24].

In this section, we propose novel symmetrical molecules having a large $\chi^{(3)}$. The key-point behind this novel molecular design is transition moment control between excited states[25], which is enhancement of the transition moment between the Bu and Ag excited state by symmetrical donor-substitution on short π-conjugation system. This idea is different from the well-known donor-acceptor substitution based on a two-level model. To embody this molecular design, SBA and SBAC were newly synthesized[14]. We use the abbreviation SBA, to refer to symmetrically substituted benzylidene aniline. SBAC is an abbreviation of SBA dichloride.

Table 2. Molecular Design for $\chi^{(3)}$ Materials

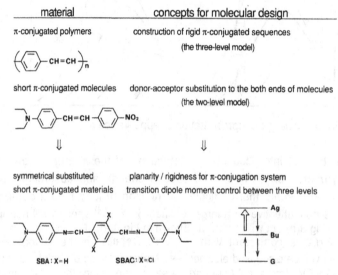

material	concepts for molecular design
π-conjugated polymers	construction of rigid π-conjugated sequences (the three-level model)
short π-conjugated molecules	donor-acceptor substitution to the both ends of molecules (the two-level model)
symmetrical substituted short π-conjugated materials	planarity / rigidness for π-conjugation system transition dipole moment control between three levels

SBA : X = H   SBAC : X = Cl

Transition dipole moments contributing to γ

3.2.2 $\chi^{(3)}$Spectra

Figure 4 shows the $\chi^{(3)}$ spectra and the absorption spectra of SBA and SBAC thin films made by vacuum deposition . A strong $\chi^{(3)}$ enhancement is observed when incident wavelengths are three times the wavelengths of the lowest energy peaks in the absorption spectra. These large $\chi^{(3)}$ can be identified as three-photon resonance. The assignments for structure of absorption spectra is discussed later. The three-photon resonant $\chi^{(3)}$ values reach around 10^{-10} esu[14]. Even in the off-resonant regions, the $\chi^{(3)}$ value is 4×10^{-11} esu. These values are comparable to those of non-crystalline or non-oriented π-conjugated polymers.

3.2.3 Vibronic Coupling Effect for $\chi^{(3)}$

Figure 5 shows the concentration-dependence of $\chi^{(3)}$ for SBAC in a PMMA matrix at a fundamental wavelength of 1.9 μm, which is an off-resonant

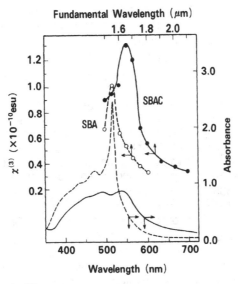

Fig. 4 $\chi^{(3)}$ spectra of deposited SBA and SBAC films

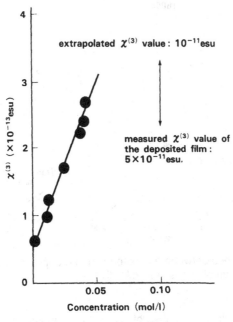

Fig. 5 Concentration dependence of $\chi^{(3)}$ for SBAC doped PMMA.

wavelength region. The $\chi^{(3)}$ values increased in proportion to the concentration. The extrapolated $\chi^{(3)}$ value of a 100% SBAC state is about 1.0 X 10^{-11} esu provided molar density of the deposited SBAC film is about 2 mol/l, analogue with SBAC molecular crystals. The measured $\chi^{(3)}$ value of the deposited SBAC film is 5 X 10^{-11} esu at a fundamental wavelength of 1.9 μm as shown in Fig. 4 Thus the $\chi^{(3)}$ of the deposited SBAC thin film is 5 times as large as that of SBAC-doped PMMA.

The difference between the deposited film and the doped film was also observed in their absorption spectra. As shown in Fig. 6, the spectrum of the deposited film has a structure while the spectrum of the doped film has not. The distance between neighboring peaks in the deposited SBAC film is approximately 1600 cm^{-1}. This value is ascribed to the C=C and C=N stretching vibrations in an excited state. A corresponding vibrational mode for the ground state is observed at 1585 cm^{-1} in a Raman spectrum (λ_{ex}=457.9 nm) for the deposited SBAC film. The structure of the absorption spectrum for the deposited SBAC film is due to vibronic coupling of electronic transition. On the absorption spectra of π-conjugated polymers such as PT[26] or PPV[27], vibronic structures are well assigned. A similar phenomenon is observed with deposited SBA film. For these molecules, vibronic coupling in electronic excitation produces some excited states contributing to $\chi^{(3)}$ as intermediate states[25], so it should be noted that this coupling enhances $\chi^{(3)}$. On the other hand, the absorption spectrum for SBAC molecules in the PMMA matrix is not as broad as that of the deposited film. In this case, the coupling is weak and the $\chi^{(3)}$ is not so large.

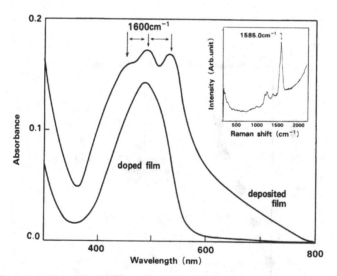

Fig. 6   Spectra difference between deposited film and doped film of SBAC, and Raman spectrum of deposited SBAC film.

Usually, vibronic coupling in the absorption spectrum appears when a planarity of π-conjugated molecules is relatively high. For example, it was reported that vibronic structure in absorption spectra for retinal and related molecules became clear as the planarity of π-conjugation for the molecules increased[28].

We consider that the $\chi^{(3)}$ enhancement by the planarity of π-conjugation system is attributed not only to vibronic coupling but also to the contribution of transition moments between excited states.

4. CONCLUSIONS

The $\chi^{(3)}$ spectra of MO-PPV films for various conversion levels have been measured for first time. It has been clearly demonstrated that the $\chi^{(3)}$ increases with the development of π-conjugation length. In particular, $\chi^{(3)}$ increases significantly due to three-photon resonance at the final conversion level where long rigid π-conjugated sequences are formed. The position of three-photon resonance is 30 nm longer than the absorption peak wavelength, suggesting the existence of a certain dominant resonant level for $\chi^{(3)}$ which possibly results from the distribution of long rigid π-conjugated sequences[13].

We have synthesized new symmetrical π-conjugated molecules, named SBA and SBAC. The magnitudes of $\chi^{(3)}$ for SBA and SBAC films reach 10^{-10} esu at their three-photon resonant regions. This value is comparable to those of π-conjugated polymers. One reason for these large $\chi^{(3)}$ is the planarity of their π-conjugated structure. The positions of three-photon resonance lie at the lowest energy peaks in the vibrational structures of their absorption spectra[14].

Considering these, we summarize the requirements to enhance $\chi^{(3)}$ through an enlargement of the transition moment from Bu to Ag, as follows:

1. Constructing long rigid π-conjugated sequences
2. Achieving high-planarity of the π-conjugation system
3. Symmetrical substitution of donors at both π-conjugation ends

A novel polymer material with a large $\chi^{(3)}$ should be able to be developed by satisfying all these requirments.

ACKNOWLEDGEMENTS

The authors wish to thank Yuhei Mori, Naoki Oba, Satoru Tomaru, Hiroaki Hiratsuka and Kei Murase of NTT. Furthermore, they wish to thank Hideyuki Murata, Professors Shogo Saito and Tetsuo Tsutsui of Kyusyu University for thier kind support and discussion.

REFERENCES

1. "Nonlinear Optical and Electroactive Polymers" Eds. P.N.Prasad and D.R.Ulrich, (Plenum Press, New York, 1988).
2. C.S.Sauteret, J..Hermann, R.Frey, F.Pradere, J.Ducuing, R.H.Baughmann, R.Chance, *Phys. Rev. Lett.* **36**, 956 (1976).
3. W.S.Fann, S.Etemad, G.L.Baker and F.Kajar, *Phys.Rev.Lett.* **62**, 492 (1989).
4. T.Kanetake, K.Ishikawa, T.Hasegawa, T.Koda, K.Takeda, M.Hasegawa, K.Kubodera and H.Kobayashi, *Appl. Phys. Lett.* **54**, 2287 (1989).
5. T.Kaino, K.Kubodera, S.Tomaru, T.Kurihara, S.Saito, T.Tsutsui and S.Tokito, *Electronics Letters* **23**, 1095 (1987).
6. T.Kaino, K.Kubodera, H.Kobayashi, T.Kurihara, S.Saito, T.Tsutsui, S.Tokito and H.Murata, *Appl. Phys. Lett.* **53**, 2002 (1988).

7. T.Kaino, H.Kobayashi, K.Kubodera, T.Kurihara, S.Saito, T.Tsutsui and S.Tokito, *Appl. Phys. Lett.* **54**,1619 (1989).

8. N.E.Schlotter, J.L.Jackel. P.D.Townsend and G.L.Baker, *Appl. Phys.Lett.* **56**, 13 (1990).

9. T.Yoshimura, *Opt. Commun.* **70**, 535 (1989).

10. S.A.Jenekhe, S.K.Lo and S.R.Flom, *Appl. Phys. Lett.* **54**, 2524 (1989).

11. H.Nakanishi, H.Matsuda, S.Okada, A.Masaki and M.Kato, Proceedings of International Workshop on Crystal Growth of Organic Materials, pp.103-112 (1989).

12. A.F.Garito, J.R.Heflin, K.Y.Wong and O.Z.Khamiri, Special Publication of Royal Society of Chemistry **69**, "Organic Materials for Nonlinear Optics", pp16-27 (1989).

13. T.Kurihara, Y.Mori, T.Kaino, H.Murata, N.Takeda, T.Tsutsui and S.Saito, *Chem. Phys. Lett.* (to be published).

14. T.Kurihara, N.Oba, Y.Mori, S.Tomaru and T.Kaino, *J. Appl. Phys.* (to be published).

15. T.Momii, S.Tokito, T.Tsutsui and S.Saito, *Chem.Lett.* 1201 (1988).

16. H.Kobayashi, K.Kubodera, T.Kurihara and T.Kaino, Extended Abstracts (The 36th Spring Meeting,1989); The Japan Society of Applied Physics and Related Societies, 2pG8 (1989).

17. G.R.Meredith, B.Buchalter and C.Hanzlik, *J. Chem. Phys.* **78**, 1533 (1983).

18. K.Kubodera and T.Kaino, Nonlinear Optics of Organics and Semiconductors,edited by T.Kobayashi, pp.163-170, (Springer Proceedings in Physics 36, Berlin, 1989)

19. P.O.Nilsson, *Applied Optics* **7**, 435 (1968).

20. Y.Mori, T.Kurihara, T.Kaino, S.Tomaru, (unpublished).

21. Y.Mori and Y.Okano, Material Research Society Symposium Proceedings **173**, "Advanced Organic Solid State Materials", pp. 665-670 (1989).

22. J.F.Ward, *Rev. Mod. Phys.* **37**, 1 (1965).

23. T.Kurihara, H.Kobayashi, K.Kubodera and T.Kaino, *Chem. Phys. Lett.* **165**,171 (1990).

24. T.Kurihara, H.Kanbara, H.Kobayashi, K.Kubodera, S.Matsumoto and T.Kaino, *Opt. Commun.* (to be published).

25. Y.Mori, T.Kurihara, T.Kaino, S.Tomaru, (unpublished).

26. T.Danno, J.Kurti and H.Kuzmany, *Phys. Rev. B* **43**(6), 4809 (1991).

27. S.Lefrant and J.P.Buisson, *Synth. Met.* **37**, 91 (1990).

28. A.Warshel and M.Karplus, *J. Am. Chem. Soc.* **96**, 5677 (1974).

NONLINEAR OPTICAL SPECTROSCOPY IN CONJUGATED POLYMERIC THIN FILMS

F. Kajzar

Commissariat à l'Energie Atomique
Direction des Technologies Avancées
LETI, DEIN/LPEM
CEN Saclay
91191 Gif Sur Yvette Cedex, France

INTRODUCTION

Conjugated polymers with quasi-one dimensional π electron delocalization represent a great potential for applications in nonlinear optical devices. This is due to their enhanced, electronic hyperpolarizability and connected with that fast, under picosecond response time. On the other hand the rich electronic structure of these polymers gives rise to different resonance enhancements in cubic susceptibility . A systematic study of such resonances is interesting from both fundamental and practical points of view as it will be discussed later.

The bulk polarization of a medium under an external electric forcing field (e.g. optical field) can be developed into its power series and its i-th component is given by

$$P_i(t) = P_i^{\circ} + \chi_{ij}^{(1)} E_j(t) + \chi_{ijk}^{(2)} E_j(t)E_k(t) + \chi_{ijkl}^{(3)} E_j(t)E_k(t)E_l(t) + \ldots \tag{1}$$

where P° is the permanent static polarization, $\chi^{(n)}$ are (n+1) rank tensors describing the nonlinear optical response of the system and E_j is the electric field component along the j-th direction. For the sake of simplicity we use hereafter the Einstein's notation.

The conjugated polymers are intrinsically centrosymmetric and all odd rank tensors are vanishing ($\chi^{(2n)} \equiv 0$) as well as there is no, a priori, macroscopic static polarization ($P^{\circ} \equiv 0$). For this reason our discussion here will be limited to four photon processes described by cubic susceptibilities.

The quantum mechanical formula for cubic susceptibility describing any four photon process with three input and one output (or vice versa) photons and derived by the time dependent perturbation calculations reads[1]:

Frontiers of Polymer Research, Edited by P.N. Prasad and
J.K. Nigam, Plenum Press, New York, 1991

$$\chi^{(3)}(-\omega_4;\omega_1,\omega_2,\omega_3) = WFNK\left[\frac{\Omega_{gl}\Omega_{lm}\Omega_{mn}\Omega_{ng}}{(E_{lg}-\omega_4)(E_{mg}-\omega_1-\omega_2)(E_{ng}-\omega_1)} + \right.$$

$$\frac{\Omega_{gl}\Omega_{lm}\Omega_{mn}\Omega_{ng}}{(E_{lg}+\omega_3)(E_{mg}-\omega_1-\omega_2)(E_{ng}-\omega_1)} + \frac{\Omega_{gl}\Omega_{lm}\Omega_{mn}\Omega_{ng}}{(E_{lg}+\omega_1)(E_{mg}+\omega_1+\omega_2)(E_{ng}-\omega_3)} +$$

$$\frac{\Omega_{gl}\Omega_{lm}\Omega_{mn}\Omega_{ng}}{(E_{lg}+\omega_1)(E_{mg}+\omega_1+\omega_2)(E_{ng}+\omega_4)} - \left\{\frac{\Omega_{gm}\Omega_{mg}\Omega_{gn}\Omega_{ng}}{(E_{mg}-\omega_4)(E_{mg}-\omega_3)(E_{ng}-\omega_1)} +\right.$$

$$\frac{\Omega_{gm}\Omega_{mg}\Omega_{gn}\Omega_{ng}}{(E_{mg}-\omega_3)(E_{ng}+\omega_2)(E_{ng}-\omega_1)} + \frac{\Omega_{gl}\Omega_{lm}\Omega_{mn}\Omega_{ng}}{(E_{mg}+\omega_4)(E_{mg}+\omega_3)(E_{ng}+\omega_1)} +$$

$$\left.\left.\frac{\Omega_{gm}\Omega_{mg}\Omega_{gn}\Omega_{ng}}{(E_{mg}+\omega_3)(E_{ng}-\omega_2)(E_{ng}+\omega_1)}\right\}\right] \tag{2}$$

where W is a numerical factor depending on units used, K is a factor depending on polarization and electric field definitions as well as on the four photon process under consideration, F is the local field factor and N is the number of molecules per unit volume. Following electric and polarization fields definitions are used throughout this paper

$$E_\omega(t) = \frac{1}{2}\{E_\omega e^{i\omega t} + E_\omega^* e^{-i\omega t}\} \tag{3}$$

$$P_\omega^{NL}(t) = \frac{1}{2}\{P_\omega^{NL} e^{i\omega t} + P_\omega^{NL*} e^{-i\omega t}\} \tag{4}$$

E_{ij} in Eq. (2) is the energy difference between i and j states in ℏ units, g is the fundamental and l m, n are excited states (cf. Fig. 1), $\Omega_{ij} = \langle i|er|j\rangle$ are transition matrix elements between $|i\rangle$ and $|j\rangle$ state. The sum in Eq. (2) is over all excited states (excluding the ground g). Equation (2) shows occurence of possible resonance enhancements in cubic susceptibilities when the sum, difference or any linear combination of interacting photon energies will approach the energy difference between fundamental and one of the excited states with corresponding symmetry (selection rules). Different multiphoton resonance enhancements as well as nonlinear optical processes leading to their observation will be discussed in the next Chapter.

MULTIPHOTON EXCITATIONS

An example of a four photon process is shown in Fig. 1. This process is going through virtual states represented schematically by broken lines. As mentioned before, if one of the virtual states matches with an excited state of the unperturbed system a resonance enhancement will occur

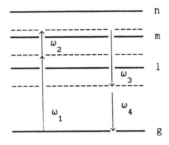

Fig. 1. Schamatic representation of a four photon process. Solid lines represent fundamental and excited states of unperturbed system whereas broken lines show virtual states. A resonance enhancement in $X^{(3)}$ occurs when one of the virtual states matches with an excited state of unperturbed system.

in corresponding cubic susceptibilty describing this process(cf.Eq.(2)). In this case one of the denominators in Eq.(2) will tend to zero and consequently the $X^{(3)}$ susceptibility will be complex. Depending on the number of photons involved in the resonant transition it will be called one, two and three photon resonances.

One photon resonances

Typical one photon resonant processes observed in conjugated polymers are saturation absorption (SA) and photoinduced absorption (PIA). They were observed in typical pump-probe experiments (cf. Fig. 2). In the first case the process is degenerate (the same pump and probe frequencies) and is descibed by the Kerr susceptibility $X^{(3)}(-\omega;\omega,-\omega,\omega)$ ($\omega_1 = \omega_2 = \omega_3 = \omega_4 = \omega$ and $K = 3/4$). The photoinduced absorption is desc/ ribed by another Kerr susceptibility $X^{(3)}(-\omega_2;\omega_1,-\omega_1,\omega_2)(\omega_4=\omega_2;\omega_3=\omega_1$ and K=3/2). The one photon resonance may be also observed in degenerate four wave mixing experiments like phase conjugation and optical Kerr effect. They are described by the same cubic susceptibility as the saturation absorption process.

In the SA and PIA processes the corresponding Kerr susceptibilities describe the nonlinear polarization of the medium created by an intense pump beam.

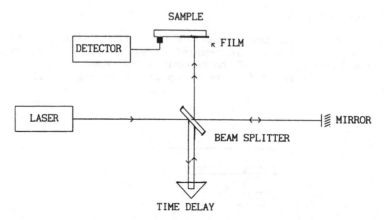

Fig.2. Schematic representation of an experimental arrangement for saturation absorption measurements by a photoacoustic technique
(after ref. 2).

This is directly connected with the nonlinear index variation

$$n = n_0 + n_2 I \qquad (5)$$

where I is the pump intensity.
The nonlinear index of refraction is linked to the cubic susceptibility
through the following relation

$$n_2 = N_n X^{(3)}/c\, n_0^2 \qquad (6)$$

where $N_n = 12\pi^2$ in esu and $N_n = 3/4\varepsilon_0$ in rational units. $X^{(3)}$ is the Kerr
susceptibility describing saturation or photoinduced absorption.
In both cases one measures relative change of transmission which is
directly linked to the imaginary part of nonlinear refractive index κ
$(n = n^r + i\kappa)$:

$$\frac{\Delta T}{T} = \frac{4\pi l}{\lambda} \kappa_2 I \qquad (7)$$

where I is the pump intensity, l is the thin film thickness and λ is the
probe wavelength. Because of absorption and consequently possible thermal
damage to the material experiments are possible on thin films only with
ultrashort laser pulses. Figure 3 shows an example of such a saturation
experiment performed in transmission on 11 Langmuir-Blodgett monolayers
thick thin film of polydiacetylene, obtained by a transfer of polymer
film prealably polymerized on water subphase[2]. The response time,
unresolved in this experiment, is below 3ps, the last number being the

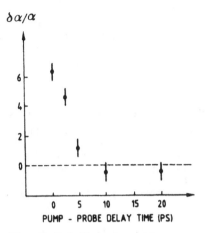

$\delta \alpha / \alpha$

PUMP - PROBE DELAY TIME (PS)

Fig. 3. Relative absorption co-
efficient variation $\delta \alpha / \alpha$ as a
function of pump - probe delay
time in the saturation absorp -
ption measurements
(after ref. 2).

Larger values are observed in photoinduced absorption, with, however a
much slower response time.[5]

Two photon resonances

As we mentioned before, the conjugated polymers are intrisically
centrosymmetric. In that case the excited levels can be classified into
gerade (g) and *ungerade* (u) states with corresponding selection rules
(one photon transitions forbidden between two states of the same symme -
try). Consequently if the fundamental state is a g state as it is the case in these polymers all excited g states are not seen in the optical
absorption spectrum. If g» g' transitions are forbidden with odd number
of photons they are allowed with even (two in lowest case) number of
photons. These g' states are called for this reason two photon states.

The two photon transitions between g and g' states can be realized
in multiple photon processes including at least four interacting photons
(we do not consider three photon processes possible in
noncentrosymmetric media where the resultant electric field on the
molecule breaks the selection rules discussed above) and they manifest
by a resonance enhancement in corresponding nonlinear susceptibility
describing the given process. Obviouslv the associated cubic
susceptibility will be complex. Therefore it is important to dispose an
experimental technique allowing not only the modulus of corresponding

cubic susceptibility determination but also its phase Φ ($X^{(3)}$ = $|X^{(3)}|exp(i\Phi)$). Such an opportunity is offered by harmonic generation techniques in thin films or in solutions.

Harmonic generation is a coherent process; the resultant harmonic field being in coherence with the exciting fundamental field. For these reason the harmonic generation measurements give the electronic susceptibility with very fast ($\propto 10^{-15}$ s) response time. For conjugated polymers and four photon processes under consideration here two harmonic generation processes are possible: third harmonic generation (THG) and electric field induced second harmonic generation (EFISHG).

Third harmonic generation

The THG process (K=3/4, $\omega_1 = \omega_2 = \omega_3 = \omega$, $\omega_4 = 3\omega$) takes place in every material medium and consequently also in surrounding air. It leads, if THG experiments are given in air, to a significant change in harmonic intensities and their variation as a function of e.g. incidence angle.[5] For these reasons, if the THG measurements are performed in air, the air contribution has to be taken into acoount exactly, using e.g. the procedure and formulas derived by Kajzar and Messier[5] otherwise the measurements have to be done in vacuum. We note here that for every four photon process, if experiments are not done in vacuum, there exists a less or more important air contribution whose importance, in principle has to be checked out consequently. For the sake of simplicity we consider here the THG experiments performed in vacuum.

In general a thin film has to be supported by a substrate which contributes also to the measured harmonic intensity. This drawback offers at the same time an opportunity for the $X^{(3)}$ phase determination. In fact, in the thin film case the optical pathlength variation between the free and bound waves is realized by a rotation of the substrate supporting thin film along an axis perpendicular to the beam propagation direction. The harmonic intensity for a single side coated substrate and as a function of incidence angle θ is given by[6]

$$I_{3\omega}(\theta) = N_f \left(\frac{X^{(3)}}{\Delta\varepsilon}\right)_s^2 |e^{i(\phi_\omega^s + \phi_\omega^p)}\left[T_1(1-e^{-i\Delta\phi_s}) + \rho T_2 e^{i\Phi}(e^{i\Delta\phi_p}-1)\right]|^2 I_\omega^3$$

$$(8)$$

where

$$N_f = \begin{cases} \dfrac{64\pi^4}{c^2} & \text{in e.s.u.} \\[3mm] \dfrac{1}{4\varepsilon_0 c^2} & \text{in SI units} \end{cases}$$

$$\rho = \left(\frac{X^{(3)}}{\Delta\varepsilon}\right)_p / \left(\frac{X^{(3)}}{\Delta\varepsilon}\right)_s \qquad\qquad (9)$$

where θ_ω, $\theta_{3\omega}$ are propagation angles at $\omega, 3\omega$ frequency, respectively. and l is the nonlinear medium thickness. The nonlinear medium absorption is taken into account by introducing complex refractive indices : $n = n^r + i\kappa$.

Equation (8) shows that if one knows the substrate cubic susceptibility, its thickness and refractive indices at fundamental and harmonic frequencies one can determine the modulus and the phase of of polymer thin film hyperpolarizability. It can be done in two ways. First one consists on THG measurements on the assembly of polymer film and substrate. Consequently the polymer film is removed in situ and measurements are done on the substrate alone.[6-9] Alternatively, if one dispose a substrate with a perfect planarity and surface quality one can determine its characteristics before the thin film deposition and introduce them after when reducing THG mesurements from the polymer film and substrate assembly[10]. An example of such a $X^{(3)}$ modulus and phase determination[7] is shown in Fig. 4.

In the case when the harmonic field generated in the polymer film is much larger than that generated in substrate itself Eq. (8) reduces to a much simpler form[11]

$$I_{3\omega} = N_f \left(\frac{X^{(3)}}{n_\omega^P + n_{3\omega}^P}\right)^2 \left(\frac{l}{\lambda_\omega}\right)^2 \mid A_1 \mid^2 \mid t_{os}^\omega \mid^6 \mid t_{sp}^\omega \mid^6 I_\omega^3 f_a \qquad (11)$$

where $N_f = \begin{cases} \dfrac{2304\ \pi^6}{c^2} \\ \dfrac{9\pi^2}{\varepsilon_0^2 c^2} \end{cases}$

l is the polymer thin film thickness, t's are transmission factors be - tween vacuum and substrate (os)a and substrate and polymer film (sp), respectively and under an assumption that the polymer film is on the detector side. The coefficient f in Eq. (11) takes account of absorption. For a thin film absorbing at harmonic r frequency only it is given by

89

$$f_a = \frac{1}{36\pi^2} \frac{\{1 - \exp(-\alpha^P_{3\omega} l/2)\}^2 + (\Delta\phi)^2 \exp(-\alpha^P_{3\omega} l/2)}{\{(n^P_{3\omega})^2 - (n^P_{\omega})^2 - (\kappa_{3\omega})^2\}^2 + 4(n^P_{3\omega}\kappa^P_{3\omega})^2} \left(\frac{\lambda_\omega}{l}\right)^2 (n^P_\omega + n^P_{3\omega})^2 \qquad (12)$$

where $\alpha^P_{3\omega}$ is the thin film linear absorption coefficient at harmonic frequency and the phase mismatch $\Delta\phi$ is given by Eq. (10). The factor A in
Eq. (11) arises from boundary conditions[5] In that case one can determine
only the modulus of $X^{(3)}$ susceptibility.

The two photon resonance in THG was first observed in
Langmuir-Blodgett films of blue form of polydiacetylene with
aliphatic chain sidegroups[11]. A strong resonance enhancement was seen
at 1.35 μm fundamental wavelength falling at the polymer transparency
range, with harmonic frequency lying in the deep of polymer optical
absorption spectrum. This result was interpreted in terms of a two
photon resonance with two photon state lying below one photon state.
This very important observation is in good agreement with fluorescence
spectroscopy in polyenes[12-13] and in short diacetylene oligomers[14]. Such
an order of excited states evidences importance of Coulomb correlations
in these materials. In fact, as it show calculations of Hayden and
Mele[15] for short polyenes with the increasing Coulomb correlations
strength a crossover between two and one photon states occurs; the first
lying below the second one at larger Coulomb correlations. This result
has been confirmed by other calculations on conjugated
polyenes[16-18] showing importance of Coulomb correlations in these
materials.

Since that time a lot of wave dispersed THG measurements have been
done on different conjugated polymers like: red form polydiacetylenes[19],
polyacetylenes[20-21] spun-p-4BCMU thin films[9,22-23] polythiophenes[8,10,23-24],
epitaxied blue form polydiacetylene[25], p-DCH[26], blue form p-4BCMU[27],
thiophene oligomers[28]. In all these materials the THG data are in favour
of two photon state lying below the one photon state. The same
conclusion has been also drawn from the $X^{(3)}$ phase measurements in
polyphenylacetylenes[29] and poly(decyl)thiophene[30].

Interesting is the case of the stable red form of polydiacetylene.
The early measurements[19] show a shoulder on three photon resonance
curve at around 1.22 μm (cf. Fig. 5) fundamental wavelength. The recent
THG measurements by Torruellas[23] (see also Torruellas et al.[9]) on spun
p-4BCMU red films (with a very similar linear optical absorption
spectrum) allowed a better resolution of two and three photon
resonances;with the first weeker than the second one. The confirmation
of this result was obtained by Torruellas[23] in a direct two photon absor-
tion measurements on this material. In fact Torruellas has found two two
photon state, one lying below one photon state exactly at the same
position as reported by Kajzar and Messier for PDA red form[19] and the
second one with significantly stronger two photon absorption intensity
above the one photon state cf. Fig. 6). This result reconciliate the THG
data with early wave dispersed four wave mixing experiment by Chance and
cow.[31] on polydiacetylene solution. Although in a general case the elect-

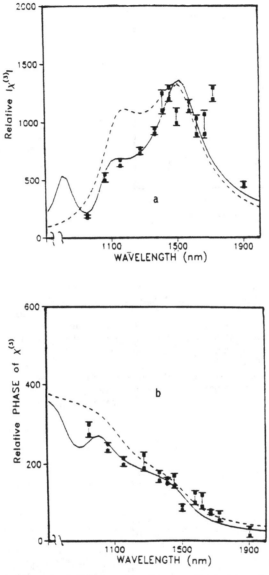

Fig. 4. Relative modulus and phase of cubic susceptibility $\chi^{(3)}$ $(-3\omega:\omega,\omega,\omega)$ (with respect to that of silica substrate) as determined from THG measurements on a thin film of poly(alkyl)thiophene. Broken and solid lines represent calculated values within a three and four level model, respectively (after Torruellas et al[8]).

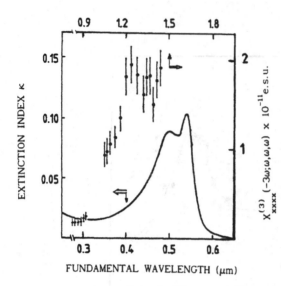

Fig. 5. Wavelength dependence of cubic susceptibility $X^{(3)}_{xxxx}(-3\omega;\omega,\omega,\omega)$ for a red form of polydiacetylene (after ref. 19).

ronic structure of molecules in solution is not the same as in thin film however in the case of PDA's red form they are very similar. In fact the wave dispersed $|X^{(3)}|$ and phase Φ spectra for p-4BCMU[9,23] and for poly(alkyl)thiophenes[8,23] (see also Fig.4) fit much better within a four level model with one two photon state below u state and one above than within a three level model with only one two photon state taken into account. This electronic structure agrees very well with the recent theoretical calculations by Dixit et al[16] who found that the oscillator strength corresponding to the transition between u and above lying g' state is significantly larger than between g and u state. It is also in good agreement with recent electroabsorption measurements on monooriented p-DCH thin films.[32]

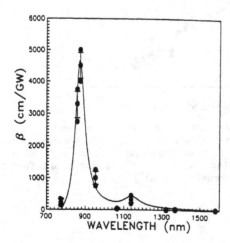

Fig. 6. Wavelength dependence of two photon absorption coefficient in spun p-4BCMU thin film obtained from direct measurements (reprinted from ref. 23 with author's permission).

The electric field induced second harmonic generation (EFISHG) (K = 3/2; $\omega_1 = \omega_2 = \omega$; $\omega_3 = 0$; $\omega_4 = 2\omega$) measurements on polymeric thin films offer another opportunity for nonlinear spectroscopy studies in polymeric thin films.

The conjugated quasi 1D polymers are intrinsically centrosymmetric and no second harmonic generation is observed. By applying an external fields one breaks the centrosymmetry and one observes the second harmonic generation whose intensity is given by

$$I_{2\omega} = N_f \left| \frac{X^{(3)}(-2\omega;\omega,\omega,0)}{(n_\omega^P)^2 - (n_{2\omega}^P)^2} \right|^2 \; |T|^2 I_\omega^2 \; E_0^2 \; \left| e^{i\phi_{2\omega}} (e^{i\Delta\phi} - 1) \right|^2 \tag{13}$$

where $X^{(3)}(-2\omega;\omega,\omega,0)$ is the cubic suseptibility responsible for EFISHG process, T is an overall transmission and boundary conditions factor, and E_0 is the applied DC field intensity.
The phase mismatch $\Delta\phi$ in this case reads:

$$\Delta\phi = \phi_\omega - \phi_{2\omega} = \frac{4\pi}{\lambda} (n_\omega \cos\theta_\omega - n_{2\omega}\cos\theta_{2\omega})l \tag{14}$$

and

$$N_f = \begin{cases} \dfrac{288\pi^3}{c} & \text{in esu} \\ \\ \dfrac{9}{2\varepsilon_0 c} & \text{in SI} \end{cases} \tag{15}$$

In Eq.(13) we have neglected a priori the substrate contribution to the harmonic intensity as it generally takes place (thin film response much larger than that of substrate itself). Otherwise a more complicated expression has to be used, similar to Eq. (8) with normalization factor N_f given by Eq. (15) $\Delta\phi$ given by Eq. (14) and I_ω^3 has to be replaced by $I_\omega^2 E_0$. Other symbols have the following meaning:

$$\rho = \left(\frac{X^{(3)}(-2\omega;\omega,\omega,0)}{\Delta\varepsilon} \right)_p \Big/ \left(\frac{X^{(3}(-2\omega;\omega,\omega,0)}{\Delta\varepsilon} \right)_s \tag{16}$$

Φ is the relative phase of polymer $X^{(3)}$ with respect to that of substrate and

$$T_1 = A_s (t_\omega^{Os})^2 t_{3\omega}^{PO} \tag{17}$$

$$T_2 = A_p (t_\omega^{Os} t_\omega^{sp})^2 \tag{18}$$

where A is a factor arising from boundary conditions (cf. ref. 5) and $t_{\omega(3\omega)}^{ij}$ are transmission factors between i and j media at $\omega(3\omega)$ frequency.

In the case of a nonabsorbing thin film and neglecting harmonic field generated in the substrate Eq. (13) reduces to a much simpler form

$$I_{2\omega} = N_f \left| \frac{X^{(3)}(-2\omega;\omega,\omega,0)}{(n_\omega^P + n_{2\omega}^P)} \right|^2 \left(\frac{1}{\lambda_\omega}\right)^2 |T|^2 I_\omega^2 E_0^2 B^2 \tag{19}$$

where

$$B = \frac{n_\omega^P \cos\theta_\omega - n_{2\omega}^P \cos\theta_{2\omega}}{n_\omega^P - n_{2\omega}^P} \tag{20}$$

and

$$N_f = \begin{cases} \dfrac{4608\pi^5}{c} & \text{in e.s.u.} \\[2ex] \dfrac{72\pi^2}{\varepsilon_0 c} & \text{in SI units} \end{cases}$$

At normal incidence B=1 (cf.Eq.(20)) and the EFISHG intensity in thin film does not depend on the refractive index dispersion. For practical convenience (multiple reflection effects) it is better to work at a small incidence angle θ ($\propto$ 5°) where the dependence of harmonic intensity on the refractive index dispersion is still negligible.

The EFISHG intensity from polymer thin film is generally calibrated with α-quartz single crystal second harmonic generation intensity measured at the same condition as that for the studied thin film. Under the above conditions one obtains a very simple expression for $I_{2\omega}$

$$X^{(3)}(-2\omega,\omega,\omega,0) = \frac{2}{3E_0} \frac{I_c^Q}{I_p} \frac{T_Q}{T_p} \left(\frac{I_p^{2\omega}}{I_Q^{2\omega}}\right)^{1/2} \frac{n_\omega^P + n_{2\omega}^P}{n_\omega^Q + n_{2\omega}^Q} X_Q^{(2)}(-2\omega;\omega,\omega) \tag{21}$$

94

where l_c^Q is the quartz single crystal coherence length and $I_Q^{2\omega}$ is the maximum Maker fringes intensity from quartz reference. The subscripts or superscripts p and Q refer to polymer film and quartz single crystal respectively.

The EFISHG experiments on conjugated polymer thin films have to be performed carefully because of internal polarization observed in these materials.[33-34] This polarization is due to electron-hole pairs separa - ation by the applied electric field and created by two photon absorption (direct harmonic photons absorption or absorption of incident photons from a two photon state). The created in this way internal polarization cancels the applied external field and results in the decrease of se - cond harmonic intensity

The principle of observation of two photon states in EFISHG is dif- ferent than in THG. As we mentioned already the applied electric field breaks the selection rules and the two photon resonance takes place in centrosymmetric polymers with both u and g states (cf. Fig. 7). An examp- le of a two photon resonance observation is shown in Fig. 8. In that case, because the u and g states lie closely, both resonances with u

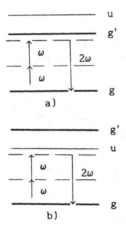

Fig. 7. Possible mechanisms leading to two photon resonances in EFISHG : with two photon (g') state (a) and with one photon (u) state (b).

and g state overlap. For this reason it is practically impossible to conclude the origin of two photon resonance if no other information is avaiable. Therefore in nonlinear spectroscopy studies both techniques can be considered as complementary. In the case when one and two photon states are close only THG can resolve them .The EFISHG technique will be useful in the case of well separated u and g levels and especially in the case when two and three photon levels overlap in THG (e.g. the g' state in the center of optical gap).

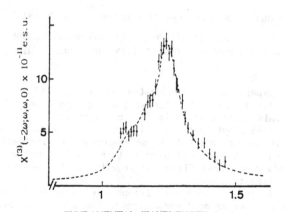

FUNDAMENTAL WAVELENGTH (μm)

Fig. 8. Wavelength variation
of EFISH susceptibility in
blue form of poly-4BCMU.
(after ref.33).

THREE PHOTON RESONANCES

Three photon resonances in a four photon process can be observed only by third harmonic generation technique. It takes place when the triple of fundamental photon energy is equal to the transition energy between fundamental and excited state of opposite symmetry (cf. Eq. (2)). The whole spectrum of three photon resonances have been obtained for polyacetylenes[21] (cf. Fig. 9) using tunable dye and free electron lasers, in polydiacetylene red[9] and blue[25] forms, thiophene oligomers[28] poly(alkyl)thiophenes[10,24], polythieno-thienyl and poly -di-thieno benzene[36], Langmuir-Blodgett films of hemicyanine[37]. If generally one expects a narrowing of three photon resonance curve with respect to the one photon spectrum sometimes one observes a broadening due either to two-photon resonance or to vibronic contributions[36], the last result, if confirmed for other materials, may reveal the three photon resonance study as an interesting tool for the molecule vibronic structure investi - gation in thin films. We note here that all older[38] and recent THG measurements on polydiacetylene blue form and on polyacetylene[39-40] at 1.91 μm fundamental wavelength give three photon resonant value of cubic susceptibility$X^{(3)}$ $(-3\omega;\omega,\omega,\omega)$.

DISCUSSION

In this paper we have shown how nonlinear optical techniques give information about the electronic structure of conjugated polymers through the observed multiphotonresonances in cubic susceptibilities describing corresponding processes. The one photon resonances may be observed in saturation and photoinduced absorption experiments which gives not only the modulus of cubic susceptibility but also its response time. The harmonic generation measurements are useful techniques for the study of the width and the position of two photon resonances. We limited our discussion here to harmonic generation experiments on thin films only where the reduced thickness and consequently the absorption allows to

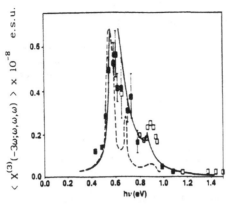

FUNDAMENTAL PHOTON ENERGY (eV)

Fig. 9. Fundamental photon
energy dependence of cubic
susceptibility $\langle\chi^{(3)}$ (-3ω;
ω,ω,> as measured for un-
oriented polyacetylene.
Closed rectangles are the
free electron laser data
(after ref.21). Broken
and solid lines show cal-
culated values.

do the measurements over a large fundamental wavelength range, including
material absorption band, at least for harmonic frequencies. The nonli -
near spectroscopy study is possible also in solution with both techni -
ques: EFISHG[41] and THG[42] where one can also determine the modulus and the
phase of corresponding $X^{(3)}$ susceptibility. The avaiable wavelength range
is however strongly reduced compared to the thin film experiments because
of larger optical interaction lengths needed in these measurements
(several coherence lengths of solvent).

Practical importance of two and three photon resonances

The above mentioned practical importance of the two photon state
lying below the one photon state consists on the fact that the resonance
occurs at the polymer transparency range for fundamental photons (in
the one photon resonance case the incident photons are absorbed). Thus
the use of high intensity laser beams is possible. Moreover the
$X^{(3)}$ susceptibility is complex. Complex is also the Kerr susceptibility
responsible for refractive index variation (cf. Eq.(6)). Exactly at the
resonance the imaginary part of refractive index will be purely imaginary (cf. Fig. 10) due to the fact that in this case n is real and real
part of $X^{(3)}$ changes sign. Thus with increasing light intensity the nonlinear medium refractive index increases or decreases, depending on the
pump wavelength (below or above a two photon resonance or *vice versa*).
This can find application in different optical switches. Similarly, the
complex character of cubic susceptibility may lead to the active optical
bistability. Of course, the $X^{(3)}$ value at two photon resonance is strongly enhanced.

The three photon resonance, observed in THG only, if one limits to four photon processes, have limited practical applications. Very thin polymeric films with enhanced $X^{(3)}(-3\omega;\omega,\omega,\omega)$ values may be useful in frequency tripling of ultrashort large instantenous power pulsed lasers. In fact the tripling efficiency of a free standing film is given by (cf. Eq. (11))

$$\eta = I_{3\omega}/I_{\omega} = N_f \left(\frac{X^{(3)}(-3\omega;\omega,\omega,\omega)}{1 + n^P_{3\omega}} \right)^2 \left(\frac{1}{\lambda_{\omega}} \right)^2 \left(\frac{2}{1 + n^P_{\omega}} \right)^6 I^2_{\omega} \qquad (22)$$

where the N coefficient is defined in Eq. (11).

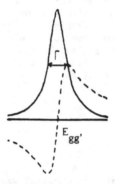

Fig. 10. Expected variation of real (broken line) and imaginary part (solid line) of Kerr susceptibility $X^{(3)}(-\omega;\omega,\omega,-\omega)$ around a two photon resonance.

For the three photon resonant value reported by Krausch et al[39] for an oriented polyacetylene ($X^{(3)}(-3\omega;\omega,\omega,\omega) = 2.7\times10^{-8}$ e.s.u.) and assuming reasonable values for refractive indices: $n^P_{\omega} = 2$, $n^P_{3\omega} = 5$, the polymer thin film thickness of 0.2 μm one gets a 50% conversion efficiency at 10 GW/cm^2 fundamental beam intensity. Such instantenous power is realizable with femtosecond lasers where because of time broadening due to the group velocity dispersion (cf ;) only thin crystals can be employed for the frequency conversion. In the above estimation we did not take into account the polymer film absorption at harmonic frequency which may reduce the above η value by maximum 75%.

REFERENCES

1. Orr B. J. and J. F. Ward, Perturbation theory of the nonlinear optical polarization of an isolated system, Mol. Phys., **20**, 513(1971).

2. F. Kajzar, L. Rothberg, S. Etemad, P. A. Chollet, D. Grec, A. Boudet and T. Jedju, Saturation absorption and Kerr susceptibility in poly - diacetylene Langmuir-Blodgett film, Opt. Commun., **66**, 55(1988).

3. B. I. Greene J. Orenstein,R. R. Millard and L. R. Williams, Nonlinear optical response of excitons confined to one dimension, Phys.Rev. Lett., **58**, 2750(1987).

4. G. Ruani , A. J. Pal, R. Zamboni, C. Taliani and F. Kajzar, Photoinduced absorption and nonlinear optical response in a polycondensed thiophene based polymer PTT, in Conjugated polymeric materials: opportunities in electronics, optoelectronics and molecular electronics, J. L. Bredas and R. R. Chance (Eds), NATO ASI series, Series E, vol. 182, Kluwer Academic Publ., Dordrecht 1990, pp. 429-441.

5. F. Kajzar. and Messier J.,Third harmonic generation in liquids, Phys. Rev. A, **32**, 2352(1985).

6. F. Kajzar, J. Messier and C. Rosilio, Nonlinear Optical Properties of Polysilane Thin Films, J. Appl. Phys., **60**,3040(1986).

7. J. C. Baumert, G. C. Bjorklund, D. H. Jundt, M. C. Jurich, H. Looser, R. D. Miller, J. Rabolt, R. Sooriyakumaran, J. D. Swalen and R. J. Twieg, Temperature dependence of the third-order nonlinear optical susceptibilities in polysilanes and polygermanes, Appl. Phys. Lett., **53**,1147(1988).

8. W. E. Torruellas, D. Neher, R. Zanoni, G. I. Stegeman, F. Kajzar and M. Leclerc, Dispersion measurement of the third order nonlinear susceptibility of nonlinear susceptibility of polythiophene thin films, Chem. Phys. Lett.,**175**,11(1990).

9. W. E. Torruellas, K. B. Rochford, R. Zanoni, S. Aramaki and G. I. Stegeman, The cubic susceptibility dispersion of poly(4-BCMU) thin films: third harmonic generation and two-photon absorption measure - ments, Opt. Commun. (in press).

10. F. Charra.,J. Messier, C. Sentein, A. Pron and M. Zagorska, Influence of conformation on two photon-spectra of polyalkylthiophene, in Organic molecules for nonlinear optics and photonics, J.Messier, F.Kajzar and P. Prasad (Eds), NATO ASI Series, Series E., Kluwer Academic Publ., Dordrecht 1991, pp. 263-272.

11. F. Kajzar. and J. Messier, Resonance enhancement in cubic suscepti - bility of Langmuir-Blodgett multilayers of polydiacetylene, Thin Sol. Films, **132**, 11(1985).

12. B. S. Hudson, B. E. Kohler and K. Shulten, in Excited States, E. C. Lim ed., Academic Press, New York 1982, vol. 6, pp. 1-52.

13. B. E. Kohler, The polyene 2^1A state in polyacetylene photoinduced absorption and thermal isomerization, J. Chem. Phys., **88**, 2788(1988).

14. B. E. Kohler and D. E. Schilke, J. Chem. Phys., Low-lying singlet states of a short polydiacetylene oligomer, J. Chem. Phys., **86**, 5214-5215(1987).

15. G. W. Hayden and E. J. Mele, Phys. Rev. B, **34**, 5484(1986)

16. S. N. Dixit, D. Guo and S. Mazumdar, Mol. Cryst. Liq. Cryst. **194**, 33 (1991).

17. Z. G. Soos, P. C. M. McWilliams and G. W. Hayden, Coulomb correlations and two-photon spectra of conjugated polymers, Chem. Phys. Lett. **171**, 14(1990).

18. D. Baerisswyl, D. K. Campbell and S. Mazumdar, An overview of the theory of π-conjugated polymers, in The physics of conducting poly - mers, H. Kiess ed., Springer Verlag, Berlin 1991.

19. Kajzar F. and J. Messier, Cubic nonlinear optical effects in conjugated polymers, Polymer J., **19**, 275-284(1987).

20. Kajzar F., S. Etemad, G. L. Baker and J. Messier, χ^3 of trans - (CH) experimental observation of 2A^x excited state, Synthetic Met., **17**,^{x} 563-567(1987).

21. W. S. Fann., S. Benson, J. M. J. Madey, S. Etemad, G. L. Baker and F. Kajzar, Spectrum of $\chi^{(3)}(-3\omega;\omega,\omega,\omega)$ in polyacetylene: an appli - cation of the free electron laser in nonlinear optical spectroscopy, Phys. Rev.Lett., **62**, 1492(1989).

22. S. Etemad, P. D. Townsend, G. L. Baker and Z. Soos, Two-photon ab - sorption in polydiacetylene based waveguides, in Organic molecules for nonlinear optics and photonics, J. Messier, F. Kajzar and P. Prasad eds, NATO ASI Series, vol. 194, Kluwer Akademic Publ., Dordrecht 1991, p. 489.

23. W. E. Torruellas, The optical cubic susceptibility dispersion of some transparent thin films, Thesis, University of Arizona at Tucson, December 1990.

24. F. Kajzar, J. Messier, C. Sentein, R. L. Elsenbaumer and G. G. Miller, Cubic susceptibility of polythiophene solutions and films, in Nonlinear optical properties of organic materials II, G. Khanarian (Ed.), Proc. SPIE vol. 1147, 36(1990).

25. T. Kanetake, Ishikawa K., Hasegawa T., Koda T., Takeda K., Hasegawa M., Kubodera K. and Kobayashi M., Nonlinear optical properties of highly oriented polydiacetylene evaporated films, Appl. Phys. Lett., **54**, 2287(1989).

26. J. Le Moigne, A. Thierry, P. A. Chollet, F. Kajzar and J. Messier, Morphology, linear and nonlinear optical studies of poly (1,6-di(N-carbazolyl) -2,4 hexadiyne) thin films (pDCH), J. Chem. Phys., **88**, 6647(1988).

27. P. A. Chollet, Kajzar F. and Messier J., Frequency and temperature variations of cubic susceptibility in polydiacetylenes, in Nonlinear optical and electroactive polymers, P. N. Prasad and D. R. Ulrich (Eds), Plenum Publ. Corp., New York 1988, pp.121-135.

28. D. Fichou , F. Garnier, F. Charra, F. Kajzar and J. Messier, Linear and nonlinear optical properties of thiophene oligomers, in Organic materials for nonlinear optics, R. A. Hann and D. Bloor (Eds), The Royal Society of Chemistry Special Publication, n°69, London 1989, pp. 176-182.

29. D. Neher, A. Wolf, C. Bubeck and G. Wegner, Third harmonic generation inpolyphenylacetylene: exact determination of nonlinear optical susceptibilities in ultrathin films, Chem. Phys. Lett., **163**, 116(1989).

30. D. Neher , A. Wolf, M. Leclerc, A. Kaltbeitzel, C. Bubeck C. and G. Wegner,Optical third harmonic generation in substituted poly(phenyl-acetylene)s and (poly(3-decylthiophene)s, Synth. Met., **34**, 249(1990).

31. R. R. Chance, M. L. Shand, C. Hogg and R. Silbey, Three-wave mixing in conjugated polymer solutions: Two-photon absorption in polydiace-tylenes, Phys. Rev. B, **22**, 3540(1980).

32. Y. Kawabe, F. Jarka, N. Peygambarian, D. Guo, S. Mazumdar, S. N. Dixit and F.Kajzar, Roles of band states and two-photon transitions in the electro-absorption of a polydiacetylene, submitted for publi-cation.

33. P.A. Chollet, F. Kajzar and J. Messier, Electric field induced opti-cal second harmonic generation and polarization effects in polydiace-tylene films, in Polydiacetylenes. Synthesis, Structure and Electro-nic Properties, D. Bloor and R. R. Chance (Eds), NATO ASI Series, vol. E102, Martinus Nijhof Publ., Dordrecht 1985.

34. P. A. Chollet, F. Kajzar and J. Messier, Electric field induced op - tical second harmonic generation and polarization effect in polydia- cetylene Langmuir-Blodgett multilayers, Thin Sol. Films, **132**, 1(1985).

35. Chollet P. A., F. Kajzar and J. Messier, Resonances of cubic nonli - near optical susceptibilities in polydiacetylenes, in Nonlinear op- tics of organics and semiconductors, T. Kobayashi (Ed.), Springer Verlag, Berlin 1988, pp. 171-179.

36. F. Kajzar, G. Ruani, C. Taliani, and R. Zamboni, Frequency variation of cubic susceptibility in new conjugated polymers: PTT and PDTB, Synth.Metals, **37**, 223(1990).

37. Kajzar F., I. R. Girling and I. R. Peterson,Third-order hyperpola - rizability of centrosymmetric Langmuir-Blodgett films of stilbazo - lium dyes, Thin Sol. Films, **160**, 209-215(1987).

38. Sauteret C., J. P. Hermann, R. Frey, F. Pradere, J. Ducuing, R. M. Baughman and R. R. Chance, Optical nonlinearities in one- dimensional conjugated polymer crystals, Phys. Rev. Lett., **36**, 956(1976).

39. Krausch F., Winter E. and Leising G., Optical third harmonic genera- tion in polyacetylene, Phys. Rev. B, **39**, 3701-3710(1989).

40. M. R. Drury, Observation of third harmonic generation in oriented Durham polyacetylene, Sol. State Commun., **68**, 417(1988).

41. F. Kajzar, I. Ledoux and J. Zyss, Electric-field induced optical se- cond harmonic generation in polydiacetylene solutions, Phys. Rev. A **36**, 2210(1987).

42. F. Kajzar.,Cubic susceptibility of organic molecules in solution,in Nonlinear optical effects in organic polymers, J. Messier, F. Kajzar, P. Prasad and D. Ulrich (Eds), NATO ASI Series, Series E, vol. 162, Kluwer Academic Publ., Dordrecht 1989, pp.225-245.

43. W. H. Glenn, Second harmonic generation by picosecond optical pulses, IEEE J. Quantum Electron., **QE-5**, 248(1969).

THIN FILMS SYSTEMS FOR SEPERATION AND OPTO-ELECTRONIC APPLICATIONS

Naoya Ogata

Department of Chemistry, Sophia University
7-1 Kioi-Cho, Chiyoda-Ku, Tokyo 102, Japan

INTRODUCTION

Thin films systems of multi-functionalities have been developed in terms of optical resoutions and non-linear optics (NLO). Optical isomers such as α-amino acids or pharmaceutical materials are usually very difficult to separate by means of conventional methods such as distillation or recrystallization, because their chemical or physical properties are same. Optical resolutions of these isomers are normally carried out by chromatographic methods, using small differences of absorption-desorption behaviors of these isomers on chromatographic columns. If the optical resolutions could be successfuly done through membranes, wide applications are expected in terms of industrial purposes. The optical resolutions through thin membranes require an effective molecular recognition so that only one isomer can permeate through the membrane and the other isomer is rejected to permeate.

Non-linear optical (NLO) materials have been attracting a great interest in opto-electronic applications. Particularly, second harmonic generation (SHG) for frequency doubling and switching has been widely studied. SHG activities require a good orientation of polar groups within thin films when guide waves are designed. Organic materials are much superior to in-organic materials in terms of SHG activities and thin films of polymers are much advantageous for practical applications because of easy processing. Thin films of SHG active polymers having high thermal stabilities are required when a strong raser beam is applied, because of heat accumultions within the films. Therefore, it has been desired for the NLO active polymers to make thin films of high temperature polymers.

This paper deals with thin films systems for optical resolutions and NLO active high temperature polymers.

OPTICAL RESOLUTIONS THROUGH MEMBRANES [1,2]

Effective separations of optical isomers through thin membranes request molecular recognitions by membranes so that only one-side of isomers can permeate through the membranes. Chiral interactions of α-helix of poly(α-aminoacids) with optical isomers provide a great possiblity to separate through the membranes. In case when water-soluble substrates such as α-aminoacids are selected, it would be necessary to arrange hydrophilic domains along through the helix of poly(α-aminoacid). On the other hands, hydrophobic domains would be required for self-organizations

of the helix of poly(α -aminoacid) within the membranes.   Based on this
molecular design of the membranes, following structure was proposed:

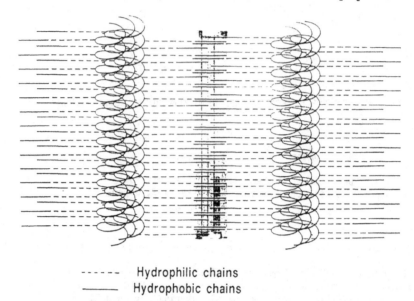

------   Hydrophilic chains
———   Hydrophobic chains

This molecular design of the membrane can be achieved by poly(L-glutamate) having following amphiphilic side chains, which can be synthesized
by an ester exchange reaction of poly(methyl L-glutamate) with NONIPOL 6
(a commercially available surfactant):

$$\{COCHNH\}_n$$
$$(CH_2)_2$$
$$COO(CH_2CH_2O)_6\text{—}\langle\bigcirc\rangle\text{—}C_9H_{19}$$

A 15% solution of poly(methyl L-glutamate) in dichloroethane was heated at 65°C for 60 hr with slightly excess amount of NONIPOL 6 in the presence of p-toluenesulfonic acid as a catalyst, with eliminating methanol
out of the solution.   Polymer, NON 6 PLG, was obtained by pouring into
methanol.   FT-IR spectrum of NON 6 PLG is shown in Fig. 1 and NMR spectrum is indicated in Fig. 2.

$$\{COCHNH\}_n \quad + \quad HO\{CH_2CH_2O\}_m\text{—}\langle\bigcirc\rangle\text{—}C_9H_{19}$$
$$(CH_2)_2$$
$$COOCH_3 \qquad\qquad NON6$$
$$PMLG$$

$$\xrightarrow[\text{in EDC}]{PTS}$$

$$\{COCHNH\}_n$$
$$(CH_2)_2$$
$$COO\{CH_2CH_2O\}_m\text{—}\langle\bigcirc\rangle\text{—}C_9H_{19}$$
$$NON6\text{-}PLG$$

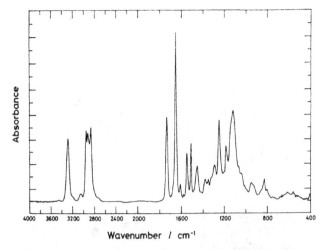

Fig. 1   FT-IR spectum of NON6-PLG film. NON6-PLG film cast
on a KBr plate from EDC solution was measured

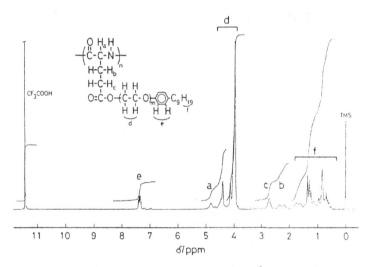

Fig. 2   NMR spectrum of NON 6 PLG

Cross-polarized microscopic picture shows a cholesteric texture as
indicated in Fig. 3 which supports the existence of helical structure of
NON 6 PLG film.   CD spectrum caused by phenoxy groups in NON 6 PLG was
measured as shown in Fig. 4, which indicates temperature dependences
of the CD spectrum.   The cholesteric texture of NON 6 PLG was observed
even at 136°C, while the CD spectrum disappeared above 53°C.
The temperature dependences of the CD spectrum are clearly seen in
Fig. 5 in which the CD measurements were carried out for the dry and wet
samples.   The dry sample showed a complete disappearance of the CD spectrum above 53°C, while the wet sample showed the CD spectrum   even at 70°C.
A glass transition temperature of NON 6 PLG appeared at around 50°C by means
of DSC analysis.   It is presumed that   isoprotic and anisotropic transitions might take place for the pendant amphiphilic groups at around 50°C.
Hydration by water at hydrophilic domains might stabilize the transition
caused by temperatures, so that the wet sample might keep the self-organized structure of NON 6 PLG at elevated temperatures.

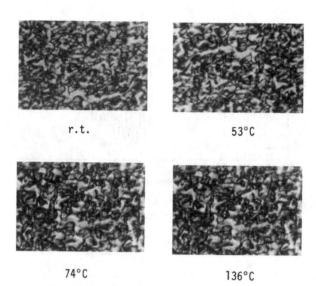

r.t.   53°C

74°C   136°C

Fig. 3   Cross-polarized microscopic pictures of NON 6 PLG

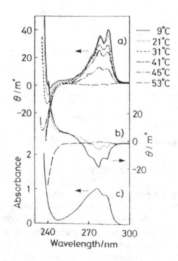

Fig. 4   CD Specrtum of NON 6 PLG film at various temperatures
(a)   directed to light (b)   quartz surface
(c)   UV spectrum

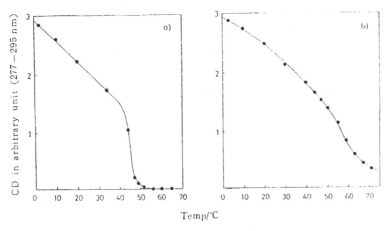

Fig. 5   CD spectrum changes at various temperatures

(a)   dry sample   (b) wet sample

These results strongly suggest that NON 6 PLG has a self-organized structure after casting, as shown below:

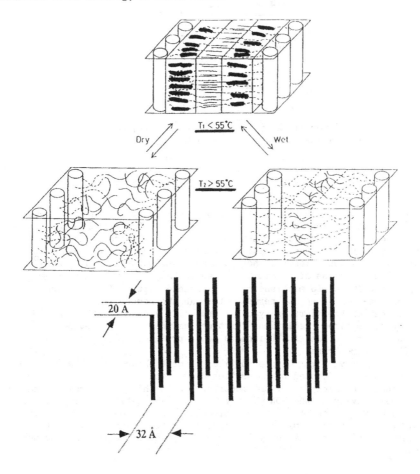

Helix distances between two helices are 20Å, while the lamellar distances are 32Å, which were determined by X ray analysis of the NON 6 PLG membrane.

NON 6 PLG was a soft polymer and was not mechanically strong for self-standing. Therefore, the solution of NON 6 PLG in dichloroethane was cast on micro-porous ultra-filter as a support and the optical resolutions of α –aminoacids were carried out by using a cell as shown in Fig. 6.

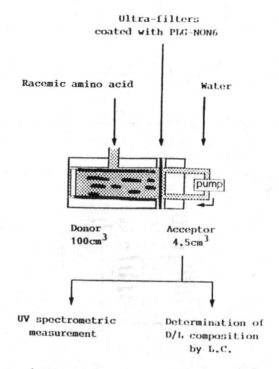

Fig. 6   Separation cell for optical resolutions

Fig. 7 shows separation behaviors of racemic tryptophan in aqueous solution.   It is clearly seen in Fig. 7 that only D-tryptophan could permeate through the membrane and L-tryptophan was completed rejected to permeate through the membrane.   The permeation rate of D-isomer was proportional to permeation times and increased as temperature went up from $1°$ to $34°$C.   The thickness of the membrane was 40 μ.   Therefore, the selectivity of the NON 6 PLG membrane might be related with molecular recognitions of one-side isomer to permeate through the membrane.

Permeation behaviors of tyrosine and serin are shown in Figs. 8a and 8b. The complete optical resolutions for these amino acids did not occur as the case of tryptophan, but D-isomers could preferentially permeate through the NON 6 PLG membrane.

The selective separation of optical isomers through the NON 6 PLG membrane might not be caused by absorption-desorption behaviors of the isomers, since the thickness of the membrane was so thin.   The real mechanism of the selectivity of the NON 6 PLG membrane is not still clear yet, but it would be very possible that the molecular recognition of one side isomers, that is, L-isomers might take place owing to the helix of poly(L-glutamate) chains, so that only D-isomers could permeate through the membrane.

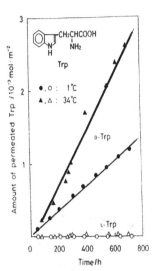

Fig. 7   Optical resolution of tryptophan through
         NON 6 PLG membrane (Try concn.= 4.5 mmol/l)

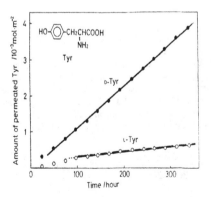

Fig. 8a Otical resolution of tyrosine by NON 6 PLG
         membrane.   (Tyr concn.= 2 mmol/1, 40°C)

NLO THIN FILMS OF HIGH TEMPERATURE POLYMERS

Non-linear optical (NLO) polymers have been attracting a great interest in opto-electronic applications.   Second harmonic generation (SHG)
of NLO active polymers has been widely studied in terms of frequency
doubling and optical switching.   Molecular designs for the SHG active
polymers require a well-ordered structure of donor-acceptor groups having
a non-centrosymmetric structure.   It is preferable for these SHG active
polymers to have a high thermal stability when a strong power of raser
beam is applied.   Therefore, there are demands to make NLO active polymers with a high thermal stability.

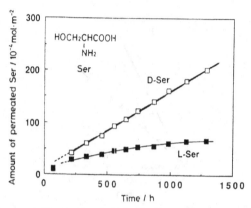

Fig. 8b Optical resolution of serin by NON 6 PLG membrane
(serin concn = 6.8 mmol/l, 40°C)

It was previously reported[3,4] that thin films of high temperature
polymers such as poly(benzimidazole) (PBI), poly(benzoxazole) (PBO) or
poly(benzothiazole)(PBT) were obtained at air/water interface by using
Langmuir-Blogdett (LB) method.   When an electron-accepting group such
as nitro group is incorporated into PBT with an well-ordered structure,
it is expected that the built-up film of the PBT might exhibit a strong
SHG activity.   Based on this expectation, PBT having nitro group (NO_2-
PBT) was synthesized at air/water interface as shown in following scheme:

$$OHC-\langle O \rangle-CHO \xrightarrow[30°C, 24hr]{HNO_3/H_2SO_4} OHC-\langle O \rangle-CHO$$
$$NO_2$$

$$OHC-\langle O \rangle-CHO + 2CH_3(CH_2)_5NH_2$$
$$NO_2$$

$$\xrightarrow[80°C,12hr]{C_6H_6} CH_3(CH_2)_5N=CH-\langle O \rangle-CH=N(CH_2)_5CH_3 + 2H_2O$$
$$NO_2$$

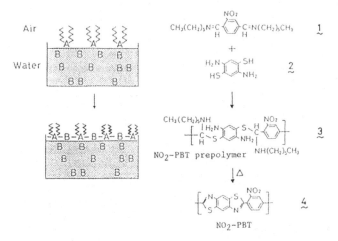

2-Nitro-dihexylterephthalaldimine (1) was synthesized by refluxing nitroterephthalaldehyde, which was obtained by nitration reaction of terephthalaldehyde, with 2 moles of hexlamine in benzene at 80°C for 12 hr. The thin film of NO_2-PBT was made at air/water interface as follows: (1) in benzene (0.1 wt%) was spread dropwisely onto the aqueous solution of 2,5-diamino-1,4-benzenedithiol (2) (0.001 wt%). Polymerization proceeded with standing the solution at room temeprature. After 8 hr, the ultra-thin film of the NO_2-PBT prepolmer was formed on water, which was transfered onto a substrate such as quartz.

The NO_2-PBT prepolymer (3) film, however, could not be deposited onto a quartz plate by a vertical lifting method by using a conventional LB trough, owing to a friction between the film on water surface and the wall of the LB trough. Therefore, a modification of the wall to a moving-wall type trough was made in order to avoid the friction. It was also important to modify the surface of the quartz plate by previously depositing acetalized poly(vinylalcohol) in order to increase the affinity between the surface and the NO_2-PBT prepolymer film.

Fig. 9 indicates results of the deposition of the NO_2-PBT prepolymer film by using the modified moving-wall trough. The film was deposited with the deposition ratio of 1. The deposition was occured by only up-stroke so that a Z-type deposition was attained. This result indicates that the multilayer film of the NO_2-PBT prepolymer has a non-centro-symmetric structure which is abso lutely required for the SHG activity.

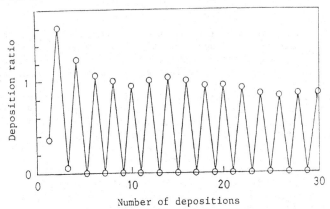

Fig. 9   Deposition of NO_2-PBT prepolymer on quartz

The SHG activity of the multilayer film of the NO_2-PBT prepolymer (20 layers) was evaluated by using Nd:YAG raser beam (1064 nm) which was irradiated to the sample film on a rotational stage and by measuring the SHG light intensity at 532 nm with a photomultiplier after filtering off remaining fundamental light.

Fig. 10 shows a result of the SHG measurement, which exhibits a fringe pattern caused by the SH light. This result suggests that the multilayer film of the NO_2-PBT prepolymer has a well-ordered structure with the orientaton of molecular chains out of the surface of the substrate with a constant angle.

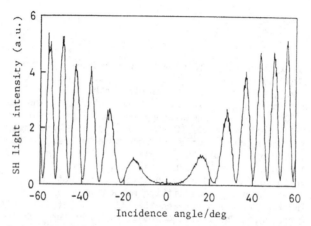

Fig. 10   Fringe pattern of the multilayer (20 layers) film of the NO_2-PBT prepolymer

The SH light intensity is given by following equation:

$$I^{2\omega} = \frac{2\omega^2 d_{eff}^2 l^2}{c^3 \varepsilon_0 (n^\omega)^2 (n^{2\omega})} \, (I^\omega)^2 \, \mathrm{sinc}^2 \left(\frac{\Delta k l}{2} \right)$$

where sincX denotes (sinX)/X, l is the sample thickness, d_{eff} is the effective nonlinear optical susceptibility, n is refractive index of the film and c is the light speed, and $k = k_2 - 2k_1$ is the phase mismatch between the fundamental and the SH light with wave vectors k_1 and k_2, respectively.

As shown in Fig. 11, the measured light intensity increased quadratically with the number of layers, that is , the thickness of the multilayer films. Therefore, it was concluded that the observed SH light was produced by the multilayer film of the NO_2-PBT prepolymer.

The multilayer films of the NO_2-PBT were subjected to a heat-treatment at 300°C in order to convert to corresponding PBT. Results of the SHG measrement of these films showed fringe patterns of SH light, but the intensity decreased to one fourth of that of the prepolymer, as can ce seen in Fig. 12. Presumably, the decrease in the SH light intensity might be attributed to the disordered structure of the multilayer films which was caused by the heat-treatment.

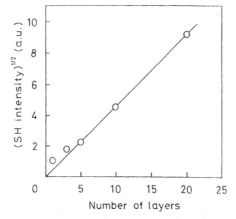

Fig. 11   Square root of the SH intensity as a
function of the number of layers of
NO$_2$-PBT prepolymer

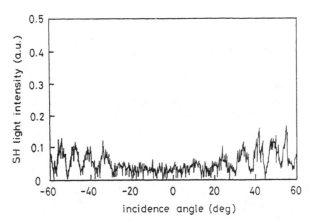

Fig. 12   Fringe pattern of the heat-treated PBT
multilayer film (20 layer, 300^{0}C for 10
min.)

Since PBI, PBO, PBT or PVBT are having fully extended conjugated
structure in main chains, it is expected that they might exhibit also
a third harmonic generation (THG) for those thin films which were obtained
at air/water interface by the LB method.   Their chemical structures are
shown as follows:

PBT   PBI

PVBT   PBO

THG values of these thin films of high temperature polymers were evaluated and results are summarized in Table I.

Table I   THG values of thin films of high temperature polymers

Substrate	Film	$\chi_3 (\times 10^{12} \text{ esu})$
quartz	PBO	1.85
quartz	PBI	6.00
CaF_2	PBT	9.56
CaF_2	PBVT	1.30

χ_3 values are in the order of 10^{-12} esu.   It was difficult to measure the molecular weights of these high temperature polymers because of their solubility problems, but they may not have enough high molecular weights which could be assumed from these small χ_3 values.

CONCLUSION

Multi-functional thin films have been developed in terms of optical resoluion and NLO activities for SHG & THG. Membranes from poly(glutamate) having amphiphilic side chains could permeate selectively D-somoers of α-amino acids and molecular recognitions of optical isomers were achieved.

Thin films of high temperature polymers which were formed at air/ water interface by LB method, indicated NLO activities.   PBT thin films containing electron donor-acceptorgroups exhibited a strong SHG actvitiy.

REFERENCES

1) A. Maruyama, N.Adachi, T.Takatsuki, M.Torii, K.Sanui, and N.Ogata, Macromol.,23, 2748(1990).
2) A.Maruyama, N.Adachi, T.Takatsuki, K.Sanui, and N.Ogata, Nihon-kagakukaishi, 1178(1990).
3) A.Angel, T.Yoden, K.Sanui, and N.Ogata, Polym. Mater. Sci. & Eng., 54, 119(1986).
4) T.Ueda, S.Yokoyama, M.Watanabe, K.Sanui, and N.Ogata, J.Polym.Sci., Part A, Polym.Chem., 28, 3221(1990).

STATIC AND DYNAMIC PHOTO-INDUCED INDEX OF REFRACTION CHANGES

Larry R. Dalton, Linda Sapochak, and Malcolm R. McLean

Department of Chemistry
University of Southern California
Los Angeles, CA

Luping Yu and Mai Chen

Department of Chemistry
University of Chicago
Chicago, IL

Charles W. Spangler

Department of Chemistry
Northern Illinois University
DeKalb, IL

INTRODUCTION

This article deals upon three aspects of the interaction of light with matter. First of all, we consider exploitation of photo-induced chemical and structural changes for fabrication of integrated optical circuits and realization of phase matching in second harmonic generation. The systematic generation of spatial variations in index of refraction, which we demonstrate, can be a crucial factor in the development of nonlinear optical devices for commercial application. We then examine development of dynamic second order optical nonlinearities and the exploitation of these for frequency generation, electro-optic modulation, and beam steering. Our particular focus in this article is the stabilization of poling-induced order and second order nonlinear optical activity. The final discussion focuses upon third order optical nonlinearities. This area of research is less advanced and attention is directed toward definition of mechanisms of optical nonlinearity and upon identifying research avenues for significantly improving optical nonlinearities for device applications such as sensor protection, dynamic holography, and optical computing.

PHOTO-INDUCED CHEMICAL AND STRUCTURAL CHANGES FOR DEVELOPMENT
OF CHANNEL WAVEGUIDES AND GRATINGS

Photo-induced configurational changes are well-known in chemistry. For example, lasers have frequently been used to initiate addition to multiple bonds, e.g., photo-induced

Frontiers of Polymer Research, Edited by P.N. Prasad and
J.K. Nigam, Plenum Press, New York, 1991

crosslinking and polymerization reactions involving polyalkenes and polyalkynes. Also, conformational changes, such as cis-trans isomerization about double bonds, which proceed via excited triplet states, are well-known photo-induced processes. Finally, it can be observed that changes of state (solid-liquid-gas and between various liquid crystalline phases) can be driven by laser irradiation and local heating effects. We have undertaken the systematic exploitation of these processes to control ultrastructure for the development of precise patterns or circuits for optical signal transport and processing.

In a recent article[1], we demonstrated the use of photo-induced isomerization and cross-linking reactions to develop integrated optical circuits (e.g., channel waveguides). In this work, an index of refraction change on the order of 0.3 is induced by a photo-stimulated isomerization about the azo (-N=N-) linkage which is locked in place by a simultaneous photo-induced cross-linking reaction involving either double or triple bonds. This work involved synthesizing polymers containing three functional segments (see Figure 1 below)

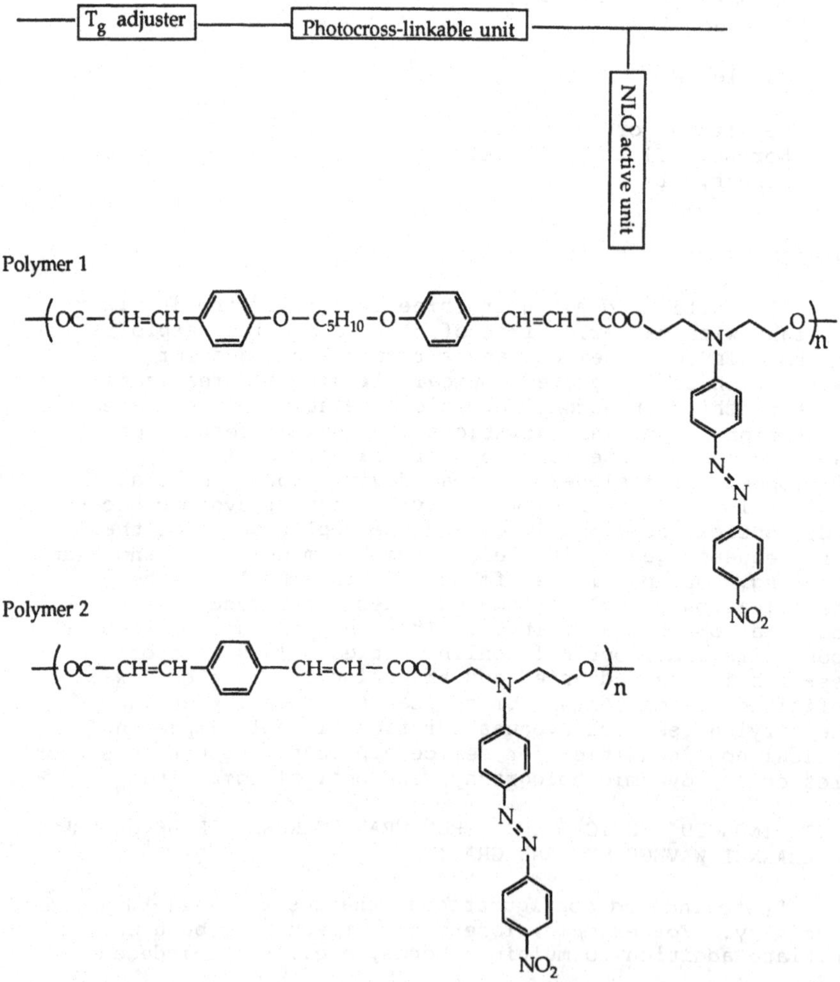

Polymer 1

Polymer 2

Table 1. SHG results for three polymer thin films

Poly-mers	λ (nm)	$N/10^{20}$ $(1/cm^3)$	T_g (°C)	E (MV/cm)	d_{13} (pm/V)	d_{33} (pm/V)	τ (days)
1	1320	3.4 ± 0.2	132 ± 3	0.2 ± 0.03	7 ± 1	47 ± 5	> 90
2	1320	6.3 ± 0.2	125 ± 4	0.2 ± 0.03	6 ± 0.5	52 ± 4	> 90
3	1320	5.6 ± 0.2	127 ± 3	0.2 ± 0.03	6 ± 0.5	52 ± 4	> 90

In addition to development of integrated optical circuits this particular polymeric structure is convenient for realizing polymers with large second harmonic generation efficiencies. Second harmonic generation requires the material to exhibit macroscopic as well as molecular non-centrosymmetric symmetry. For polymeric materials containing a noncentrosymmetric electroactive moiety as a side chain pendant, the required macroscopic order is typically achieved by electric field poling of the material in a liquid or fluid state followed by cooling of the material to a solid state to lock-in the poling-induced order. Unfortunately, dynamic modes are not completely quenched in the solid state at finite temperatures and induced order has typically been observed to relax with time. We, and many other groups, have endeavored to stabilize the poling induced macroscopic order by crosslinking the material exploiting either photo-induced or thermally-induced processes. Some representative results are given in the accompanying table (Table 1).

One of the requirements for successful exploitation of second harmonic generation in non-centrosymmetric materials is phase matching of the first and second harmonic signals. We have recently achieved this feat for polymers in general by use of laser assisted poling techniques to generate a diverging grating pattern in polymeric materials. As shown in figure 2, local laser heating is used to generate regions of local order through electric field poling.

Polymer 3

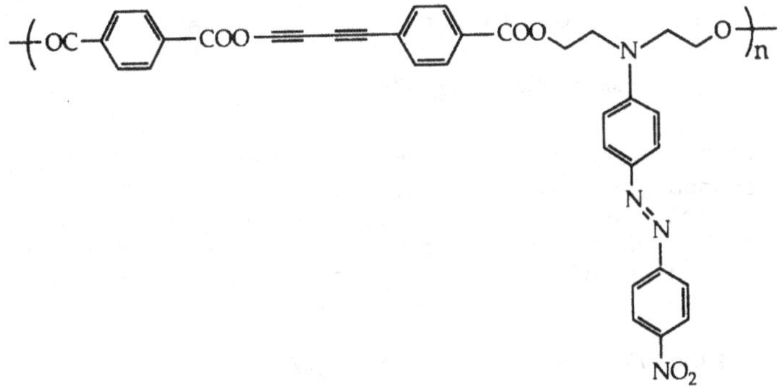

Fig. 1. Schematic and three examples of photo-crosslinkable polymeric systems.

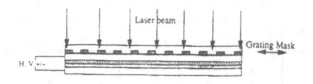

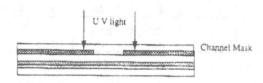

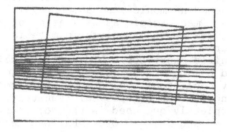

Fig.2. Illustrating the fabrication of a waveguide through laser annealing / poling followed by UV induced crosslinking.

It is clear from the figures that only the regions melted by laser heating will exhibit molecular ordering under the force of the poling field.

The condition for phase matching is given by

$$\Delta k = 4\pi [N^{2\omega} - N^{\omega}]/\lambda = 2\pi/\Lambda$$

where N is the mode index in the waveguides and Λ is defined in the accompanying figure (Fig. 3).

The second harmonic signal is plotted versus grating period in the accompanying figure and the relationship of phase matching length L_{pm} to grating period is given as follows:

$$P(2\omega)/P_o(2\omega) = Sinc^2(\Delta k L_{pm}/2) = 1/2$$

$$L_{pm} = 1.4/\Delta k = 1.4\Lambda^2/(2\pi\Delta\Lambda) = 2.3 \text{ mm}$$

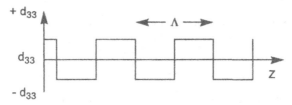

Fig. 3. Quasi-phase matching SHG in polymer waveguides.

In the accompanying figure (Fig. 4), a plot of the second harmonic power versus input power is given, demonstrating an absolute conversion efficiency, $P(2\omega)/P(\omega)$, of approximately 10^{-4}. The normalized conversion efficiency is approximately 30%.

While the values demonstrated here are not adequate for device development the improvement over previous values are encouraging as is the demonstration of modest efficiency in second harmonic generation. Note that the measurement wavelength for the data in Table 1 is 1320 nm. Measurements in the near infrared were performed to better estimate the non-resonant d_{ij} values.

The preceding discussion emphasizes that several stages of research are crucial for the development of devices utilizing second order optical nonlinearities. Initially, molecules must be screened for appropriate microscopic or molecular optical nonlinearities (hyperpolarizibilities, β).

Then, these molecular moieties must be incorporated into polymers in a manner which permits electric field poling to be carried out.

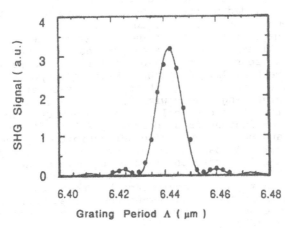

Fig. 4. Second harmonic generation vs. grating period.

This is not a simple matter with highly rigid units such as ladder oligomers as these are particularly sensitive to local intermolecular potentials and reorient, under the influence of a poling field, only with difficulty. Finally, the poling-induced order must be stabilized by crosslinking, hydrogen bonding, or some other strong interaction. A convenient polymer structure for accomplishing the necessary latter steps has been illustrated herein (Fig. 1), particularly when modifications include connecting the pendant electroactive moieties to the polymer backbone by variable length flexible chain spacer segments. However, because many features must be simultaneously realized, an ideal material has yet to be achieved.

MECHANISTIC STUDIES OF THIRD ORDER OPTICAL NONLINEARITIES

A variety of mechanisms can contribute to third order optical nonlinearities including exciton phase space filling, structural relaxation, multi-photon processes, thermal and acoustic processes, etc. We have recently noted that when more than one mechanism is active (multiple relaxation pathways exist) observed nonlinear optical responses (e.g., the amplitudes and temporal behavior of phase conjugate signals generated in four wave mixing) will depend upon pulse characteristics such as pulse widths[2]. Indeed, because of the "pulse integration" effect, different conclusions may be reached using picosecond pulses and using femtosecond pulses. Because of the complications introduced by competing mechanisms, realistic assessment of optical nonlinearity may require an extensive series of measurements employing different pulse widths, different laser wavelengths, pulse powers, polarizations, and beam delays. In addition, it may be appropriate to determine the real and imaginary components of the optical nonlinearity using interferometric degenerate four wave mixing (IDFWM) techniques or by a combination of pump-probe (PP) and dynamic Kerr effect (DKE) experiments as discussed elsewhere[2]. It is also often appropriate to investigate measured optical nonlinearities as a function of intermolecular separation of nonlinear optical moieties; such studies frequently permit identification of intermolecular excitonic interactions and the effects of lattice-induced polarization[2]. An illustration of the elucidation of mechanism is provided by the accompanying data deriving from IDFWM measurements of a ladder oligomer[2].

For the data shown in the figure 5, it is clear that the observed optical nonlinearity is determined by photo-generated bipolarons. Strong evidence for this mechanism is provided by the correlation of the wavelength dependence of the imaginary component of the third order susceptibility shown in the preceding figure with changes in the optical spectrum observed upon chemical or electrochemical doping. Moreover, the assignment is supported by EPR studies and by the temporal and pulse width responses exhibited by the phase conjugate signal. A possible mechanism is shown in figure 6.

This work illustrates the problem of attempting to define optical nonlinearity by performing a single measurement in a region of one-phonon transparency. The assumption has been that measurements performed sufficiently far from one-phonon transitions will yield the non-resonant optical nonlinearity.

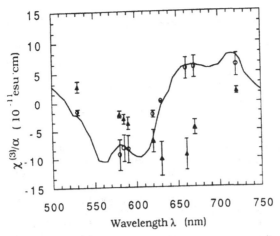

Fig. 5. Wavelength dependence of the real (--Δ--) and imaginary (--0--) part of $\chi^{(3)}$ $(-\omega, \omega, \omega, -\omega)/\alpha$ in a ladder copolymer. The solid line is the normalized absorption spectra change during electro-chemical doping.

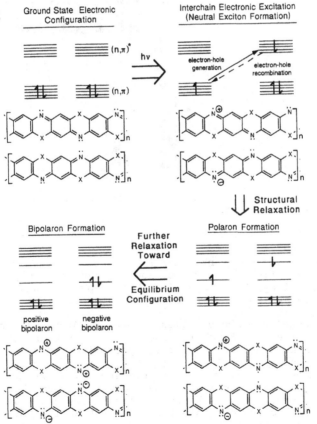

Fig. 6. A hypothetical mechanism for the photo-generation of bipolarons by interchain electronic excitation is illustrated.

Obviously, photo-induced processes such as bipolaron generation can be quite effective even exciting the tail of one phonon transitions. Moreover, multi-phonon processes can be quite effective. It is clear that heating effects can be associated with non-linear absorption and thus the interpretation of results can be quite erroneous unless the additional nonlinear processes are taken into account.

Thus, detailed studies of third order optical nonlinearities are of great importance in identifying the most promising candidates for device development. To this point in time, surveys have been largely limited to studies employing picosecond and longer pulses by the absence of lasers capable of delivering sub 200 femtosecond pulses with pulse powers approaching 1 GW/cm^2. Fortunately, two commercial systems have recently become available which taken together facilitate studies in the frequency range 590 to 900 nm. The first of these is illustrated at figure 7 and is based upon laser components from Coherent and Continuum.

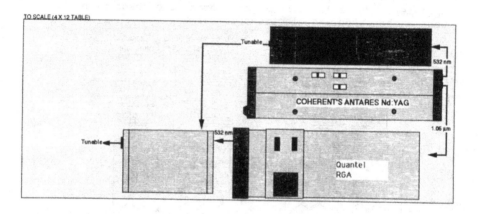

Fig. 7. Coherent / Continuum - high repetition rate, tunable femtosecond amplification system

A mode-locked Coherent Antares YAG laser is used as the primary source which is frequency doubled and used to pump a Coherent Satori Dye Laser system. This output is then fed to Continuum PTA and RGA amplifiers to produce high power femtosecond pulses.

The second approach utilizes a Coherent Argon Ion laser to pump a Coherent Ti-Sapphire laser. The output is then amplified using Continuum amplifiers producing sub 100 femtosecond pulses.

We are currently carrying out preliminary measurements with both of these systems and should ultimately be able to obtain third order susceptibilities measured over the visible and near infrared regions as a function of time for both femto and picosecond pulses.

A number of interesting third order nonlinear optical materials such as mixed valence metal complexes, metallomacrocylic materials, and metallated ladder polymers have been identified.

122

Intervalence and metal-ligand charge transfer events can extend over significant distances and contribute to optical nonlinearity. However, the amount of structural relaxation that is associated with each of these specific events is largely unknown at this time and a meaningful discussion of third order nonlinearities must await more detailed studies.

ACKNOWLEDGEMENTS

This work was supported by the Air Force Office of Scientific Research under contracts F49620-87-0100 and F49620-88-0071 and by the National Science Foundation under grant DMR-88-15508. The authors particularly wish to thank Drs. X. F. Cao, J. P. Jiang, R. W. Hellwarth, and W. H. Steier of the University of Southern California and to thank Dr. R. Norwood of Hoechst-Celanese for various nonlinear optical characterizations. The authors wish to acknowledge the role of Dr. Donald R. Ulrich of the Air Force Office of Scientific Research in motivating this research.

REFERENCES

1.   Y. Shi, W. H. Steier, L. P. Yu, M. Chen and L. R. Dalton, Large Stable Photoinduced Refractive Index Change in a Nonlinear Optical Polyester Polymer with Disperse Red Side Groups, Appl. Phys. Lett., 58: 1131 (1991).

2.   A. N. Bain, L. P. Yu, M. Chen, L. S. Sapochak and L. R. Dalton, Time-Dependent Nonlinear Spectroscopy, Spectros. Int. J., 8, 73 (1990).

NONLINEAR OPTICAL PROPERTIES OF THIOPHENE BASED

CONJUGATED POLYMERS

C. Taliani, G. Ruani and R. Zamboni

Istituto di Spettroscopia Molecolare, CNR
Via de' Castagnoli, 1 - 40126 Bologna (Italy)

L. Yang, R. Dorsinville and R.R. Alfano
Department of Electrical Engeenering and IUSL
The City College of CUNY, New York 10031, (USA)

INTRODUCTION

Investigation of the nonlinear optical (NLO) properties of organic systems possessing delocalized π electrons has attracted considerable interest because of the large and fast NLO response [1-4]. The most studied systems have been polyacetylene [5-10] and polythiophene [11-14]. Even though the NLO response in these macromolecules derives from the high hyperpolarizability of π delocalized electrons in the carbon polymer backbone, the origin of their large optical nonlinearity is not yet clear and therefore it is not yet possible to take full advantage of the chemical tailoring of new systems in order to make more advanced NLO materials. One way to contribute to an understanding of the origin of the NLO response is to study the electronic cubic susceptibility $\chi^{(3)}$ as a function of frequency for different conjugated polymers with different electronic structures. In fact in the absence of a spectrum it is difficult to distinguish between the resonant and nonresonant values of $\chi^{(3)}$ because in principle, resonance may occur not only in spectral regions where the material is absorbing but also in transparent regions by resonance with forbidden electronic states which do not contribute to the absorption. Up to now, only a limited set of data have been available on the electronic $\chi^{(3)}$ and many of these are limited to a single fundamental frequency. Wavelength dispersed $\chi^{(3)}_{THG}$ by third harmonic generation (THG) has been studied in polyacetylene [15] and in two thiophene based polyconjugated systems [16].

In the first part of this paper, we report on the investigation of a series of thiophene based conjugated polymers with rather different backbone structures and energy gaps in order to explore the effect of their molecular architecture on electronic and NLO properties.

In particular, we report on the dispersion of $\chi^{(3)}$ $(-\omega;\omega,-\omega,\omega)$ in the spectral range covering the rising slope of the optical absorption as well as in off resonance conditions.

In the second part we report on the experimental investigation of photoinduced infrared and near infrared absorption (PA). In fact the electronic excitations of conjugated macromolecules are controlled by a strong e-ph coupling resulting in the formation of self-localized excitations such as charged solitons, polarons (P) and bipolarons (BP). These excitations are associated with new electronic states that are formed in the subpicosecond timescale inside the energy gap. In polythiophene (PT), which has a nondegenerate ground state, as well as the thiophene based conjugated polymers considered in this work, P and BP are predicted [17-18] as dominant nonlinear excitations which can be produced either by photoexcitation or doping.

In the third part we examine to what extent the transient photogeneration of these nonlinear excitation may alter nonresonant optical linearities. A new pump and probe degenerate four wave mixing (DFWM) technique is discussed which allows measurements of $\chi^{(3)}$ when nonlinear excitations are transiently generated by photoexcitation above the energy gap, and we show that in PT the $\chi^{(3)}$ measured at 1064 nm is enhanced by resonance with photogenerated transient electronic states.

In our opinion this discovery opens up the possibility of interesting new developments in the area of NLO properties of conjugated systems.

ELECTRONIC PROPERTIES AND NLO MEASUREMENTS

Electronic properties

We have investigated a homologous family of sulphur containing heteroaromatic compounds, with monomer sizes varying from one up to three

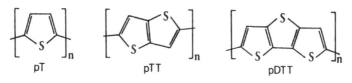

Fig. 1. Molecular structure of polythiophene (PT), polithieno [3,2-b]thiophene
(PTT) and polydithieno[3,2-b;2',3'-d]thiophene (PDTT).

fused thiophene rings. The structures of corresponding polymers, polythiophene
(PT), polythieno(3,2-b)thiophene (PTT) and polydithieno(3,2-b;2',3'-d)thiophene
(PDTT) are illustrated in fig.1.

PTT and PDTT were prepared by electrochemical oxidation from an
acetonitrile/LiClO₄ solution containing the monomer. Experimental details have
been reported previously [19-20]. PT was prepared by the same method starting
from the dimer which requires lower oxidation potential than thiophene itself
[21]. The electronic absorption and emission experiments were carried out on
neutral (undoped) thin films deposited on Indium Tin Oxide (ITO) glass and
silicon respectively. Excitation of the emission was performed both by c.w. laser
beam at 2.54 eV and by the pulsed second harmonic of a Nd:YAG pumped dye
laser at 4.0 eV. The UV-Visible absorption and emission spectra of undoped PT,
PTT and PDTT are show in fig.2.

The maxima in the absorption spectra are located at 2.7 eV for PT and
PDTT and at 2.85 eV for PTT, whereas the band gaps, which are estimated from
the intercept of the low energy rising edge of the optical absorption with the
baseline, are at about 2 eV, in PT and PTT and 1.9 eV in PDTT. These
differences are too small to be related to the existence of significant
modifications in the electronic structure of the three neutral polymers. The large
width of the electronic absorption envelope may be the result of the electronic
absorption bands due to a rather wide distribution of conjugation lengths. The
short chains would contribute to the absorption at higher energy, while chains
with longer conjugation would give rise to the lower energy absorption; the
overall result is the large inhomogeneous broadened envelope. The fact that the
bandwidth of the emission spectra is much narrower is in agreement with this
hypothesis. In fact, excitation within the electronic absorption envelope would
give rise to a cascade of energy transport processes involving chains with
progressively longer conjugation lengths and radiative emission would take place
from the low energy distribution of long conjugated segments. This is a usual
process widely observed in the electronic spectra of disordered molecular
crystals and glasses [22]. Optical measured band gaps have been compared with
theoretical band structure calculations which have been performed using the
Valence Effective Hamiltonian (VEH) technique [23]. This method has been

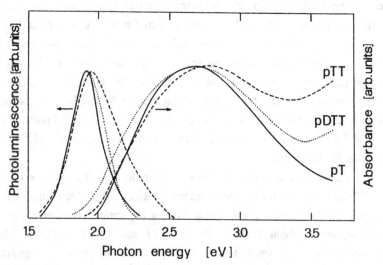

Fig. 2. Emission and absorption spectra of PT (—), PTT (---) and PDTT (•••) (from ref. 24).

shown to give reliable results, even for band gap values which are generally poorly estimated by ab-initio Hartree-Fock calculations. From the calculated band structures, it is possible to determine the band gap between the top of the highest occupied band (HOMO) and the bottom of the lowest unoccupied band (LUMO) and the HOMO bandwidth.

The calculated values of the band gap (Eg) and the width of the highest occupied band (BW) for PT, PTT and PDTT along with the optical measured band gap are summarized in table 1.

Table 1. Experimental results and theoretical data (from ref. 24).

Polymer	$E_g(eV)$[a]	$E_g(eV)$[b]	$BW(eV)$[b]
PT	2.0	1.9	2.5
PTT	2.0	1.8	1.7
PDTT	1.9	1.8	2.2

[a]From optical measurements on the neutral polymer.
[b]From theoretical calculations on the neutral polymer.

The calculated band gaps are in very good agreement with the optical data and in spite of the different molecular structure of the starting monomers, the neutral polymers show very similar band gaps.

NONLINEAR OPTICAL PROPERTIES

A prerequisite for organic materials with large and fast NLO responses is a conjugated electron backbone structure. In the previous paragraph we described a series of such polyconjugated systems: PT, PTT and PDTT. These are centrosymmetric systems therefore the third order electrical susceptibility, $\chi^{(3)}$, is the first non vanishing nonlinear term. It is well known that there are different methods to measure $\chi^{(3)}$, and each of them have different meanings. $\chi^{(3)} (-3\omega;\omega,\omega,\omega)$ is the expression of $\chi^{(3)}$ measured via THG which addresses the purely electronic optical response. On the other hand $\chi^{(3)} (-\omega;\omega,-\omega,\omega)$ measured by DFWM concerns the optical response determined by the energy migration cascade from excited electronic species, to vibrational excitation and eventually thermal phonons. Therefore the optical response at a given time may be affected by several excitations. Nevertheless the DFWM configuration is technologically more interesting because it is closer to the hypotetical envisaged device $\chi^{(3)}$ $(-\omega;\omega,-\omega,\omega)$ design. We therefore investigated the wavelength dependence of $\chi^{(3)}$ $(-\omega;\omega,-\omega,\omega)$ in resonance and nonresonance conditions in the series. The dispersion of $\chi^{(3)}$ $(-3\omega;\omega,\omega,\omega)$ of PTT has been reported in a previous paper [16].

Typical thicknesses of the undoped films were in the 0.5-2 μm range. The $\chi^{(3)}$ was measured using a folded boxcar four wave mixing configuration. The details of the experimental set-up are given in ref. [11].

The DFWM signal as a function of the delay time was measured for all three different samples, indicating a response time limited by the laser pulse duration (15 psec) as measured by a 2 psec streak camera. The third order nonlinearity is obtained from measurements of the four wave mixing signal by using the formula [25]:

$$\chi^{(3)} = A \frac{n^2}{\omega L} \left(\frac{I_s}{I_1 I_2 I_3} \right)^{1/2} \left[\frac{\alpha L}{[1-\exp(-\alpha L)]} \right] \exp\,(\alpha L)$$

where α and n are the absorption coefficient and the refractive index of the sample. I_s is the four wave mixing signal intensity, I_1, I_2, and I_3 are the intensities of the three interacting beams, and A is a constant which depends on the laser pulse duration and on the geometry of the experiment. The $\chi^{(3)}_{SP}$ of the sample was determined by comparing the sample signal with the CS_2 signal namely:

$$\chi^{(3)}_{SP} = \chi^{(3)}_{CS_2} \gamma \left(\frac{I_s(SP)}{I_s(CS_2)} \right)$$

where γ takes into account the transmission coefficient of the sample and thickness and refractive index of the CS_2, and $\chi^{(3)}(CS_2)$ = 8.8x10^{-13} esu [26]. Refractive indexes of samples were measured by Brewster angle technique at different wavelengths, values varied from 1.9 to 2.1. For all measurements, interacting beams had parallel polarizations, and no significant change in four wave mixing signal was observed when the samples were rotated, indicating that films were highly isotropic. Under these conditions the observed four wave mixing signal was generated by the $\chi^{(3)}_{1111}$ component of the NLO susceptibility.

The values of $\chi^{(3)}$ as a function of the wavelength are given in table 2 and in fig.3 one-photon absorption and $\chi^{(3)}(-\omega;\omega,-\omega,\omega)$ spectra are shown.

Table 2. Nonlinear coefficient $\chi^{(3)}$ vs wavelength (from ref. 11)

Wavelength (nm)	PT $\chi^{(3)}(\times 10^{-9}\text{esu})$	PTT $\chi^{(3)}(\times 10^{-9}\text{esu})$	PDTT $\chi^{(3)}(\times 10^{-9}\text{esu})$
532	6.6 ± 1	5.9 ± 1	11.3 ± 2
585	5 ± 1	4.4 ± 1	7.7 ± 1
590	3.6 ± 1	3.0 ± 1	5.8 ± 1
595	3.8 ± 1	3.6 ± 1	6.1 ± 1
600	3.2 ± 1	2.9 ± 1	4.5 ± 1
605	3.0 ± 1	3.0 ± 1	5.5 ± 1
630	0.7 ± 0.5	0.7 ± 0.5	1.3 ± 0.6
1064	0.03 ± 0.02	0.03 ± 0.02	0.03 ± 0.02

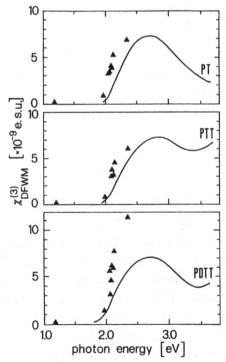

Fig. 3. One-photon absorption spectra (full line) and nonlinear coefficient $\chi^{(3)}_{\text{DFWM}}$ (triangles) of PT, PTT and PDTT.

We observe that above the gap the values are larger than 10^{-9} esu for all three samples. In addition there is a large increase in the value of $\chi^{(3)}$ (up to 10^{-8} esu for PDTT at 532 nm) by going into the rising one photon absorption edge. Below the band gap, far away from the first π-π^* transition in a spectral region where the one-photon absorption spectra is negligible, (at 1064 nm), all the three polymers have approximately the same value of $\chi^{(3)} \approx 10^{-11}$ esu (see fig. 3 and table 2). This latter value is anyway one order of magnitude larger than in CS_2.

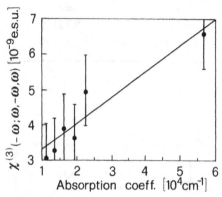

Fig. 4. Nonlinear optical coefficient $\chi^{(3)}_{DFWM}$ versus absorption coefficient of PT (from ref. 11).

It is interesting to note how for devices applications the nonresonant value of $\chi^{(3)}$ should be higher than 10^{-8} esu, in particular for wave guided devices. In general, we may define a figure of merit: $\chi^{(3)}/\alpha\tau$ where α is the absorption coefficient and τ is the time response of the $\chi^{(3)}$ process. For application the figure of merit has to be maximized in the following way [2]: large $\chi^{(3)} \approx 10^{-4}$ esu ; low $\alpha \leq 10^3$ cm^{-1}; small $\tau \leq 1$ psec. Fig. 4 shows a plot of the nonlinear coefficient versus the one-photon absorption coefficient for PT.

Similar results were obtained for PTT and PDTT. The solid line is a theoretical fit assuming $\chi^{(3)}$ proportional to α. The plot is compatible with the experimental curve indicating that the $\chi^{(3)}$ $(-\omega;\omega,-\omega,\omega)$ is determined by a population grating effect. The high values of $\chi^{(3)}$ are obtained at the expence of large absorption losses in the media making these systems not yet suitable for device applications.

As one can immediately see from fig. 4 the figure of merit for the investigated polymers is rather a long way from the maximized one necessary for guided device applications. One possible way out of this "catch 22" effect is to consider resonances with transient excited states. We will address this aspect in the last section.

PHOTOGENERATED NONLINEAR EXCITATIONS IN THIOPHENE BASED CONJUGATED POLYMERS

In this section we report investigation on the generation and stability of nonlinear excitations and the relative mobility of P and BP (P, BP) under different average interchain distances. The role of interchain coupling on the stability of P and BP has been theoretically studied by D. Baeriswyl and K. Maki [27] and by Y.N.Gartstein and A.A. Zakhidov [28]. The interchain transfer integral, which is the key parameter that controls the stability of P and BP [27-28], is expected to increase monotonically between PT and PDTT. Photoinduced absorption (PA) in PT has been studied by several authors [29-30]. In the framework of a BP charge transport model the dynamic mass of BP has been estimated from a PA experiment [30]. In fact the correspondence between doping induced bands (DIB) and photoinduced (PA) bands in the NIR region [31] indicates the identical nature of charged defect states created by doping and by photoexcitation. In the following we report on the PA experiment on PT, PTT and PDTT.

The infrared PA spectra of PT, PTT and PDTT at 6 K are shown in fig.5 and the phonon region is shown in the inset of each figure.

The PA spectrum of PT reported here is virtually identical to spectra reported in the literature [29-30]. PA spectra of PTT and PDTT are quite similar to that of PT both in the mid-IR and in the NIR spectral range.

In all three systems (see Table 3) the prominent phonon bands which are observed in the mid-IR (IRAV modes) coincide with the bands that appear by doping [31] (DIB).

Photogenerated charges induce a local perturbation of the polymer backbone geometry. The correspondence between DIB and IRAV bands reflects

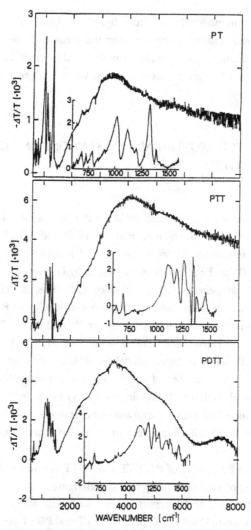

Fig. 5. Photoinduced absorption spectra at 6 K of PT, PTT and PDTT.
Magnification of the phonon range is show in the inset (from ref. 32).

Table 3. Frequencies (cm^{-1}) of DIB and IRAV bands in PT, PTT and PDTT (from ref. 32).

PT		PTT		PDTT	
DIB	PA	DIB	PA	DIB	PA
671w	688m	455w		467m	
	727m	618m	631w	695m	698m
794w	783m	691m	695m	892w	893w
1031s	1015s	1055s		1080s	1118s
1100m	1116s	1108s	1111s		1203s
1205m	1200m	1182m	1197s	1262s	1259s
1331s	1321s	1282s	1264s	1291w	1303m
1392s	1382m	1470w	1481m	1402s	1407m
					1494m

the identical nature of the charged defect states which are generated by doping or by photoinjection of electron-hole pairs. In the near IR a broad electronic absorption is observed in PT, PTT and PDTT with maxima at 0.42 eV, 0.51 and 0.45 eV respectively. The similarity of IRAV and DIB modes and the simultaneous presence of two subgap electronic excitations constitute the most convincing evidence of the presence of BP in polythiophene [33]. The similar spectral behaviour of PT, PTT and PDTT indicates that BP are stable excitations in PTT and PDTT as in the case of PT. Then the closer interchain packing and the consequent larger interchain transfer integral in PTT and PDTT do not inhibit the formation of BP in these polymers.

DIRECT EVIDENCE OF $\chi^{(3)}_{DFWM}$ ENHANCEMENT BY TRANSIENTLY PHOTOGENERATED NONLINEAR EXCITATIONS

The electronic excitations of conjugated organic systems, which exhibit large and fast NLO response, are controlled by a strong e-ph coupling resulting in the formation of self-trapped excitations such as charged solitons, P and BP with their associated electronic states inside the gap. Consequently, the redistribution of oscillator strength, produced in less than a picosecond, is proposed as the possible cause for the larger NLO effects in these polymers. In nondegenerate ground state systems like PT, PTT and PDTT we have shown in the previous section that BP are the dominant charge carriers after photoexcitation or doping. The dispersion of $\chi^{(3)}_{DFWM}$ was obtained at the edge of the absorption region. Below gap, measurements at 1064 nm gave a $\chi^{(3)}$ value of 10^{-11} esu. We here report preliminary measurements of non resonant $\chi^{(3)}$ at 1064 nm when the PT sample is photoexcited by above gap excitation. The pump-probe DFWM technique monitors the picosecond dynamics of nonlinear photoexcitations.

PT thin films (1000 - 2500 Å) were electrochemically polymerized on ITO. The pump-probe DFWM technique is outlined in fig. 6.

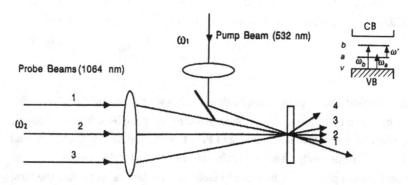

Fig. 6 Schematic of the geometry of "pump-probe" DFWM technique. In the inset polaron levels are outlined.

The fundamental output of a 10-Hz mode-locked Nd:YAG laser (25 ps) pulses at 1064 nm is divided into three beams and used in a DFWM geometry. A fourth beam at 532 nm having the same polarization direction as the 1064 nm beams is used for the photoexcitation. DFWM beams at 1064 nm probe the system excited at different delay times.

Three DFWM beams at 1064 nm are spatially and temporally overlapping on the sample. The photoexciting beam at 532 nm is delayed as shown in the inset of fig. 7.

In the absence of the 532 nm beam there is a background non resonant DFWM signal as previously reported. Fig. 7 illustrates the increase of the

DFWM signal at 1064 nm as a function of the delay of the 532 nm beam. An evident resonant enhancement is observed when the photoexcited 532 nm beam overlaps with the 1064 nm DFWM beams. In order to determine the response time of the enhanced DFWM signal we performed a second experiment in which the photoexciting 532 nm beam was illuminating the sample simultaneously with two 1064 nm beams, the third one being delayed. This experiment is outlined in the inset of fig. 8.

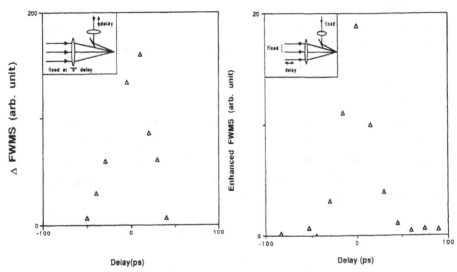

Fig.7. The enhanced DFWM signal at 1064 nm of PT. The delay is between the 532 nm and DFWM at 1064 beams.

Fig.8. The time resolved enhancement of DFWM signal at 1064 nm. The 532 nm beam is fixed at 0 delay while one of the 1064 nm beams is delayed.

The results clearly show that the response time is still limited by the laser pulse resolution (<25 ps) indicating that the enhanced DFWM signal is ultrafast. When PT is photoexcited above the energy gap by 532 nm (2.33 eV) photons, midgap states are formed inside the gap [30a]. In a recent photomodulation experiment it has been shown that a broad midgap photoinduced absorption attributed to P appears in the subpicosecond timescale [30b]. In this case the beams at 1064 nm are in resonance with the transitions involving the photogenerated gap states. Now $\chi^{(3)}$ obtained from DFWM measurements has a triple resonant term [34]:

$$\chi^{(3)}_R = \frac{NA^4}{(\omega_{ij} - \omega_2 - i\Gamma_{ij})^3}$$

.where, $ij = vb$ *(or ab)*, A includes the dipole matrix element for the transitions. Γ_{ij} is the phenomenological damping factor, and N is the number of P or BP per unit volume.

The full expression of $\chi^{(3)}$ is $\chi^{(3)} = \chi^{(3)}_{NR} + \chi^{(3)}_R$. The enhancement of DFWM signal at 1064 nm is expected when the system is photoexcited at 532 nm. As shown in fig. 7, we have observed a photoinduced enhancement of the DFWM signal at 1064 nm because of the resonance of $\chi^{(3)}$ about the photoinduced polaronic bands in PT.

The two time resolved experiments that we performed (figs. 7 and 8) indicate that both the formation of photoinduced states and the formation of the resonant (with polaron transitions) transient grating are faster than the time resolution of the experiment (25 ps). The short lifetime of the photoinduced states detected in the present work could be indicative of the fact that P rather than BP are responsible for the enhancement of $\chi^{(3)}$ [30b]. The dependence of the enhanced DFWM signal on pump fluence (I) at 532 nm is shown in fig. 9.

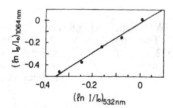

Fig. 9. The enhanced DFWM signal at 1064 nm vs the 532 nm pump fluence.

This corresponds to a sublinear dependence: $\chi^{(3)}_R = k\, I^{0.7}$ where k is a constant. The sublinearity may arise from saturation effects such as volume or phase filling as seen in *trans* $(CH)_x$ [6]. In conclusion the resonance enhancement of $\chi^{(3)}$ due to photogenerated polaronic states has been observed for the first time in polythiophene on picosecond time scale using a novel "pump-probe" DFWM technique. Intrachain P are most likely to be responsible for this enhancement at times less than 25 ps.

CONCLUSIONS

In this paper we have studied the NLO and electronic properties of a homologous series of thiophene based polyconjugated systems. The one-photon absorption spectra of the investigated compounds show an inhomogeneous broadened envelope giving rise to an extended absorption tail. This residual absorption in the low energy edge has, up to now, been common to chemically or electrochemically synthesized conjugated polymers and has prevented optimization of the $\chi^{(3)}$ response, in wave guided devices. In fact the enhancement of $\chi^{(3)}$ due to resonance with stationary electronic states is adversely compensated by absorption losses.

One way to get round this difficulty is to take advantage of $\chi^{(3)}$ response enhancement due to resonance with transiently photogenerated subpicosecond polaronic states which we have shown for the first time to be effective in polythiophene. This, together with highly pure and aligned conjugated systems will offer new possibilities for development of an all optical wave guided device. Along this line we would like finally to mention our recent discovery [35] of the liquid crystalline phase of an oligomer of thiophene: α-sexithienyl (T6). The conjugation length of T6 is long enough to show large $\chi^{(3)}$ response [36-37]. The nematic order of T6, as well as the possibility to synthetise a variety of rigid rod-like conjugated molecules, opens a new prospective for organic materials for molecular electronics.

ACKNOWLEDGMENTS

We are very grateful to profs. R. Tubino and J.L. Bredas and Drs. R. Lazzaroni and A.J. Pal for the invaluable contribution that they gave to part of this work. We would like to thank Progetto Finalizzato "Materiali speciali per tecnologie avanzate" of CNR for financial support.

REFERENCES

1. "Nonlinear Optical Properties of Polymer", ed. by A. Heeger, J. Orenstein and D.R. Ulrich, MRS Symp. Proc. Vol. 109, (1988).
2. "Nonlinear Optical Effects in Organic Polymers", ed. by J. Messier, F. Kajzar, P. Prasad and D. Ulrich; Kluwer Acad. Publ., Dordrecht (1989); P.N. Prasad in "Nonlinear Optical effects in Organic Polymers" pag. 377, ed. by J. Messier, F. Kajzar, P. Prasad and D. Ulrich; Kluwer Acad. Publ., Dordrecht (1989).

3. "Conjugated Polymer Materials: Opportunities in Electronic, Optoelectronics and Molecular Electronics", ed. by J.L. Bredas and R.R. Chance, NATO ASI Series, Vol. 182, Kluwer Acad. Publ., Dordrecht (1990).

4. "Organic Molecules for Nonlinear Optics and Photonics", ed. by J. Messier, F. Kajzar and P. Prasad, NATO ASI Series, Vol. 194, Kluwer Acad. Publ., Dordrecht (1991).

5. C.V. Shank, R. Yen, R.L. Fork, J. Orenstein and G.L. Baker, Phys. Rev. Lett. 49, 1566 (1982); Phys. Rev. B. 28 6095 (1983).

6. L. Rothberg, T.M. Jedju, S. Etemad and G.L. Baker, Phys. Rev. Lett. 57, 3229 (1986); Phys. Rev. B 36 7524 (1987).

7. F. Kajzar, S. Etemad, G.L. Baker and J. Messier, Solid State Commun 63, 1113 (1987).

8. M. Sinclair, D. Moses, K. Akagi and A.J. Heeger, Phys. Rev. B 38 10724 (1988).

9. D.M. Mackie, R.J. Cohen and A.J. Glick, Phys. Rev. B 39 3442 (1989).

10. W.S. Fann, S. Benson, J.M.J. Madey, S. Etemad, G.L. Baker and F. Kajzar, Phys. Rev. Lett. 62, 1492 (1989).

11. L. Yang, R. Dorsinville, Q.Z. Wang, W.K. Zou, P.P. Ho, N.L. Yang, R.R. Alfano, R. Zamboni, R. Danieli, G. Ruani and C. Taliani, J. Opt. Soc. Am. B 6 753 (1989); R. Dorsinville, L. Yang, R.R. Alfano, R. Zamboni and C. Taliani, Opt. Lett. 14, 1321 (1989).

12. R. Worland, S.D. Philips, W.C. Walker and A.J. Heeger, Synth. Met. 28, D663 (1989).

13. B.P. Singh, M. Samoc, H.S. Nalwa and P.N. Prasad, submitted.

14. D. Neher, A. Wolf, M. Leclerc, A. Kaltbeitzel, C. Bubeck and G. Wegner, Synth. Met. in press.

15. W.S. Fann, S. Benson, J.M.J. Madey, S. Etemad, G.L. Baker and F. Kajzar; Phys. Rev. Lett. 62, 1492 (1989); S. Etemad, W.S. Fann, P.D. Townsend, G.L. Baker and J. Jackel in "Conjugated Polymer Materials: Opportunities in Electronic, Optoelectronics and Molecular Electronics", ed. by J.L. Bredas and R.R. Chance, NATO Series, Vol. 182, Kluwer Acad. Publ., Dordrecht 341-352, (1990).

16. F. Kajzar, G. Ruani, C. Taliani and R. Zamboni, Synth. Met. 37, 223 (1990).

17. W.P. Su and J.R. Shrieffer, Proc. Natl. Acad. USA 77, 5626 (1980).

18. R. Ball, W.P. Su and J.R. Shrieffer, J. Phys. (Paris) Colloq 44, c3 (1983).

19. P. Di Marco, M. Mastragostino and C. Taliani, Mol. Cryst. Liq. Cryst. 118, 241 (1985).

20. R. Danieli, C. Taliani, R. Zamboni, G. Giro, M. Biserni, M. Mastragostino and A. Testoni, Synth. Met. 13, 325 (1986).

21. M.A. Druy and R.J. Seymour, 44, C3-595 (1983).

22. M. Pope and C.E. Swenberg, "Electronic Processes in Organic Crystals" Clarendom, Oxford (1982) (and reference therein).

23. J.M. Andrè, L.A. Burke, J. Delhalle, G. Nicolas and P. Durand, Int. J. Quantum Chem. S13, 283 (1979); J.L. Bredas, R.R. Chance, R. Silbey, G. Nicolas and Ph. Durand, J. Chem. Phys. 75, 255 (1981).

24. C. Taliani, R. Zamboni, R. Danieli, P. Ostoja, W. Porzio, R. Lazzaroni and J.L. Bredas, Physica Scripta 40, 781 (1989).

25. G.M. Carter, M.K. Thakur, Y.J. Chen and J.V. Hryniewicz, Appl. Phys. Lett. 47, 457 (1985).

26. S.L. Shapiro and H.P. Broida, Phys. Rev. 154, 129 (1967).

27. D. Bariswyl and K. Maki, synth. Met. 28 D507 (1989).

28. Y.N. Gartstein and A.A. Zakidov, Sol. St. Commun. 213 (1987).

29. F. Moraes, H. Schaffer, M. Kobayashi, A.J. Heeger and F. Wudl, Phys. Rev. B. Rap. Comm. 30, 2948 (1984).

30. a) Z. Vardeny, E. Ehrenfreund, O. Brafman, A. Heeger and F. Wudl, Synth. Met. 18, 183 (1987); b) G.S. Kanner and Z.V. Vardeny, Synth. Met. (1991) in press.

31. C. Taliani, R. Danieli, R. Zamboni, P. Ostoja and W. Porzio, Synth. Met. 18, 177 (1987).

32. A.J. Pal, G. Ruani, R. Zamboni, R. Danieli and C. Taliani, Synth. Met. 41, 579 (1991).

33. A.J. Heeger, S. Kivelson, J.R. Shrieffer and W.P. Su, Rev. Mod. Phys. 60, 781 (1988).

34. Y.R. Shen, "The Principle of Nonlinear Optics", John Wiley & Sons, Inc. 1984, Chap. 14.

35. C. Taliani, R. Zamboni, G. Ruani, S. Rossini and R. Lazzaroni, J. Mol. Elctr. 6, 225 (1990).

36. M.T. Zhao, B.P. Singh and P.N. Prasad, J. Chem. Phys. 89, 5535 (1988).

37. D. Fichou, F. Garnier, F. Charra, F. Kajzar and J. Messier, Spec. Publ. Royal Soc. Chem. 69, 176 (1989).

OPTICAL STUDIES OF POLYSILANES

Jonathan R. G. Thorne and Robin M. Hochstrasser
University of Pennsylvania, Philadelphia, PA

John M. Zeigler
Silchemy, Inc., Albuquerque, NM

Andreas Tilgner and H. Peter Trommsdorff
University of Grenoble, France

Roger H. French and Paul J. Fagan
DuPont Co., Wilmington, DE

Robert D. Miller
I.B.M., Almaden, CA

The polysilane polymers $[RR'Si]_n$ are the silicon analogues of the alkanes (see [1] for a review). They have a number of important applications arising largely from their photochemical activity. Of particular interest, however, are their electronic excited states which, unlike the alkanes, show evidence of delocalized sigma bonding along the polymer backbone, and lead to a host of interesting linear and non-linear optical phenomena.

We give a review of optical experiments that have been performed on the polysilane high polymers to investigate the consequences of sigma conjugation and the nature of excited states and excited state dynamics.

1. Super Emission

The radiative lifetime of polysilane polymers has been determined to be ~700 ps. This is ~25 times shorter than the absorption strength per Si-Si bond would predict and suggests a chromophore length of ~25 silicon atoms over which excitations are delocalized (2). The behavior is thus characteristic of a super emissive molecule. This property is further confirmed by the saturation and other experiments described below.

The process of spatial energy transfer in the polymers has been followed by the technique of time-resolved fluorescence depolarization [2,3] which reveals relaxations occurring on several different timescales from picosecond to nanosecond and longer. We have concluded that energy migration along a twisted chain is a principal cause of the anisotropy loss that occurs prior to any possible motion of the whole polymer chain in the laboratory frame.

2. Phase transition effects

Many polysilanes show interesting conformational phase changes both in solution and as thin films near to room temperature. These phase changes are frequently accompanied by large absorption spectral shifts and have been interpreted as changes of the conformation of the silicon backbone leading to greater electron delocalization. The prototypical, poly(di-n-hexylsilane), $[(C_6H_{13})_2Si]_n$ is the most widely studied by optical techniques of the sigma-conjugated polysilane polymers. The absorption spectrum was proposed to consist of an inhomogeneous distribution of substates having a range of excitation energies [4]. Emission would then originate from the lower energy group following energy transfer. The origin of the segment length inhomogeneous distribution is however not understood.

We have shown (3) that the fluorescence lifetime of the polysilane is significantly increased in crystalline forms as a result of a decrease in the rate of non-radiative processes accompanied by an increase in the radiative lifetime. This lengthening of the lifetime is not seen in a glass at 77K and

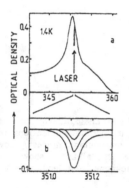
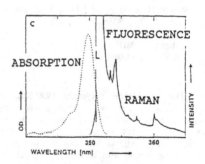

Figure 1. Hole burning in poly(di–n–hexylsilane) glass

(a) Linear Absorption
(b) Enlarged Hole Region
(c) Absorption and Fluorescence

we regard this as evidence of the existence of an isolated ordered all-trans rodlike form of the polymer in this medium. The fluorescence anisotropy was shown to be comparatively independent of phase and to decay with a characteristic time of 2ns attributed to motion of the excitation among the polymer segments having different orientations in the laboratory frame.

3. Photochemistry and Hole burning

The first observation of hole–burning in the optical spectra of delocalized polymer backbone states was observed. It was shown that narrow photochemical holes ($\sim$1cm^{-1}) could be burnt in the inhomogeneous spectrum of a glassy polymer at low temperature (4). We have studied the temperature

dependence and energy dependence of the holewidth (5).   There is a residual
zero temperature width of ca. 0.4 cm^{-1} and an Arrhenius-like line broadening
having an activation energy of 37 cm^{-1}.   We find the hole width to increase
with increasing separation between the excitation energy and the absorption
edge.

Narrow-band   laser   excitation   of   low-temperature   solid   films   of
poly(di-n-hexylsilane) showed photochemically induced absorption changes that
indicated that extensive energy transfer occurred after light absorption
[6,8].   It was concluded that the transfer occurred between adjacent polymer
segments but not between the highly internally ordered microcrystallites that
compose the structure.

The low temperature glassy spectrum of the polymer can be modelled by
introducing a small random Gaussian energy disorder at silicon sites in a
linear chain, analogous to the continuous disorder caused by curvature of the
skeleton, proposed for polydiacetylene [7]   leading to the conclusion that the
lowest energy states of the .chain have extents of about 10 silicon atoms.
This model explains the effects of energy transfer prior to photochemistry on
the hole burning spectrum (8-10).

Evidence from these hole-burning studies and the very small (<10 cm^{-1})
shift of the fluorescence and absorption peaks found in low-temperature
glasses shows that an individual chromophore upon excitation does not strongly
relax due to coupling to the lattice.   Negligible phonon structure is seen in
absorption and emission spectra, and the excitations in polysilanes do not
have the character of strongly bound polarons.   By careful sample manipulation
we have obtained spectra that directly display weak phonon coupling as shown
in Figure 1c.

Fluorescence emission has been shown to be highly sensitive to sample
preparation.   The effects of dissolved molecular oxygen on the low temperature
spectra are particularly significant (9,10).   It is these observations that
necessitated very careful sample manipulation.

We have observed transient absorptions at 450 nm and 370 nm in room
temperature solutions of polyphenylmethylsilane (11) using picosecond laser
pulses.   We attribute a persistent transient absorption at 450 nm to
photogenerated phenylmethylsilylene.   The precursor state is likely to be a
relaxed triplet state.   Photochemical selection from the length distribution
occurs from the solution-phase polymer at ambient temperature (the first
example of room temperature hole-burning in solution) (11).

4.   Multiphoton   spectroscopy   and   excited   state   spectroscopy

Two-photon excitation studies have identified a dipole forbidden even
parity excited state ~leV above the first excited odd parity state [12].   The
2-photon cross-section is large indicative of high non-linear polarizability:
i.e. $\delta = 10^{-48}$ cm^4 sec^{-1}.

This work has stimulated a theoretical reexamination of the problem of the
energetic ordering of covalent/ionic states of 'alternated' linear chain
polymers and the effects of electron correlation (13).

We have shown that the excitations in these polymers are best regarded as
delocalized excitons and are   similar to those found in the π-conjugated
polydiacetylenes.   Observation of 1,2 and 3 photon resonances (the latter by
third harmonic generation [14] has clarified this picture of excited states
(see Fig. 2).

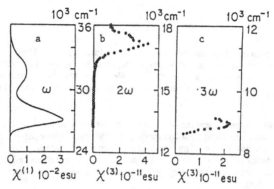

Figure 2.   Multiphoton
             resonances in poly
             di-n-hexylsilane
             Films

(a) Linear Absorption
(b) Two-Photon Absorption
(c) Third Harmonic Generation

We have also observed excited states of poly(di-n-hexylsilane) in femtosecond transient absorption [13,15]. The state in the near infrared (~1eV) correlates well with that seen in two-photon spectroscopic measurements and is attributed to an exciton having charge transfer character. The even parity state in the ultra-violet (~3.5eV) is possibly a biexciton. Ab initio calculations present an alternative view of these g-states (16).

We examined the femtosecond time resolved excited state dynamics occurring in linear-chain poly (di-n-hexylsilane) following optical excitation (15). The process of energy transfer in the polymers has been followed by the technique of time-resolved excited state absorption depolarization (15) which reveals highly dispersive kinetics characteristic of relaxations occurring on many different timescales from subpicosecond to nanosecond and longer. Fast energy transfer in the linear chain polymer is observed to occur on a time scale of 700 fs after excitation at short wavelength. We believe this process populates a distribution of lower energy states through phonon assisted relaxations. In Figure 3a, $\bar{\theta}$ is the average change in dipole direction after time t.

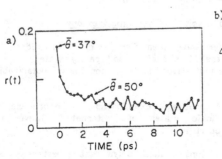
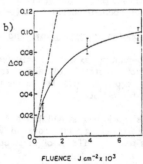

Figure 3.   a) Anisotropy of
             Polysilane Absorption

b) Poly(di-n-hexylsilane)
   Saturation Curve

In other experiments saturation spectroscopy [15] was used to obtain an upper limit for the average excitation conjugation length. In this method the number of photons needed to bleach the absorption is measured (using fs pulses) and used to calculate the number of absorbing centers, hence the number of silicons per excitation. The experiment yielded an upper limit of 25 silicon atoms and a most probable value of 10-15 atoms for each absorbed photon. This is in reasonable agreement with the extent of delocalization obtained from fluorescence lifetimes [2] (see Fig. 3b).

5. Absorption spectra in the deep UV (2-30eV)

Polysilane properties depend on sidechains. To evaluate sidechain-backbone coupling spectra at high energies are needed: We have accomplished this with a plasma source reflectance spectrometer. Some results for a series of poly(di-n-alkyl)silanesare shown in Figure 4 [(a) shows absorption, (b) dispersion]. Excitonic transitions are seen at 3-4 eV: the side chains show up in the 10-16 eV region. The intermediate region requires further interpretation.

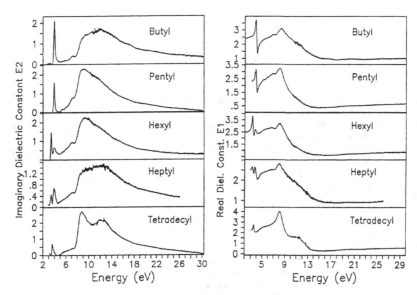

Figure 4.   Reflectance spectra of poly(di-n-alkyl silanes)

6. Oligomeric Silanes

Figure 5 shows absorption and fluorescence spectra of silanes, $Cl(SiMe_2)_2(SiPh_2)_5(SiMe_2)_2Cl$ and $Cl(SiMe_2)_2(SiPh_2)_5(SiMe_2)_2(SiPh_2)_5(SiMe_2)_2(SiPh_2)_5(SiMe_2)_2Cl$ with chain lengths, n = 9 and 23, silicon atoms. As the silicon chain length decreases, the quantum yield of fluorescence drops, the emission becomes broad, structured and red shifted and the radiative lifetime increases. This demonstrates the change in character of the excitation from large super-emissive exciton-polaron to small exciton-polaron vibronically coupled molecular transition [17].

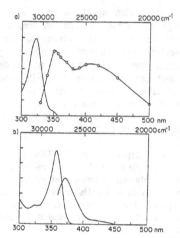

Figure 5. Absorption and fluorescence spectra
a) Si_9 b) Si_{23} chains

References

1. R. D. Miller and J. Michl, Chem. Rev. **89**, 1359 (1989).
2. Y. R. Kim, M. Lee, J. R. G. Thorne, R. M. Hochstrasser and J. M. Zeigler, Chem. Phys. Lett. **145**, 75 (1988).
3. J. R. G. Thorne, R. M. Hochstrasser and J. M. Zeigler, J. Phys. Chem. **92**, 4275 (1988).
4. H. P. Trommsdorff, J. M. Zeigler and R. M. Hochstrasser, J. Chem. Phys. **89**, 4440 (1988).
5. A. Tilgner, H. P. Trommsdorff, J. M. Zeigler, R. M. Hochstrasser, J. Chem. Phys., submitted.
6. H. P. Trommsdorff, J. M. Zeigler and R. M. Hochstrasser, Chem. Phys. Lett. **154**, 463 (1989).
7. G. Wenz, M. A. Muller, M. Schmidt, G. Wegner, Macromolecoles **17**, 837 (1984).
8. A. Tilgner, H. P. Trommsdorff, J. M. Zeigler and R. M. Hochstrasser, J. Lumin. **45**, 373 (1990).
9. A. Tilgner, J. P. Pique, H. P. Trommsdorff and R. M. Hochstrasser, Polymer Preprints **31**, 244 (1990).
10. A. Tilgner, J. R. G. Thorne, J. P. Pique, J. M. Zeigler, H. P. Trommsdorff and R. M. Hochstrasser, J. Lumin. **46**, 000 (1991).
11. Y. Ohsako, J. R. G. Thorne, C. M. Phillips, J. M. Zeigler and R. M. Hochstrasser, J. Phys. Chem. **93**, 4408 (1989).
12. J. R. G. Thorne, Y. Ohsako, J. M. Zeigler and R. M. Hochstrasser, Chem. Phys. Lett. **162**, 455 (1989).
13. Z. G. Soos and G. W. Hayden, Chem. Phys. **143**, 199 (1990).
14. J. R. G. Thorne, J. M. Zeigler and R. M. Hochstrasser, Proc. NATO ASI (1990); in press.
15. J. R. G. Thorne, S. T. Repinec, S. A. Abrash and R. M. Hochstrasser, Chem. Phys. **146**, 315 (1990).
16. V. Balaji and J. Michl, "Singlet Excitation in Polysilanes: Ab Initio Calculations on Oligosilane Models" (preprint).
17. J. R. G. Thorne, S. A. Williams, R. M. Hochstrasser and P. J. Fagan, Chem. Phys. Lett., submitted.

NEW COPOLYMERS FOR NONLINEAR OPTICS APPLICATION WHICH INCORPORATE

ELECTROACTIVE SUBUNITS WITH WELL DEFINED CONJUGATION LENGTHS

Charles W. Spangler, Pei-Kang Liu and Eric G. Nickel

Department of Chemistry
Northern Illinois University
DeKalb, IL

David W. Polis, Linda S. Sapochak and Larry R. Dalton

Department of Chemistry
University of Southern California
Los Angeles, CA

INTRODUCTION

Over the past five or six years, rapid advances have been made in the field of photonics, or the manipulation of light by light. Although organic polymers have been generally regarded as insulators in the electronics industry, research advances in the area of electroactive materials have shown that polymers such as polyacetylene can achieve metallic conductivity approaching that of copper when oxidized or reduced chemically or electrochemically (doping). More recently, it has been recognized in our laboratory and others that electroactive materials also have exceptionally high optical nonlinearities. Research on these materials holds the promise that they indeed may be suitable for the eventual design of many nonlinear and electro-optic devices, including an all-optical computer.

THE DESIGN OF NEW POLYMERS FOR NONLINEAR OPTICS

Conjugated pi-electron polymers, including such diverse structures as polyacetylene, polythiophenene and ladder polymers, have all been shown to have large third order optical nonlinearities and extremely fast switching times (femtosecond regime).[1] However, there have been problems associated with processibility, the magnitude of the third-order response so that low switching energies can be used, and the fabrication of high-quality, low-loss optical films. In this paper we will focus our attention on polymers containing structural repeat units related to polyacetylene, poly[p-phenylene vinylene], poly[2,5-thienylene vinylene] and the ladder polymer POL.

polyacetylene
PA

$$\left[\!\!\left\langle\bigcirc\right\rangle\!-C\!=\!C\right]_n$$

poly [p-phenylene vinylene]
PPV

$$\left[\!\!\left\langle\!\!\stackrel{\displaystyle \fbox{}}{S}\!\!\right\rangle\!-C\!=\!C\right]_n$$

poly [2,5-thienylene vinylene]
PTV

POL

Although large third-order response has been observed for all of the above polymers, they all absorb over a wide rang of the UV and VIS spectrum thus resulting in large absorption losses and subsequent low $\chi^{(3)}/\alpha$ values. PPV and PTV can be fabricated into optical quality films with $\chi^{(3)}/\alpha$ values around 10^{-13} esu-cm. It is much more difficult to fabricate PA and POL into optical quality films due to their inherent insolubility and intractability. In order to circumvent both problems, we have recently proposed that copolymers in which high-NLO activity oligomeric segments alternate with various low-NLO activity spacers, might yield materials with enhanced solubility due to the flexible spacer repeat units, while at the same time narrowing the absorption band characteristics due to the well-defined and invarient conjugation length of the NLO-active oligomeric repeat units. In addition, the absorption windows associated with these repeat units can be shifted by the mesomeric interaction of the functionalities used to link the active and non-active spacers. A generic formulation of this type of copolymer is illustrated in Figure 1.

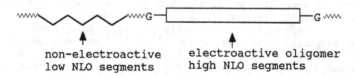

non-electroactive   electroactive oligomer
low NLO segments   high NLO segments

G = Mesomerically interactive functional group

Figure 1 General formulation of NLO-active copolymer

THE RELATIONSHIP OF THIRD ORDER SUSCEPTIBILITY WITH CONJUGATION LENGTH

As was pointed out repeatedly at the recent American Chemical Society Symposium, Materials for Nonlinear Optics (Boston, April, 1989), the second hyperpolarizability γ, is related to the extent of electron delocalization in molecules with extensive pi-electron clouds that can be easily polarized. Several workers have attempted to model the consequences of increasing conjugation (or delocalization) in relation to third-order properties. Beratan, et al.[2], have shown that the second order hyperpolarizability γ, increases rapidly for trans-polyenes as conjugation increases to 10-15 repeat units, and then more slowly up to 40 repeat units. This suggests that very long conjugation sequences (e.g, a fully conjugated high molecular weight pi-electron polymer) may not be required for high NLO activity, and that oligomeric segments could be intermixed with nonactive segments to maximize both NLO response and desirable physical properties.

Hurst and coworkers[3] have also calculated second hyperpolarizability tensors _via ab initito_, coupled-perturbed Hartree-Fock Theory for a series of polyenes up to $C_{22}H_{24}$. They found that γ_{xxxx} was proportional to chain length, with a power dependence of 4.0, but that this dependence tapered off as N increased. More recently, Garito and coworkers[4] calculated a power law dependence of γ_{xxxx} on chain length on the order of 4.6 ± 0.2. They also suggest that large values of $\chi^{(3)}$ should be attainable with conjugation sequences of intermediate length (100Å). Prasad[5] concurs with the conclusion that γ/N levels off with increasing N. In addition, Prasad has measured γ for a series of polythiophene oligomers by degenerate four wave mixing (DFWM) in solution and found a power law dependence for γ of 4. $\chi^{(3)}$ measurements for poly(3-dodecylthiophene) prepared by either chemical or electrochemical means were approximately the same, even though the molecular weights and number of repeat units were substantially different. Prasad concludes that one should consider what the effective conjugation length is compared to the overall length, and that in most cases, the effective conjugation length for NLO purposes probably does not extend much beyond ten repeat units. Further evidence for this interpretation was obtained by Prasad when similar DFWM measurements for poly[p-phenylene] showed a leveling off of $\chi^{(3)}$ at the terphenyl level (N = 3), presumably due to the gradual twisting of the phenyl rings out of conjugation as the number of repeat units increased.[6] These theoretical predictions and experimental results all seem to indicate that synthetic efforts should be concentrated on relatively short, highly conjugated molecules (oligomers) rather than long conjugation sequences.

SYNTHESIS OF NLO-ACTIVE COPOLYMERS

We have previously described[7,8] a general approach to the design and synthesis of the type of copolymer illustrated in Figure 1. Since even low molecular weight oligomers of PPV and PTV are relatively insoluble, we chose a condensation reaction approach that is not dependent on exact stoichiometry. Bis-carboxylic acid derivates of PA, PPV and PTV have been synthesized, converted to bis-acyl halides by reaction with thionyl chloride and the resulting bis-acyl halides reacted with an aliphatic diamine _via_ interfacial polymerization conditions. Initial attempts to form good optical films from these polyamides were thwarted by the relative insolubility of the high molecular weight (ca. 100,000 by GPC) of these polyamides. To alleviate this problem, we adopted a copolyamide approach wherein a more soluble aliphatic bis-acyl halide was mixed in with the NLO moiety to produce copolyamides illustrated below for PTV-dimer and trimer.
In this fashion, copolyamides with 10-20% incorporation were formed for PA (x = 3,4), PPV (x = 2,3) and PTV (x = 2,3) oligomers. In all cases the yields of purified polymer ranged from 80-100%. The full synthetic details of the copolyamide preparations and spectroscopic characterization have been published elsewhere.[7,8] Thermal analysis of these copolyamides show that they are at least as stable as the fully saturated Nylon 6,12, which may be regarded as a "parent" structure for comparison sake.

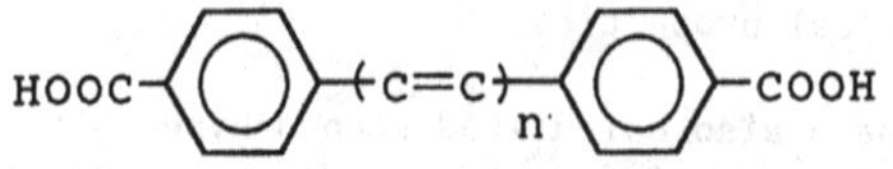

HOOC—⬡—(C=C)ₙ—⬡—COOH n= 3, 4

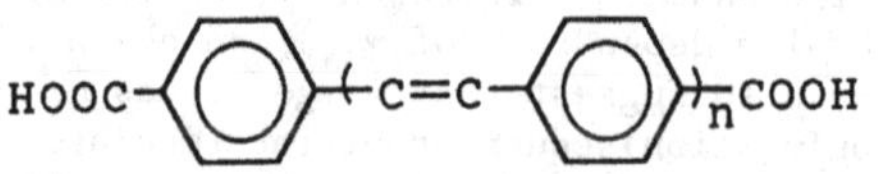

HOOC—⬡—(C=C—⬡)ₙ—COOH n= 2, 3

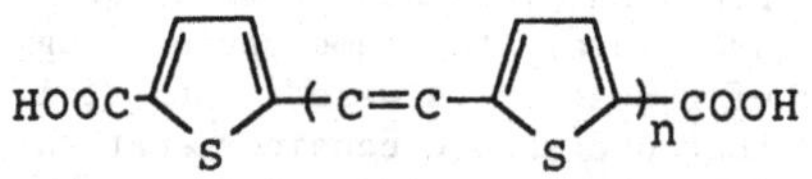

HOOC—[thiophene]—(C=C—[thiophene])ₙ—COOH n= 2, 3

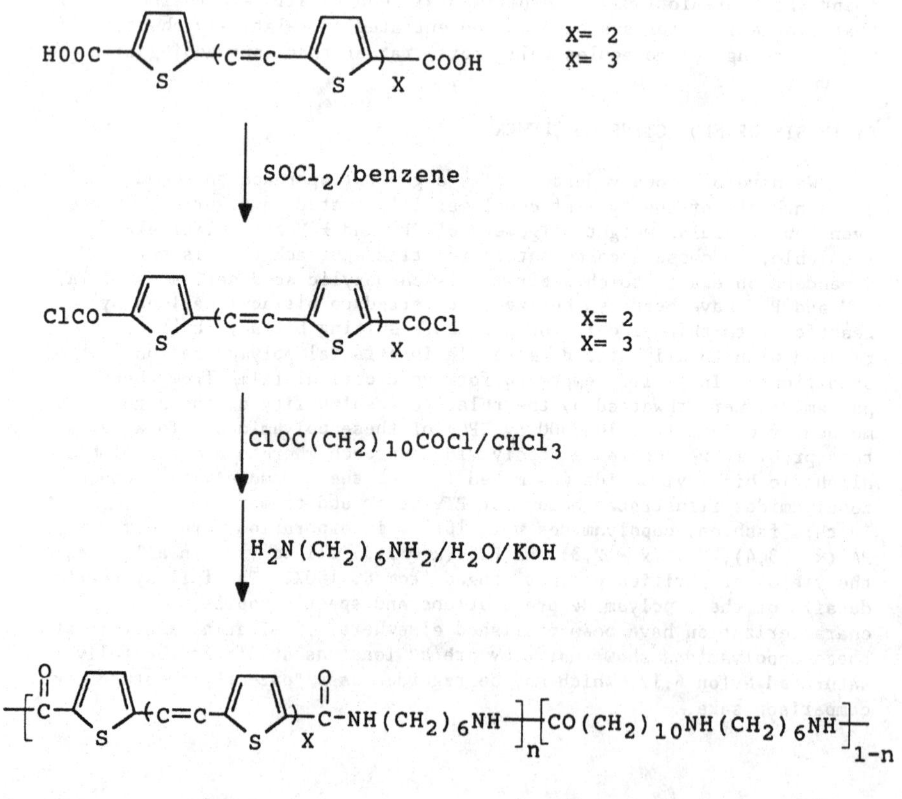

HOOC—[thiophene]—(C=C—[thiophene])ₓ—COOH X= 2, X= 3

↓ $SOCl_2$/benzene

ClCO—[thiophene]—(C=C—[thiophene])ₓ—COCl X= 2, X= 3

↓ $ClOC(CH_2)_{10}COCl/CHCl_3$

↓ $H_2N(CH_2)_6NH_2/H_2O/KOH$

$\left[C(=O)-[thiophene]-(C=C-[thiophene])_X-C(=O)-NH(CH_2)_6NH\right]\left[CO(CH_2)_{10}NH(CH_2)_6NH\right]_{1-n}$

More recently we have expanded our investigations to include ladder
moieties as the NLO-active segment of the copolyamide structures.[9,10]
Ladder polymers are well known for their thermal stability, and the pi-
conjugated segments may be regarded as essentially planar. Dalton and
coworkers[11] have previously shown by ENDOR and ESE spectroscopic studies
that electron delocalization in ladder polymers is limited to one repeat
unit. Thus we have attempted to incorporate POL and PTL repeat units
into copolyamides _via_ ladder bis-carboxylic acids as illustrated below:

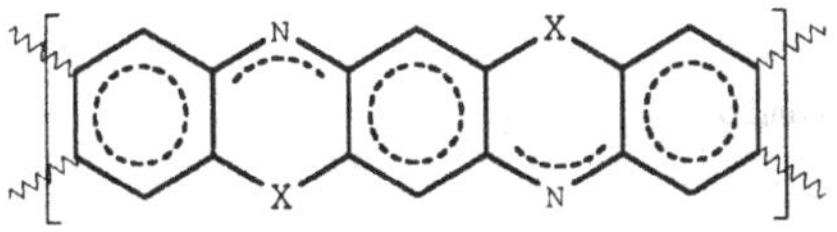

proposed delocalized unit in ladder polymers

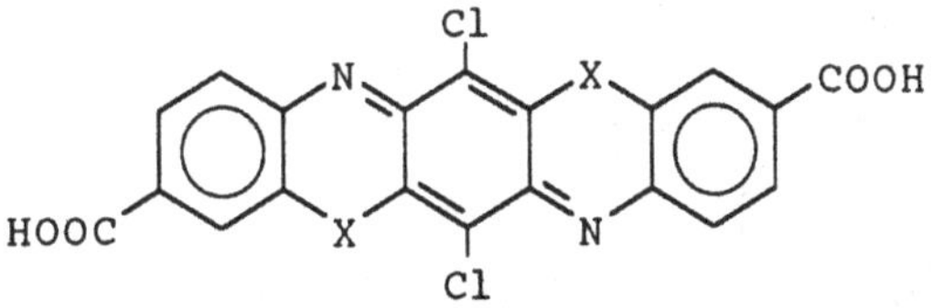

POL (X=O) and PTL (X=S) monomer
for copolyamide synthesis

Copolyamides with a 10-20% incorporation of the ladder subunit were
prepared by the standard interfacial polymerization approach. These
polymers, however, proved to be much less soluble than the corresponding
PPV and PTV copolyamides. Films have only been produced by static
evaporative casting techniques since we have not yet identified a
solvent which yields a solution viscous enough for spin-coating.

PREPARATION OF GUEST-HOST COMPOSITES

Guest-host composites of the NLO-active oligomers can be prepared
by dissolving weighed samples of the oligomer and optical-grade
polycarbonate in a suitable solvent (dichloroethane or NMP) and spin
coating on a glass or silica surface. The optical absorption
characteristics of the monomers, polymers and selected composites are
shown below in Table 1.

PRELIMINARY $\chi^{(3)}$ EVALUATION

There have been relatively few $\chi^{(3)}$ measurements on polymers in
which one repeat unit has a relatively large contribution to the optical
nonlinerity, while the other repeat unit contributes little. Copolymers
of the type illustrated in Figure 1 are characterized by sharp
absorptions from the well-defined conjugation length of the NLO active
repeat unit, which does not vary significantly from the monomer unit.

Table 1. Absorption Spectra of NLO-Active Repeat Units.

Repeat Unit	π-π^* absorption	
	λ_{max} (nm)	band edge (nm)
POLM (soln)[a]	504, 470[d]	550
POLM (10% copolyamide)[b]	539, 491	577
POLM composite (1%)[c]	512, 478	690
PTV dimer (soln)[a]	411	494
PTV dimer (20% copolyamide)[b]	444, 468	510
PTV dimer (10% composite)[c]	458	550
PTV trimer (soln)[a]	457	400
PTV trimer (20% copolyamide)[b]	479, 504	550
PTV Trimer (10% composite)[c]	465, 493	550

[a]DMF solution. [b]NMP solution. [c]Thin Film. [d]If more than one absorption peak, peak of maximum absorptivity is underlined.

The composites do not seem to follow any particular trend compared to the parent and covalently bound copolymer. Dalton and Yu[12] formed a ladder copolymer via a double ring closure in a soluble open-chain

precursor, however, elemental analysis showed that the ring closure was incomplete. However, the absorption characteristics were those of a 5-ring ladder unit. DFWM measurements indicated a $\chi^{(3)}$ of 4.5 x 10^{-9} and a $\chi^{(3)}/\alpha$ in the range of 10^{-12}-10^{-13} esu-cm. Two of the polymers listed in Table 1 show similar $\chi^{(3)}$ values.[7,8]

$$\chi^{(3)} = 2.0 \times 10^{-12} \text{ esu}$$

$$\chi^{(3)} = 1.5 \times 10^{-12} \text{ esu}$$

154

A $\chi^{(3)}/\alpha$ value of 0.6×10^{-13} esu-cm is quite comparable to the $\chi^{(3)}$ value for the Dalton and Yu ladder copolymer. More recently a copolymer incorporating a PPV dimer segment as a repeat unit has been characterized by THG, and shows similar nonlinear behavior.[13]

$$\left[\begin{array}{c} \overset{O}{\underset{\|}{C}}O-\bigcirc-C=C-\bigcirc-C=C-\bigcirc-O\overset{O}{\underset{\|}{C}}(CH_2)_5 \end{array} \right]_n \qquad \chi^{(3)} = 8.0 \times 10^{-13} \text{ esu}$$

We have also prepared a copolyamide incorporating a PPV dimer segment, but the lack of solubility prevented incorporation above 20%, and DFWM studies showed little if any, activity. It is possible that this low level of incorporation coupled to a lower third order susceptibility compared to PTV-dimer accounts for the inability of our measurements to detect significant activity. At the current time, DFWM studies are underway for 10% composites of PTV dimer, trimer, tetramer and pentamer. PPV and ladder oligomers are not soluble enough to allow for such composites at the 10% incorporation level. These studies should yield further information regarding the effect of increasing delocalization length on $\chi^{(3)}$.

CONCLUSION

Electroactive repeat units can be incorporated in formal copolymer backbones as copolyamides. These polymers show considerable nonlinear susceptibility with $\chi^{(3)}/\alpha$ in the range 10^{-13} esu-cm. This is quite respectable when one considers their relative low-level of incorporation (10-20%), and the relative sizes of the copolyamide nonactive segments. Obviously, if the solubility of the PPV, PTV, or POL monomers and polymers could be increased, the need for the nonactive copolyamide segments would be eliminated and we could expect at least an order of magnitude incease in $\chi^{(3)}$. Similar arguments would also apply to the composite structures. Recent developments in our laboratories in the incorporation of solubilizing substituents in α,ω-diphenylpolyene series demonstrates that such solubility enhancement in electroactive materials is certainly possible, and we hope to be able to report on this approach to enhance third-order susceptibilities in the near future.

ACKNOWLEDGEMENTS

This work was supported by the Air Force Office of Scientific Research under grant #AFOSR-90-0060 and contract AFOSR-F49620-88-071. The authors would also like to thank Drs. J. P. Jiang and X. F. Cao (USC) and Dr. Robert Norwood (Hoechst-Celanese) for preliminary DFWM characterization of the polymers.

REFERENCES

1. For a general review of nonlinear activity in pi-conjugated electoactive polymers, see:

    a. D. J. Williams, ed., "Nonlinear Optical Properties of Organic Polymeric Materials", Amer. Chem. Soc. Symp. Ser. 233, New York (1983).

    b. D. S. Chemla and J. Zyss, eds., "Nonlinear Optical Properties of Organic Molecules and Crystals", Vol. 2, Academic Press, New York (1987).

c. P. N. Prasad and D. R. Ulrich, eds., "Nonlinear Optical and Electroactive Polymers", Plenum Press, New York (1988).

d. S. R. Marder, J. E. Sohn and G. D. Stucky, eds., "Materials for Nonlinear Optics", Amer. Chem. Soc. Sym. Ser. 455, Washington, DC, (1991).

2. D. N. Beratan, J. N. Onuchic and J. W. Perry, "Nonlinear Susceptibilities of Finite Conjugated Organic Polymers, J. Phys. Chem. 91:2696 (1987).

3. G. J. B. Hurst, M. Duplis and E. Clementi, Ab Initio Analytic Polarizability, First and Second Hyperpolarizibilities of Large Conjugated Organic Molecules: Applications to Polyenes C_4H_6 to $C_{22}H_{24}$, J. Chem. Phys. 89:385 (1988).

4. A. F. Garito, J. R. Heflin, K. Y. Wong and O. Zamani-Khamiri, Enhancement of Non-linear Optical Properties of Conjugated Linear Chains through Lowered Symmetry, in "Organic Materials for Non-Linear Optics", R. A. Hann and D. Bloor, eds., Roy. Soc. Chem., London (1989).

5. P. N. Prasad, Studies of Ultrafast Third-Order Non-linear Optical Processes in Polymer Films, in Ref. 1(c), p. 264.

6. P. N. Prasad, Third-Order Nonlinear Optical Effects in Molecular and Polymeric Materials, in Ref. 1(d), p. 50.

7. C. W. Spangler, T. J. Hall, K. O. Havelka, D. W. Polis, L. S. Sapochak and L. R. Dalton, New Copolymers for Nonlinear Optics Applications which Incorporate Delocalized π-Electron Subunits with Well Defined Conjugation Lengths, Proc. SPIE, 1337: 125 (1990).

8. C. W. Spangler, P.-K. Liu, and T. J. Hall, The Design of New Copolymers for $\chi^{(3)}$ Application, Polymer (In Press) (1991).

9. C. W. Spangler, T. J. Hall, P.-K. Liu, L. R. Dalton, D. W. Polis and L. S. Sapochak, The Design of New Copolymers for $\chi^{(3)}$ Applications: Copolymers Incorporating Ladder Subunits, "Organic Materials for Non-Linear Optics:, R. A. Hann and D. Bloor, eds., Roy. Soc. Chem. London (1991) (In Press).

10. C. W. Spangler, M. L. Saindon and Eric G. Nickel, Synthesis and Incorporation of Ladder Polymer Subunits in Copolyamides, Pendant Polymers and Composites for Enhanced Nonlinear Optical Response, Proc. SPIE (In Press) (1991).

11. L. R. Dalton, J. Thomson and H. S.Nalwa, The Role of Extensively Delocalized π-Electrons in Electrical Conductivity, Nonlinear Optical Properties and Physical Properties of Polymers, Polymer, 28:543 (1987).

12. L.-P. Yu and L. R. Dalton, Synthesis and Characterization of New Polymers Exhibiting Large Optical Nonlinearities III: Rigid Rod/Flexible Chain Copolymers, J. Amer. Chem. Soc., 111:8699 (1989).

13. T. E. Mates and C. K. Ober, Model Polymers with Distrylbenzene Segments for Third-Order Nonlinear Optical Properties, in Ref. 1(d),P. 497.

SECOND HARMONIC GENERATION IN LANGMUIR-BLODGETT FILMS OF

N-DOCOSYL-4-NITROANILINE

T. Geisler

Danish Institute of Fundamental Metrology
Lundtoftevej 100
DK-2800 Lyngby, Denmark

S. Rosenkilde

Department of General and Organic Chemistry
The H.C. Ørsted Institute, Universitetsparken 5
University of Copenhagen, DK-2100 Copenhagen, Denmark

W.M.K.P. Wijekoon, P.N. Prasad

Photonics Research Laboratory, Department of Chemistry
State University of New York at Buffalo, Buffalo, NY 14214 USA

P.S. Ramanujam

Risø National Laboratory
Optics and Fluid Dynamics Department
DK-4000 Roskilde, Denmark

INTRODUCTION

In recent years there has been an increasing interest in the development of organic materials showing large nonresonant nonlinear optical response [1],[2]. Some organic materials have already shown off-resonance nonlinear optical properties that make them promising candidates for various applications in optical signal processing [3].

In order to be efficient for second-order effects such as second harmonic generation (SHG), the molecules should be arranged in a noncentrosymmetric structure, since in the electric-dipole approximation the second-order susceptibility, $\chi^{(2)}$, vanishes in the bulk for centrosymmetric structures. Furthermore, the microscopic first order hyperpolarizablity β of the molecule should be large. In most cases, the molecules in question feature electron rich donor substituents in resonance with acceptor substituents through a conjugated molecular π-electron system [4]. Thin organic films can be made by the Langmuir-Blodgett (LB) technique where a molecular layer first is prepared at an air-water interface and then transferred layer by layer onto a solid substrate. The LB-technique offers the possibility for controlling the structure and thickness of the films precisely and thus provide unlimited opportunities in designing materials in the form of waveguides [5] with extensive nonlinear optical properties.

Frontiers of Polymer Research, Edited by P.N. Prasad and
J.K. Nigam, Plenum Press, New York, 1991

In this paper, we report on measurements of SHG from the small chromophore, para-nitroaniline (pNA), derivatized with docosanoic (Behenic) acid (BA) to form N-docosyl-4-nitroaniline (DCpNA) in LB films, see **1.**. A comparison is made with LB films of 2-docosylamino-5-nitropyridine (DCANP) [6], see **2.**.

$$O_2N-C \overset{C-C}{\underset{C-C}{\bigcirc}} C-\overset{H}{\underset{}{N}}-C_{22}H_{45}$$

1 DCpNA

$$O_2N-C \overset{C-C}{\underset{C-N}{\bigcirc}} C-\overset{H}{\underset{}{N}}-C_{22}H_{45}$$

2 DCANP

EXPERIMENTAL

N-Docosanoyl-4-nitroaniline was prepared from docosanoic acid (Aldrich Chem. Co., 98%) *via* docosanoyl chloride and 4-nitroaniline (Riedel-de Haön, purum 99% - purified from a solution of acetone treated with charcoal and filtered through celite). In order to reduce N-docosanoyl-4-nitroaniline to N-docosyl-4-nitroaniline, we have applied the method of Kuehne and Shannon [7] to reduce unsubstituted amides and lactams by reaction with phosphorus oxychloride and sodium borohydride. Under such mild conditions, the carbonyl group was selectively reduced to a methylene group without conversion of the nitrogroup.

The compounds were spread from stock chloroform solutions (3 mg in 10 ml) on a pure water subphase (distilled, deionized (resistivity > 18 MΩ/cm), pH = 5.7) in a LB-trough. The solvent was allowed to evaporate for 15 minutes before the pressure *versus* molecular area (π-A) isotherm was measured. Also the stability of the compressed monolayers was checked by measuring the decrease in monolayer area for a constant held pressure after slow compression.

The solvent was allowed to evaporate for at least 10 minutes, and the monolayers were allowed to relax for 30 minutes at constant held pressure before any transfer was attempted. The deposition speed was 3 mm/minute. The substrates used were microscope slides, which after cleaning were rendered hydrophobic with octadecyltrichlorsilane.

UV-visible absorption spectra for the DCpNA compound in both chloroform solution, and LB film form were measured with a Perkin-Elmer 137 UV spectrophotometer. For the films UV-visible spectra as a function of time and number of layers were measured.

The experimental setup for the SHG measurements on LB films is shown in figure 1. A Nd:YAG laser at 1064 nm 1000 Hz Q-switched with a 100 nsec pulse width and approximately 0.1 mJ per pulse is used. The beam is weakly focused (f = 500 mm) on the film. The polarization of the fundamental beam can be rotated through the use of a half-wave plate and a polarizer inserted in front of the sample to be tested. A Personal Computer (PC) controls a rotational stage for changing the angle of incidence, θ. In order to measure the second harmonic intensity in the transmission mode, the remaining fundamental beam is removed by IR-blocking filters, and after passing a monochromator around 532 nm the signal is detected by a photomultiplier and counted by a photon-counter/personal computer system. Detuning

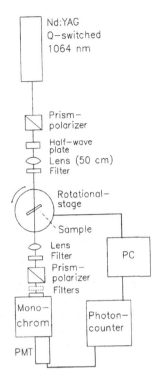

Figure 1. The experimental setup for SHG
measurements on LB films.

the monochromator away from 532 nm confirms that it is the second harmonic and not some broad band flouroscence we are measuring.

ISOTHERMS

The pressure *versus* molecular area (π–A) isotherms at 17 °C of DCpNA and DCpNA mixed with BA are shown in figure 2. For the pure DCpNA the mean molecular area is large, 42.5 Å^2 at the collapse pressure of 18 mN/m, indicating a large tilt angle with respect to the surface normal. Extrapolating the steepest slope to zero pressure gives a molecular area of 47 Å^2. As expected the mixed monolayers possess a smaller mean molecular area.

In order to check the stability of the monolayers, the decrease in monolayer area at constant held pressure, 12 mN/m, after a slow compression was measured, and the time dependence for the first 60 minutes is shown in figure 3. It is seen that the effect of adding BA produces a film which is more stable. The small decrease in area for the mixed monolayers indicates that they indeed are stable monolayers, even though the collapse pressure is low. The DCpNA film transferred well, nearly 100% onto the hydrophobic substrates, in both the down- and up-stroke at 17 °C and a surface pressure of 12 mN/m. We therefore have a Y-type LB film. Lower temperature and higher pressure seemed to disturb the transfer.

UV-VISIBLE ABSORPTION SPECTRA

UV-visible absorption spectra showed that the absorption maxima for DCpNA is redshifted from $\lambda_{max} = 372$ nm in chloroform solution to $\lambda_{max} = 425$ nm in LB film form. The linear dependence in the optical density at λ_{max} with the number of deposited layers for

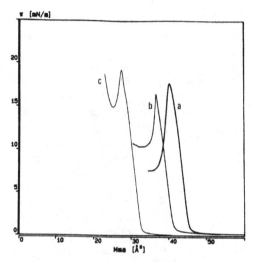

Figure 2. Pressure **versus** molecular area (π-A) isotherm for N-docosyl-4-nitroaniline (DCpNA) mixed with Behenic acid (BA) at 17°C on a pure water subphase (pH = 5.7). Molar ratios of DCpNA:BA are a) 1:0, b) 4:1, and c) 2:1, respectively.

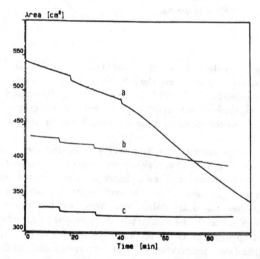

Figure 3. Stability of the mixed monolayers. The monolayers were first compressed to 7.5 mN/m then 10 mN/m and finally 12 mN/m, this is seen as steps on the curves. Molar ratios are a) 1:0, b) 4:1, and c) 2:1, respectively.

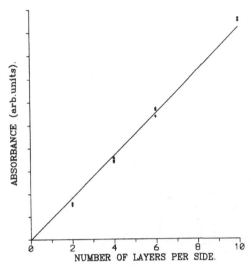

Figure 4. Optical density as function of number of deposited layers.

DCpNA is shown in figure 4, together with the decrease in monolayer area during transfer confirmed that the transfer was succesful, i.e., nearly 100% in both down- and up-stroke. The absorption spectra from the LB films changed with time after deposition even for unexposed (to laser light) samples. The appearance of the films changed from transparent yellowish in color to less transparent milky.

SECOND HARMONIC GENERATION IN LB FILMS OF DCpNA

The second harmonic signal also revealed that the films change with time, in the sense that only freshly made samples gave good second-harmonic signal. A monolayer sample kept in nitrogen and argon atmospheres and darkness for several days maintained most of the ability to generate the second harmonic. We have observed that the maximum intensity of the signal decreases and that the fringes become less contrasty, after the film has been exposed to air for several hours. The reason for the decay of the signal could be due to oxidation of the film.

Multilayer films of DCpNA showed a faster decay than monolayer films. Second harmonic fringes from a fresh and a three-day-old film are shown in figure 5. This could be due to a reorientation of the chromophores towards antiparallel alignment of the dipoles. In a multilayer film, this will cause a structural cancellation of the bulk susceptibility. The degradation of the films was not due to laser induced damage, as confirmed by moving to unexposed places on the films.

SYMMETRY PROPERTIES AND SECOND HARMONIC GENERATION

In figures 6a and 7a, the second harmonic signal from a film with ten Y-type layers on each side are shown for the s-in/s-out and p-in/p-out configurations of the polarization with the rotation axis being parallel and perpendicular to the dipping axis. If we rotate around the dipping axis, the intensity is larger for s than that for p-polarization. For rotation around the axis perpendicular to the deposition axis, the p-polarized fundamental gives the largest signal. This dependence on the polarization and rotation axis is the same as for DCANP. A similar structure of the film is therefore believed in the case of DCpNA. The dependence on

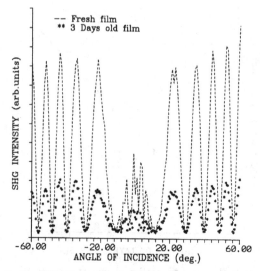

Figure 5. The SHG fringes from a fresh sample and a 3 days old sample
kept in air. The input-output polarization combination is s-
in/s-out, rotation is performed around the symmetry axis.

polarization makes us suggest that the symmetry of the structure for a monolayer is C_s, where
the xz-plane forms a mirror plane as the only symmetry element and C_{2v}, for even number
of layers. Theoretically simulated envelope functions for the second harmonic intensity for a
monolayer with C_s symmetry and for an even number of layers with C_{2v} symmetry for the
s-in/s-out configuration and for an even number of layers with C_{2v} symmetry for the p-in/p-
out configuration, are shown in figures 6b and 7b. The corresponding theoretical curve for
the $C_{\infty v}$ symmetry (figure 8) clearly shows that this symmetry can be ruled out.

We have compared the signal level from DCpNA monolayers and DCANP monolayers as
this compound has been investigated in the literature [6], [8]. Mono and multilayer films of
DCANP have been found to be stable over a period of several months. The total SHG from
fresh monolayers of DCpNA is about three times larger than signals from fresh monolayers of
DCANP, when the fundamental beam was s-polarized, which in both cases gave the largest
signal. In both DCpNA and DCANP, the molecules are oriented along the dipping direction.
However, in DCpNA, the chromophores lie at an angle of only about 10 degrees with respect
to the substrate, whereas in DCANP, they make an angle of about 30 degrees [8]. This could,
in part, explain the larger signal for the former. DCpNA can be considered a one-dimensional
molecule in the sense that the β_{xxx} is much larger than any other component of the β-tensor
due to the charge transfer axis, [1] whereas in DCANP the nitrogen atom in the pyridine ring
will cause other components to be non-negligible at the expense of the β_{xxx} component [9].
Powder measurements on DCpNA show that the SHG efficiency is weak, much less than 1%
that of urea, whereas DCANP has a strong powder efficiency.

CONCLUSION

In conclusion, we have shown that it is possible to form a monolayer of DCpNA at an
air-water interface, and that this can be transferred to a hydrophobic glass slide and thereby
form a homogeneous well-ordered LB film, with a strong anisotropy in the orientation of the
chromophores with respect to the surface normal, confirmed by SHG measurements. However,
the film is unstable and the SHG intensity is found to decay with time.

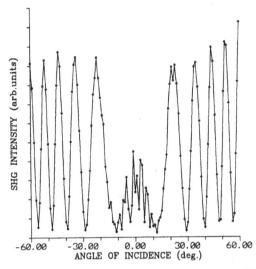

Figure 6a. The SHG fringes from a fresh 10 layer sample. The input-output polarization combination is s-in/s-out, rotation is performed around the symmetry axis.

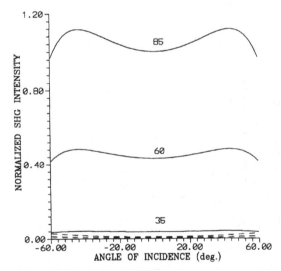

Figure 6b. The theoretical envelopes for the transmission mode. The film possess C_s- or C_{2v}-symmetry, rotation is performed around the symmetry axis. The input-output polarization combination is s-in/s-out. The numbers next to the lines denote the assumed molecular tilt angle, and solid and dashed lines are the envelopes for the maxima and minima, respectively.

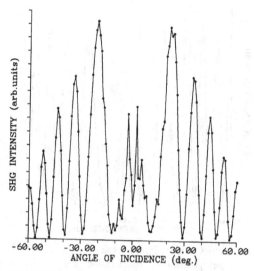

Figure 7a. The SHG fringes from a fresh 10 layer sample. The input-output polarization combination is p-in/p-out, rotation is performed around axis perpendicular to the symmetry axis.

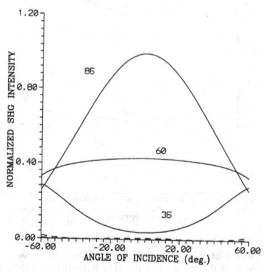

Figure 7b. The theoretical envelopes for the transmission mode. The film possess C_{2v}-symmetry, rotation is performed around the axis perpendicular to the symmetry axis. The input-output polarization combination is p-in/p-out.

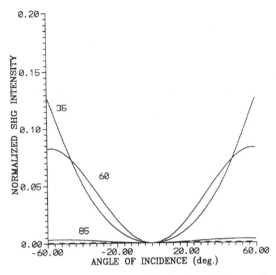

Figure 8. The theoretical envelopes for the transmission mode. The film possess $C_{\infty v}$-symmetry. The input-output polarization combination is p-in/p-out.

ACKNOWLEDGMENTS

Mr. Jørgen Stahl is acknowledged for his patient work with the preparation of the LB films. The Danish Research Council for Scientific and Industrial Research (STVF) is acknowledged for financing part of this work.

REFERENCES

[1] D.S. Chemla, and J. Zyss, Eds., "Nonlinear optical Properties of Organic Molecules and Crystals," Vol. I,II, Academic Press, New York (1987).

[2] J. Messier, F. Kajzar, P. Prasad, and D. Ulrich, Eds., "Nonlinear Optical Effects in Organic Polymers," NATO ASI series, Applied Science, Vol.162, Kluwer Academic Publishers, London (1989).

[3] P.N. Prasad, and D.R. Ulrich, Eds., "Nonlinear Optical and Electroactive Polymers," Plenum, New York (1988).

[4] D.J. Williams, Organic Polymeric and Non-Polymeric Materials with Large Optical Nonlinearities, Angew. Chem. Int. Ed. 23:690 (1984).

[5] Ch. Bosshard, M. Küpfer, P. Günter, C. Pasquir, S. Zahir, and M. Seifert, Optical waveguiding and nonlinear optics in high quality 2-docosylamino-5-nitropyridine Langmuir-Blodgett films, Appl. Phys. Lett. 56:1204 (1990).

[6] G. Decher, B. Tieke, Ch. Bosshard, and P. Günter, Optical Second-harmonic Generation in Langmuir-Blodgett Films of 2-Docosylamino-5-nitropyridine, J. Chem. Soc. Chem. Commun. 933 (1988).

[7] M.E. Kuehne and P.J. Shannon, Reduction of Amides and Lactams to Amines by Reactions with Phosphorus Oxychloride and Sodium Borohydride, J. Org. Chem. 42:2082 (1977).

[8] Ch. Bosshard, G. Decher, B. Tieke, and P. Günter, Linear and nonlinear optical properties of Y-type Langmuir-Blodgett films of 2-docosylamino-5-nitropyridine, in "Nonlinear
Optical Materials," SPIE, 1017:141 (1988).

[9] M.Nakano, K.Yamaguchi, and T.Fueno, CNDO/S-CI Calculations of Hyperpolarizabilities. I: Substituted Benzenes, Polyaza Compounds and Related Species, in: "Nonlinear Optics of Organics and Semiconductors," T.Kobayashi ed, Springer Proceedings in
Physics 36, Springer-Verlag, Berlin (1989).

SECOND HARMONIC GENERATION BY THERMOPLASTIC POLYMER

DISPERSED LIQUID CRYSTAL FILMS

Cheng-Tsung Huang and Hwa-Fu Chen

Materials Research Laboratories
Industrial Technology Research Institute
Hsinchu, Taiwan, Republic of China

INTRODUCTION

Polymer dispersed liquid crystal (PDLC) films are made of micron-sized nematic liquid crystal droplets dispersed in a polymer matrix [1]. PDLC films are normally opaque, due to the refractive index mismatch between the liquid crystal and the polymer matrix. However, the opaque state can be switched into a transparent state by applying electric fields [2], or laser heating [3-4]. Recently, second harmonic generation (SHG) by thermosetting PDLC films at their opaque state was reported [5]. The SHG intensity reached a maximum after the PDLC film was fully cured. However, this SHG intensity decreased when a 60 Hz voltage was applied to turn it transparent.

While the nature of SHG of bulk nematic liquid crystals was complicated [6,7], the origin of the SHG by PDLC films was attributed to the presence of surface and interface in PDLC films [7,8]. A mechanism based on dielectric constant gradient [5] was proposed. The reduction of SHG, when applying a 60 Hz voltage, was then explained by the reduction of dielectric constant mismatch. Since the dielectric constants are frequency dependent, it would be of interest to determine the influence of the frequencies of the applied voltages on the SHG by PDLC films.

The goals of this paper are 1) to study the SHG by a thermoplastic PDLC film and 2) to investigate the effect of applied voltages with different frequencies on the SHG by PDLC films. The frequencies of 0, 60, 1 k, and 1 M Hz, were used. Our results showed that the same reduction of

SHG intensity was found for thermoplastic PDLC films and that in the case of applying d.c. (0 Hz) voltages, the threshold voltages for SHG reduction as well as for transmittance switching are higher than those in the case of applying a.c. voltages.

EXPERIMENTAL

A commercial liquid crystal mixture E7 (BDH) was mixed with the same amount of polyvinyl acetate (Janssen, medium molecular weight) in chloroform (Merck). This solution was spin-coated onto a conductive ITO glass (Ashai). After solvent evaporation by heating at 60°C for 30 minutes, a phase-separated PDLC film of about 5 um thick was obtained. The experimental apparatus for measuring the transmittance characteristic of PDLC films was shown in Fig. 1. A pulsed Nd-YAG (1.06 um) laser system, as shown in Fig. 2, was used to measure the SHG intensity.

RESULTS AND DISCUSSIONS

According to the model of dielectric constant gradient, SHG by PDLC films results from the second harmonic polarization,

$$P(2w) = X^{(3)}E_{(ind)}(0)E(w)E(w) \tag{1}$$

where $X^{(3)}$ is the third order susceptibility tensor, and $E_{(ind)}(0)$ is the induced electric field and is, to the lowest order, proportional to the dielectric constant gradient [5]. This dielectric constant gradient decreases when the applied voltage aligns the liquid crystal molecules within the droplets. This effect was used to explain the reduction of SHG intensity by PDLC films [5].

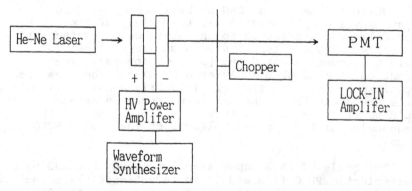

Fig. 1 Experimental set-up for measuring light transmittance of polymer dispersed liquid crystal films.

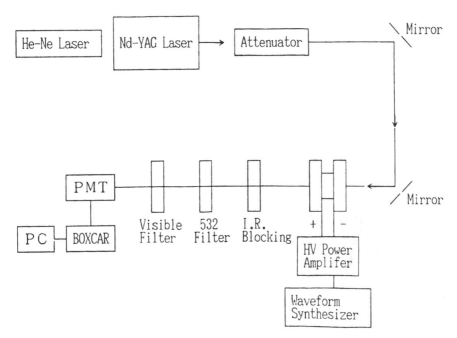

Fig. 2 Experimental set-up for measuring second harmonic
generation of polymer dispersed liquid crystal
films.

Fig. 3 shows the curves of SHG intensity vs. applied
voltage. In both d.c. and a.c. (up to 1 M Hz) regimes, the
SHG intensity decreases to a constant level when the
applied voltages are greater than threshold voltages. This
indicates that thermoplastic PDLC films exhibit the same
behavior of SHG as that of thermally cured PDLC films.
Furthermore, from Eq. 1, it is noted that the dielectric
constant plays a key role in the SHG by PDLC films. The
dielectric constant is, in general, frequency dependent.
However, as shown in Fig. 3, the reduction of SHG exhibits
only two different threshold voltages; one is for d.c.
regime and another is for a.c. regime. The significantly
lower threshold voltages for a.c. regime await further
explanation.

Fig. 4 shows the curves of the transmittance vs.
applied voltage. It is noted that the threshold voltages
for transmittance coincide with the threshold voltages for
SHG intensity. This indicates that the alignment of liquid
crystal molecules within the droplets affects both
phenomena. This is again consistent with the results of
thermally cured PDLC films [5].

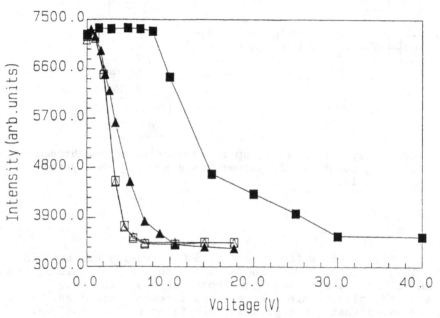

Fig. 3 Second harmonic generation intensity vs. applied
voltage of a polymer dispersed liquid crystal film,
consisting of 50% E7 and 50% polyvinyl acetate,
5 um thick (■ : 0 Hz; △ : 60 Hz; □ :1 k Hz; ▲ : 1 M
Hz; except for the case of 0 Hz, the applied
voltages given were V_{rms}.).

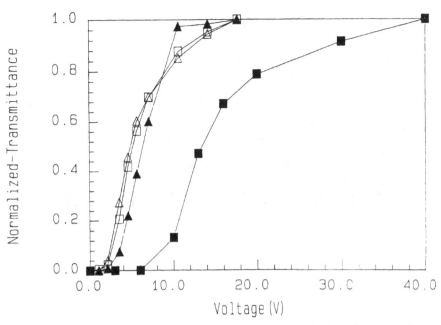

Fig. 4 Normalized transmittance vs. applied voltage of a
polymer dispersed liquid crystal film, consisting
of 50% E7 and 50% polyvinyl acetate, 5 um thick
(■: 0 Hz; △ : 60 Hz;□:1 k Hz; ▲ : 1 M Hz; except
for the case of 0 Hz, the applied voltages given
were V_{rms}.).

ACKNOWLEDGMENTS

The authors wish to thank Dr. Jong-Ming Liu of
Materials Research Laboratory for his helpful discussions.
The authors also thank the financial support from the
Ministry of Economic Affairs, Republic of China, under the
contract number 34P2100 to the Industrial Technology
Research Institute.

REFERENCES

1. For a review see: G. P. Montgomery, Jr., " Polymer-
   Dispersed and Encapsulated Liquid Crystal Films", in "
   Large-Area Chromogenics: Materials and Devices for
   Transmittance Control", C.M. Lampert and C.G. Granqvist
   ed., SPIE, Bellingham, Washington, (1990).
2. J.W. Doane, N.A. Vaz, G.G. Wu, and S. Zumer,"Field
   Controlled Light Scattering From Nematic
   Mircrodroplets", Appl. Phys. Lett., 48(4):269 (1986).
3. P. Palffy-Muhoray, B.J. Frisken, J. Kelly and H.J. Yuan,
   "Nonlinear Optical Response of Polymer Dispersed Liquid
   Crystal Films", SPIE Proc., 1105:33 (1989).
4. G. Cipparrone, C. Umeton, G. Arabia, G. Chidichimo, and
   F. Simoni, "Nonlinear Optical Effects in Polymer

Dispersed Liquid Crystals", Mol. Cryst. Liq. Cryst., 179:269 (1990).

5. L. Li, H.J. Yuan, and P. Palffy-Muhoray, "Second Harmonic Generation by Polymer Dispersed Liquid Crystal Films", Mol. Cryst. Liq. Cryst., 108:239 (1991).

6. M.I. Barnik, L.M. Blinov, A.M. Dorozhkin, and N.M. Shtykov, "Optical Second Harmonic Generation in Various Liquid Crystalline Phases", Mol. Cryst. Liq. Cryst., 98:1 (1983).

7. P. Palffy-Muhoray, "The Nonlinear Optical Response of Liquid Crystals", in "Liquid Crystals: Applications and Uses, Vol. 1", B. Bahadur ed., World Scientific, Singapore (1990).

8. Y.R. Shen, "Surface Properties Probed by Second-Harmonic and Sum-Frequency Generation", Nature 337:519 (1989).

THIRD-ORDER OPTICAL NONLINEARITY AND CONDUCTIVITY OF 2,5-DISUBSTITUTED

POLY(1,4-PHENYLENE VINYLENE) COMPOSITES WITH SOL-GEL PROCESSED INORGANIC

GLASSES

K.-S. Lee, C. J. Wung and P. N. Prasad

Photonics Research Laboratory, Dept. of Chemistry
State University of New York at Buffalo, Buffalo, New York 14214

C. K. Park, J. C. Kim and J.-I. Jin

Dept. of Chemistry, Korea University, Seoul, Korea

H.-K. Shim

Dept. of Chemistry, Korea Advanced Institute of Science and
Technology, Taejeon 305-701, Korea

ABSTRACT

Composites of poly(1,4-phenylene vinylene) (PPV) and some of its derivatives such as poly(2-bromo-5-methoxy phenylene vinylene) (BrMPPV) and poly(2-buthoxy-5-methoxy phenylene vinylene) (BuMPPV) with sol-gel processed silica and V_2O_5 glasses were prepared as materials for optical waveguide and photonics applications. Optical nonlinearities of the pure polymers and the polymer/sol-gel composites were measured by the degenerate four-wave mixing technique. The $\chi^{(3)}$ values are ~ 3 x 10^{-10}, ~ 9 x 10^{-10}, and ~ 10^{-9} esu for the PPV, BrMPPV and BuMPPV polymers, respectively. The composites exhibited the $\chi^{(3)}$ value scaled by the number density, but were of enhanced optical quality for waveguide application. The maximum conductivity of the PPV/sol-gel film doped with AsF_5 was 0.023 S/cm.

INTRODUCTION

Conjugated polymers with extensive π-electron delocalization have emerged as an important class of third-order nonlinear optical (NLO) materials because of the large π-electrons contribution to optical nonlinearity.[1] In order to operate photonic devices at low energy pulses, materials must have large nonlinearity, $\chi^{(3)}$.[2] In addition, they must have mechanical strength, environmental stability, high optical damage threshold, and low optical losses. Many conjugated polymers have all these properties except poor optical quality. To get good optical quality polymers, we have succeeded to improve the optical quality by preparing composites of conjugated polymers with a silica glass by sol-gel processing.

Inorganic glass like silica or V_2O_5 are excellent photonic media and can be made into high quality fibers and films with extremely low optical losses. However, the nonlinear optical coefficients of these glasses are extremely low. Therefore, if an NLO polymer and an inorganic glass can be mixed homogeneously, both the NLO and optical property can be optimized. Furthermore, these composites can be made as thick as 1 μm up to several μm and can be used as optical wave guide materials[2]. In general, the advantages of the sol-gel processed composite are as follows; (i) they provide potential to introduce multifunctionality for desired device applications; (ii) they can be made into various guided-wave forms such as planar waveguides, channel waveguides and fibers; (iii) these systems can achieve good dipolar alignments during gelation in electric field because of the increased molecular mobility of the structure; (iv) they can also be used to prepare composites with unusual electronic and optical properties by using heterostructures consisting of mixed valent inorganic semiconductors and p-type organic semiconductors. However, there are two main difficulties in making composites of this kind. One concern is that high temperature conventional processing of glasses cannot be used since most polymers decompose around 300-400°C. Another consideration is that there are incompatibilities between the organic polymeric and inorganic glass structures; phase separation often occurs at higher compositions. Furthermore, there is additional problem in mixing two components; that is poor solubility and poor fusibility of conjugated polymers resulting from polymer backbone stiffness. This problem can be solved by making conjugated polymers which have flexible side chains. This fact was observed in alkoxy-substituted polyazomethines, polyphthalocyaninatosiloxanes, etc. Recently, several researchers have found a way to circumvent this processability problem by the synthesis of soluble polymeric precursors[3-6]. PPV precursor polymers can be cast as a film and subsequently converted by thermal or chemical treatment to yield the desired conjugated polymers. The precursor polymers, before the thermal elimination, are very useful materials for making sol-gel composites because they are soluble in water or common organic solvents. Therefore, these precursors can be processed into composite forms at low temperature.

In this paper we report the preparation, characterization, optical and electrical properties of composites of the poly(p-phenylene vinylene) and some of its 2,5-disubstituted derivatives with sol-gel processed silica glasses.

EXPERIMENTAL

(a) Polymer synthesis

Poly(p-phenylene vinylene)[3](PPV), poly(2-bromo-5-methoxy-p-phenylene vinylene[4] (BrMPPV) and poly(2-methoxy-5-butoxy-p-phenylene vinylene)[5] (BuMPPV) were prepared by the precursor technique as reported elsewhere.

(b) Sol-gel process

The BrMPPV or BuMPPV/sol-gel composite was made by mixing the sulfonium polyelectrolyte precursor polymer for BrMPPV or BuMPPV, and tetramethyl orthosilicate (TMOS) (from Aldrich) in methanol. The processing sequence for BrMPPV/sol-gel composite is as follows: Equal amount by volumes of TMOS, methanol and precursor polymer are mixed with stirring and added proper amount of 0.1 N of HCl to control the pH. For sol formation, this mixture was heated at 60 $^{\circ}$C for about 30 min, cooled down to room temperature and diluted to be 10 times by volumes of TMOS in solution by addition of methanol. Consequently, equal volumes of above mixture and BrMPPV precursor solution (5% by weight) are mixed. After storing at room temperature for 24 hrs, the final BrMPPV/sol-gel precursor solution is casted on a suitable substrate (silicate glass or quartz plate) using a spin coater. The resulting films were thermally treated at 210 $^{\circ}$C for 20 hr. As a result, the BrMPPV/so-gel precursor was converted into the BrMPPV/sol-gel composite film with good optical quality. The preparation of the BuMPPV/sol-gel composite is similar to that of BrMPPV/sol-gel composite; the preparation of PPV/sol-gel is published elsewhere[6].

(c) Measurements

IR spectra were obtained by an infra-red (Alpha Centauri FT-IR) spectrometer. A Shimadzu UV-VIS-NIR scanning spectrophotometer (model UV-3101 PC) was used for UV-visible spectral analysis. The band gap was determined from the onset between the baseline and extrapolation of the absorption position in the transmittance spectra. A prism coupler (Metreicon PC-2000) was employed for the refractive index and film thickness measurements. Differential scanning calorimeter and thermogravimetric analyzer (Du pont 9900) were used to study the thermal elimination process of precursor polymers. Electrical conductivities of

175

the AsF_5-doped films were monitored by the four-probe method. The third-order optical nonlinearities of samples were evaluated by the degenerate four-wave mixing technique[1] using 400 fs pulses at 602 nm. The laser system consists of a Nd-YAG laser (Spectra Physics, model 3800 mode-locked) with a fiber optic pulse compressor (Spectra physics, model 3690), a synchronously pumped dye laser (Spectra Physics, model 375 B), and a three-stage amplifier (Quanta-Ray PDA-1) pumped at 30 Hz with a Quanta-Ray DCR Nd-YAG laser.

RESULTS AND DISCUSSION

In the polymer/sol-gel composites preparation, the organic precursor polymer and the silica sol-gel (obtained from a spontaneous polymerization of tetramethyl orthosilicate (TMOS) are mixed in common solvents and converted to the final conjugated polymer composites by heat treatment above 200 °C. This process involves the following two reactions.

$$n\ Si(OCH_3)_4 + 4n\ H_2O \xrightarrow[-4n\ CH_3OH]{H^+} n\ Si(OH)_4 \xrightarrow{-2n\ H_2O} (SiO_2)_n \qquad (1)$$

Polymer precursor   Polymer

For PPV: X=H, Y=H
BrMPPV: X=Br, Y=OCH_3
BuMPPV: X=O(CH_2)_3CH_3, Y=OCH_3

Using the sol-gel technique one can prepared conjugated polymer composites in which two components are homogeneously mixed in large compositions up to 50 % by weight without any phase separation. The result of homogeneous mixing maybe due to a synergistic effect occurring during the chemical reaction of each component. We do not have any evidence for molecularly mixing, but the good optical quality indicates that if micro-domains are formed in sol-gel process or in elimination step of precursor polymers, they are much smaller than the wavelength of

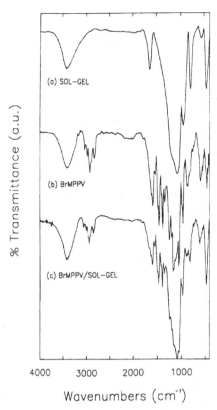

Figure 1. The IR transmission spectra of (a) thermally treated (210°C) sol-gel, (b) BrMPPV, and (c) BrMPPV/sol-gel composite.

light. Actually, the average pore diameters of the glass created in the sol-gel process were reported 1.5-10 nm[7] which is smaller than the near-UV or visible radiation wavelength. The composite films described here can be formed in the thickness rage from 0.5 μm up to tens of micrometers without cracking. The films showed good mechanical strength and environmental stability. The structure of the resulting BrMPPV/sol-gel composite was identified by FT-IR spectroscopy (Figure 1 (a) sol-gel, (b) BrMPPV, and (c) BrMPPV/sol-gel composite). The strong absorption band at 1100 cm^{-1} is mainly due to the silica network, and the 452 cm^{-1} band corresponds to the deformation vibration of Si-O-Si-O bonds. A trans-CH out-of-plane bending vibration of BrMPPV appeared at 960 cm^{-1}. No absorption band is observed in the vicinity of 630 cm^{-1} which is associated with the cis-CH bending vibration mode. This indicates that the structure of the conjugated polymers in the silica-matrix composite does not change by the thermal elimination reaction. Multiabsorption peaks in the frequency range 2800-3000 cm,$^{-1}$ correspond to the C-H stretching vibration of BrMPPV. Very similar spectral features were also observed for the BuMPPV/sol-gel composite.

Figure 2, curves a, b, c, and d show the UV-visible spectra of BrMPPV precursor, BrMPPV/sol-gel precursor, polymer BrMPPV, and the BrMPPV/sol-gel composite, respectively. The precursor films showed very weak absorptions in the visible range. The 261 nm absorption band is associated with the sulfonium salt. The appearance of absorption bands between 285 and 370 nm are evidence of trans-stilbene groups in the polymer chain by partial elimination.

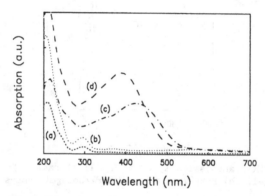

Figure 2. The UV-visible absorption spectra of (a) BrMPPV precursor, (b) BrMPPV/sol-gel precursor, (c) BrMPPV, and (d) BrMPPV/sol-gel composite.

Differential scanning calorimetry (DSC) was used to measure physical, chemical and enthalpic changes which accompanied gel or composite densification. Figure 3 shows the DSC trace of BrMPPV precursor, silica gel precursor and the BrMPPV/sol-gel precursor which had been dried in a desiccator for 3 months. In the case of BrMPPV (figure 3 (a)), we found that between 60°C and 175°C is a endothermic effect which may be related to a combination of the loss of organic volatiles and E1cB elimination. In the case of sol-gel (figure 3 (b)), we observed an endothermic peak in the range of 50-150°C. This endotherm is a contribution of dehydration and viscous sintering. Figure 3 (c) shows a broadened endothermic response between 50°C and 130°C and peak maximum is lower than that of BrMPPV in figure 3 (a) (from 120°C to 90°C). The shifted peak position provides further evidence for the interaction of the two components (BrMPPV and silica)in the system. All these characteristic features are attributed to a lowering of the

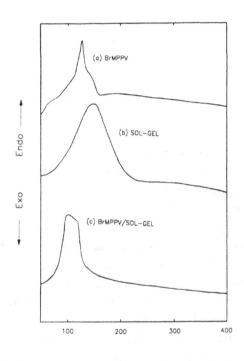

Temperature (°C)

Figure 3. Differential scanning calorimetry (DSC) curves for (a) BrMPPV
precursor, (b) sol-gel precursor, and (b) BrMPPV/sol-gel
precursor.

activation energy during the condensation-polymerization and sintering of the BrMPPV/sol-gel composite. Furthermore, the repeat scan shows that the process associated with the endothermic reaction is irreversible and no glass transition is detected.

The $\chi^{(3)}$ value was evaluated by comparing the strength of the conjugate signal at low incident photon flux with that of CS_2 according to the following relationship[1]:

$$\frac{\chi_s^{(3)}}{\chi_c^{(3)}} = \left(\frac{n_s}{n_c}\right)^2 \frac{\ell_c}{\ell_s} \left(\frac{I_s}{I_c}\right)^{1/2} X \frac{\alpha \ell_s}{\exp(-\alpha\ell_s/2) \ [1-\exp(-\alpha\ell_s)]} \tag{3}$$

Where n is the refractive index, ℓ is the interaction length, and α is the linear absorption coefficient. The subscript c and s refer to CS_2 and the sample, respectively. The CS_2 sample is nonabsorbing at the wavelength used in our experiments. The value of $\chi^{(3)}= 6.8 \times 10^{-13}$ esu was used as the reference value for CS_2[9]. However, in the case of BrMPPV the absorption term in Eq.(3) is ignored because the sample is nearly nonresonant at the wavelength 602 nm (see figure 2). The measured effective $\chi^{(3)}$ values for BrMPPV and BuMPPV are ~9 $\times$ 10^{-10} and ~10^{-9} esu, respectively. The subpicosecond DFWM response of BrMPPV/sol-gel is shown in figure 4. These values can be compared with ~4 $\times$ 10^{-10} esu reported by Singh[10] et al for PPV at 602 nm. The $\chi^{(3)}$ values for the polymer/sol-gel composites show slightly smaller value than the corresponding pure polymer. This can be attributed to two reasons: (1) the onsets of the absorption spectra in the composites are shifted to shorter wavelength indicating a reduction of effective configuration (2) the number density of the polymeric units responsible for $\chi^{(3)}$ is reduced in the composite.

Organic polymeric structures with extensive π-conjugation have a relatively large nonresonant third-order susceptibility $\chi^{(3)}$ which is derived from the π-electron contribution[11] The trend of our measurement reflects the contribution of the electron donor substituents at the 2, 5 positions of PPV. The buthoxy and methoxy electron donor groups at 2, 5 positions enhance the π-electron density in the conjugated polymer BuMPPV. Thus, the energy band gap is smallest and the effective $\chi^{(3)}$ value is the largest for BuMPPV. The structures of PPV and its derivatives, band gap energies, and the third order non-linear susceptibilities $\chi^{(3)}$ are presented in Table 1.

Uniaxial Orientation of films are known significantly to improve electrical conductivity along the stretching direction.[12-13] Doping studies of PPV show that conductivities approaching those obtained for doped polyacetylene are achievable. The maximum conductivities[3] of fully eliminated PPV films with the dopant AsF_5 is 10 S cm.[-1] On the other hand, the only known conductivities of BrMPPV are iodine doped stretched films. The stretching ratio ($1/1_0$) of 1 and 6 reached conductivities at 7.4×10^{-5} and 2.5×10^{-3} S cm^{-1} respectively.[13] Our results show that the AsF_5 doped PPV/sol-gel film has reached maximum conductivity of 0.023 S cm^{-1} and the AsF_5 doped unstretched BrMPPV and BrMPPV/sol-gel films have maximum conductivities of 1.4×10^{-2} and 8.3×10^{-4} S cm^{-1} respectively. Because the polymer/glass composites contain silica glass,

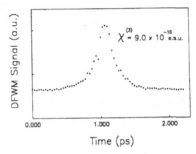

Figure 4. Degenrate four wave mixing (DFWM) signal observed for BrMPPV/sol-gel composite as a function of the forward beam delay. The wavelength is 602 nm and the pulses are ~400 fs.

Table 1. The structures of PPV and its derivatives, band gap energies, and the third order non-linear susceptibilities $\chi^{(3)}$.

Structures	Materials	Band gap energies (eV)	$\chi^{(3)}$ (esu)
	PPV film	2.46	~ 4 x 10^{-10}
	PPV/sol-gel composite	2.70	~ 3 x 10^{-10}
	BrMPPV film	2.37	~ 9 x 10^{-10}
	BrMPPV/sol-gel composite	2.52	~ 8 x 10^{-10}
	BuMPPV film	2.19	~ 10^{-9}
	BuMPPV/sol-gel composite	2.23	~ 10^{-9}

the conductivity value will be expected to be smaller than that of the pure polymer film.

CONCLUSION

The BrMPPV and BuMPPV polymers have been mixed successfully with silica using the sol-gel technique. The characterization of these materials has been made using UV-visible, FT-IR spectra and thermal analyzer. The blue shift in the UV-visible spectra and lowered peaks position in DSC response provide strong evidence of interaction between the polymer and the silica components. The $\chi^{(3)}$ values of PPV and its derivatives as well as polymer/sol-gel composites have been determined using degenerate four-wave mixing technique. The trend of increasing third order nonlinearities shows important contribution derived from the electron donor substituents at the 2, 5 positions in PPV. Even though the $\chi^{(3)}$ values are slightly lower in the polymer/sol-gel composites, initial study of interference patterns with multiple beam interferometry show better film homogeneous and smooth surfaces in this composites. This evidence supports that sol-gel processed polymer/silica films can produce better optical quality film than its pure polymer.

References

1. P.N. Prasad and D.J. Williams "Introduction to Nonlinear Optical Effects in Molecules and Polymers", Wiley, New York (1991).
2. R. Burzynski, D.N. Rao, and P.N. Prasad, unpublished result.
3. D.R. Gagnon, J.D. Capistran, F.E. Karasz, R.W. Lenz and S. Antoun, Polymer, **28**, 567 (1987).
4. Chi-Kyun Park, Hong-Ku Shim, and Jung-Il Jin, unpublished results.
5. J.I. Jin, C.K. Park, H.K. Shim, and Y.W. Park, J. Chem. Soc., Chem. Commun.,no. 17, 1205 (1989).
6. C.J. Wung, Y. Pang, P.N. Prasad, and F.E. Karasz, Polymer, **32**, 605, 1991.
7. M. Yamane, S. Aso, and T. Sakaino, J. Mater. Sci., **13**, 865 (1978). 8.
G. Carturan, V. Gottardi, and M. Graziani, J. Non-Crystalline Solids **29**, 41, (1978).
9. N.P. Xuan, J.L. Ferrier, J. Gazengel, and G. Rivorie, Opt. Commun. **51**, 433 (1984).
10. B.P. Singh, P.N. Prasad, and F.E. Karasz, Polymer, **29**, 1940, (1988).
11. "Nonlinear Optical and Electroactive Polymers" Eds. P.N. Prasad and D.R. Ulrich, Plenum Press (New York, 1988).
12.C.C. Han, R.W. Lenz, and F.E. Karasz, Polym. Commun., **28**, 26, (1987).
13 K.Y. Jen, L.W. Shacklettle, and R. Elsenbaumer, Synth. Met., **22**, 179, (1987).

OPTICAL PROPERTIES OF POLYDIACETYLENES WITH π-CONJUGATING SUBSTITUENTS IN

SOLUTION AND SOLID STATE

K. Nagendra Babu, Abhijit Sarkar, Lalita P. Bhagwat and
Satya S. Talwar

Department of Chemistry, Indian Institute of Technology
Powai, Bombay 400 076, INDIA

INTRODUCTION

Solid state 1,4-addition polymerisation of crystalline disubstituted diacetylenes(DAs) is known to give crystalline polydiacetylenes(PDAs[1]) which exhibit very interesting optical and electronic properties because of the extended conjugation of π-electrons along the backbone[2,3]. The high and fast third order nonlinear optical response (χ^3) shown by PDAs projects them as alternate materials for potential applications in electro-optics and all optical communication[4] system in place of conventional inorganic materials. Efforts are being made to enhance the polarizability and conjugation along the PDA backbone to get enhanced χ^3 response as it is related to the π-conjugation of the system.

Although the optical properties of these polymers are primarily dominated by the π-conjugated backbone, the substitutent pendant groups markedly influence the electronic, optical and other physical properties of the polymer. As the optical properties of backbone can not be electronically tuned with aliphatic sidegroups, it is believed that formally conjugating sidegroups could influence the electronic properties of the backbone through conjugative interaction between the backbone and such sidegroups. However, relatively few PDAs with substituent groups in formal π-conjugation with backbone have been reported. This has been partly due to lack of solid state reactivity in many precursor diacetylenes[5,6]. Recently, some success in synthesis of such PDAs and study of their properties has been reported[7,8]. However, depending on the extent of substituent-backbone interaction (orientation of π-system of the conjugating group with respect to planar backbone), π-conjugating substituents may lead to perturbation of electronic structure of backbone and enhanced nonlinear optical response due to enhanced polarizability.

We have been involved in the synthesis of DAs and PDAs with π-conjugating substituents and study of properties of such PDAs. In this paper, we report the reactivity of several such DAs towards polymerisation and the optical spectra of a few PDAs in solid state and in solution.

EXPERIMENTAL METHODOLOGY

All the diacetylenes were synthesized in our laboratory. The details of synthesis will appear elsewhere[9]. Thermal polymerisation was carried out

Frontiers of Polymer Research, Edited by P.N. Prasad and
J.K. Nigam, Plenum Press, New York, 1991

by heating the samples in sealed glass tubes in a constant temperature oil bath at appropriate temperatures. Monomer free polymer samples were prepared by extracting the partially polymerised samples with appropriate solvent. Reflectance spectra of polymerised polycrystalline samples were recorded on Shimadzu UV260 UV-visible spectrophotometer with diffuse reflectance accessory. Raman spectra of solid samples were recorded on Spex Ramalog 1401 Raman spectrometer using Argon laser excitation frequency at 5145 A°.

RESULTS AND DISCUSSION

We prepared a series of diacetylenic monomers (R–C≡C–C≡C–R[1]) with conjugated sidegroups. Their reactivity towards polymerisation is given in the Table 1.

Table 1. Reactivity of different dialetylene monomers towards solid state polymerization

Monomer : R–C≡C–C≡C–R'

No	R	R'	Reactivity towards Polymerization			λ_{max}(nm) of PDA
			Δ	hϑ	ɣ	
1	–HC=CH—⬡—CH₃	R	X	X	X	–
2	–HC=CH—⬡—OCH₃	R	X	X	X	–
3	(quinolyl)	R	+ + +	+ + +	+ + +	756
4	(pyridyl)	R	+ +	+ +	+ +	a
5	(thienyl)	R	+ +	+ +	+ +	a
6	(thienyl)	R	+	+	+	a
7	(naphthyl)	R	b	b	b	b
8	(quinolyl)	– CH₂OH	+ + +	+ + +	+ + +	a
9	(quinolyl)	–CH₂OCONH—⬡	+ + +	+ + +	b	730
10	(quinolyl)	(thienyl)	+	+	+	a
11	(quinolyl)	(thienyl)	+	+	+	a

Δ : thermal, hϑ : photochemical, ɣ : ɣ-radiation, R : reflectance

X : unreactive, + + + : high reactivity, + + : moderate reactivity, + : low reactivity.

a : featureless and broad reflectance in the visible region, b : to be studied in detail.

Although the absorption spectra of these PDAs are expected to be influenced by the sidegroups, only PDQ(3) and PQPU(9) showed dramatic changes in the optical spectra. The reflectance spectra of partially polymerised (thermal) DQ and monomer free PDQ are shown in Fig. 1 whereas those of PQPU are illustrated in Fig. 2. Partially polymerised DQ and QPU absorbed at 756 nm and 730 nm respectively with the visible absorption edges at 830 nm. These two polymers show longest wavelength optical transition of all known PDAs so far. This red shift in the λmax of these compounds, we feel, is due to conjugative interaction of quinolyl groups with the backbone causing an extension of backbone conjugation. The reflectance spectra of monomer free polymer samples of PDQ and PQPU are different from that of partially polymerised samples. The long wavelength absorption peaks of PDQ (756 nm) and

PQPU (730 nm) blue shifted to 730 nm and 675 nm respectively. A larger blue shift observed in case of PQPU is probably due to the disorder created along the backbone by the flexible methylene phenyl urethane sidegroup as consequence of reduced conjugative interaction between quinolyl group and backbone. Detailed raman spectral and ^{13}C-NMR studies are in progress to understand this phenomenon.

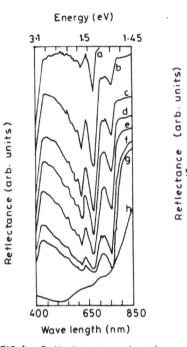

FIG.1. Reflectance spectra of partially polymerised (thermal, 160°C) samples of DQ (a–g:0,2,5,1,4, 7hr) and monomer free poly DQ (h)

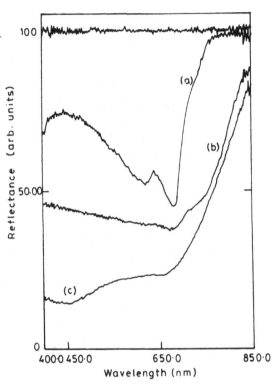

FIG.2. Reflectance of partially polymerized crystals of QPU
(a): Ambient heat and light
(b): Thermal (130°C, 2 hr)
(c): Monomer free polymer

Laser raman spectrum of partially polymerised DQ recorded with 514.5 nm excitation showed two bands at 1430 cm^{-1} and 2100 cm^{-1} for the double bond and triple bond stretching frequencies respectively. The double bond frequency (1430 cm^{-1}) observed for the PDQ backbone is the lowest value of all known PDAs reported so far[11]. The lowering of raman frequency $\nu_{(-C\equiv C-)}$ for PDQ substantiates the suggestion that the conjugating quinolyl sidegroups cause extension of backbone conjugation.

SOLUTION PROPERTIES

These polymers were found to be soluble in acidic solvents. This is the first report on soluble polydiacetylenes with π-conjugating sidegroups.

PDQ furnishes a blue solution in conc.H_2SO_4. Its absorption spectrum showed long wavelength absorption at 634 nm which red shifted to 680 nm on dilution with water with a cut off at 800 nm. Phenolic solution of PDQ absorbed at 656 nm and showed solvato and thermochromism. The absorption characteristics of PDQ and PQPU in different solvents are presented in Table 2. Details of the solution properties of PDQ will be reported else-where[12].

Table.2. Absorption characteristics of solutions of PDQ and PQPU.

Sample	Solvent	Colour	λ_{max}(nm)
PDQ	(a) Conc. H_2SO_4	blue	634
	Conc. $H_2SO_4 + H_2O$ (20%)	blue	670
	Conc. $H_2SO_4 + H_2O$ (>50%)	blue	680
	(b) $F_3C\,COOH + O_2N$ —⬡	blue	637
	$F_3C\,COOH + O_2N$ —⬡ + ⬡	blue	641
	(c) Phenol	blue	656
	Phenol + Toluene (>95%)	blue	692
			643
PQPU	Phenol	red	530
	Phenol + Toluene	red	570

In the literature[13], it is reported that PDAs form yellow solutions (470 nm, coiled conformation of individual chains) in good solvents such as chloroform and DMF and turn red or blue (λmax: 550 or 630 nm, rodlike individual chains or aggregates of coiled chains) only upon changing the quality of the solvent by adding a poor solvent such as hexane or by lowe-ring the temperature of the solution. The long wavelength absorption of solutions of PDQ (656 nm) and PQPU (530 nm) in phenol and other solvents is in contrast to the reported λmax (470 nm) of yellow solutions of different PDAs in good solvents. The long wavelength absorption maxima and the insen-sitivity of the absorption spectral profile of PDQ solutions with respect to its concentration indicate greater conjugation along PDQ chains in good solvents. These chains may resemble isolated rod like structures.

The rigidity of the backbone skeleton decreases in PQPU with the replacement of a quinolyl group by a flexible methylene phenylurethane group which manifests in the lowering of the λmax of the corresponding solution. These results point to modulation of optical properties of PDA solutions depending on the sidegroups.

In summary, we have reported the solution and solid state optical pro-perties of two polydiacetylenes with π-conjugating sidegroups which are oriented in a manner causing enhanced conjugation along the backbone. These two PDAs show longest wavelength absorption of all the PDAs reported so far both in solution and solid state.

ACKNOWLEDGEMENTS

Three of us (KNB, AS and LPB) thank CSIR, New Delhi, for providing senior research fellowship. We thank Dr. V.B. Kartha (BARC, Bombay) for his kind help in recording laser raman spectra.

REFERENCES

1. G. Wegner, Z. Naturforsch 24b, 824 (1969).
2. Polydiacetylenes (ed. D.Bloor and R.R. Chance), Nato ASI series, Series E, Applied Science No. 102, Martinus Nijhoff, (1985).
3. Adv. Polm. Sci. 63, (1984),; W. Huntsmann, The Chemistry of Functional Groups, supplement C, edited by S. Patai and Z. Rapport, (John Wiley and Sons Ltd., (1983), p. 917-981.
4. G. M. Carter, Y. J. Chen, M. F. Rubner, D. J. Sandman, M. K. Thakur and S. K. Tripathy in "Nonlinear optical properties of organic molecules and crystals", vol.2, edited by D.S. Chelma and J. Zyss, (Academic press, New York, 1987), p.85
5. G. Wegner, J. Polym. Sci., Part B 9, 133 (1971).
6. P. K. Khandelwal and S. S. Talwar, J. Polym. Sci., Polym. Chem. Ed. 21, 3073 (1983).
7. Y. Tokura, T. Koda, A. Itsubo, M. Miyabashi, K. Okuhara and A. Veda, J. Chem. Phys. 85, 99 (1986).; H. Matsuda, H. Nakanishi, T. Hosomi and M. Kato, Macromolecules 21, 1238 91988).; K. Ichimura, T. Kobayashi, H. Matsuda, H. Nakanishi and M. Kato, J. Chem. Phys. 93, 5510 (1990). H. Matsuda, H. Nakanishi, S. Kato and M. Kato, J. Polym. Sci. Polym. Chem. Ed. 25, 1663 (1987).
8. Satya S. Talwar, M. Kamath, K. Das and U.C. Sinha, Polymer Communications, 31, 198 (1990).
9. A. Sarkar, M.B. Kamath, P. Sekher, L.P. Bhagwat, K.N. Babu, K. Rajlakshmi and S.S. Talwar, Ind. J. Chem. 30B, 360 (1991),; P. K. Khandelwal, Ph.D. Dissertation, I.I.T. Bombay (1983).; M. B. Kamath, Ph.D. Dissertation, I.I.T. Bombay (1989).
10. Satya S. Talwar, M. B. Kamath and K.N. Babu, Mat. Res. Soc. Symp. Proc., vol. 173, 583 (1990).
11. D. N. Batchelder and D. Bloor in "Advances in Infrared and Raman Spectroscopy", vol. II, edited by R. J. H. Clark and R. E. Hester (Wiley Heyden, New York, 1984) p.133.
12. K. N. Babu and S.S. Talwar (submitted for publication).
13. C. Rosenblact and M. F. Rubner, J. Chem. Phys. 91, 7896 (1989).; R. Xu and B. Chu, Macromolecules 22, 4523 (1989).; N.A. Taylor, J.A. Odell, D. N. Batchelder and A. J. Campbell, Polymer 31, 1116 (1990).; M. Ramiso, J. P. Aime, J. C. Fawe, M. Schott, M. A. Muller, M. Schmidt, H. Baumgardtl and Wegner, J. Phys. France 49, 861 (1988).; K.C. Lim and A.J. Heeger, J. Chem. Phys. 82, 522 (1985).

APPLICATION OF LIQUID CRYSTALS IN SPACE ACTIVITIES

S. K. Gupta

Polymers and Special Chemicals Division
VSSC, Trivandrum-695 022
India

INTRODUCTION

Space activities are becoming more and more demanding with respect to development of new materials which should have high thermal stability, low density and ultra high strength. These activities also need thorough inspection of the components used for their different systems whereas these components vary from tiny integrated electronic circuits to big size assembled rocket motors, heat shield etc. Space scientists also want to note the stresses on the space craft structure during its flight. In this article, the recent achievements in the area of material fabrication out of liquid crystal polymers have been briefly mentioned. Nondestructive testing of electronic packages used in space programme by thermal analysis using liquid crystals has been reviewed. Aero-dynamic testing of models of rocket motors in wind tunnel using liquid crystals helps in the evaluation of the drag, shock waves and boundary layer flow of the rocket or aircraft. It assists for studying the discontinuities in adhesively bonded composite structures e.g. heat shield. Applications of side chain liquid crystal polymers for space activities have also been reviewed.

Testing of Electric and Electronic Components

Hundreds of cholesteric materials, both pure and mixture, are known and have been used used to measure temperatures ranging from -20° to 250°C. For a given material each colour corresponds to an exact temperature. Reference may be had to British Patent 1,041,490, lines 5.102 of page 4, for a comprehensive list of compounds suitable for use in making such compositions. Based on the generation of temperature pattern by dissipation of power, liquid crystals can be employed to detect faults in electronic devices. The liquid crystalline material is painted on the surface of the device and the device is then switched on. The defective areas are clearly indicated by the change in colour of the liquid crystal film at the point of excessive thermal stress. This technique is particularly useful for locating faults in solid state devices, especially integrated circuits which are used in satellite and in different parts of the rocket.

Frontiers of Polymer Research, Edited by P.N. Prasad and
J.K. Nigam, Plenum Press, New York, 1991

Acoustic Level Determination

Cholesteric liquid crystal can be used to display sound fields.[2] Attempts are being made to estimate acoustic level in heat shield while launching a rocket.

Application of Liquid crystals for Aerodynamic Testing

The remarkable sensitivity of liquid crystals have made it possible to observe some of the finer details of slight temperature rise produced by weak shock waves impinging on the surface of the rocket or aeroplane. This information is needed to estimate drag and other characteristics of the full scale aeroplane or rocket from wind tunnel results[3]. Another interesting application, made possible by the reversibility of the liquid crystals color indication would be the investigation of slow transient surface phenomena. Liquid crystal can presently be considered as a valuable tool in aerodynamic testing of launch vehicles.

Determination of Discontinuties in Aerospace Components and Structures by Liquid Crystal System

Liquid crystal have been used to evaluate the efficiency of heat exchangers, to detect nonuniformities in electrically resistive coatings[4] on rocket parts. Discontinuties in adhesively bonded composites structures may be located particularly beneath thin sheets with low thermal diffusivity, by rapidly heating or cooling a liquid crystal coated surface. Bond irregularities which locally alter heat losses from the test surface of honey comb structure may be defined by the contrasting colours seen over these regions due to transcient localised temperature gradients.

Polymer Liquid Crystals As Materials of Construction For Aerospace Industry

The weight of different parts of rockets, satellite or aircrafts and of components used in then can be reduced by using liquid crystalline materials for their fabrication because liquid crystal polymers have high strength and high modulus. Liquid crystalline polymeric materials which find applications for fabrication of space components can be divided into four categories.

1) Heterogeneous composites based on liquid crystalline fibers.
2) Thermotropic polymers
3) Liquid crystal blends
4) Molecular composite

Heterogeneous Composites Based On Liquid Crystal Fibers

Out of liquid crystalline fibers, carbon fibre and polyaramide fibre have been used in space activities due to their very high mechanical properties. Therefore they are discussed below.

Carbon fibre: Carbon fibers may be used for different aerospace applications. However carbon fibers[5] made from mesophase pitch have the best mechanical properties. Therefore they are captivating. They are used as a reinforcement in resin, carbon and metal matrices to form composites having high specific strength and stiffness. These composites find applications in space connected activities for fabrication of: 1. Rocket nozzle, 2. Heat shield, 3. Rocket motor cases 4. Gas bottles,

5. Satellite structures etc. Mesophase pitches are also used for making carbon-carbon composites required for rocket nozzles.

Polyaramide Fibre: Fibres made from polyparaphenyiene terephthalamide[6] have been available commercially since the early 1970's under the Du Pont trade mark Kevlar aramid. Recently Du Pont has developed a new aramid fibre called Kevlar-149 with a different combination of properties including inherent low moistures pick up and a modulus of elasticity up to 40% greater than that of previous high modulus aramid fibre.

Due to higher specific strength due to liquid crystalline nature, polyaramid fibres are used in making composites used for rocket motor cases.

Thermotropic Polymer Liquid Crystal

Thermotropic polymer liquid crystals can be processed with ordinary thermoplastic processing equipment[7,8]. Because of low viscosities and low extrudate swells, this is easier to process than processing of ordinary thermoplastics. Out of thermotropic fibres, polyethylene based composites have found many applications in aerospace programme. They also find application in balistic protection system. Its utility can also be envisaged in automotive and aircraft composites structures which have to be designed to dissipate the kinetic energy during crash impacts.

Blends of Liquid Crystal Polymer

Blends where thermoplastic resins are reinforced by thermotropic liquid crystalline polymers mixed in the melt, upon cooling spontaneously form in situ composites[6]. Liquid crystallaine polymers from ordered fluids with few entanglement and thus fascilitate the processing of the resin by lowering the melt viscosity. Upon cooling, the liquid crystalline polymer solidifies forming fibrilliar structures which reinforce the thermoplastic resin. Thus blends of liquid crystalline polymers and thermoplastic form light weight high strength materials. A useful property of the thermotropic polymers is the absence of shrinkage in the mould. This property is a consequence of the low co-efficients of thermal expansion of these polymers.

Addition to a polymer that forms an isotropic melt of a relatively small amount of the order 10% of a thermotropic polymer that forms a nematic phase at the processing temperature employed leads to a reduction in melt viscosity which may be employed to reduce the processing temperature, improve mould filling or enable fibres to be incorporated at higher concentration.

For example addition of a small amount (4.5%) of PLC[7] (polyhydroxy benzoic acid) in polystyrene produces a 40% increase of the tensile modulus as compared to polystyrene. These polymeric blends can find applications in reducing the weight of electric and electronic housing required for pay loads in satellite and different fitting in aircraft industry. This can also be used for moulding turbine blades and different components required for space craft. Blends should find applications in electrical, electronic, chemical, aircraft, aerospace and automobile industries and this list of material industries is not complete either.

Molecular Composites

Molecular composites[7] are made by complete dispersion of liquid crystalline rod like molecules in the matrix of an amorphous homologue that contains a sufficient number of flexible bonds to adopt a random-coil configuration. This helps in eliminating delamination and concentration of stress at discontinuities such as fibre ends. They can be used for making mission – adaptive wings that is wings with shapes which can be changed during a flight. The other advantage of these systems is that the component made of it are radar invisible which has special military applications for stealth[7].

Applications Of Side Chain Liquid Crystal Polymers (SCLCP)

SCLCP Based Elements Of Integrated Circuits: Calamitic groups, discotic groups and amphilic groups have been incorporated by Flinkleman[9] in the side chain of elastomers. He further developed the technology so that it may find use in elements for integrated circuits.

SCLCP Based Fixed Wavelength Filters Or Reflecters:

A cholesteric SCLCP could be heated above Tg to a temperature at which it was selectively reflecting desired wavelength of light, and then quenched below Tg, thereby locking in both the helical structure of the cholesteric mesophase and the required optical characteristics[10].

The required optical characteristics of the structural helicity and the light reflecting charcteristics of the polymer can be controlled by the chirality of the side chains, the incorporation of non-chiral or racemic side chains, the ratio of non-chirals to racemic side chains etc. Recently mesogenated poly (Hydrogenmethyl cyclo-siloxanes)[11] is used for optical polymer film and coatings on paper. This film reflects beautiful iridecent colors, varying with viewing angle. By the addition of pigments, it was demonstrated that the colour vs, viewing angle dependence could be reversed from blue to red with decreasing angle.

SCLCP Based Laser Addressed Data Storage Device: Aradle[12] demonstrated the use of SCLCP in an analogue optical data store with an unoptimised sensitivity of 12 nJ u/m^2 at 24°C with light of λ = 632.8 nm. The device was used to portray the fine details of an area of an ordance survey map.

Photochromic Effects Of SCLCP For Imaging Technology: Cabbera, Kongrauza and Ringsdorf[13] have studied mesogenated polysilaxane side chain copolymers.

A pale pink (at room temperature) cast film becomes yellow (spiropyran) on irradiation with light λ = 500 nm and then deep red on irradiation with light of λ = 365 nm. Aggregation effects involving dimerisation of the merocyanine moities and crosslinking of the macromolecules explain the deep red colour. This system have potential for applications in imaging technology.

Conclusion

Liquid crystals for inspecting electrical and electronic devices are having a great utility for space applications. Use of liquid crystal for studying aerodynamic aspects of rocket and aeroplane is very important. Integrity of adhesively bonded composite structures can be scrutinised using cholestric liquid crystals. Fabrication of different space components with heterogeneous composite based on liquid crystalline fire, liquid crystalline polymers, blends and molecular composites is of vital importance for space industry. Side chain liquid crystalline polymers have great potential for their application in integrated circuits, fixed wavelength filters or reflectors, data storage devices, optical polymer films and coatings, and in imaging technology. In a nutshell liquid crystals have been used and will be utilised much more in space activities.

Acknowledgement

The author is thankful to Miss. P.B.Latha, scientist PU & E Section for her valuable help in course of preparation of this article. He is grateful to Dr.S.C.Gupta, Director VSSC for encouragement and kind permission to publish it. He is also thankful to the organising committee of ICFPR, New Delhi for providing the opportunity to present the paper.

Reference

1. G.Vlulkianoff, 1967, 8, 389.

2. G.Meier, E.Sackmann, J.G.Grabmaier Applications of Liquid Crystal Springer verlag Berlin Heidelberg, New York, 1975 page 87.

3. E.J.Kliem, "Applications of Liquid Crystals" proceedings of the 2nd International Liquid Crystal conference held at Kent state University, August 12-16, 1968 part - I Gordon and Breach Science Publishing, New York page 27.

4. AW.E.Wordmanse, aid page 2 29.

5. S.K.Gupta, Chemical Age of India 1976, 27, 11, 961.

6. D.C.Prevotsek, "Recent Advances in High Strength Fibers and Molecular Composites" Polymer Liquid Crystal Academic Press New York, 1982 page 3 34.

7. Witold Brostow, Polymer 1990, 31.6.979.

8. M.G.Dobb and J E Melntyre, Advances in Polymer science vol 60-61, Springer-Verlag, Berlin Heidelerg New York 1984 page 95.

9. H.Finkelmann, Angew, Chem. Int. Ed. Engl. 1988, 27, 987.

10. H.Finklemann and H.J.Kork, Dispp. Technol 1985, 1, 87.

11. H.J.Eerie, A Milter and F.H.Kreuzer, Proc. 12th Intl Liq. Crystal Conf., Freiberz, FRG, 1988, poster CHZD.

12. C..Mc Ardle, MG.Clark, C.M.Haws, M.C.Wiltshire, A.Parker, G.Nester, G.W.Gray, D.Lacey and K.Toyne, J.Liq.Cry. 1987, 2. 573.

13. I Cabera, V.Kongrauz and H.Ringstorf Angrew. Chem., Intl. Edn. Engl. 1987-26, 1178.

DEVELOPMENT OF MULTILAYER AR COATINGS DERIVED FROM

ALKOXIDE BASED POLYMERIC SOLS

Arup K. Atta, Prasanta K. Biswas and Dibyendu Ganguli

Central Glass & Ceramic Research Institute
Calcutta 700032, India

INTRODUCTION

Various types of antireflective (AR) coatings deposited on a wide range of substrates are reported in the literature.[1-3] Most of the designs and techniques involve a combination of oxide and non-oxide materials. All-oxide AR coatings promise good durability unless they are porous beyond a certain degree; however, information on all-oxide AR coatings is scanty.[4,5]

The polymeric route in sol-gel processing is known to give rise to hard, durable oxide coatings.[6,7] Usually the process consists of (i) preparation of a sol in which alkoxide-derived polymeric particles are homogeneously dispersed (ii) application of thin layers of the sol on the substrate (iii) drying of the sol layer, its conversion to a porous gel layer and (iv) thermal conversion of the gel to denser oxide layer.

In the present work a variety of sols were synthesized via the sol-gel polymeric route by exploiting the chemical reactivities of the precursor aloxides; different types of all-oxide AR coatings were made by depositing these sols onto soda lime glass substrates (n_s, refractive index =1.52) and baking at elevated temperatures following suitable designs.

EXPERIMENTAL

Preparation of Sols of SiO_2, TiO_2 and ZrO_2 Systems

The methods of sol preparation and the chemicals used for SiO_2, TiO_2 and ZrO_2 coatings have been mostly described earlier.[8-10] The solvent and chelating agent used for TiO_2 system were 1-propanol and acetylacetone (acacH) respectively; the mole ratio of tetraisopropyl orthotitanate : 1-propanol : acacH : water was 1.0 : 27.0 : 0.5 : 2.2.

Preparation of Sols of ZrO_2-SiO_2 Systems

Three sols of ZrO_2-SiO_2 system were prepared. The chosen wt%

Frontiers of Polymer Research, Edited by P.N. Prasad and
J.K. Nigam, Plenum Press, New York, 1991

ratios of ZrO_2 : SiO_2 were 95 : 5 (Sol A), 80 : 20 (Sol B) and 40:60 (Sol C). Tetraethyl orthosilicate (TEOS) was dissolved in a mixture of 1-propanol and 2-butanol in a suitable proportion. The required amounts of water and 1M hydrochloric acid were added to it and the mixture was refluxed for 6h for partial hydrolysis. The required amount of zirconium (IV)n-propoxide mixed with 2-butanol was added to the partially hydrolyzed SiO_2 sol system and refluxed again for 6h. The refluxed sol was aged for two days followed by the addition of suitable amounts of water, alcohol and acacH. The mole ratio of zirconium n-propoxide : TEOS : 1-propanol : 2-butanol : 2-methoxy-ethanol : acacH : water : hydrochloric acid was 1.0 : (0.1-1.6) : (5.6-27.3):(13.8-17.7):(0.0-18.2):(0.8-3.4):(2.2-6.8):(0.015-0.150).

In all cases the viscosities of the sols were kept in the range of 2-4 cps and monitored by a viscometer.

Preparation and Characterization of Coatings
_____

Cleaned[11] soda lime glass substrates were dipped at different speeds of withdrawal (2-20 cm/min) followed by drying and baking at 450°C. All the coatings were found to be X-ray amorphous. The refractive indices and physical thicknesses of the oxide films were determined ellipsometrically at 6328 Å. The film thicknesses were calibrated against lifting speeds of the dipping unit. Reflection spectra (5° angle of incidence) of the films were recorded spectrophoto-metrically.

OPTICAL DESIGN

AR coatings were prepared by using a 2-layer design (G/H/L/air) and two 3-layer designs : (i) G/M/H/L/air and (ii) G/M/2H/L/air where M,H,L stand for quarterwave optical thicknesses corresponding to the chosen wavelength (λ_o) of the medium, high and low refractive index components respectively; G stands for the substrate glass.

For the 2-layer design, the L layer was either SiO_2 (n_1=1.46) or a ZrO_2-SiO_2 composition (from Sol C), with n_1 =1.55 ; the H layer was ZrO_2 (n_2 =1.92). For the 3-layer designs the H layer was TiO_2 (n_2 =2.04), the M layer was a ZrO_2- SiO_2 composition, from either Sol A or Sol B with n_3 = 1.84, used for design (ii), or 1.68, used for design (i). The L layer was SiO_2 (n_1=1.46). The symbols n_1,n_2 and n_3 indicate the refractive indices of the layers from air to the substrate.

RESULTS AND DISCUSSION

Basic Chemistry for the Formation of Polymeric Sols
_____

Generation of a sol is based on mainly two types of reactions, namely, hydrolysis of the alkoxide:

$$M(OR)_n + H_2O \rightarrow M(OR)_{n-1} OH + ROH \qquad \text{.... (i)}$$

and polycondensation of the hydrolysed products:

$$2M(OR)_{n-1} OH \longrightarrow 2M(OR)_{n-1} O_{\frac{1}{2}} + H_2O$$

or/and $M(OR)_{n-1}$ OH + $M(OR)_n \longrightarrow 2M(OR)_{n-1}$ $O_{\frac{1}{2}}$ + ROH   ...(ii)

Where M = Si, Ti, Zr and R = $-CH_3$, $-C_2H_5$,$- C_3H_7$ etc.

The rates of hydrolysis of zirconium and titanium alkoxides are much faster than that of silicon alkoxide. Hence, to control the rates of their hydrolysis, a chemical modification has been employed by incorporating a suitable proportion of acetic acid in the sol of ZrO_2 system and of acacH in the sol of TiO_2 system. Both the species act as bidentate ligands[12] and stabilize the system by controlling the rates of hydrolysis and polycondensation.

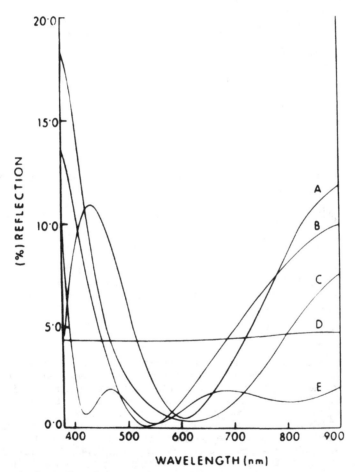

Fig.1. Reflection spectra in the visible region of two and three layer antireflective coatings. A,2-layer system, refractive indices n_1=1.46 and n_2=1.92; B, 2-layer system, refractive indices n_1=1.55 and n_2=1.92; C, 3-layer system, refractive indices n_1=1.46, n_2=2.04 and n_3=1.68; D, uncoated glass substrate; E, 3-layer system, refractive indices n_1=1.46, n_2=2.04 and n_3=1.84.

Table 1. Comparison of Spectral Performance of Different Multilayer
AR Coatings

Design Followed (λ_0 in nm)	Refractive Indices of the Materials used	%R min (λmin in nm)	Bandwidth (nm) with % R $\leqslant$ 1	Bandwidth (nm) with % R $\leqslant$ 1.5
2-Layer (600)	n_1=1.46 n_2=1.92	0.74 (600)	637-574 =63	655-553 =102
2-Layer (540)	n_1=1.55 n_2=1.92	0.15 (541)	592-499 =93	617-482 =135
3-Layer Design-(i) (600)	n_1=1.46 n_2=2.04 n_3=1.68	0.45 (602)	675-550 =125	725-511 =214
3-Layer Design-(ii) (550)	n_1=1.46 n_2=2.04 n_3=1.84	0.67 (423) 0.21 (550)	434-412 =22 607-506 =101	900-405 =495

The reaction products with acetic acid (Eq. iii) and acetylacetone (Eq. iv) are as follows :

$$\text{Zr }(OC_3H_7^n)_4 + CH_3COOH + H_2O \longrightarrow \text{Zr }(OC_3H_7^n)_3 \ (OOCCH_3)(H_2O) \quad \ldots (iii)$$

$$\text{Ti}(OC_3H_7^i)_4 + acacH \longrightarrow \text{Ti}(OC_3H_7^i)_3 \ acac + C_3H_7OH \quad \ldots (iv)$$

For preparing sols of ZrO_2 - SiO_2 system initially TEOS was allowed to undergo partial hydrolysis as shown in equation (i); subsequent condensation occurred with the partially hydrolyzed or unhydrolyzed $Zr(OC_3H_7)_4$ as shown in equation (ii).

Optical Properties of AR Film
_____

Reflection spectra of the 2-layer and 3-layer coatings are shown in Fig.1; comparative data on their spectral performance are given in Table 1. For the 2-layer system, the relationship of refractive indices for having AR effect is $n_2/n_1 = n_s^2$. If the glass substrate be a soda lime glass then n_2/n_1 =1.23 and this ratio should be maintained for having good AR effect. However, as has been indicated earlier, in a sol-gel polymeric system, the available substances are ZrO_2 (n_2=1.92) and TiO_2 (n_2=2.04) as high index material and SiO_2 (n_1=1.46) as low index material. Of these two possible combinations, i.e. $TiO_2(n_2)/SiO_2(n_1)$ and $ZrO_2(n_2)/SiO_2$ (n_1),the latter, i.e. 1.92/1.46=1.31 was chosen as it is close to 1.23. Using this combination the bandwidth with % R$\leqslant$1 is 63 nm and % R min at the chosen λ_0(600 nm) is 0.74 (A in Fig. 1). On the other hand, to maintain the ratio n_2/n_1=1.23, the lower index component, SiO_2 (n_1=1.46) was replaced by a mixed oxide component in the system ZrO_2-$SiO_2(n_1$=1.55) maintaining the high index component as constant (n_2=1.92). The AR effect was thus improved and the spectrum is shown in Fig. 1(B). The bandwidth with % R$\leqslant$1 is 93 nm and % R min at the chosen λ_0 (540 nm) is 0.15 and the bandwidth with % R$\leqslant$1.5 covers 135 nm. In the 3-layer systems, a large bandwidth of 495 nm (Fig.1,E) covering the whole visible region with % $R \leqslant 1.5$ and $\lambda_0 = 540$ nm was obtained in case of design (ii), which was much better (Table 1) in performance than that obtained via design (i). Design (ii) is therefore preferable for achieving broad band AR effect in the visible region.

ACKNOWLEDGEMENT

The authors thank Dr. B.K. Sarkar, Director for his kind permission to publish this paper.

REFERENCES

1. H.A. Macleod, "Thin Film Optical Filters" Adam Hilger Ltd., London (1969).
2. H.K. Pulker, "Coatings on Glass", Elsevier, New York (1984).
3. J. T. Cox and G.Hass, Antireflection coatings for optical and infrared optical materials in :"Physics of Thin Films", G.Hass and R.E. Thun, eds., Academic Press, New York(1969).
4. M.Buehler, J.Edlinger, G.Emiliani, A.Piegari and H.K.Pulker, All-oxide broad band antireflective coatings by reactive ion plating deposition, Appl. Opt. 27:3359(1988).
5. A.Piegari, Antireflection coatings on glass for the visible spectrums, Rivista della Staz Sper, Vetro n. 1 :127 (1989).
6. P.K. Biswas, D. Kundu and D. Ganguli, A. sol-gel derived antireflective coating on optical glass for near-infrared applications, J. Mat. Sci.Letts. 8 : 1436 (1989).
7. H. Schroeder, Oxide Layers deposited from organic solutions, in : "Physics of thin Films" G.Hass and R.E.Thun, eds. Academic Press, New York (1969).
8. D. Kundu, P.K. Biswas and D. Ganguli, Alkoxide-derived amorphous ZrO_2 coatings, Thin Solid Films 163:273(1988).
9. D.Kundu, P.K. Biswas and D. Ganguli, Sol-gel preparation of wavelength-selective reflecting coatings in the system ZrO_2-SiO_2, J.Non-Cryst. Solids 110 :13 (1989).
10. A. Atta, P.K. Biswas and D. Ganguli, Preparation of multilayer colour-separating filters by sol-gel processing, J.Mat.Sci. Letts. (1991) in press.
11. A. K. Atta, P.K. Biswas and D. Ganguli, A. sol-gel derived yellow-transmitting coating on glass, J.Non-Cryst. Solids 125 : 202 (1990).
12. C.Sanchez, F. Babonneau, S. Doeuff, and A. Leaustic, Chemical modifications of titanium alkoxide precursors, in : "Ultrastructure Processing of Advanced Ceramics" J.D. Mackenzie and D.R. Ulrich, eds., Wiley-Interscience, New York(1986).

THIRD ORDER NONLINEAR OPTICAL RESPONSE OF PURE AND DOPED POLYPHENYL

ACETYLENE: A WAVELENGTH DEPENDENCE STUDY

R. Vijaya, Y.V.G.S. Murti, G. Sundararajan[*] and
T.A. Prasada Rao

Department of Physics, Indian Institute of Technology
Madras 600 036, India
*Department of Chemistry, Indian Institute of Technology
Madras 600 036, India

INTRODUCTION

Polyphenylacetylene [PPA] is a π-electron conjugated polymer with carbon atoms on the backbone and having hydrogen and phenyl ring for sidegroups attached to alternative carbon atoms. This paper reports nonlinear optical studies on this material which is recognised to be important due to its reasonably high values of the third order suscepti-bility $\chi^{(3)}$ in transparent regions. The wavelength dependence study helps us to get the dispersion in $\chi^{(3)}$. Such a dispersion is brought about by the varying contributions from the different denominators in the expression[1] for $\chi^{(3)}$ given below:

$$\chi^{(3)}(\omega;\omega,\omega,-\omega) = \frac{2N}{3\hbar^3\epsilon_0}\left[\sum_{lmn}{}' \frac{\vec{\epsilon}^*\cdot\vec{Q}_{gl}\ \vec{\epsilon}^*\cdot Q_{lm}\ \vec{\epsilon}\cdot Q_{mn}\ \vec{\epsilon}\cdot Q_{ng}}{(\Omega_{lg}-\omega)(\Omega_{mg}-2\omega)(\Omega_{ng}-\omega)}\right.$$

$$\left. - \left(\sum_{m}{}' \frac{|\vec{\epsilon}\cdot\vec{Q}_{mg}|^2}{(\Omega_{mg}-\omega)}\right) \times \left(\sum_{n}{}' \frac{|\vec{\epsilon}\cdot\vec{Q}_{ng}|^2}{(\Omega_{ng}-\omega)^2}\right)\right] \qquad (1)$$

This is the form of $\chi^{(3)}$ as obtained in a degenerate four wave mixing (DFWM) process, g is the ground state and l, m, n are the different states in the gap. $\vec{Q}_{ij}$ are the matrix elements $<i|Q|j>$ of the electric dipole moment operator between stationary states $|i>$ and $|j>$ of the unperturbed atom, N is the number density of atoms and $\vec{\epsilon}$ is the unit polarization vector, ω is the operating frequency, Ω_{1g}, Ω_{mg} and Ω_{ng} are the frequencies of transition between the ground state and the gap states. Primes indicate

Frontiers of Polymer Research, Edited by P.N. Prasad and
J.K. Nigam, Plenum Press, New York, 1991

that g is to be omitted in the summations over intermediate states.

When either the operating frequency ω or its second harmonic 2ω becomes comparable to any of the transition frequencies, it can be seen that the value of $\chi^{(3)}$ will show a large resonance enhancement in contrast to the condition when the operating frequency is very small compared to the transition frequencies.

PRESENT STUDIES

PPA, both in its pristine and iodine-doped forms, has been studied at the three wavelengths of 532 nm, 694 nm and 1064 nm. These are obtained respectively at the second harmonic of a Nd: YAG laser, fundamental of a ruby laser and fundamental of a Nd: YAG laser. In the present studies, PPA which is present in amorphous form is taken in solution state in the solvent 1,2 dibromoethane. PPA in this solvent has an absorption edge near 400 nm and is almost transparent throughout the visible region. Iodine-doped polymer, however, has an absorption band peaking at 490 nm, apart from the edge at 400 nm. Both the samples have negligible absorption at 694 nm and 1064 nm.

Doping the polymer can affect the electronic properties[2]. If doping increases the disorder of the chains through local geometric modifications, thus lowering the delocalization length, then it will have a deleterious effect on the polymer's characteristics. However, if doping introduces new states in the band gap, the oscillator strengths on the various transitions get changed and this can lead to an improvement in the properties over the undoped polymer. Doping studies in the context of nonlinear optics in conjugated polymers are scarce. Hence we have studied the effect of acceptor doping with iodine on the nonlinear optical characteristics of PPA.

The third order optical response of this polymer is probed via optical phase conjugation[3] experiments in a degenerate four wave mixing geometry[4]. The experimental arrangement is described in detail elsewhere[5,6]. Three input laser beams, designating two of them as pump beams and the third as the probe beam, interact in a medium having a nonzero nonlinear polarization and third order susceptibility $\chi^{(3)}$ to generate an output beam with special properties. In the process of DFWM, the frequencies of all the mixing fields are the same. The parameter that is of interest in a phase conjugation experiment is the phase conjugate reflectivity R. The value of $\chi^{(3)}$ is evaluated[7] from the quantities R, n(linear refractive index of the medium), I_1 (intensity of the first pump beam) and α (linear absorption coefficient of the medium) whenever the samples absorb at the operating wavelength.

$$\chi^{(3)} = \frac{2c^2 n^2 \varepsilon_0 \alpha}{3\omega \, e^{-\alpha L'}(1 - e^{-\alpha L})} \frac{R^{\frac{1}{2}}}{I_1} \tag{2}$$

L' is the interaction length (= L sec θ) and L is the sample thickness. θ is the angle between pump and probe beams. ε_0 is the permittivity of free space and c is the speed of light. The measured values of reflectivity are least-squares-fitted to the following quadratic relation with the pump intensity I_1:

$$R = C \, I_1^2 \tag{3}$$

C is a constant. This relationship is valid when the pump depletion is negligible and the applied fields are such that the reflectivity is small. The slope C of the above-mentioned fit is used in (2) for the determination of $\chi^{(3)}$.

RESULTS AND DISCUSSION

In the studies at all the three wavelengths, the undoped and iodine-doped polymer samples gave considerable values of reflectivity. At the wavelength of 532 nm and at a pump energy of 8 mJ, PPA gave a reflectivity of 1.5×10^{-3} and PPA: iodine (20 wt. % doping) gave a reflectivity of 3.5×10^{-3}. At 694 nm, undoped PPA gave a reflectivity of 2.5×10^{-3} and 20 wt. % iodine-doped PPA, a value of 7.0×10^{-3}, both at a pump intensity of 30 MW/cm^2. However at the wavelength of 1064 nm, iodine-doped sample gave reflectivities (of the order of 10^{-2} at a pump energy of 120 mJ) lower than that of undoped PPA.

At 532 nm, very low doping of 1% to 5% did not enhance the conjugate signal above that for the undoped sample. Dopant level of 20 wt. % gave the highest conjugate energy followed by a fall for higher levels of doping. This is due to the combined effect of absorption and doping. The sample of 20 wt. % iodine-doped PPA has an absorption of $\alpha L = 0.68$, which is close to the value (~ 0.70) for obtaining the maximum reflectivity from (2).

At 694 nm, a doping concentration of 20 wt. % again gives the maximum reflectivity with a fall for higher dopant concentrations. Inspite of the samples having negligible absorption at 694 nm, this effect is observed. Hence it can be concluded that very high doping levels probably distort the polymeric chains thus reducing their efficiency for nonlinear optical studies.

The evaluated values of $\chi^{(3)}$ of the samples at all the three wavelengths are listed in table I.

Table I

Third order susceptibility of PPA

No.	Sample		Wavelength (nm)	$\chi^{(3)}$ (m^2/v^2)
1.	PPA		532	1.6×10^{-18}
2.	PPA: I_2	22.0 wt. %	532	1.9×10^{-17}
3.	PPA: I_2	43.5 wt. %	532	3.2×10^{-17}
4.	PPA: I_2	63.5 wt. %	532	3.7×10^{-17}
5.	PPA: I_2	85.5 wt. %	532	4.4×10^{-17}
6.	PPA		694	2.1×10^{-19}
7.	PPA: I_2	5.0 wt. %	694	2.1×10^{-19}
8.	PPA: I_2	10.0 wt. %	694	2.7×10^{-19}
9.	PPA: I_2	20.0 wt. %	694	6.0×10^{-19}
10.	PPA: I_2	40.0 wt. %	694	5.4×10^{-19}
11.	PPA: I_2	60.0 wt. %	694	4.5×10^{-19}
12.	PPA		1064	5.1×10^{-19}
13.	PPA: I_2	20.0 wt. %	1064	3.5×10^{-19}

The higher value of $\chi^{(3)}$ in undoped PPA at 532 nm in comparison to that at 694 nm and 1064 nm where the values are of the same order of magnitude is due to the proximity of the operating wavelength to the absorbing region of the sample.

All these studies are conducted with the polarization direction of the mixing fields being parallel to each other. This combination gives the highest values of the signal intensity. When the polarization of one of the pump beams is parallel to that of the probe but the other pump has orthogonal polarization, the conjugate signal is lower in intensity. Thermal mechanisms will contribute to a large extent in these polarization combinations for nanosecond pulsed experiments. However, when both the pump beams have polarizations parallel to each other but orthogonal to that of the probe, no absorption-related effects can contribute[8]. We find from our studies that this last combination of polarization of the mixing beams also gives a considerable amount of conjugate signal (about half of the second-mentioned combination). This leads us to conclude that there are nonthermal contributions to the nonlinearity in our polymer samples.

REFERENCES

1   D.C. Hanna, M.A. Yuratich and D. Cotter, "Nonlinear optics of free atoms and molecules", Springer-Verlag, Berlin (1979).
2   T.A. Skotheim (Ed.), "Handbook of conducting polymers", Marcel-Dekker, New York (1986).
3   R.A. Fisher (Ed.), "Optical phase conjugation", Academic Press, New York (1983).
4   A. Yariv and D.M. Pepper, Amplified reflection, phase conjugation and oscillation in degenerate four wave mixing, Opt. Lett. 1:16 (1977).
5   R. Vijaya, Y.V.G.S. Murti, G. Sundararajan and T.A. Prasada Rao, Degenerate four wave mixing in solutions of pure and iodine-doped polyphenylacetylene, Opt. Commun. 76: 256 (1990).
6   R. Vijaya, Y.V.G.S. Murti, G. Sundararajan and T.A. Prasada Rao, Nonresonant third-order optical response of polyphenylacetylene, J.Appl. Phys. in press .
7   R.G. Caro and M.C. Gower, Phase conjugation by degenerate four wave mixing in absorbing media, IEEE J. Quant. Electr. QE-18: 1376 (1982).
8   R. Trebino, C.E. Barker and A.E. Siegman, Tunable-laser- induced gratings for the measurement of ultrafast phenomena, IEEE J. Quant. Electr. QE-22: 1413 (1986).

PHOTOPOLYMER TRENDS IN OPTICS

Y. P. Kathuria

Shriram Institute for Industrial Research

19, University Road, Delhi - 110 007 (INDIA)

INTRODUCTION

In late 1960's and early 1970's there has been a great interest among researchers in photopolymer material as new entrant for holographic recording media possessing very high resolution. During this period many research articles were published by several authors[1-4] exhibiting unusual behaviour of these materials. But they ended up with its non-binding and degradation properties. However, in the recent past, with the exhibit of 3rd order non-linearities in organic polymers, interest has been reviewed for its application in HOE, phase conjugate mirrors[5], wavelength multiplexer/demultiplexer & more recently in non-linear grating/directional coupler and fast switching devices where the feedback gratings using PMMA films are used[6]. This communication describes few characteristic properties such as surface as well as index modulation in a phase hologram with a volume effect[4] recorded in photopolymer media. This is designed for grating coupler (Fig.1) in coupling a light beam from a laser into a thin film waveguide, where the diffraction efficiency play an effective role.

MATHEMATICAL BACKGROUND

It has been well established that the mechanism of hologram gratings made from photo-polymer comprises mainly of two effects: namely, surface modulation where the typical layer thickness is < 25 um that results in the formation of thin phase hologram and secondly the volume effect due to index modulation, which becomes dominant when the layer thickness ranges in excess of 100 um.

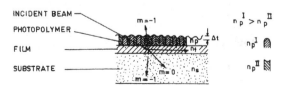

Fig.1. Block diagram of the grating coupler

To calculate the diffraction efficiency one need to derive the expressions for phase and index modulation as follow:

(a) Phase modulation[7,8]

Starting from the fundamental ray tracing in the geometrical optics, one can write for the path difference in the photopolymer and the glass plate

$$P = \frac{t}{\cos\psi} \qquad\qquad G = \frac{T}{\cos\xi} \qquad\qquad (1)$$

t ; T being the polymer layer and glass thickness whereas ψ and ξ are the respective angle of refraction.

The total optical path difference w.r.t. air is

$$\text{path difference} = (n-1)\frac{t}{\cos\psi} + (n_g-1).\frac{T}{\cos\xi} \qquad\qquad (2)$$

n; n_g : air & glass refractive indices

$$\text{Phase difference} = 2\frac{\pi}{\lambda}\ .\ \text{path difference} \quad \text{and}$$

$$\underset{(\Delta\phi)}{\text{Phase modulation}} = \frac{2\pi n}{\sqrt{n^2-\sin^2\theta}}\ [\ \Delta t(n-1)+t\Delta n] \qquad\qquad (3)$$

$$n\cos\psi = \sqrt{n^2-\sin^2\theta} \qquad \theta: \text{Angle of incidence}$$

(b) Index modulation[2]

The refractive index of the photopolymer is given by

$$\frac{n^2-1}{n^2+1} = \frac{4\pi}{3}\ N\alpha$$

α : Mean polarizability;   N   : No. of molecules per volume

and αN : Molecular density which varies inversely as the volume or approximately thickness 't' of the material.

i.e. $\alpha N \sim 1/t = \beta/t$ therefore,

$$n^2 = \frac{1 + \dfrac{4\pi}{3}\cdot\dfrac{2\beta}{t}}{1 - \dfrac{4\pi}{3}\cdot\dfrac{\beta}{t}}$$

with the index/thickness modulation as

$$\Delta n = -\Delta t\ \frac{(n^2-1)(n^2+2)}{6nt} \qquad\qquad (4a)$$

208

$$\& \quad \Delta t \; = -6t \; \frac{\Delta n.n}{(n^2-1)(n^2+2)} \qquad (4b)$$

Knowing the phase modulation ($\Delta\phi$) and thickness/index modulation from equation (3) and (4) one can calculate the diffraction efficiency by[1,3,8]

$$\eta_t \; = \; Sin^2\!\left(\frac{\pi\Delta nt}{\lambda\,Cos(\theta/2)}\right); \eta_p \; = \; J_1^2 \; (\Delta\phi) \qquad (5)$$

The generated plots of these diffraction efficiencies versus exposure agree very well with the experimental results of many authors.

EXPERIMENTAL SET UP

The experimental setup is basically a two beam interference (Fig.2) in which a beam from a He-Ne/He-Cd laser source is made spatially filtered and collimated by a microscope objective, pinhole and lens combination system. The collimated beam after being equally split up by the beam splitter, traverse the equal path length and are made to interfere at the recording plane. A spin coat/few drops of a photopolymer with monomer to catalyst ratio of 6:1 and thickness ranging from 25 um to 100 um was sandwitched between the two glass plates. Several samples in this thickness range were prepared. These were exposed to the two beam interference for varying degree of angular separation. The data is tabulated in Table I.

Table I. Details of the Sample

S.No.	Frequency (lines/mm)	Exposure (mJ/Cm^2)
1.	80	–
2.	100	10.35
3.	334	9.80
4.	580	8.17
5.	700	8.17
6.	863	9.80
7.	1100	8.17

The photopolymer holographic grating so recorded were fixed with the following process.

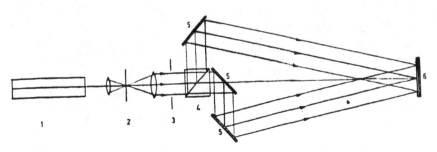

Fig. 2. Experimental set up for the hologram recording.

FIXING PROCESS

For fixing of the photopolymer hologram one can use one of the following process. (1) Xenon lamp emission (2) UV illumination : With corning glass filter No. 7-54, which allows only those radiations to be absorbed by photopolymers. (3) Thermal process : This process can only be used for permanent fixing, when the photopolymer hologram becomes visible through either of the above two process.

In this experiment we have fixed the photopolymer holograms by (2) and (3) process. But however, for better or increased diffraction efficiency, its advisable to keep the holograms in darkness for two or three days.

RESULTS AND DISCUSSION

A complete surface and diffraction efficiency study of the above samples was carried out to yield the following interesting results.

(a) The photographs of the surface of such hologram grating (Fig.3) taken with scanning electron microscope show clearly a depth modulation of 0.6 um and 3 um in the hologram grating for two set of interfering angles. The varying degree of cross linking in the dark and bright region may be due to refractive index modulation of the medium.

(b) The diffraction efficiency in the frequency range of 5001/mm - 600 1/mm was found to be higher ($\eta \sim 30\%$) than that at lower or higher frequency with monomer to catalyst ratio of 6:1 only (Fig.4). This may be due to the fact that for frequencies below 100 1/mm the diffraction efficiency is primarily due to surface relief and above 10001/mm it is primarily due to refractive index variation. But however inbetween it may be due to surface as well as refractive index modulation.

(c) With the present photopolymer it became possible to resolve 1100 1/mm without any difficulty, but it became difficult to resolve more than 1500 1/mm. With the optimum ratio of monomer to catalyst, the resolution and the diffraction efficiency however, could be improved upon.

 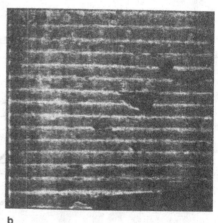

a

b

$\Delta t = 0.6 \mu m$ $\Delta t = 3 \mu m$

Fig. 3. SEM photographs of the interference patterns on photopolymer at two interference angles with (a) $\Delta t = 0.6 \mu m$ (b) $\Delta t = 3 \mu m$.

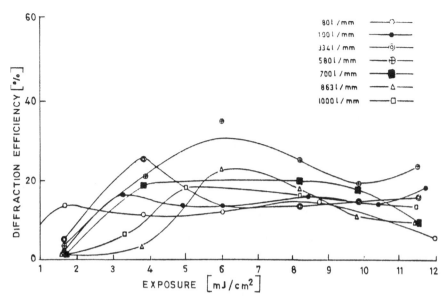

Fig. 4. Plot of Diffraction Efficiency (η) versus Exposure (E) for different line frequencies.

(d) The aging of the sample in the adverse conditions have shown only slight degradation. Few of the samples were as old as 10 yrs.

CONCLUSION

Construction and study of the photopolymer holographic gratings has revealed us its surface and index modulation behaviour which results in higher diffraction efficiency in the frequency range of 580 lines/mm. The aging of the sample showed a remarkable improvement.

REFERENCES

1. J.A.Jenney,J.O.S.A. 60, 1155 (1970).
2. W.S.Colburn, K.A.Haines, Appl.Opt. 10, 1636(1971).
3. J.A.Jenney, Appl. Opt. 6, 1371 (1972).
4. B.L.Booth, Appl. Opt. 11, 2994 (1972).
5. Simon Gray, Photonics Spectra, 125 (Sept. 1989)
6. R.A.Hahn, D.Bloor, Organic Materials for non-linear optic, Royal Society of Chemistry, London, 369 (1989).
7. S.Martelluci, A.N.Chester, Integrated Optics:Physics and application, ASI 90, 411 (1983).
8. N.Sadlej, B.Samolinska, Optics and Laser Tech., 175 (1975).

EMERGING DEVELOPMENTS IN PLASTIC OPTICAL FIBERS

J.K. Nigam, Amita Malik and G.L. Bhalla

Shriram Institute for Industrial Research
19, University Road, Delhi-110 007, India

INTRODUCTION

The bridging element between the transmitter and receiver is
generally known as transmission channel. Any improvement in this
transmission channel aims at either to improve the transmission fidelity
or to increase the data rate or to increase the transmission distance
between the two stations. Also, the amount of information transmitted is
directly related to the frequency range over which the carrier wave
operates, increasing the carrier frequency increases the available
transmission band width and consequently provides larger information
carrying capacity. Therefore, great interest in communication at optical
frequencies ($\simeq$ 5 x 10^{14} Hz) was created with the advent of lasers in
1966 and hence the birth of optical fibers (working on the principle of
total internal reflection). The first silica optical fiber showed the
losses in the transmission signal over 1000dB/km. The attenuation in the
signals is caused by absorption, scattering, radiative, distortions,
pulse broadening, mode couplings etc[1]. However, the attractive
advantages, (low weight, wider band width, hair size dimensions, immunity
to the electric interference, low fiber to fiber cross talks, high degree
of data security, etc), over the conventional modes of transmission led
researchers to reduce the attenuation in glass fiber around 0.2 dB/km in
the 1100 to 1600 nm by early 1980.

Though the achieved attenuation in glass fibers are low yet the
plastic optical fibers has good world market potential, particularly, in
local area network (LAN)[2] because of the various advantages, viz improved
ductility, large fiber core diameter (100-1400 μm) as compared to 0.1 μm
for glass fiber, high numerical aperture, etc. Large core diameter and
high numerical aperature allow the design of simple connection system
that reduces the connectors and also has economical installation cost.
However, plastic optical fiber has certain limitations of higher
attenuation, water absorption, low temperature, lack of available high
speed LED's etc [3-5]. Nevertheless, attempts have been made to overcome
the drawbacks by substitution as well as by using different polymers as
core and sheath as described in the following sections. The plastic
optical fibers with an attenuation of 125 dB/km are available[2]. Though
in laboratory the attenuation losses upto 20 dB/km have been achieved.
Plastic optical fiber, even with high attenuation losses, finds
application in short data links, dash board, automobiles lights,
decorative sign, medical application, video displays etc.

STRUCTURE OF OPTICAL FIBER

Optical fibers/waveguides of different configurations have been discussed in literature[6]. However, accepted optical fiber consists of a single solid dielectric cylinder (core) with index of refraction n_1. Core is surrounded by a solid dielectric, known as cladding, of refractive index n_2, ($n_1 > n_2$). In addition, cladding is further encapsulated in an elastic, abrasion-resistant plastic material. Further, optical fibers may be classed as single or multimode step-index or multimode graded index fibers, depending upon the refractive index variation in the material composition of the core.

OPTICAL SOURCES

Generally optical power is launched from LED (incoherent source), consisting of GaAlAs for wavelength 600-900 nm and InGaAsP for wavelength 1100-1600 nm and laser diodes (coherent source) for the optical fiber. The type of the power launching system used depends upon various factors such as bit rates, emission response time, quantum efficiency, bandwidth, type of the optical fiber used, etc. [7,8].

POLYMERS IN PLASTIC OPTICAL FIBER (POF)

The polymers used as POF's are mainly polycarbonates, PMMA and their copolymers. Table 1 gives the losses observed in case of polycarbonates used as cores, and the losses observed in the near IR region vary from 630-1000dB/km, while with PMMA 91dB/km loss has been observed in the visible region. Table 2 gives the copolymers used as sheath and polymers as cores. The use of fluorinated copolymers as sheath with PMMA have reduced the losses by 2dB/km in the visible region, while with the increasing wavelength the losses increase to 134dB/km and and 226dB/km (at 650 nm). The use of polycarbonate as POF shall find advantage over PMMA due to its unique properties, viz high temperature capability, curability, and low moisture absorption, and the most important is the availability of high power high speed LEDs for launching optical power, although the losses are three times higher at 650 nm[5]. Table 3 gives few references concerning the copolymers used as cores and polymer as sheath. The losses observed are again towards the higher side in case of copolymers used as cores and sheath (Table 4). The losses observed with acrylate copolymers are higher as compared to the copolymer of acrylate with styrene. These losses further increase with the copolymers of polycarbonate. Thus the use of PMMA and styrene copolymers as cores and fluorinated copolymer as sheath have shown lowest loss.

TABLE 1

POLYMERS USED AS CORE AND SHEATH

Sl.No.	Polymers		Wavelength nm	Losses (dB/km)	Ref. No.
	Core	Sheath			
1.	Polycarbonate	Poly(4-methyl-1-pentene)	–	1000	10
2.	Polycarbonate (Bisphenol-A)	Bisphenol AF polycarbonate	780	630	11
3.	Polycarbonate	Poly(4-methyl-pentene)	765	800	12
4.	PMMA	Silicone polymers	510	91	13

TABLE 2

POLYMERS USED AS CORE AND COPOLYMERS AS SHEATH

Sl.No.	Polymers Core	Sheath	Wavelength nm	Losses (dB/km)	Ref. No.
1.	PMMA	Hexafluoro acetone -hexafluoroethylene -Vinylidene fluoride copolymers	570 650	87 134	14
2.	PMMA	Vinylidene fluoride polymers and tetrafluo ethylene	570 650	89 226	14

TABLE 3

COPOLYMER USED AS CORE AND POLYMER AS SHEATH

Sl.No.	Polymers Core	Sheath	Wavelength nm	Losses (dB/km)	Ref. No.
1.	Co-polycarbonate	Poly (4-methyl-1-pentene)	770	1400	15
2.	α-methyl styrene -styrene copolymer	PMMA	-	15% loss	16

TABLE 4

CORE AND SHEATH AS COPOLYMERS

Sl.No.	Polymers Core	Sheath	Wavelength nm	Losses (dB/km)	Ref. No.
1.	Polycarbonate + isophthaloyl chloride - tere- phthaloyl dichlo- ride copolymers	2,2'-bis (4-hydroxy phenyl) 1,1,1,3,3,3,- hexafluoropropane - phosgene copolymer	-	2000	17
2.	Bu-methacrylate- Me- acrylate	Eth-methacrylate - Me- -fluoroacrylate 2,2,2-trifluoroethyl - fluoroacrylate	650 570	270 350	18
3.	Polystyrene and Me-methacrylate styrene copolymer	PE or ethylene copolymer or fluoropolymer	630	500- 1000	19
4.	diethylene glycol bis(allyl carbo- nate methylacry- late	Heated in C_2F_4- C_3F_6 pipe	660	1380	20
5.	Styrene - Me- methacrylate	C_2F_4-CH_2 = CF_2 Copolymer + PMMA	580 650	144 180	21

MODIFIED POLYMERS IN PLASTIC OPTICAL FIBRES

Modifications in POF have been made to minimize signal losses by substitution of hydrogen of CH by fluorine or deuterium.

TABLE 5

MODIFICATION OF POLYMERS WITH LOWER LOSSES

Sl.No.	Polymers Core	Sheath	Wavelength nm	Losses (dB/km)	Ref. No.
1.	Polystyrene d-5	EVA cladding	800	370	22
2.	Polystyrene d-8	EVA cladding	804	160	22
3.	PMMA d-8	Fluoroalkyl methylacrylate	565 / 640	41 / 55	22
4.	PMMA d-8	Fluoroalkyl methylacrylate	760 / 780 / 850	180 / 25 / 50	22
5.	$C_6F_mD_nCD=CD_2$	Poly (tetrafluoro-propyl methacrylate)	846	175	23
6.	$C_6F_aD_bCD=CD_2$ $CD_2 = CDR'CO_2R_2$ $(R'= R^2 = D,CD_3)$	Poly tetra fluoro propyl methacrylate	650	87	23
7.	PS-8F and alkyl methacrylate	Tetrafluoro ethylene Vinylidene fluoride	630	120	24
8.	PS-8F- PMMA-8F	Vinylidene fluoride copolymer	640 / 830	110 / 290	25
9.	Alkyl (meth) acrylate - 5D-3F styrene - Me-methacrylate	Tetraluoroethylene - Vinylidene fluoride	640 / 830	120 / 290	26
10.	5D-3F styrene - 3F-5D methyacrylate	$CF_2=CF_2$ - Vinylidene fluoride	640 / 830	110 / 290	27

Table 5 gives the losses observed in case of modified polymers. It can be observed from the table that the losses with copolymers of acrylates and fluorinated styrene have shown lower losses as compared to the use of only fluorinated styrene as cores. But with increasing wavelength these losses increase due to the presence of CH group.[9] Thus, substitution of H by D and use of 5D-3F styrene with fluoroalkyl pentadeuteroalkyl methacrylate reduces the loss by 50%. These losses are further reduced when deutrated PMMA is used with fluoroalkyl methacrylate sheathing. Thus these find applications where LED's can be used as light source i.e. 840 nm. But with the increasing wavelength (1100 nm) the losses increase, thereby permitting use for short distance optical signal transmission.

REFERENCES

1. Gerd Keiser, "Optical Fibre Communication," McGraw-Hill Book Co., Singapore, 1984.
2. C.T.Troy, "New Spectra, Plastic FO market said set to soar," Photonics spectra, 24(7): 38 (1990)
3. P. Avakian, W.Y. Hsu, P. Meakin and H.L. Synder, "Optical absorption of predeuterated PMMA and influence of Water, " J. Polym. Sci. Polym. Phys. Ed., 22: 1607 (1984).
4. T. Kaino, "Ultimate loss limits in plastic optical fibers", Appl. Opt. 24:4192 (1985).
5. F.L. Sanford, "Polycarbonate : The next optical fiber," Photonics Spectra, 23(10): 83 (1989).

6. S.E. Miller, E.A. Marcatili and T. Li, "Research towards optical fiber transmission system," Proc. IEEE, 61:1703 (1973).

7. H. Kressel and J.K. Butler, "Semiconductors Lasers and Heterojunction LED's," Academic, New York, 1977.

8. G.H.B. Thompson, "Physics of Semiconductor Laser Devices," Wiley, New York, 1980.

9. T. Kaino, "Preparation of plastic optical fibers for near IR region transmission, "J. Poly. Sci, Part A, Polym. Chem., 25:37(1987).

10. O. Masaya and S. Koji, "Heat-resistant optical fibers from polycarbonates," Jpn. Kokai Tokkyo Koho JP 0119, 307 (1989).

11. F. Hiroshi and K.Toshimasa, "Clad optical fibers with polycarbonate cores," Jpn. Kokai Tokkyo Koho JP 0161,706 (1989).

12. T. Akira, S. Hisashi, T.Takehisa and W. Noboru, "New plastic optical fiber with polycarbonate core and fluorescence doped fibers for high temperature use," Proc SPIE-Int. Soc. Opt. Eng. 840:19 (1987).

13. M. Shiruyoshi, D.S.Katsuhikoi and U. Yosihiro, "Silicone acrylate polymers for optical fibers clads," Jpn. Kokai Tokkyo Koho JP 63,243,110 (1988).

14. T. Shinichi, K. Shigeki, M. Kazuhiko, Y. Taku and K. Toshio, "Heat resistant cladding material for plastic optical fiber and plastic optical fiber using the same," Eur. Pat. Appl. EP 307,164 (1989).

15. N. Takashi and S. Kazuyoshi, "Heat resistant polycarbonate optical fibers," Jpn. Kokai Tokkyo Koho JP 6402,006 (1989).

16. Theodore L. Parker and Donald J. Parettie, "Vinyl aromatic core polymeric clad optical fiber," PCT. Int Appl. WO 8705,117 (1987).

17. U. Naoya, "Aromatic polyester cores for optical fibers", Jpn. Kokai Tokkyo Koho JP 0129,805 (1989).

18. N. Tsuneyuki, T. Yoshiharu, M. Setsuo and I. Hitoshi, "Manufacture of core sheath optical fibers," Jpn. Kokai Tokkyo Koho JP 01,13,102 (1989).

19. Liu. Hanming and Wong. Zhiyuan, "Preparation of R_1 and R_2 optical fibers," Shiyou-Huayong, 17(11): 714 (1988).

20. A. Fumito, F. Hiroshi and H.Hisako, "Flexible diethylene glycol bis (allyl carbonate) copolymer optical fiber, "Jpn. Kokai Tokkyo Koho JP 63,146, 004(1988).

21. T. Seishiro and S.Heiroku, "Optical fibers having methyl methacrylate styrene copolymer cores with low transmission loss," Jpn. Kokai Tokkyo Koho JP 63,101,803 (1988).

22. T. Kaino and Y. Katayuma, "Polymers For Opto-electronics," Polym. Eng. Sci., 29(17): 1809 (1989).

23. T. Kaino, T. Fukuda and T. Matsunaga, "Optical fibers", Jpn. Kokai Tokkyo Koho JP 61,223,805 and 61,223,806 (1986).

24. T.Masayuki, M. Tsuneaki, Y. Shotaro and H. Shoici, "Manufacture of acrylate polymer optical fibers," Jpn. Kokai Tokkyo Koho JP 63,214,705 (1988).

25. T. Masayuki, M. Tsuneaki, Y. Shotaro and H. Schoici, "Manufacture of fluorostyrene-methacrylate copolymer optical fibers," Jpn. Kokai Tokkyo Koho JP 63, 214, 704 (1988).

26. T. Masayuki, M. Tsuneaki, Y. Shotaro and H. Schoici, "Optical fibers with deuterated fluoropolymer cores," Jpn. Kokai Tokkyo Koho JP 63,182,607 (1988).

27. T. Masayuki, "Optical fibers with deuterated fluoropolymer cores," Jpn. Kokai Tokkyo Koho JP 63,182,608 (1988).

SPECTROSCOPIC ANALYSIS OF SOLUBLE AND PROCESSABLE

ELECTROACTIVE POLYMERS

T. Danno*, J. Kuerti** and H. Kuzmany

Institut für Festkörperphysik, Universität Wien
Strudlhofgasse 4, A-1090 Wien, Austria

INTRODUCTION

For the large scale application of electroactive polymers solubility, melt processability and evironmental stability are prerequisites. Poly-(alkylthiophenes) have attracted considerable attention and represent a new class of conjugated polymers characterized by their suitable processabilities[1-3] and by their interesting physical properties ,e.g., the thermochromic phase transition due to the thermally induced inter-ring torsions of the polymer backbone.[4-6] We report the spectroscopic properties, especially the optical absorption and the resonance Raman scattering, of oriented or unoriented poly(octylthiophene)(POT) in the neutral and doped states.

EXPERIMENTAL

POT was synthesized chemically from octylthiophene. From toluene solution thin films with a thickness of ca. 0.5 μm were formed on the substrates using spin-coating. The thin films on the polyethylene and polyester sheet could be stretched at room temperature up to 500 %. Doping was carried out chemically by immersing the POT film in 0.1 M $FeCl_3$ / nitromethane solution. Optical absorption was measured using a Hitachi U-3410 spectro-photometer between 15 K and 480 K under reduced pressure of $5X10^{-4}$ mbar. Raman measurements were carried out after excitation with Kr or Ar ion lasers using a Spex-double monochromator and a photon counting system between 100 K and 480 K also under the reduced pressure. Anisotropy of optical absorption and Raman scattering were determined for the oriented samples by a Glan-Thomson prism.

RESULTS

The optical absorption of unoriented POT at lower temperature showed three characteristic features: The absorption maximum was located at 2.4 eV, the oscillator strength of energies beyond the absorption maximum was still strong and some structure with a spacing of 0.18 eV was observed around the absorption maximum. This energy corresponds very well to the frequency of the Raman line of the C=C stretch mode at 1475 cm^{-1} as shown in Figure 1, which dominates the Raman spectrum. Thus, the structures in the optical absorption can be assumed to originating from vibronic transitions. The double and

---

* On leave from Mitsui Petrochemical Industries, LTD, Tokyo Japan
** Permanent address: Department of Atomic Physics, Roland Eötvös University Budapest H-1088, Puskin u. 5-7. Hungary

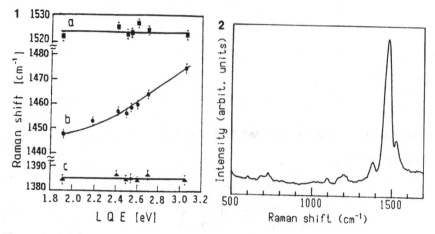

Figure 1. Raman spectrum of unoriented POT at 100 K taken with 408 nm laser excitation.

Figure 2. Change of Raman shift at 1530 cm^{-1} (a), 1470cm^{-1} (b), and 1380 cm^{-1} (c) for POT as a function of laser energy.

triple overtones of $\nu_{c=c}$ were also observed at 2949 and 4432 cm^{-1}, respectively. Figure 2 shows the change of the Raman shift with the laser quantum energy (LQE) for three differnt Raman lines. There is a clear dispersion of the Raman line from the $\nu_{c=c}$ with the LQE, as compared to the others, but the shift is smaller than in trans-(CH)$_x$. The peak position decreased linearly with decrease of the LQE down to 2.4 eV but after that changed only gradually. No dramatic change of line shape for the various laser excitations was observed for POT. However, the widths of Raman lines increased with increasing LQE similar to trans-(CH)$_x$.[7]

The thermochromic phase transition temperature T_c was determined as 390 K from the shift of the absorption maximum with temperature. The intensity of the Raman line of $\nu_{c=c}$ increased up to T_c with temperature and after that decreased.

In the oriented POT, polarized Raman scattering of $\nu_{c=c}$ showed a strong anisotropy (Figure 3). The intensity ratio of the Raman scattering $I_{///}/I_{\perp}$ was 5.5. The downwards peak shift of about 5 cm^{-1} as compared to unoriented POT was observed in the spectrum of parallel orientation.

Figure 4(a) shows the optical absorption spectra of the oriented POT. The diffrence of the position of the absorption maximum between parallel and perpendicular geometries is 0.43 eV which agrees well with the shift by the thermochromic phase transition (0.55 eV). For both orientations the absorption maxima shifted also towards higher energy with temperature. However, for the parallel direction a clear phase transition was observed at 415 K. This temperature is higher than the one for unoriented POT. On the other hand, the phase transition observed for the perpendicular geometry at 370 K was lower than for the unoriented POT.

Figure 4(b) shows the optical absorption spectra of oriented POT after chemical doping. Even after doping the spectra show a clear anisotropy. In the parallel geometry absorptions at 0.5, 1.4, 2.3 eV are observed while they are located at 0.7 and 1.5 eV in the perpendicular geometry. The strong absorption beyond 3 eV is from FeCl$_3$.

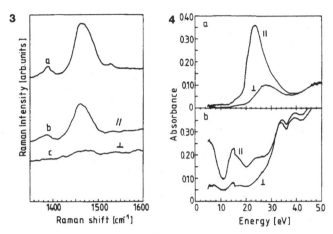

Figure 3. Raman spectra of $\nu_{C=C}$ of POT. (a) Unoriented POT; (b) 500 % stretched POT with parallel scattering geometry; (c)same sample as (b) with perpendicular scattering geometry. The laser wave length was 457 nm.

Figure 4. Optical absorption of 500 % stretched POT. (a) Undoped; (b) $FeCl_3$ doped.

DISCUSSION

The behaviour of the Raman intensity during the thermochromic phase transion is explained by the change of the resonance condition in connection with the change of the optical absorption. A similar result was observed for poloy(di-n-silane) due to the order-disorder transition by the defects formation along the polymer backbone to increase the σ-σ^* gap energy.[8]

In the case of oriented POT the phase transition temperature T_c for the parallel direction was higher than the one for unoriented POT. This can be understood from an increased order and an increased conjugaton length due to stretching of polymer backbones. This interpretation is in agreement with the behaviour of the temperature dependence of the absorption maximum for perpendicular polarization, since for this geometry T_c was observed to be reduced. This result also suggests that disordered or short conjugated regions of the polymer are easier thermally affected than the ordered region, and at higher temperatures the states of polymer chains are more disordered than those at lower temperature.

As mentioned above, the structures on the optical absorption can be assumed as the vibronic structures. Thus, an analysis of the optical absorption experiments within the vibronic theory of optical absorption is suggestive. The simple formulation of the optical-absorption line shape within the adiabetic/Condon approximation and a harmonic-oscillator model[9] was improved by coupling the explicit evaluated transition moments and a distribution function of the conjugation length to represent the total oscillator strength from the chanis with a given conjugation length[10]. Transition moments were calculated, based on the one-electron energy eigenvalues and eigenfunctions, which were obtained from a Longuett-Higgins, Salem-type Hückel Hamiltonian including geometry optimization. A bimodal Gaussian distribution was chosen to consider the contribution from high-energy regions, i.e., the contribution from the short conjugated chains. The use of a bimodal distribution is also suggested from the viewpoint that the ordered and disordered regions coexist in the real system. From the calculation, it becomes clear that in the system of POT, conjugation length is about 4 thiophene rings which corresponds to 8 C=C double bond. This value is a factor of 4 smaller than that obtained for trans-(CH)$_x$.[11] This may originate from the higher flexibility of the chains in POT, since the long side chains interrupt the inter-chain interaction and make the thermal motion easier.

The linear relation between laser energy and vibrational frequency of $\nu_{c=c}$ has been interpreted by the conjugation length model, which is lead from the relation between the conjugation length and the gap energy or optical frequency of the system, respectively.[7,12]

The 5 cm^{-1} downwards shift of the Raman line of $\nu_{c=c}$ for parallel geometry of oriented POT is explained by an extension of the conjugation length with molecular orientation. Correspondingly, the line shape for parallel excitation at frequencies beyond the peak looks more slim as compared to that for unoriented POT. On the contrary the Raman spectrum for perpendicular polarization looks broader and especially the contribution from higher frequencies within one Raman band is stronger than for parallel polarization. These results suggest that short conjugated chains with a broad distribution exist already in the starting material and these disordered chains cannot be further oriented by stretching. Thus, the spectrum of perpendicular geometry characterizes the disordered chains. This interpretation agrees well with experimental results of the optical absorption for stretched POT and with the results of the calculation for the optical absorption in which the bimodal distribution function of the conjugation length was introduced as described above. From the results discussed above it is concluded that both effects of temperature and molecular orientation are fully interpreted by means of a change of the conjugation length.

The optical anisotropy was preserved even after the doping, and the spectrum for the parallel geometry changed more drastically than the one for perpendicular. The spectrum of doped POT is dominated by the well known biporalon transitions in the near-IR region.[13]

Acknowledgement- This work was supported by the Fonds zur Förderung der wissenschaftlichen Forschung in Austria. T.D. acknowledges Mitsui Petrochemical Industries Ltd. for his visit at the Universität Wien.

REFERENCES

1. K.Y.Jen, G.G.Miller and R.L.Elsenbaumer, J.Chem.Soc.,Chem.Commun., 1346 (1986).
2. M.Sato, S.Tanaka and K.Kaeriyama, J.Chem.Soc.,Chem.Commun., 873 (1986).
3. K.Yoshino, S.Nakajima, M.Onoda and R.Sugimoto, Synth.Met.,28:C349 (1989).
4. O.Inganäs, G Gustafsson, W.R.Salaneck, J.E.Österholm, and J.Laakso, Synth.Met., 28:C377 (1989).
5. B.Themans, W.R.Salaneck and J.L.Bredas, Synth.Met., 28:C359 (1989).
6. C.X.Cui and M.Kertesz, Phys.Rev., B40:9661 (1989).
7. H.Kuzmany, phys. stat. sol., B97:521 (1980).
8. H.Kuzmany, J.F.Rabolt, B.Farmer and R.D.Miller, J.Chem.Phys., 85:7413 (1986).
9. H.Kuzmany, "Festkörperspektroskopie, Eine Einführung", Springer-Verlag, Berlin (1989).
10. T.Danno, J.Kürti and H.Kuzmany, Phys. Rev., B43(6):4809 (1991).
11. H.Kuzmany, Pure & Appl.Chem.,57(2):235 (1985).
12. H.Kuzmany, Makromol.Chem.. Macromol.Symp., 37:81 (1990).
13. G.Harbeke, E.Meier, W.Kobel, M.Egli, H.Kiess and E.Tosatti, Solid State Comm., 55(5):419 (1985).

OPTICAL PROPERTIES OF ORIENTED CONJUGATED POLYMERS AND COPOLYMERS

Serge Lefrant[a], Thierry Verdon[a], Evelina Mulazzi[b] and
Gunther Leising[c]

[a]Laboratoire de Physique Cristalline, Institut des Matériaux de Nantes
(UMR CNRS-Université de Nantes n°110)
2, rue de la Houssinière, 44072 Nantes Cedex 03 (France)
[b]Dipartimento di Fisica dell'Università, via Celoria 16, 20133 Milano
(Italy)
[c]Institut für Festkörperphysik, Technische Universität, Petergasse 16
Graz (Austria)

INTRODUCTION

In recent years, optical properties of polymers and copolymers have played an important role in the design of new materials for applications in electronics, non-linear optics and photonics. In fact, the new route for the synthesis of polymers, which involves a precursor polymer step, has proved to be rather powerful to obtain stretch-oriented samples offering good processability and higher purity in comparison with those obtained by using standard methods. In addition, most of these compounds exhibit anisotropic optical properties which are of real interest to built new optical and electro-optical devices.

In that respect, we have studied the optical properties of oriented trans-polyacetylene and oriented triblock copolymers made of polyacetylene and polynorbornene $(CH)_x$-$(NBE)_n$-$(CH)_x$, in which x and n can be varied in different ways. In order to achieve an investigation as complete as possible, we have used different optical techniques such as optical absorption, infrared absorption, Resonance Raman Scattering (RRS) and Photoinduced Infrared Absorption (PIA). Each of these spectroscopic techniques provides complementary information since they probe either the bulk or the surface properties of the samples. The experimental data have been analyzed and interpreted by using different models and among them, the one based on the electronic and vibrational properties of conjugated segments of different lengths which form the polymeric chains. The contribution coming from the electronic and vibrational states of the different conjugated segments entering the various optical functions studied are then weighted by a bimodal distribution of conjugated segments. The final theoretical results depends on the properties of long (double bonds N > 30) and short (N < 30) segments involved in the calculations through the G factor which weights the long segment distribution with respect to the short one. With such an approach, one can obtain a consistent interpretation of the optical properties derived from the different experimental data.

The aim of this paper is to report some of the optical data obtained from oriented trans-$(CH)_x$ and triblock copolymers $(CH)_x$-$(NBE)_n$-$(CH)_x$ and to give the theoretical interpretation of the most important features of the experimental results.

Frontiers of Polymer Research, Edited by P.N. Prasad and
J.K. Nigam, Plenum Press, New York, 1991

EXPERIMENTAL RESULTS

Preparation of samples

Polyacetylene samples have been prepared according to the Durham - Graz procedure using the poly-(BTFM-TCDT) as a precursor[1,2]. This precursor polymer, purified by several washing and precipitation cycles, is then dissolved in acetone and films are consequently cast onto cleaned glass substrates and dried under pure argon gas. By a thermal treatment at 80°C (approximately 40 minutes), one obtains cis-$(CH)_x$ films and if the treatment is longer (1h30-2 h) and performed at higher temperature (140°C), trans-$(CH)_x$ films are directly synthetized. In order to obtain oriented polymer films, a stretching of the precursor is to be applied during the thermal transformation and stretching ratios ($\Delta l/l$) of the order of 15 can be achieved. Such a synthesis method leads to highly dense polyacetylene films.

The triblock copolymers $(CH)_x$-$(NBE)_n$-$(CH)_x$ have been synthetized by the use of a bifunctional metathesis catalyst following a procedure described in details in Ref. 3. Films were cast from prepolymer solutions in toluene or THF on a substrate suitable for spectroscopic measurements. Let notice that prior to the thermal treatment which transform the prepolymers to the final triblock copolymers, the average molecular mass of each prepolymers was determined by means of size exclusion chromatography. The structure of the triblock copolymers is shown Fig. 1. The samples we have studied are characterized by $x = 88$ and $n = 415$; $x = 75$ and $n = 100$. The copolymer films were also stretch-oriented to $\Delta l/l$ ratios of the order of $\sim$ 3-6.

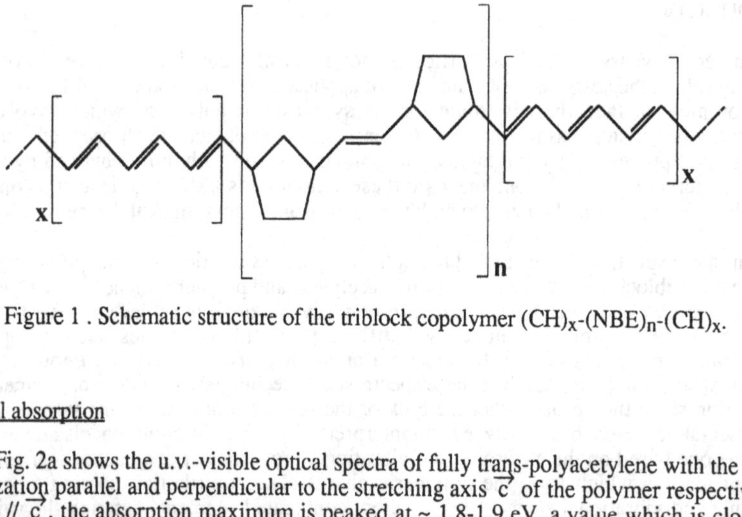

Figure 1 . Schematic structure of the triblock copolymer $(CH)_x$-$(NBE)_n$-$(CH)_x$.

Optical absorption

Fig. 2a shows the u.v.-visible optical spectra of fully trans-polyacetylene with the light polarization parallel and perpendicular to the stretching axis $\vec{c}$ of the polymer respectively. For $\vec{E} \parallel \vec{c}$, the absorption maximum is peaked at $\sim$ 1.8-1.9 eV, a value which is close to the absorption of standard polyacetylene, whereas the spectrum for $\vec{E} \perp \vec{c}$ is very weak and rather flat all over the visible range.

From the strong difference in the absorbance in these optical spectra, it is clear that the oriented trans-$(CH)_x$ exhibits a very strong anisotropy, as a proof of a very good orientation of the polymeric chains. Let recall that this way of synthesis of $(CH)_x$ allows to achieve stretching ratios $\Delta l/l$ as high as $\sim$ 15. Fig. 2b show the u.v.-visible absorption spectra of the copolymer $(CH)_{88}$-$(NBE)_{415}$-$(CH)_{88}$. The maximum of the absorption curve is peaked $\sim$ 2.70 eV for $\vec{E} \parallel \vec{c}$ and at $\sim$ 2.90 eV for $\vec{E} \perp \vec{c}$. This shift of the absorption maximum towards higher energy is related to the length of the conjugated segments in the polyacetylene parts of the polymer. It is consistent with the average value of $x = 88$ as determined by means of size exclusion chromatography and will be corroborated by the Raman data analysis. This average value is significantly smaller than in standard Durham-Graz polyacetylene. Also, as previously reported by Stelzer et al[3], the absorption for $\vec{E} \perp \vec{c}$ is further shifted towards higher energy, reflecting in the $(CH)_x$ sequences which are not oriented along the stretching axis a high disorder and a signature of shorter conjugated segments. Notice in addition that the three different copolymers studied with x = 75, 88 and 115 respectively showed very small differences in their absorption curves.

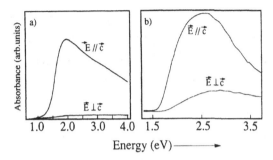

Figure 2 . Absorption spectra of : a) fully trans-$(CH)_x$, b)$(CH)_{88}$-$(NBE)_{415}$-$(CH)_{88}$.

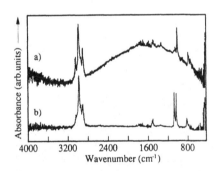

Figure 3 . Polarized infrared spectra of the triblock copolymers $(CH)_{88}$-$(NBE)_{415}$-$(CH)_{88}$ a) $\vec{E} \,/\!/\, \vec{c}$; b) $\vec{E} \perp \vec{c}$.

Infrared absorption

We present in Fig. 3 the infrared absorption spectra of the triblock copolymers $(CH)_x$-$(NBE)_n$-$(CH)_x$ (x = 88, n = 415) stretch-oriented with a stretching ratio $\Delta l/l = 5.3$, for light polarization parallel to the stretching axis (Fig. 3a) and perpendicular to the stretching axis (Fig. 3b). It appears first that the different absorption bands correspond to the superposition of those coming from trans-$(CH)_x$ segments and from polynorbornene segments.

For $\vec{E}$ // $\vec{c}$, one observes also a broad feature which is due to interferences effects due to the geometric parallelism of the two sides of the polymer. This feature can be used to estimate the thickness of the film if one knows the refraction index value $n_{//}$. In this case, taking a value $n_{//} = 2.7$ as determined in ref.4, the film has been found to be 0.5 µm thick. Details are given elsewhere[5].

Table 1 . Proposed assignment for the infrared vibrations of polynorbornene

Infrared peaks in cm^{-1}	proposed assignment
2994	C-H stretching (sp^2 carbon)
2946	C-H stretching (sp^3 carbon)
2865	C-H$_2$ stretching
1464	in-plane C-H deformation
1446	in-plane C-H$_2$ deformation
1042	ring vibration
966	out-of-plane C-H deformation (trans)
740	out-of-plane C-H deformation (cis)
729	out-of-plane C-H$_2$ deformation

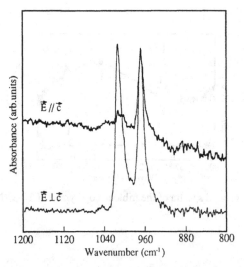

Fig. 4 . Polarized infrared spectra of (CH)$_{88}$-(NBE)$_{415}$-(CH)$_{88}$ in the 800-1200 cm^{-1} region for $\vec{E}$ // $\vec{c}$ and $\vec{E}$ $\perp$ $\vec{c}$.

A proposed assignment of the infrared bands in polynorbornene is given in ref.5 and reported in Table 1. It appears that the cis conformation of polynorbornene is present in the polymer since a mode at 740 cm^{-1} can be assigned to an out-of-plane C-H deformation on the C=C bond in the cis conformation. If we look carefully on the polarization of the infrared bands peaked at 966 cm^{-1} and 1010 cm^{-1} which are attributed to the out-of-plane C-H deformation in (NBE)$_n$ and trans-(CH)$_x$ respectively in the copolymer (see Fig.4), one can clearly observe that the 1010 cm^{-1} vibrational mode has a weak contribution for $\vec{E} \,/\!/ \, \vec{c}$, whereas the 966 cm^{-1} is much more depolarized. Therefore, one can deduce that the polyacetylene segments are better oriented than the polynorbornene ones in the copolymer.

Resonance Raman Scattering (RRS)

The RRS technique has been widely used in the past to study $(CH)_x$ films in terms of electronic and vibrational properties since a peculiar behavior in the Raman band shapes and peak positions has been observed when different excitation wavelengths are used[6-8]. Generally speaking, if we use an excitation line in the red range, λ_{exc} = 676.4 nm for example, one observes two main and assymetric bands at 1070 and 1460 cm^{-1}. When the excitation wavelength is tuned to the violet range, satellite bands develop at ~ 1120 cm^{-1} and ~ 1510 cm^{-1} respectively, resulting in a double peak structure for both features. Among the different models, one of them considers a double distribution of conjugated segments[9] (long and short) and therefore, the RRS technique can be used as a probe to establish which kinds of conjugated segments compose the polymeric chains[10]. In addition, different measurements have been made in polarized light in stretched oriented samples[11-13].

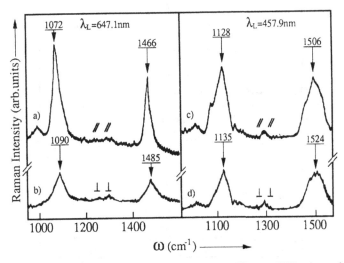

Fig. 5 . Experimental RRS spectra of oriented trans-$(CH)_x$ at 77K taken with two different wavelengths in the $/\!/\,/\!/$ and the $\perp\perp$ configurations.

A typical result is shown Fig. 5 in oriented trans-$(CH)_x$. The main features can be described as follows : i) for λ_L = 647.1 nm, the main Raman bands in the $/\!/\,/\!/$ configuration are peaked at 1072 and 1466 cm^{-1} respectively, while they are shifted to 1090 and 1485 cm^{-1} in the $\perp\perp$ configuration ; ii) for λ_L = 457.9 nm, the $/\!/\,/\!/$ polarized spectrum exhibits the double peak structure with the higher intensity components peaked at 1128 and 1506 cm^{-1} respectively for each band, whereas the $\perp\perp$ polarized spectrum shows only the high frequency bands shifted to 1135 and 1524 cm^{-1} respectively. The low frequency components, in spectrum d, have a negligible intensity. Since the two main vibrational modes are totally symmetric, they should be observed only in the $/\!/\,/\!/$ configuration. A similar experiment carried on stretch-oriented cis-rich $(CH)_x$ prepared by the Durham-Graz method[13] shows clearly that this behavior is specific to the trans-$(CH)_x$ segments. As a matter of fact, the vibrational modes due to the "cis" segments (totally symmetric) are completely polarized with a contribution only in the $/\!/\,/\!/$ configuration whereas those due to the trans-$(CH)_x$ segments show a strong anisotropy.

227

We have recorded RRS spectra in polarized light of stretch-oriented triblock copolymers $(CH)_x$-$(NBE)_n$-$(CH)_x$. Figure 6 shows these spectra for $\lambda_{exc} = 647.1$ nm and 457.9 nm respectively. For $\lambda_{exc} = 647.1$ nm, the bands peaked at 1086 and 1473 cm^{-1} for the // // configuration are slightly shifted to 1098 and 1477 cm^{-1} respectively. For $\lambda_{exc} = 457.9$ nm, the shift looks smaller. Notice that no double peak structure is observed in the // // configuration as a consequence of the smaller concentration of long conjugated segments, in good consistency with the optical absorption data. The theoretical analysis, which will be described in the next section, leads also to results in agreement with this result.

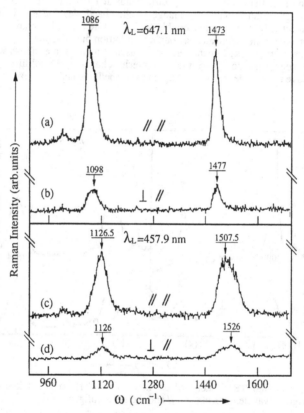

Fig. 6 . Raman spectra of the triblock copolymer $(CH)_{88}$-$(NBE)_{415}$-$(CH)_{88}$ for two different wavelengths in the // // and the $\perp\perp$ configurations.

Infrared Photoinduced Absorption (PIA)

In trans-$(CH)_x$, fundamental information has been gained concerning the photogenerated charges with are pinned at defect sites[14-16]. One observes a photogenerated electronic absorption band peaked at 3500-4200 cm^{-1} together with photoinduced vibrational bands peaked at 1370, 1282 and 500-600 cm^{-1}. Most of the studies have been carried out on disordered samples[14-16], but more recently, further experiments have been reported by Pichler and Leising[17] on oriented $(CH)_x$. As an example, we show on Fig. 8a the PIA absorption of fully trans-$(CH)_x$, with the pump beam (514.5 nm) polarized perpendicular to the stretching axis whereas the probe beam has been set parallel to this axis. A systematic study was carried out on oriented $(CH)_x$ films with different cis-trans ratios[17].

In the case of the triblock copolymers $(CH)_x$-$(NBE)_n$-$(CH)_x$, the PIA spectra show features which are dependant upon the x and n values. Figs.7b and c show such features in the range 400-2000 cm^{-1}. We observe, as in trans-$(CH)_x$ the sharp vibrational infrared modes at 1370 and 1287 cm^{-1} in both copolymers $(CH)_{75}$-$(NBE)_{100}$-$(CH)_{75}$ and $(CH)_{88}$-$(NBE)_{415}$-$(CH)_{88}$. The main difference is the broad band peaked at ~ 500-600 cm^{-1} in the frequency region 700-800 cm^{-1}.

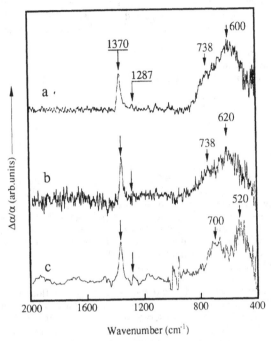

Fig. 7 . Photoinduced Infrared Absorption spectra taken at 77 K with the pump beam perpendicular and the probe beam parallel to the stretching axis:
a) fully trans-(CH)x, b) $(CH)_{88}$-$(NBE)_{415}$-$(CH)_{88}$,
c) CH_{75}-$(NBE)_{100}$-$(CH)_{75}$.

In the case of $(CH)_{75}$-$(NBE)_{100}$-$(CH)_{75}$, we observe a well resolved structure composed of a double peak at 700 and 520 cm^{-1} different from the shape observed in curve 7b which is approximately similar to trans-$(CH)_x$. Notice also that the peak of the LE (Low Energy) band is recorted at 0.51 eV in trans-$(CH)_x$ and at 0.57 eV in the triblock copolymer characterized by n=100 and x=75.

DISCUSSION AND CONCLUSION

We have reported in this paper a series of optical experiments carried out on oriented polyacetylene and oriented triblock copolymers containing polyacetylene segments, $(CH)_x$-$(NBE)_n$-$(CH)_x$. The purpose is to study the anisotropic properties of such polymers and an interpretation can be proposed by using a model which is based on the electronic and vibrational properties of conjugated segments of different lengths. Most of these properties have been explained in the case of oriented trans-$(CH)_x$, for which the analysis of the Raman spectra leads to the introduction of a bimodal distribution of conjugated segments[9]. The distribution of long conjugated segments stands for those which contain more than 30 double bonds (N) and the distribution of short ones refers to those with N < 30. The relative concentration of long and short segments are then weighted by a factor G. In many different cases, the bimodal distribution introduced to explain the features of the RRS spectra has been really successful and in particular in the interpretation of Raman spectra recorded in polarized light in stretch oriented samples[11-13]. Let recall that this theory takes into account the electron-phonon interaction in the electronic excited state as well as the electric dipole moments M_n for the electronic transition and he relative energies of the conjugated segments of different lengths. It was shown that the electric dipole moment component parallel to the axis of the chain reaches a constant value for $N = 30$. On the other band, the component perpendicular to the chain axis has a contribution for $N < 10$ with a maximum value at $N = 6$. As a consequence, a light polarization perpendicular to the conjugated segments will lead to a non-negligible electronic transition when those segments have less than 10 double bonds. Details on the calculations have been published elsewhere and a clear interpretation of the depolarization of the Raman bands as well as of their shift in position was given [11-12].

The study of oriented copolymers $(CH)_x$-$(NBE)_n$-$(CH)_x$ was carried out with the same procedure. From the Raman data analysis, the parameters of the double distribution of conjugated segments has been determined as follows for two copolymerts :

x=88	$N_1 = 40$	$N_2 = 13$	$\sigma_1 = 20$	$\sigma_2 = 6$	$G = 0.45$
x=75	$N_1 = 40$	$N_2 = 13$	$\sigma_1 = 20$	$\sigma_2 = 6$	$G = 0.35$

N_1 is the peak of the long conjugated segments distribution and N_2 is the peak of the short ones distribution. It turns out that the three different copolymers which were studied ($x = 75$, $n = 100$; $x = 88$, $n = 415$; $x = 115$, $n = 220$) give very close results, in agreement with optical absorption data. The peaks of the distributions are found equal and only the weight factor G is slightly different. Compared to previous results obtained in fully trans-polyacetylene, these parameters are characteristic of a polymer with rather short conjugated segments and this result confirm the fact that no double peak structure is observed in the Raman spectra taken with an excitation wavelength at 457.9 nm. This is further corroborated by the results obtained in polarized light since the shift in the Raman bands recorded in crossed polarization is much less than usually observed.

In the case of the photoinduced infrared absorption spectra, this model leads also to consistent results. Two points can be mainly discussed. First of all, the photogenerated electronic absorption (the LE band) is observed at 0.51 eV in trans-$(CH)_x$ and shifted to 0.57 eV in the triblock copolymers. This band is attributed to transitions towards a narrow level inside the band gap of the polymer and is therefore directly related to the $\pi \rightarrow \pi^*$ electronic transition value and the shorter the conjugated segments, the higher this value.

The second feature is the infrared photoinduced vibrational mode at 500-600 cm^{-1}. It is known that in this region, sharp features may also observed in cis-trans $(CH)_x$ films[15-17]. Different models have been proposed to explain these data, including thermal modulation effects of the infrared absorption of the remnant cis-$(CH)_x$ due to the (C-H) out-of-plane deformation. In another approach, i.e. in a model using the perturbed Green's function formalism[20], one can calculate the perturbation induced by the trapped charges on the lattice dynamics of the conjugated segments through a parameter Λ which represents the change of the force constant compared to the unperturbed lattice of the polymeric chain. New vibrational modes can be derived from this model and it turns out that mainly the one located in the 500-700 cm^{-1} is strongly dependant upon the parameter Λ which itself depends on the length of the conjugated segments[21]. By using the bimodal distribution and the G factor which weigths the long segments with respect to the short ones, as derived from the polarized Raman data analysis, one may be able to explain the structure at 700-750 cm^{-1} in the broad band peaked at 500-600cm^{-1} recorded in the photoinduced infrared spectra of the stretched trans-$(CH)_x$ and copolymer samples. Further experimental work is also needed in other polyacetylene type sample to confirm these data.

In conclusion, we have reported experimental data obtained by different optical techniques. Most of them have been analyzed and interpreted by taking into account the electronic and vibrational properties of conjugated segments of different lengths. The bimodal distribution model by introducing a factor G which weights the short segment distribution with respect to the long one leads to results consistent with the experimental observations.

ACKNOWLEDGMENTS

we would like to thank Dr. F. Stelzer for initiating the synthesis of the triblock copolymers on which measurements have been made. One of us (E.M.) wants to thank partial financial support from the bilateral program CNR Italy-france.

REFERENCES

1. G. Leising, Polym. Bull. **11**, 401 (1984).

2. G. Leising, Polym. Commun. **25**, 201 (1984).

3. F. Stelzer, R.M. Grubbs and G. Leising, Polym. (1991) in press.

4. G. Leising, Phys. Rev. **B38**, 10313 (1988).

5. T. Verdon, Thèse Université de Nantes, 1990 (unpublished).

6. H. Kusmany, Phys. Status Solidi b **97**, 521 (1980).

7. D.B. Fitchen, Mol. Cryst. Liq. Cryst. **83**, 95 (1982).

8. S. Lefrant, J. Phys. (Paris), Col. **44**, C3-247 (1983).

9. G.P. Brivio and E. Mulazzi, Phys. Rev. **B30**, 676 (1984).

10. E. Mulazzi, G.P. Brivio, E. Faulques and S. Lefrant
    Solid St. Comm. **46**, 851 (1983).

11. E. Faulques, E. Rzepka, S. Lefrant, E. Mulazzi, G.P. Brivio and G. Leising
    Phys. Rev. **B33**, 8622 (1986).

12. E. Mulazzi, S. Lefrant, E. Perrin and E. Faulques, Phys. Rev. **B35**, 3026 (1987).

13. E. Perrin, E. Faulques, S. Lefrant, E. Mulazzi and G. Leising,
    Phys. Rev. **B38**, 10645 (1988).

14. E. Mulazzi, G.P. Brivio, E. Faulques and S. Lefrant
    Synth. Met. **41**, 1333 (1991).

15. G.B. Blanchet, C.R. Fincher, T.C. Chung and A.J. Heeger
    Phys. Rev. Lett. **50**, 1938 (1983).

16. H.E. Schaffer, R.H. Friend, A.J. Heeger, Phys. Rev. **B36**, 7537 (1987).

17. N.F. Colaneri, R.H. Friend, H.E. Schaffer and A.J. Heeger
    Phys. Rev. **B38**, 3960 (1988).

18. K. Pichler and G. Leising, Europhys. Lett. **12**, 533 (1990).

19. S. Lefrant, E. Faulques, G.P. Brivio and E. Mulazzi
    Solid. St. Comm. **53**, 583 (1985).

20. P. Piaggio, G. Dellepiane, E. Mulazzi and R. Tubino, Polym. **28**, 563 (1987).

21. E. Mulazzi, A. Ripamonti and S. Lefrant, Synth. Met. **41**, 1337 (1991).

DIACETYLENES WITH FORMALLY CONJUGATED SIDEGROUPS :

PRECURSORS TO LIQUID CRYSTALLINE POLYMERS ?

A. Sarkar, K. Nagendra Babu, M.B. Kamath, P.K. Khandelwal
Lalita P. Bhagwat and S.S. Talwar*

Department of Chemistry
Indian Institute of Technology
Powai, Bombay 400 076, INDIA

INTRODUCTION

Recently, there has been hectic research activity in search of organic materials which show promising non-linear optical properties.[1-5] Polydiacetylenes (PDAs) attracted special attention as a candidate material for third order non-linear optical properties (X^3). This interest is due to the fact that PDAs show large third order non-linear optical susceptibility, with a fast response time. Additional supporting and important features of PDAs are that they can be obtained in single crystal form or LB films and it is possible to tune their optical properties through "Molecular and crystal engineering" through a variety of sidegroups.

PDAs are obtained by solid state 1,4-addition polymerisation of monomer crystals of disubstituted diacetylenes (R-C≡C-C≡C-R'). The polymerisation is carried out by exposing the monomer crystals to thermal, photo- or γ-radiation or even pressure. The monomers may be symmetric or unsymmetric depending upon the nature of sidegroups R and R'. Monomers with side groups formally conjugated to the diacetylene moiety are often found to be unreactive towards solid state polymerisation. However, these monomers are found to exhibit liquid crystalline properties on heating above a threshhold temperature, characteristic of each such diacetylene[5-9]. Divinyl diacetylenes (DVDAs) which are unreactive towards solid state polymerisation upon heating undergo a phase transition to a reactive nematic liquid crystalline monomeric phase to yield liquid crystalline polymers. These polymers are reported to show X^3 which are two orders of magnitude larger than quartz[2]. A variety of unsymmetrically substituted diaryl diacetylenes are also reported to show solid-liquid crystal transition which undergo polymerisation subsequently to nematic liquid crystalline polymers[5] . The monomers show enhanced properties whereas none of the respective polymers exhibited detectable second order non-linear optical response.

We have been involved in the synthesis and study of properties of PDAs with formally conjugated sidegroups with a view to tune the optical properties of PDA backbone. Even though a couple of them exhibit interesting optical properties, most of them were found to be unreactive towards solid state polymerisation. In this paper, we report the interesting thermal behaviour exhibited by this class of diacetylenes.

Frontiers of Polymer Research, Edited by P.N. Prasad and
J.K. Nigam, Plenum Press, New York, 1991

EXPERIMENTAL

Synthesis of diacetylene monomers was carried out in our laboratory following different schemes [10,11]. Shimadzu Thermal Analyser DT-30 was used to record thermogravimetric curves. DSCs were recorded using Dupont 910 Differential Scanning Calorimeter equipped with thermal analyst 2000 data processor. Optical microscopic observations were made using Orthoplan Leitz microscope to which a Chaiz Meca heating-freezing stage (-190°C to +600°C)was attached. Photographs were taken with a Orthomat-Leitz camera.

RESULTS AND DISCUSSION

The DSC and DTA/TG of all the compounds reported indicate that they (1a-e, 2a-d, 3a-b) exhibit liquid crystallinity upon heating. Typical DSCs for

three of the diacetylenes are shown in fig. 1. DSC profile of all the compounds are characterised by an endotherm followed by an exotherm. (Table 1). Thermogravimetric analysis showed only negligible (3-4%) or no weight loss at all for all the compounds through out the temperature range (30-400°) studied. All the compounds become transluscent (endotherm) upon heating and undergo irreversible polymerisation reaction (characterised by exotherm upon further thermal annealing. The area of endotherm for

Table 1 . Results from DSC measurements

No.	Sample/colour	Endotherm[a] peak/ΔH	Exotherm[a] peak/ΔH	Temp.gap endo-exo	colour of product
1.	3a / Pale Yellow	159.0/5.75	246.0/48.8	165-210	Dark reddish brown
2.	3b/ Pale Yellow	178.0/6.20	231.0/54.9	182-214	Dark reddish brown
3.	1a / Colourless	91.1/4.99	197.1/35.9	100-135	Shining Deep brown
4.	1b / Colourless	92.2/4.65	224.2/37.5	105-145	Shining Deep brown
5.	2a / Pale Yellow	125.9/4.13	232.8/26.9	130-160	Black
6.	2b / pale Yellow	133.1/6.03	234.2/39.4	140-160	Black
7.	1e / Colourless	173.6/7.7	283.0/44.7	175-240	Shining Black
8.	1d / Colourless	227.0/5.03	252.0/45.5	-	Shining Black
9.	1c / Colourless	123.0/8.5	144.0/40.5	-	Shining Black
10.	2c / Colourless	169.3/4.86	177.5/68.0	-	Shining Black
11.	2d / Colourless	159.0/18.6	167.4/56.4	-	Shining Black

[a]Endo-Exhotherm peaks are in °C,   ΔH is in Kcal/mol.

234

all diacetylenes except 2d is characterised by an enthalpy change in the range of 4.13 to 8.5 Kcal/mol and is assigned to a crystal-liquid crystal transition. Relatively high enthalpy change in case of 2d might be due to the presence of urethane groups which are capable of forming H-bonds. The enthalpy of polymerisation obtained from the area of exotherm is in the range of 35-45 Kcal/mol for most of the diacetylenes which is comparable to that for

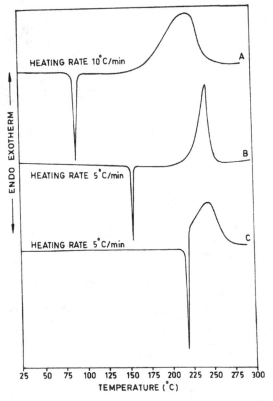

Fig. 1. DSC of (A) 1b, (B) 3a, (C) 1d.

diacetylene polymerisation in the solid state[12]. In case of diacetylene liquid crystals which show relatively large ΔH_p (45-68 Kcal/mol), the polymerisation process might be a multidimentional reaction yielding a network polymer.

The polymerisation reaction is catagorised into two types a) liquid crystalline polymerisation (endotherm followed by exotherm with an intermediate temperature gap) and b) melt polymerisation (exotherm immediately follows endotherm, fig. 1). Our preliminary results from FT-IR and UV-Vis

235

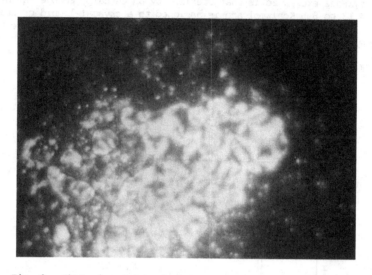

Fig. 2. Photomicrograph of the nematic liquid crystal
monomer of 3a at 190°C between crossed polarisers.
(magnification = 1 x 50)

absorption studies showed that the polymers formed from these types of reactions are different from those formed by solid state reaction.

Repeated cyclic heating of 1a and the resultant DSC's indicate that the liquid crystalline phase is stable and polymerisation occurs in liquid crystalline phase only upon prolonged thermal annealing.

The nature of liquid crystalline phase and subsequent polymerisation was examined under crossed polarisers of an optical microscope. The liquid crystalline phase of two samples 3a and 3b when examined under crossed polarisers indicated that the liquid crystalline phases were nematic. The existence of nematic liquid crystalline phases was confirmed by observation of disclination structures containing two or four brushes (fig.2). Polymerisation of monomers in liquid crystalline phase took place upon further heating to yield reddish-brown polymers. These polymers are found to be insoluble in common organic solvents. The polymers formed were not liquid crystalline unlike the other reported polymers of the class[2]. However this point is under detailed investigation with controlled annealing in the liquid crystalline phase. The response of these monomers and polymers to second order and third order non-linear optical susceptibilities is also being investigated.

CONCLUSION

Diacetylenes with formally conjugated sidegroups undergo thermotropic crystal-liquid crystal transition and undergo polymerisation even though most of these DAs are unreactive towards solid state polymerisation. Crystalline DVDAs upon heating transform into nematic liquid crystalline phases. The polymers formed from respective liquid crystalline phases are however not liquid crystalline polymers. Solid state polymerisation and liquid crystalline polymerisation of these diacetylenes are differentiated.

ACKNOWLEDGEMENTS

Three of us (AS, KNB and LPB) thank CSIR, New Delhi for the award of senior research fellowship. We are grateful to Prof. S.K. Bhatia and Prof. V. Panchapakesan for allowing us to use their differential scanning calorimeter and hot stage microscope respectively. Helpful discussions with Dr. G.N. Jadhav during optical microscopic studies is gratefully appreciated.

REFERENCES

1. D.J. Williams, Angew Chem Int. Ed. Engl., 1984, 23, 690.
2. A.F. Garito, C.C. Teng, K.Y. Wong and O. Zammani Khamiri, Mol. Cryst. Liq. Cryst., 1984, 106, 219.
3. Non linear optical properties of organic molecules and crystals, eds. D.S. Chemla and J. Zyss., Academic Press, New York, vol. 1 & 2, 1987.
4. Claudine Fouquey, Jean-Marie Lehn and Jacques Malthete, J. Chem. Soc. Chem. Commun., 1987, 1424.
5. J. Tsiboukis, A.R. Werninck, A.J. Shand and G.H.W. Milburn, Liquid Crystals, 1988, 3, 1393.
6. G.H.W. Milburn, A. Weninck, J. Tsiboukis, E. Bolton, G. Thomson and A.J. Shand, Polymer, 1989, 30, 1004.
7. G. Hardy, G.H.W. Milburn, K. Nyitrai, J. Horvarth, G. Balazs, J. Varga and A.J. Shand, New Polymeric Mater., 1989, 1, 209.
8. G.H.W. Milburn, C. Campbell, A.J. Shand and A.R. Werninck, Liquid Crystals, 1990, 8, 623.
9. G.H.W. Milburn, in Non-linear Optical Effects in Organic Polymers, eds. J. Messier et.al, Kluwer Academic Publishers, 1989, p.149.
10. S.S. Talwar, M. Kamath, K. Das and U. Sinha, Polymer Commun., 1990, 31, 198; S.S. Talwar, M. Kamath and K.N. Babu, Mat. Res. Soc. Symp. Proc. vol. 173, 1990, 583.
11. A. Sarkar, Sekher P., M.B. Kamath, L. Bhagwat, K.N. Babu, K. Rajalakshmi & S.S. Talwar, Ind. J. Chem., 1991, 30B, 360.
12. G.N. Patel, R.R. Chance, E.A. Turi and Y.p. Khanna, J. Am. Chem. Soc., 1978, 100, 6644.

POLYMERS FOR ELECTRONICS

Roland Darms

Ciba-Geigy AG
Plastics Division
Research Center Marly
1701 Fribourg, Switzerland

INTRODUCTION

The degree of success the electronic industry has had over the last decades is no doubt due to the progress and the achievements in the semiconductor technology, but without polymers and fine chemicals this success would not have been possible. These materials are vital for the manufacture of electronic components. I would like to discuss with you some examples of new polymers, which were developed in our laboratories as base materials for printed circuit boards, as photopolymers for printed circuit boards, as photopolymers for chips and as protective coatings for chips.

BASE MATERIALS FOR PRINTED CIRCUIT BOARDS

The base material most widely used today for preparing printed circuit boards is epoxy resin. A trend towards higher temperature materials can be observed. This is because the package density is getting greater and therefore the operating temperatures encountered in these boards higher. Furthermore because of higher signal speed also polymer materials with low dielectric constant are needed. In order to meet these requirements we are developing a new class of imides, the allylnadicimides as base materials.

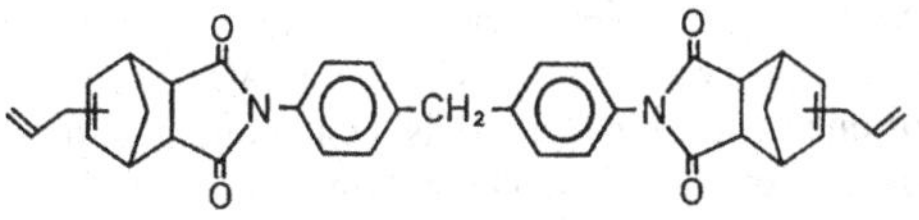

Fig. 1. Allylnadic imide

Frontiers of Polymer Research, Edited by P.N. Prasad and
J.K. Nigam, Plenum Press, New York, 1991

Figure 1 shows one representative of this class of materials. It is a low melting solid, which exhibits excellent solubility in common organic solvents. This is an important processing parameter for the fabrication of printed circuit boards. The good solubility and low melting point may be explained by the fact, that this material is a mixture of isomers. These are formed during the synthesis, which starts from cyclopentadiene. Reaction with allylchloride gives an isomeric mixture of allylcyclopentadiene, which is not separated. In a facile reaction with maleic anhydride the key intermediate allylnadic anhydride is obtained, which is reacted as an isomeric mixture with diaminodiphenylmethan in the usual fashion to give the allylnadicimide represented in Figure 1. With other diamines a large family of novel nadicimides could be prepared following the same reaction sequence. After hardening these materials at temperatures up to 260°C a Tg above 300°C is obtained. The mechanical as well as the thermal properties are excellent. Surprisingly the dielectric constant is fairly low, which is quite unusual for imides. The allylnadicimides can also be combined with other crosslinkable materials, such as e.g. bismaleinimides, which leads to novel network structures with interesting properties.

The majority of printed circuit boards today are rigid boards, but the use of flexible printed circuits is increasing because of space saving. For this application we have investigated high molecular weight polyamide-polyimide blockcopolymers.

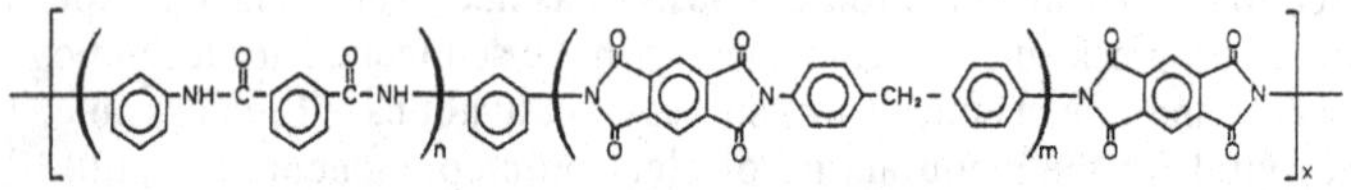

Fig. 2. Polyamide-polyimide blockcopolymer

Figure 2 shows an exemple with aromatic polyamide and polyimide blocks, which exhibits good adhesion on copper and is flame resistant. Both properties are important for printed circuits. Films can readily be cast from the soluble polyamic acid precursor. In contrast to the corresponding aromatic homopolymers the blockcopolymers can also be processed from the melt. Through variation of the chemistry and the molecular weight of the blocks many different types of blockcopolymers can easily be synthesized and tailor made for specific requirements.

PHOTOPOYMERS FOR PRINTED CIRCUIT BOARDS

Photopolymers are used as photoresists in the formation of the conductive pattern of the printed circuit. The principle of a photoresist and the technique how it is employed is well known and will not be discussed here any more. Many types of photoresists are on the market. The trend is towards materials with high photospeed and excellent resolution, because of increasing packaging density the conductive lines are getting closer and closer.

I would like to discuss two examples of photopolymers, which we have developed as photoresists. Both undergo crosslinking when irradiated with UV-light. The light sensitive component of one of these polymers is a chalcone group, which is known to dimerize under the influence of UV-light. The polymer is synthesized starting from the diglycidylether shown in Figure 3, which contains already the light sensitive group.

240

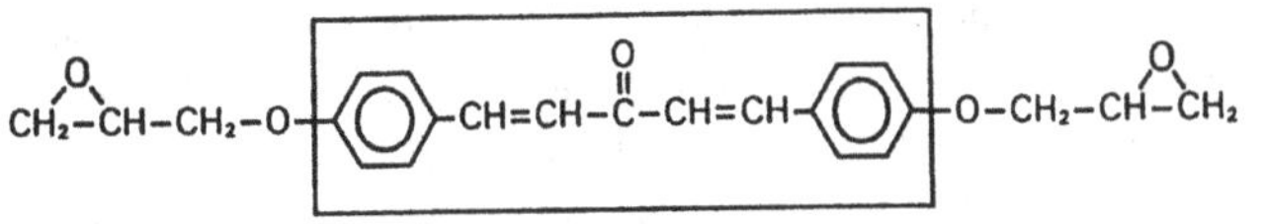

Fig. 3. Diglycidylether with photoreactive chalcone group

This compound is reacted with a diphenol (e.g. bisphenol-A) in the presence of a basic catalyst to give a linear polymer with epoxy endgroups. Molecular weights can easily be adjusted by varying the proportion of diepoxide and bisphenol. This photopolymer shows very high sensitivity in the UV-light and also an excellent resolution. After photocrosslinking the polymer still contains epoxy groups, which can be reacted by heat in a second step and in the presence of a latent hardener to give additional crosslinking. This is of technical importance in applications of this photoresist as a solder mask, where a tough and resistant layer is desired or in multilayer circuit boards, where a tight and firm bond between the conductive and insulating layers can be achieved.

The photoreactive component of the other polymer is a dimethylmaleic imide group, which undergoes a photocycloaddition reaction under UV-light. When these imide groups are attached along a polymer chain the cycloaddition leads to crosslinking of the polymer. Starting material for this polymer is the monomer shown in figure 4, which has in addition to the dimethylmaleic imide group a polymerizable methacrylate group in the same molecule.

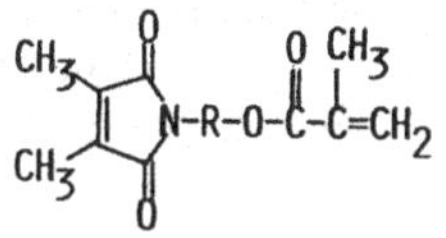

Fig. 4. Polymerizable, photoreactive dimethylmaleic imide

It can readily be polymerized or copolymerized, because the double bond of the dimethylmaleic imide group does not undergo radical polymerization. In this way many types of different copolymers can be easily synthesized depending on the type of application. One requirement, which has become increasingly important for environmental reasons, is the developability of a photoresist in aqueous solutions.

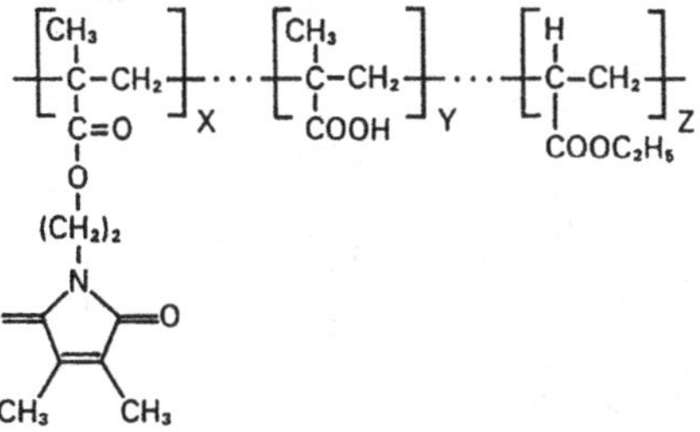

Fig. 5. Aqueous developable dimethyl maleicimide photopolymer

We therefore have copolymerized methacrylic acid with our photoreactive monomer to form the photopolymer shown in Figure 5, which not only exhibits high photosensitivity but can also be developed readily in aqueous base.

Not just photocrosslinking of polymers is used as a reaction principle in photoresists for printed circuits, but photopolymerization of monomers can also be used. We are investigating the photopolymerization of epoxy resins with the help of new UV-photoinitiators. Figure 6 shows such an initiator.

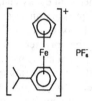

Fig. 6. Cationic photoinitiator

When irradiated with UV-light, it forms a cationic species, which initiates a cationic polymerization of the epoxy resin. On irradiation through a mask polymerization selectively takes place only in those areas, where the UV-light comes through. These systems are very versatile, because epoxy resins can readily be formulated with other polymerizing monomers.

Beside UV sensitive photoresists there is increased interest in laser sensitive materials for printed circuit boards. We are investigating acrylate formulations with photoinitiators, which absorb in the laser region. The formulations, which also contain acrylic acids for aqueous development, show high photosensitivity. With the help of CAD the desired pattern on the photoresist can readily be achieved with laser direct writing.

PHOTOPOLYMERS FOR CHIPS

Photosensitive materials are also needed for the chip manufacture. The requirements though are slightly different from those in the printed circuit field, because the resolution has to be much smaller. The trend here is towards submicron resolution. Todays most widely used photoresists in this area are based on novolac/diazoquinone systems. They are aqueous developable and work with UV-light. The trend in this field of photopolymers is towards higher resolution than achievable with these systems. In order to come way down into the submicron region newer developments aim at resists which work with deep-UV, electron beam and also X-ray as source for irradiation. We have developed a new deep-UV resist, which is shown in Figure 7.

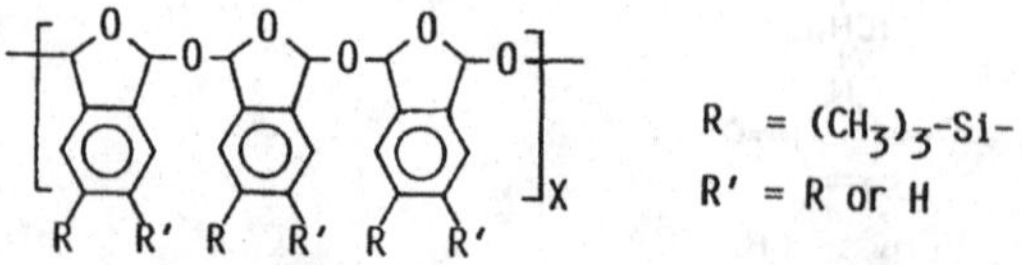

$$R = (CH_3)_3-Si-$$
$$R' = R \text{ or } H$$

Fig. 7. Dry developable, plasmaresistant phthalaldehyde photopolymer

This polyacetal, which contains silane groups, depolymerizes when irradiated in the presence of an acid generator. The reactive principle is not photocrosslinking or photopolymerization as discussed in the resists for printed circuits, it is in this case depolymerization. The acid generator, which is added to the polyacetal, produces an acid on irradiation with deep-UV, which then catalyses depolymerization on all those places, where the light hits. The polymer depolymerizes to a monomeric silane containing phthalaldehyde. Since this monomer is very volatile, it can easily be removed after irradiation of the resist simply by heating. No solvent is needed for development. It is a dry developable system. This resist is applied in combination with a polyimide, i.e. a layer of polyimide is coated onto the silicium wafer and the photoresist on top of that. After development the polyimide is etched away with oxygen plasma on those areas which are not protected by the resist. For this reason the resist has to be plasma resistant. This is achieved by the built-in silane groups. Under the action of the plasma these groups are converted into silicon dioxide which make the resist plasma resistant.

PROTECTIVE COATINGS FOR CHIPS

The use of organic polymers as protective coatings in chip manufacturing is increasing. These materials need to be thermally and dimensionally very stable as well as easily applicable from solution to form thin coatings. We have prepared the polyimide shown in figure 8 from diaminophenylindane and benzophenone dicarboxylicacid dianhydride, which fulfills the requirements for this application.

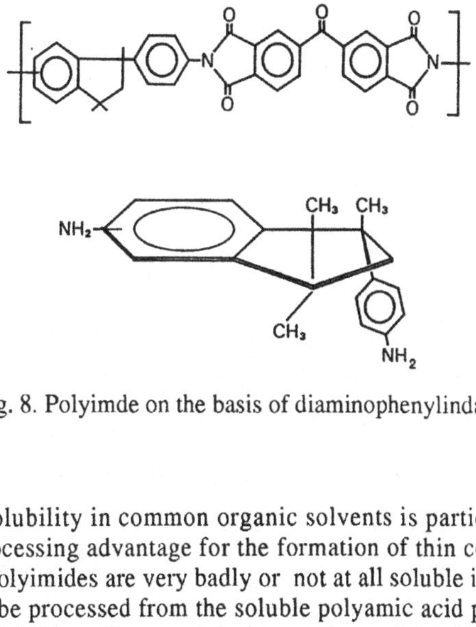

Fig. 8. Polyimde on the basis of diaminophenylindane

Its excellent solubility in common organic solvents is particularly noteworthy because it is a processing advantage for the formation of thin coatings. Most of the known aromatic polyimides are very badly or not at all soluble in the imideform and therefore have to be processed from the soluble polyamic acid precursor. We think, that the good solubility of our polyimide can be explained by the specific steric structure of the diaminophenylindane also shown in Figure 8. The non planar conformation of the diamine prevents the polymer of assuming a higher molecular ordering. In addition to that it is employed as an isomeric mixture in the polycondensation reaction. Specific applications for this polyimide are as a planarization material or as an inner-layer dielectric.

A second type of polyimide developed for use as a protective coating is shown in Figure 9.

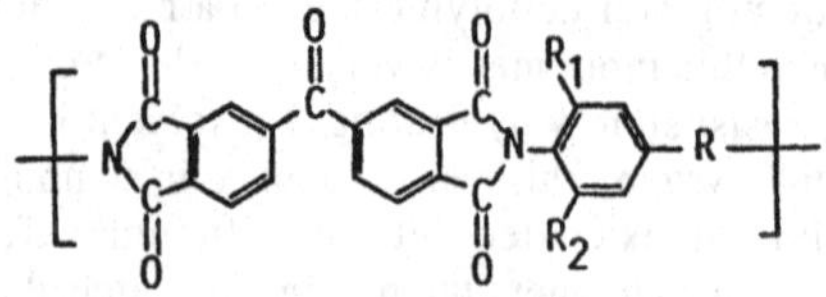

R_1, R_2 = Alkyl

Fig. 9. Photosensitive polyimide

This material is photo imageable, a feature which is important in all those applications where only some specific areas on the chip have to be protected. The photosensitivity is based on the structural element, in which alkyl groups stand ortho to the imidebond. An alkyl group in one ortho position is sufficient, with groups in both ortho positions the photosensitivity is better, without alkyl substituents in these positions there is no photosensitivity at all. The resolution on imageing is very good. This polyimide is also well soluble in the imide form and shows excellent thermal and dimensional stability.

A VIEW ON INDUSTRIALIZATION OF CONDUCTING POLYMERS

LOOKING OUT OF POLYMER BATTERY DEVELOPMENT

Tsutomu Matsunaga

Research and Development Division
Bridgestone Corporation
Kodaira-shi Tokyo 187 Japan

Present Address, 866-221 Araku Iruma-shi Saitama-ken 358 Japan

INTRODUCTION

After a plateau of several years, R&D on conducting polymers is now coming up to a new stage. In 1987, a series of polyaniline-lithium batteries was commercialized by Bridgeston and Seiko in Japan. That was the first industrial application of the conducting polymer. In 1987 Naarman with BASF in Germany developed polyacetylene with conductivity as high as copper and silver. That was a epochal break-through on the conductivity of polymers. In 1990 polyacene-lithium capacitors were on market by Kanebo in Japan. In the same year, a new type of aluminum electrolytic capacitors replacing the liquid electrolyte by polypyrrol was commercialized in Japan.

Corresponding with such a technological progress about conducting polymers, papers submitted to ICSM 1990 held at Tübingen in Germany were mainly concerned with applied research on conducting polymers. There was much difference from ICSM 1986 held at Kyoto, Japan where very fundamental things of conducting polymers were solely discussed. Additionally, the industrial products exhibition of conducting polymers were planned at ICSM 1990 for the first time. However, only two industrial companies of Japan and Germany exhibited their products. That represents correctly the state of art of the conducting polymer for the time of being. The conducting polymers are now not merely exotic materials but surely approaching the gateway of industrialization.

In this paper a review and a preview on the industrial application of conducting polymers, especially in detail on polymer batteries which have been expected to take a role of locomotive engine for the industrialization of conducting polymers.

DEVELOPMENT OF POLYMER BATTERIES

Marketing Prediction

The first polymer battery was proposed by Jozefowicz with CNRS, France in 1967. That was a polyaniline-aqueous sulfonic acid system. It was really a big event to be memorized because the conducting polymer was first proved to be electrochemically active.

However Jozefowicz's battery did not attract much attention because of no outstanding performance comparing with conventional batteries. Furthermore Jozefowicz himself did nothing on that matter after the first discovery and invention, and his pioneering work had been forgotten more than ten years until MacDiarmid et al with Pennsylvania University, USA invented rechargeable lithium batteries applying various polymer electrodes in 1981.

The invention of MacDiarmid et al was developed from the systematic research on conducting polymers and the discovery that polyacetylene prepared by Shirakawa's procedure and the many other conjugated polymers were electrochemically active in non aqueous electrolytes. Polymer industry showed intensive concern in the work of MacDiarmid et al, because they supposed polyacetylene or some other conjugated polymers would be promising materials for lithium rechargeable batteries, which had been considered as the battery of the next generation of cordless electronics era. They saw some substantial things on the polymer lithium battery more than a kind of novel batteries. Thus several industrial projects were launched in USA, Germany, Japan and so on with the objective to develop polymer lithium batteries, especially polyacetylene lithium rechargeable batteries.

The lithium batteries was originally developed as the battery of high performance for military use. In the mid of 1970s, primary lithium batteries for non military use were commercialized in Japan. They were used for power source of watches, cameras, calculators, IC memory back-up and many other electro-appliances. The market size of the primary lithium batteries have been getting bigger and bigger in these decades as shown in Fig. 1. The growth rates were 15-30% per year.

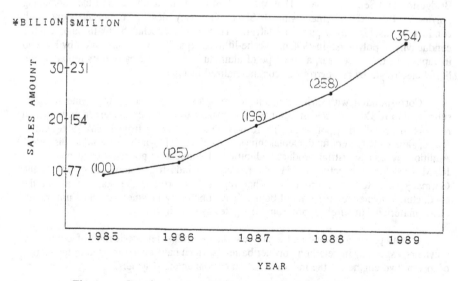

Fig. 1.   Growing market of primary lithium batteries in Japan

Thinking that battery electronic application of bigger size like word processors, micro computers, video cameras etc. equipped with big IC memories rush into market, we should take it a matter of course that requirements to batteries would be revolutionally severe.

Neither the conventional rechargeable batteries nor even lithium primary batteries may afford to such a high requirement. That is why the development of rechargeable lithium batteries of quite high performance have been seriously demanded.

Japanese market size of small rechargeable lithium batters for various electro-appliances is expected like what is shown in Fig. 2. It does not include large batteries to be applied to electric vehicles etc. According to the prediction, the market of ¥7billion or $5million is expected for the coin-type and the market of ¥30billion or $230million will be available for cylinder-type of lithium rechargeable batteries.

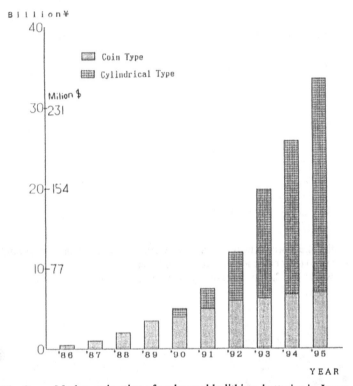

Fig. 2.   Market estimation of rechargeable lithium batteries in Japan

Assuming more than 50% of the market might be shared by polymer lithium batteries, the consumption of conducting polymers for the electrodes would be more than 300 t/year or 7million lbs/year. It depends on whether rechargeable polymer lithium batteries of bigger size than coin type ones are commercially available. If the polymer battery stays in the category of the coin type, the consumption of conducting polymers for battery electrodes would not exceed several tons per year. Then the polymer battery would be only a specialty market of conducting polymers and never would be the locomotive engine or the big driving force for the industrialization of conducting polymers.

General Performance Of Polymer Batteries

Table 1 shows the basic performance of several prototypal polymer lithium batteries comparing with conventional rechargeable batteries. Some advantages of polymer lithium batteries over the conventional ones, that is, high voltage and high energy density etc. could be seen.

Table 1. Basic Performance of Polymer Lithium Batteries

Cell	Open circuit voltage/V	Energy density/Wh Kg⁻¹
PA/LiClO₄ · PC/Li	3. 7	290
PPP/LiClO₄ · PC/Li	4. 4	320
PPy/LiBF₄ · PC/Li	3. 2 ~ 3. 5	290~360
PTh/LiBF₄ · PC/Li	3. 9	280
PAn/LiBF₄ · PC/Li-Al	3. 5 ~ 3. 7	370~400
MnO₂/LiClO₄ · PC/Li	3. 9	450
PbO₂/H₂SO₄ · H₂O/Pb	2. 1	175
Ni/KOH · H₂O/Cd	1. 3	210

Regarding the power density, polyacetylene seems to be excellent and polyaniline also seems good, due to their extended surface reaction area which comes from their fine fibrillar texture as shown in Fig. 3. On the contrary, polypyrrole is poor in that point because of the dense morphology.

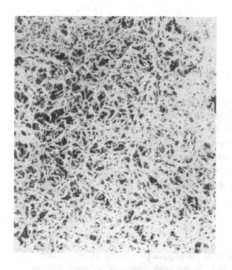 

Fig. 3. SEM pictures of polyaniline synthesized electrochemically from aqueous solution of 1 mol dm⁻³ aniline and 2 mol dm⁻³ HPF4

Coulombic efficiency or reversibility of electrode reaction is a ruling factor of the performance of batteries. All the polymer electrodes have high efficiency in the range of shallow charging and discharging as shown in Fig. 4. Especially polyaniline electrode is ideally good in wide range of the reaction.

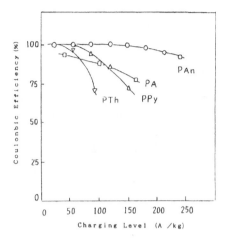

Fig. 4.   Coulombic efficiency of polymer batteries
          PAn: Polyaniline   PA: Polyacetylene   PPY: Polypyrrole
          PTh: Polythiophene

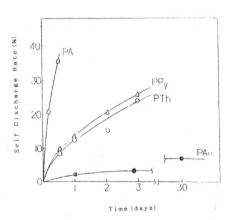

Fig. 5.   Self discharge of polymer batteries
          Charging: 30 Ah/Kg for PA, PY, PTh and 120 Ah/Kg for polyaniline

Self discharge is generally considered to be the weakest point of polymer batteries, judging from the case of polyacetylene lithium battery. However that is not the case of all the polymer batteries. For instance, self discharging of polyaniline lithium battery is very small even comparing with conventional batteries as shown in Fig. 5.

Designing And Performance Of The Commercialized Polyaniline Lithium Battery

Judging from things described above, polyaniline is surely a most suitable material for polymer electrodes. Polyaniline have been known since more than 100 years ago. However

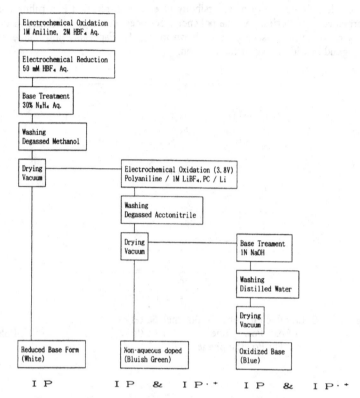

Fig. 6.　　Preparation procedure of multifarious polyanilines

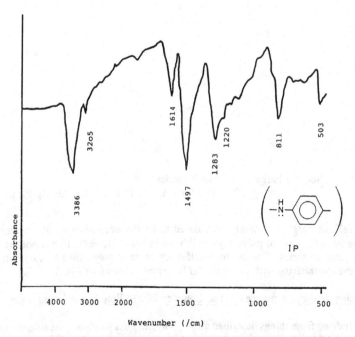

Fig. 7.　　IR Spectrum of reduced polyaniline (IP)

its molecular structure was not clearly defined because it took various structure and various colour, depending on the condition of synthesis and conditionings after synthesis, like shown by Fig. 6.

In order to design polyaniline electrodes of optimum molecular structure, identifications of polyaniline at various stages and analytical description of charge and discharge mechanism in non-aqueous electrolytes should be done. First of all, thoroughly reduced state of polyaniline was prepared by the procedure described in Fig. 6, which was opaque white and denoted as IP. Polyaniline IP was proved to be composed of imino-1,4-phenylene units by infrared spectroscopy and elemental analysis (C~78.9% H~5.5% N~15.2% O~0.4%).

Fig. 8 represents a typical voltammogram of IP and metallic lithium in a non-aqueous electrolyte.

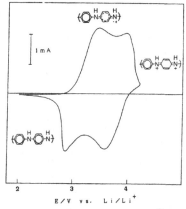

Fig. 8.   Cyclic voltammogram of polyaniline in 1 mol dm^{-3} LiClO$_4$/Propylene Carbonate
Potential sweep rate: 1 mV s^{-1}

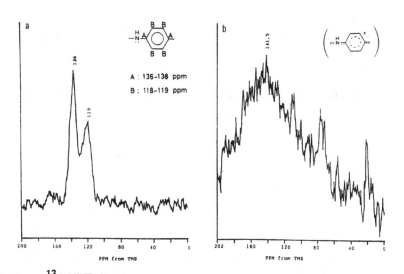

Fig. 9. ^{13}C NMR Spectra
(a) Reduced stage of polyaniline   (b) 1st oxidized stage of polyaniline

251

Molecular structural change of polyaniline from the starting point to the first oxidized stage of the voltammogram was traced spectroscopically. Fig. 9 represents ^{13}C NMR of the reduced stage and the charged stage of polyaniline. At the reduced stage, a beautiful spectrum representing imino-1,4-phenylene units was obtained. At the first oxidized stage by electrochemical charging, nuclear magnetism was shielded by scattered electrons and the spectrum was disturbed. Polyaniline at the oxidized stage was electro-conductive owing to delocalized electrons.

XPS of N_{1s} of polyaniline at reduced and first oxidized stage are shown respectively in Fig. 10 (a) and (b). By deconvolution of Fig. 10 (b), it was concluded that 50% polyaniline IP was oxidized and transformed to cation radicals.

Based on these and those analytical works, the scheme of interconversion of polyaniline and the mechanism of charge/discharge was described as shown in Fig. 11.

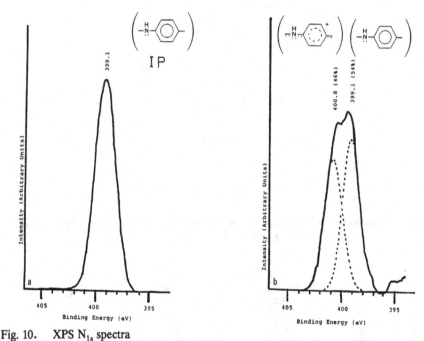

Fig. 10.   XPS N_{1s} spectra
(a) Reduced stage of polyaniline   (b) 1st oxidized stage of polyaniline

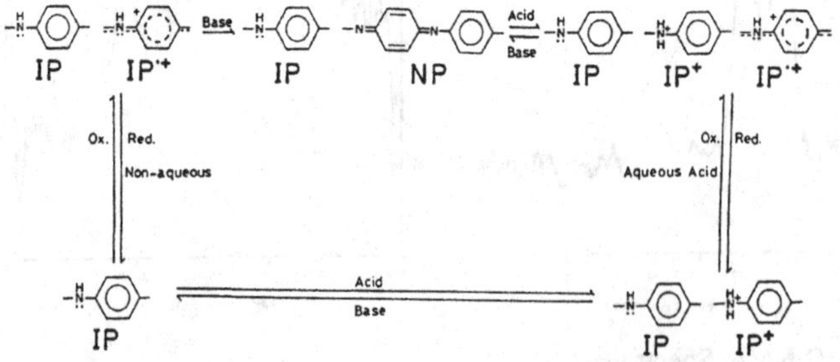

Fig. 11.   Interconversion scheme of polyaniline

According to the clarified mechanism of charge and discharge in non-aqueous electrolytes, polyaniline electrodes composed of IP units and its cation radicals, not to include quinoid units nor salt units, were prepared. Then a rechargeable polyaniline lithium battery of corn type like shown below was commercialized.

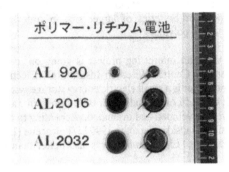

Fig. 12. Commercialized rechargeable polyaniline lithium batteries

Charge voltage was cut off at 3.0V or 3.3V. That means an ion doping level by charging is only around 30% in order to prevent decomposition of electrolytes solution which is $LiBF_4$/propylene carbonate solution. Table 2 represents general performance of commercialized grades of the battery.

Table 2. Specification of Commercialized Rechargeable Polyaniline Lithium Batteries

		AL614	AL920	AL2016	AL2032	AL2032-20
Dimension Diameter (mm)		6.8	9.5	20	20	20
Thickness(mm)		1.4	2.0	1.6	3.2	3.2
Weight	(g)	0.2	0.4	1.7	2.6	2.7
Nominal Voltage	(V)	3.3	3.0	3.0	3.0	3.3
Nominal Capacity	(mAh)	0.3	0.5	3.0	8.0	20.0
		(3.3~2V)	(3.0~2V)	(3.0~2V)	(3.0~2V)	(3.3~2V)
Cycle Life Depth	(mAh)	0.03	0.1	1	3	1
Life(cycles)		≥ 1000	≥ 1000	≥ 1000	≥ 1000	≥ 1000
Standard Current	(A)	1~50μ	1μ~1m	1μ~5m	1μ~5m	1~100μ
Recomended Charging Method		Constant Voltage Charging				
Operating Temperature		-20~+60°C				

The polyaniline lithium battery is highly evaluated by customers because of high reliability, stability and long life. The battery is mainly applied to IC memory back up where floating performance is important. The polyaniline lithium battery shows very little decay of the capacity after being kept at full charge to nominal voltage even at rather high temperature. The battery is guaranteed self proof to over discharge. After seven days of short circuit discharge, the battery recovered the capacity to the original level by recharging.

Development Of Polymer Battery Of Large Capacity

Polymer batteries of coin type are steadily growing in the market. The next target should be to develop polymer batteries of larger size in order to drive forward the industrialization of conducting polymers.

On that point of view, an interesting project is going on in Japan which is Load Conditioner Development by Central Research Institute Of Electric Power Industry in Tokyo, Japan. Load Conditioner is a small electric energy storage system for residential use. The specification of unit cells of Load Conditioner is described in Table 3. Several lithium rechargeable batteries were developed and evaluated. According to the paper presented to The 31st Battery Symposium at Osaka, Japan in 1990 Li/Polyacene (Li/PAS) was evaluated to be very promising (Table 4). Li/Polyaniline was also better than Li/BMoS2.

Table 3. Specification of Unit Cell for Lord Conditioner of Residential Use being developed by Central Research Institute of Electric Power Industry, Tokyo

Lord Conditioner : 8kwh/3kw

Requirements as to Unit Cells

Gravimetric Energy Density	~ 70Wh/kg
Volumetric Energy Density	~ 120Wh/1
Power Density	~ 20W/kg
Coulonbic Efficiency	> 80%
Cycle Life	~ 2500

Table 4. Performance of Prototypal Unit Cells for Load Conditioner of Central Research Institute of Electric power Industry, Tokyo

	Cut-off Voltage [V]	Mean Discharge Voltage [V]	Discharge Capacity [Ah/kg]	Discharge Energy [Wh/kg]
Li/PAS	4/2 (4. 2/1. 6)	2. 88 (2. 64)	102 (152)	291 (402)
Li/PAn	4/2. 5	3. 55 ~	8. 7 ~	232 ~
Li/βMoS₂	2. 2/1. 3	1. 75	97	170

With regard to capacity of polyaniline electrode a favorable thing was presented at The 30th Battery Symposium in Nagoya, Japan in 1989 by research group of Bridgestone and Tokyo University of Agriculture And Technology. Using quartz crystal microbalance, they followed mass change of polyaniline electrode by charge and discharge. In aqueous system mass of polyaniline electrode was getting increase by doping on the 1st oxidation stage and decrease by deprotonation on the 2nd oxidation stage (Fig. 13a). On the contrary in non-aqueous system, mass of polyaniline electrode monotonically increased by charging (Fig. 13b). In the latter case, the oxidation was proved to be induced by the same doping mechanism on both 1st and 2nd stages. The mass increment of the polyaniline electrode at 1st and 2nd oxidation peaks of the voltammogram were proportional to 50% doping and 100% doping per phenylen imine units.

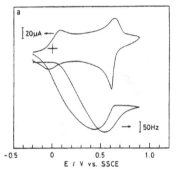 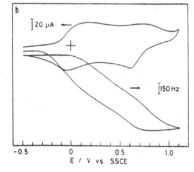

Fig. 13.   QCM frequency response and cyclic voltammogram of polyaniline electrode
(a) 0.5 mol dm^{-3} NaClO$_4$ aqueous solution (PH = 1)
(b) 0.5 mol dm^{-3} LiClO$_4$ / Acetonitrile solution

Mass Sensitivity = 13.3 x 10^{-9} g cm^{-2} Hz^{-1}

From the experimental results mentioned above, it was concluded that reversible doping to polyaniline was available up to 100% that was as twice as expected before. Therefore the highest theoretical energy density of polyaniline is estimated roughly 1000WH/Kg, that is 5 times higher than the case of commercialized polyaniline battery. Polyaniline might be now a most excellent material for electrodes of large size lithium batteries comparing with any other material.

EXTENDING SCOPE OF INDUSTRIAL APPLICATION OF CONDUCTING POLYMERS

Poor processibility had prevented the industrialization of conducting polymers. However intensive applied research on conducting polymers are now breaking off the processing barrier, and research on industrial applications of conducting polymers except the battery seems getting active like shown in Fig. 14.

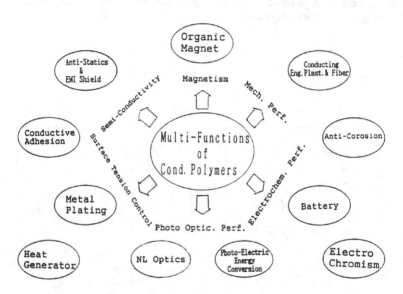

Fig. 14.   Expanded Technology Map of Conducting Polymers

Regarding applications to make the most of conductivity or semi-conductivity of the polymer, Hitachi and Shôwadenkô, two major companies in Japan, released a favorable result of their technical cooperation that the resolution of laser lithography was greatly improved by wet coating of a conducting polymer over the photoresist layer. That should be highly evaluated by micro-electronics industry. Polymer transistors are also under development elsewhere in Japan, France etc., especially F. Garnier et al with CNRS in France announced that an all-plastic transistor was developed for the first time in 1990. Many other things like EMI shielding materials etc. have been intensively investigated.

As to applications of the electrochemical function of the conducting polymers, a new type of Aluminum electrolytic capacitor was commercially developed by Marukoh-Denshi Co./Nippon Kahrito Co. in Japan in 1990. Instead of ionic conductive electrolyte was electronic conductive polypyrrole used in the capacitor which showed excellent performance in high frequency region. That was the second practical application of the conducting polymer coming after the polymer battery. Research on electrochromism induced by electrochemical doping-undoping is still active and expected some practical progress.

Fibers and films are now brought into new focus of conducting polymer research, because not only conductivity of the polymers are tremendously increased by stretching but also their mechanical and thermal properties are superior comparing with conventional liquid crystaling fibers or super engineering plastics. S. Tokito, A.J. Heeger et al and T. Ohnishi et al are independently working with fibers of arylene vinylenes. A.G. MacDiarmid, A.J. Epstein et al are mainly involved in films of polyaniline.

Optical non-linearity of conducting polymers is a most interesting subject of long range research. Several cooperative research projects for conducting polymers to be applied to non-linear optics already started in Japan. A big market for conducting polymers in that field is expected.

CONCLUSION

The market of the polymer battery of coin type is steadily growing in Japan due to the stability, reliability and other outstanding performance of batteries. However, the industrial development of polymer batteries of large capacity is required to increase consumption of conducting polymers up to some industrial scale. Recent progress of research work on electrochemical behavior of conducting polymers suggests a promising future of the polymer battery of large capacity competing well with inorganic lithium rechargeable batteries and so on.

Many industrial applications of conducting polymers other than batteries are also optimistically developing as well, because the conducting polymers are so attractive materials of multi-function.

SCIENCE AND TECHNOLOGY OF CONDUCTING POLYMERS

Alan G. MacDiarmid

Department of Chemistry
University of Pennsylvania
Philadelphia, PA 19104-6323
USA

Arthur J. Epstein

Department of Physics and
Department of Chemistry
The Ohio State University
Columbus, OH 43210-1106
USA

ABSTRACT

The applicability of the concept of "doping" is the unifying theme which distinguishes a certain class of organic polymers - "conducting polymers" - from all others. Doping results in dramatic electronic and magnetic changes with a concomitant increase in conductivity to, or approaching, the metallic regime. Doping phenomena and the chief types of dopable organic polymers are described with particular emphasis on polyaniline which is presently being commercialized on a relatively large scale and is the leading conducting polymer for technology, although closely followed by polythiophene derivatives. Polyaniline shows considerable promise for electromagnetic interference (EMI) shielding and is already used in commercial rechargeable batteries. Leading potential technological applications utilize polyaniline film membranes for gas separations and polyphenylenevinylenes as light-emitting diodes. Additional potential applications of conducting polymers such as electrochromic windows, redox capacitors, chemical sensors, etc. are also described briefly.

INTRODUCTION

An intrinsically conducting polymer (ICP), more commonly known as a "synthetic metal," is an organic polymer that possesses the electrical, electronic, magnetic and optical properties of a metal while retaining the mechanical properties, processibility, etc., commonly associated with a conventional polymer,[1]. These properties are intrinsic to the doped material. This class of polymer is completely different from "conducting polymers" which are merely a physical mixture of a non-conductive polymer with a conducting material such as metal or carbon powder. They are synthesized by "doping" an organic polymer, either an insulator or semiconductor, having a small conductivity, typically in the range 10^{-10} to 10^{-5} S/cm, to a material which is in the "metallic" conducting range (~ 1 to 10^4

Frontiers of Polymer Research, Edited by P.N. Prasad and
J.K. Nigam, Plenum Press, New York, 1991

S/cm). The concept of doping is the unique, central, underlying and unifying theme which distinguishes conducting polymers from all other types of polymers. The controlled addition of known, small ($\leq 10\%$) non-stoichiometric quantities of chemical species results in dramatic changes in the electronic, electrical, magnetic, optical and structural properties of the polymer. Doping is reversible to produce the original polymer with little or no degradation of the polymer back-bone. Both doping and undoping may be carried out chemically or electrochemically. By controllably adjusting the doping level, a conductivity anywhere between that of the undoped (insulating or semiconducting) and that of the fully doped (highly conducting) form of the polymer may be easily obtained. Conducting blends of a (doped) conducting polymer with a conventional polymer (insulator) whose conductivity can be adjusted by varying the relative proportions of each polymer can be made. This permits the optimization of the best properties of each type of polymer.

The "classical" method of doping involves the redox doping, i. e. chemical or electrochemical partial oxidation ("p-doping"), or partial reduction ("n-doping") of the π backbone of the polymer,[1]. More recently, non-redox doping which neither adds nor removes electrons from the polymer backbone has been discovered.

Since the initial discovery in 1977,[2] that polyacetylene, $(CH)_x$, now commonly known as the prototype conducting polymer, could be p- or n-doped, either chemically or electrochemically to the metallic state, the development of the field of conducting polymers has continued to accelerate at an unexpectedly rapid rate. This rapid growth rate has been stimulated not only by the field's fundamental synthetic novelty and importance to a cross-disciplinary section of investigators - chemists, electrochemists, experimental and theoretical physicists and electronic and electrical engineers - but to its actual and potential technological applications.

In the "doped" state, the backbone of a conducting polymer consists of a delocalized π system. In the undoped state, the polymer may have a conjugated backbone such as in trans-$(CH)_x$, which is retained in a modified form after doping, or it may have a non-conjugated backbone, as in polyaniline, (leucoemeraldine base form), which becomes conjugated only after p-doping. More recently it has been observed that a form of polyaniline (emeraldine base) can be doped by a non-redox process,[3]. This is accomplished by simply protonating the imine nitrogen atoms of the polymer to produce a polysemiquinone radical cation in which both charge and spin are delocalized along the polymer backbone. Protonic acid doping has subsequently been extended to systems such as poly(heteroaromatic vinylenes),[4].

For several years after their discovery conducting polymers were regarded as materials whose properties were "answers waiting for the correct question" in so far as technological applications were concerned! This is, however, no longer the case as will be described in a later section. Technological applications may be divided into two classes - (i) that in which a conducting polymer is presently used in, and sold for a given purpose, and (ii) that in which it is not presently being used and sold but which has technological potential ranging from very high to the more exotic cases, where specific applications, if any, may be in the distant future. Some of these will be described in a later section.

At the present time polyaniline is without doubt the most important conducting polymer from the point of view of large scale technological use; however, derivatives of polythiophene and poly(phenylenevinylene) and indeed, polypyrrole, polyacetylene and polyparaphenylene show considerable technological promise as specialty polymers, as described later.

The polyanilines are probably the most rapidly growing class of conducting polymers as can be seen from the number of papers and patents published (1237) during the last five years, viz., 1986 (108); 1987 (221); 1988 (236); (1989 (383); and 1990 (289) (to June 3, 1991),[5]. These figures are due in large part to the very considerable industrial interest in polyaniline. This is highlighted by the 1987 manufacture and sales of novel rechargeable batteries by Bridgestone Corp. (Japan) based on polyaniline,[6]. In February of 1990 Lockheed Corp. (USA) and Hexcel Corp. (USA) announced a joint venture to manufacture polyaniline and its blends with conventional polymers. This was followed in April, 1991,[7,8] with an announcement by Neste Oy (Finland) of the start-up of pilot plant production of polyaniline and polythiophene derivatives to support application development of conducting polymers. In the same month Uniax Corp. (USA) and Neste Oy reported a joint venture for research and development of polythiophenes and polyanilines,[7,8]. In the following month Allied-Signal Inc. (USA), Americhem Corp. (USA) and Zipperling Kessler and Co. (Germany),[9] announced a joint venture to manufacture large quantities of polyaniline and its blends with conventional polymers such as polyvinylchloride and the Nylons, especially for use in electromagnetic interference shielding. Large quantities of these materials are currently available from these companies. In view of the technological importance of polyaniline, its synthesis and properties will be emphasized in this review.

TYPES OF CONDUCTING POLYMERS

Following the discovery of the doping of both cis- and trans-$(CH)_x$,[2], other polymer systems (Table I), such as polyparaphenylene, poly(phenylenevinylene), polypyrrole, polythiophene, polyfuran, etc., and their ring- and N-substituted derivatives were found to undergo redox p-doping and/or n-doping,[1]. Polyphenylene sulfide upon doping with AsF_5 undergoes a chemical reaction to give a doped derivative of polythiophene,[1]. These discoveries were followed by the synthesis of dopable systems involving poly(heterocycle)vinylenes where Y = NH, NR, S and O which proved to be of very great importance to the whole field since, for the first time, they provided processible conducting polymers, some in the doped form, by virtue of their solubility in a number of different organic solvents; furthermore, several of them showed most promising environmental stability in the doped form,[1].

A number of other redox-dopable conducting polymer systems are also known such as those made by doping covalently-linked siloxane-phthalocyanine complexes, $[Si(Pc)O]_x$, but the above systems and their derivatives have been the most extensively investigated,[1].

Although "polyaniline" has been known for about 150 years, it was not until the mid-1980's that intense interest in it and its derivatives, as a completely different type of conducting polymer, really began,[1]. This resulted not only from its ease of synthesis and derivatization, but also from its novel non-redox doping properties and from its potential technological importance.

Table I.   Selected Examples of Redox Dopable Organic Polymers.

Polyparaphenylene	
Poly(phenylenevinylene)	
Polypyrrole	
Polythiophene	
Polyfuran	
Polyphenylene sulfide	
Poly(heterocycle)vinylenes	

REDOX DOPING

Trans-$(CH)_x$, the most extensively investigated conducting polymer, is representative, in a broad sense, of all those conducting polymers which are dopable by redox processes,[1,2]. Its chief features will therefore be outlined below.

Free-standing films of cis-$(CH)_x$ can be synthesized from gaseous acetylene at -78°C in the presence of a $Al(C_2H_5)_3/Ti(OC_4H_9)_4$ catalyst and can subsequently be isomerized at ~150°C to yield lustrous, silvery, films of the more thermodynamically stable trans-$(CH)_x$,

The bonding π system of the polymer can be readily partially oxidized, "p-doped" by a variety of reagents such as iodine vapor or a solution of iodine in CCl_4, e.g.,

$$[CH]_x + 1.5(xy)I_2 \longrightarrow [CH^{+y}(I_3)^-_y]_x \qquad (y \leq \sim 0.07) \qquad (1)$$

with a concomitant increase in conductivity from ~10^{-5} S/cm to ~10^3 S/cm. If the polymer is stretch-oriented 5 to 6 fold before doping, conductives parallel to the direction of stretching up to ~10^5 S/cm can be obtained,[1].

Analogously, the polymer backbone can be partially reduced, "n-doped" by, for example, a solution of sodium naphthalide, in THF, viz.,

$$[CH]_x + (xy)Na^+(Nphth)^- \longrightarrow [Na^+_y(CH)^{-y}]_x \qquad (y \leq \sim 0.1) \qquad (2)$$

with a very large increase in conductivity although values as large as those reported for iodine p-doping have not been obtained. The antibonding π^* system is partially populated by this process.

THE POLYANILINES

The polyanilines,[3] refer to a large class of conducting polymers which exist in three different discrete oxidation states at the molecular level, both in their doped and undoped forms. Physical mixtures of these oxidation states are also readily obtained. One of these discrete oxidation states can be doped by a non-redox process which neither adds nor removes electrons from the polymer π backbone. Until very recently, it was unique amongst all conducting polymers when similar effects, which have not yet been fully investigated, were discovered for some of the poly(hetrocyclovinylenes),[4]. The reduced form of polyaniline can also be doped by a conventional oxidation process,[3].

The interest in this conducting polymer stems from the fact that many different ring- and nitrogen- substituted derivatives can be readily synthesized and that each derivative can exist in a number of different average oxidation states (composed of the three discrete oxidation states) which can in principle be "doped" by a variety of different dopants either by non-redox processes or by partial chemical or electrochemical oxidation,[3]. These properties, combined with the relative low cost of several polyanilines, their ease of processing and satisfactory environmental stability indicate strongly their significant potential technological applicability.

The polyanilines refer to a class of polymers which can be considered as being derived from a polymer, the base form of which has the generalized composition:

and which consists of alternating

reduced, 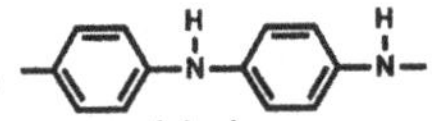 and oxidized, 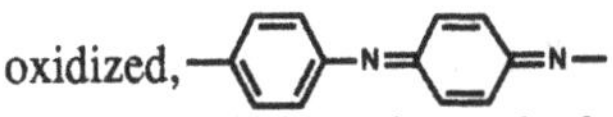 repeat units,[3,10,11].
The <u>average</u> oxidation state, (1-y) can be varied continuously from zero to give the
completely reduced polymer, 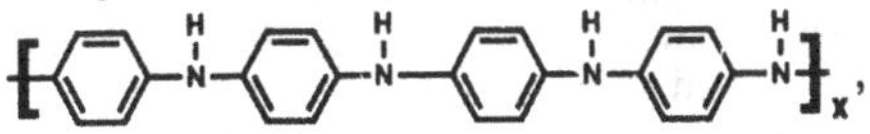 , to 0.5 to give

the "half-oxidized" polymer, 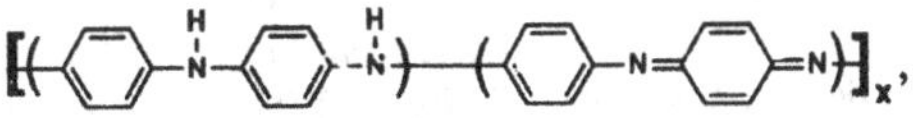 , to one to

give the completely oxidized polymer, 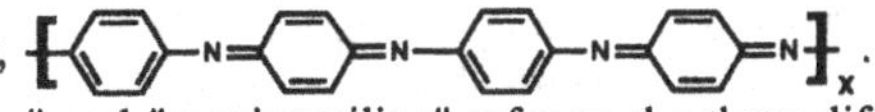 . The
terms "leucoemeraldine", "emeraldine" and "pernigraniline" refer to the three different
discrete oxidation states at the molecular level where $(1-y) = 0$, 0.5 and 1 respectively, either
in the base form, e.g. emeraldine base or in the protonated salt form, e.g. emeraldine
hydrochloride,[3,10,11]. In principle, the imine nitrogen atoms can be protonated in whole or
in part to give the corresponding salts, the degree of protonation of the polymeric base
depending on its oxidation state and on the pH of the aqueous acid. Complete protonation
of the imine nitrogen atoms in emeraldine base by, e.g. aqueous HCl, results in the
formation of a delocalized polysemiquinone radical cation,[3,11,12] and is accompanied by an
increase in conductivity of ~10^{10}.

The partly protonated emeraldine hydrochloride salt can be synthesized easily as a
partly crystalline black-green precipitate by the oxidative polymerization of aniline,
$(C_6H_5)NH_2$, in aqueous acid media by a variety of oxidizing agents, the most commonly
used being ammonium peroxydisulfate, $(NH_4)_2S_2O_8$, in aqueous HCl,[3,10-12]. It can also
be synthesized electrochemically from aniline,[3]. It can be deprotonated by aqueous
ammonium hydroxide to give an essentially amorphous black-blue "as-synthesized"
emeraldine base powder with a coppery, metallic glint having an oxidation state as
determined by volumetric $TiCl_3$ titration corresponding approximately to that of the ideal
emeraldine oxidation state,[13]. The ^{13}C,[14] and ^{15}N NMR,[15] spectra of emeraldine base are
consistent with its being composed principally of alternating oxidized and reduced repeat
units.

The emeraldine base form of polyaniline was the first well established example,[3,11,16-
18] of the "doping" of an organic polymer to a highly conducting regime by a process in
which the number of electrons associated with the polymer remain unchanged during the
doping process. This was accomplished by treating emeraldine base with aqueous protonic
acids and is accompanied by a 9 to 10 order of magnitude increase in conductivity (to 1 - 5
S/cm; 4 probe; compressed powder pellet) reaching a maximum in ~1M aqueous HCl with
the formation of the fully protonated emeraldine hydrochloride salt, viz.,

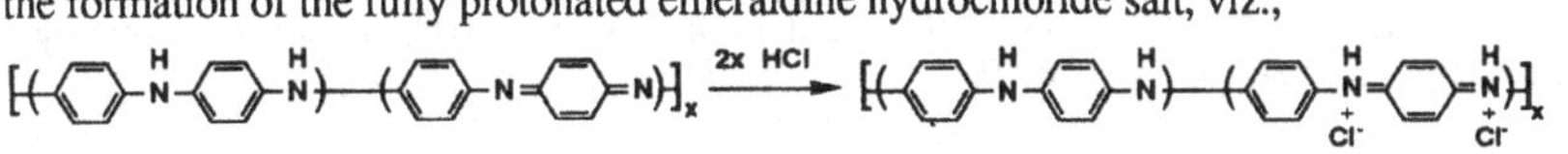

(3)

If the fully protonated i.e. ~50% protonated emeraldine base should have the above
dication i.e. bipolaron constitution as shown in equation 3, it would be diamagnetic.
However, extensive magnetic studies,[3,19] have shown that it is strongly paramagnetic and
that its Pauli (temperature independent) magnetic susceptibility increases linearly with the
extent of protonation. These observations and other earlier studies,[11,16-18] show that the
protonated polymer is a polysemiquinone radical cation, one resonance form consisting of
two separated polarons:

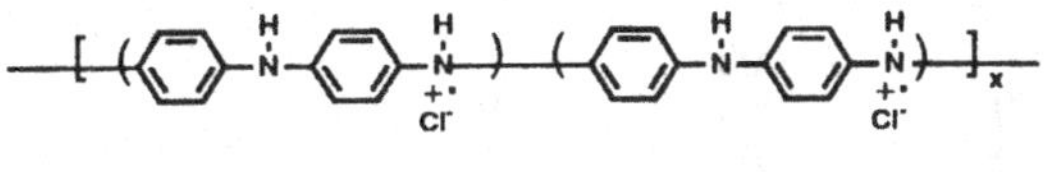

It can be seen from the alternative resonance form where the charge and spin are placed on
the other set of nitrogen atoms that the overall structure is expected to have extensive spin
and charge delocalization resulting in a half-filled polaron conduction band.

Molecular Weight of Polyaniline

As-synthesized,[12] emeraldine base in NMP solution, containing 5 wt.% LiCl exhibits a monomodal symmetrical molecular weight distribution curve by G.P.C. The monomodal symmetrical peak obtained,gives values of $M_p = 38,000$, $M_w = 78,000$ and $M_n = 26,000$; polydispersity, $M_w/M_n = \sim 3.0$. Recent light scattering results, which give absolute M_w values, show that M_w values obtained from G.P.C. studies vs. polystyrene standard are approximately double the correct value. In one study a solution of emeraldine base in NMP/LiCl was passed through a preparative G.P.C. column and six separate fractions were collected, the lowest molecular weight fraction ($M_p < 5000$) being discarded, since in a separate study it was shown that it contained oxygen-containing impurities,[20]. Each of the six fractions (Table II) were shown to be pure emeraldine base by elemental analysis, infrared and electronic spectral studies and by cyclic voltammetry,[21].

The conductivity of the doped (1M HCl) polymer rises monotonically with molecular weight up to a value of $\sim 150,000$ ($\sim 1,600$ ring-nitrogen repeat units) after which it changes relatively little. The reason for the change in dependency of conductivity on molecular weight is not clearly apparent; however, it is not caused by a change in the degree of crystallinity, since all fractions exhibited approximately the same crystallinity by x-ray diffraction studies.

Relationship Between Crystallinity and Conductivity of Polyaniline

It has been known for some time that emeraldine base is readily solution-processible,[22-24] and that it may be cast as free-standing, flexible, coppery-colored films from its solutions in NMP. These films can be doped with ~ 1M aqueous HCl to give the corresponding flexible, lustrous, purple-blue films ($\sigma \sim 1\text{-}4$ S/cm) of emeraldine hydrochloride,[22] which are partly crystalline.

Aligned Films: As will be seen below, the intrinsic properties of a conducting polymer can only be approached through processing. Uniaxially oriented, partly crystalline emeraldine base films are obtained by simultaneous heat treatment and mechanical stretching of films formed from "as-synthesized" emeraldine base containing $\sim 15\%$ by weight plasticizer such as NMP,[3,25]. Samples are observed to elongate by up to four times their original length when stretched above the glass transition temperature [$> \sim 110°C$], [3,25]. The resulting films have an anisotropic X-ray diffraction and optical response, with a misorientation of only a few degrees,[3]. The unstretched films are essentially completely resoluble in NMP. The solubility decreases with stretching (increased crystallinity);four-fold stretched films are completely insoluble in NMP. However, their crystallinity can be destroyed by repeated doping with aq. HCl and undoping with aq. NH$_4$OH whereupon they become completely re-soluble in NMP.

Table II.   Characterization of Fractions of Emeraldine Base From Preparative G.P.C. Studies (a)

	M_p	M_n	M_w	M_w/M_n	Conductivity (b) (S/cm)
Fraction					
1	15,000	12,000	22,000	1.8	1.2
2	29,000	22,000	42,000	1.9	2.4
3	58,000	40,000	73,000	1.8	7.9
4	96,000	78,000	125,000	1.6	13.1
5	174,000	148,000	211,000	1.4	17.0
6	320,000	264,000	380,000	1.4	14.9

(a)   Molecular weights relative to polystyrene standard (NMP solution).
(b)   Compressed pellet; 4-probe; after doping with 1M aq. HCl for 48 hours.

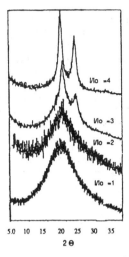

5.0   10   15   20   25   30   35
2 θ

Fig.1. X-Ray Diffraction Spectra of Ribbons of Emeraldine Base of Increasing Draw
Ratio, (l/l_0; l = final length; l_0 = original length before stretching.)

Lustrous, copper-colored ribbons of uniaxially oriented emeraldine base film up to 1.2 meters (4 feet) in length and 2.5 cm in width (thickness ~20mm) of various draw ratios can be readily fabricated by stretch-orienting emeraldine base films cast from NMP solution at ~140° C between two metal rollers rotating at different speeds,[26]. As can be seen from Figure 1 the apparent degree of crystallinity is greatly increased by processing of this type.

The tensile strength of the ribbons also increases significantly with an increase in draw ratio (and crystallinity) as shown in Table III. As expected, biaxially oriented film exhibits significantly greater tensile strength than uniaxially oriented film for the same draw ratio (l/l_0 = 2) (Table III). The conductivity of the HCl-doped uniaxially oriented ribbons increases on stretching (l/l_0 = 1, σ ~5 S/cm; l/l_0 = 4, σ ~80 S/cm). It should be noted that the conductivity of the oriented films is greatly dependent on their method of drying; conductivities of ~300 S/cm can be obtained for films which have not been dried to any great extent,[26].

The above observations show that polyaniline can be processed by methods used for commercial polymers. Even at this very early stage, its tensile strength overlaps the lower tensile strength range of commercial polymers such as Nylon 6. Unstretched Nylon 6 has tensile strengths ranging from 69 to 81 MPa, [27].

Aligned Fibers: Fibers (~30-70mm) of emeraldine base can be formed by drawing a ~20% by weight "solution" of emeraldine base in NMP in a water/NMP solution,[23,28]. If

Table III.   Tensile Strength (MPa)[a] of Emeraldine Base Ribbons as a Function
of Draw Ratio (l/l_0)

		Uniaxial Orientation:				Biaxial Orientation[b]:
		l/l_0=1	l/l_0=2	l/l_0=3	l/l_0=4	l/l_0=2
Tensile Strength	(Av.)	54.4	53.2	75.9	124.1	122.4
	(Best)	59.9	62.1	82.8	144.8	131.6

(a) gauge length = 3 inches   (b) l/l_0 = 2 in both directions

desired, the emeraldine base "solution" in NMP may also be drawn in aqueous HCl which results in direct formation of the doped fiber. Fibers can also be spun from NMP solution. The drawn fibers (containing NMP as plasticizer) can be thermally stretch oriented at ~140°C up to four times their original length in a similar manner to emeraldine base films,23. X-ray diffraction studies show directional enhancement of the Debye-Scherrer rings. A monotonic increase in apparent crystallinity with draw ratio is observed. Doping with 1M aqueous HCl results in a significant increase in the conductivity parallel to the direction of stretching (σ ~40-170 S/cm) as compared to the conductivity of the polymer powder from which the fibers are prepared (σ ~1-5 S/cm).

As can be seen from Figure 2, the conductivity of the HCl-doped drawn fibers increases monotonically with draw ratio, (l/l_o). Since the crystallinity also increases with draw ratio, it is apparent that the conductivity increases with increase in crystallinity.

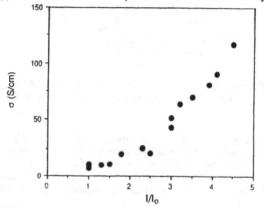

Fig.2. Conductivity of the HCl-doped Drawn Fibers vs. Draw Ratio (l/l_o); l = final length; l_o = original length before streching.

It should be stressed that the above data were obtained using "as-synthesized" emeraldine base containing low molecular weight polymer and low molecular weight impurities,23. Since l/l_o in general will increase with increasing molecular weight, e. g. by gel spinning, it is apparent that use of higher molecular weight polymer should result in greater l/l_o ratios and hence in even higher conductivities.

Preliminary studies show that emeraldine base fibers both before and after doping with 1M aqueous HCl exhibit promising mechanical properties,1,23. Values for one inch gauge length (tensile strength, MPa; initial modulus, GPa) for emeraldine base fibers stretch-oriented $(l/l_o ~ 3-4)$ at ~140°C after drawing are: 318(Av.); 366(Best) and 8.1(Av.); 8.6(Best). After doping, corresponding values are: 150(Av.); 176(Best) and 4.6(Av.); 5.0(Best). X-ray diffraction studies show some reduction in crystallinity after doping, consistent with the reduction in tensile strength. As expected, the tensile strength of the oriented emeraldine base fibers are greater than those of oriented films. As can be seen on comparing the above tensile strengths with those of, for example, fibers of Nylon 6 (200-905 MPa),29 the mechanical properties of polyaniline fibers, considering the early stage of development, are most encouraging.

TECHNOLOGY

Technological uses of conducting polymers may be classified as follows: (1) Use based on bulk conductivity of the pure conducting polymer or a blend of the conducting polymer with a conventional polymer; (2) Use based on the electrochemical redox properties of the polymer; (3) Use based on the formation of an excited state of the polymer; (4) Use based on the morphology/microstructure of the polymer. Each of these categories will be summarized below.

(1) Use Based on Bulk Conductivity

It is most unlikely that a conducting polymer will ever be used as a substitute for copper as an electrical conductor unless its conductivity and mechanical properties can be appropriately increased. However, the unique <u>combination</u> of properties which can be displayed by a conducting polymer or its blends with conventional polymers appear to show great promise for a variety of uses. A remarkable feature of several conducting polymers which has recently been brought to light is their ability to undergo percolation at very low loading levels when incorporated as a blend into an insulating polymer. When an electrically conducting material - metal or carbon powder or filaments - is mixed with an insulating polymer, essentially no increase in conductivity is observed until particles of the conducting material first touch each other and thus form a conducting pathway throughout the mixture. At this loading level "percolation threshold" (~16 vol.% for a three-dimensional network of conducting globular aggregates in an insulating matrix) the conductivity increases extremely rapidly. The percolation threshold is greatly dependent on the size and aspect ratio of the particles - whether, for example, spheres or long needles - and can vary from a few volume percent up to 30% to 40% or more in industrial composites depending on the efficiency of mixing and uniformity of size. However, in blends of doped polyaniline and also in blends of derivatives of certain substituted polythiophenes[30] in conventional insulating polymers either no or only very low (< 5%) percolation thresholds are observed. For example, a (5 wt. %) blend of doped polyaniline in Nylon 12 shows that percolation has already commenced and that the conductivity at ~ 20% loading is essentially the same as that of the pure conducting polymer,[9]. Thus the most desirable features of both the conducting and non-conducting polymers can be combined. At least in some cases it is believed that percolation commences at a very low level since the host (non-conducting) polymer forms a network upon which a very thin film of the conducting polymer is deposited, thus forming a continuous conducting film throughout the bulk material.

Doped polyaniline and blends of doped polyaniline are now commercially available in large quantities,[9]. The blends are easier to process than metal flake- or fiber-filled polymers and are non-abrasive and exhibit uniform and higher conductivities than is possible with carbon black fillers. They show superior electromagnetic (EMI) shielding properties to previous materials,[9] and have use in static dissipation. It seems highly likely that the polythiophene blends presently being developed,[7,8] may be used for similar purposes.

The bulk conductivity of films or powders of conducting polymers has also been utilized in a variety of chemical sensing devices whereby a conducting polymer is either doped or undoped by a specific chemical species whose presence is to be detected with a concomitant large change in electrical conductivity which can be recorded electronically,[31].

(2) Use Based on Electrochemical Redox Processes

The use of conducting polymers as a cathode material (with alkali metal anode) in novel rechargeable batteries is well established,[6]. Button-cell batteries containing polyaniline have been on sale since 1987. They are based on the concept of electrochemical oxidation and reduction. This is illustrated below for polyaniline/Li batteries, the fundamental electrochemical processes involved being:

Discharge Reactions

$$(4)$$

The electrons taken up by the polyaniline polysemiquinone radical cation come via an external wire from the anode.

$$\text{Anode:} \qquad 2\,\text{Li} \longrightarrow 2\,\text{Li}^+ + 2\,e^- \qquad\qquad (5)$$

The charge reactions are the reverse of the above. These batteries show superior qualities for use as a back-up power source for personal computers and for portable solar-powered calculators, etc. Somewhat similar industrial prototype rechargeable batteries involving polypyrrole have also been announced,[32,33]. Rechargeable, completely packaged batteries using a polyacetylene or polyparaphenylene/alkali metal cathode have also been developed,[34]. These have approximately twice the energy density of nickel/cadmium batteries.

Many electrochemically dopant-induced structure-property changes in conducting polymers have been described for use in, for example, electrochromic windows and displays, electrochemically controlled chemical separation and delivery systems, redox capacitors, electromechanical actuators, etc.,[31]. Some of these are apparently close to commercial application.

(3) Use Based on the Formation of an Excited State of the Polymer

The evaluation of conducting polymers for use in non-linear optical (NLO) devices has aroused and continues to arouse a very large amount of scientific activity throughout the world. Although promising properties and prototype devices have been described,[30,35] commercial use in the immediate future does not seem likely.

An extremely exciting, very recent development,[7,36] reports the use of polyphenylenevinylene and derivatives as light-emitting diodes (LED) whereby their electroluminescent properties are utilized. Relatively intense visible light is emitted with reasonable efficiency. Considering the large number of conducting polymers and their derivatives it appears that this discovery may open up a whole new technology whereby flexible, large area diodes emitting a wide range of colors may be possible.

(4) Use Based on the Morphology / Microstructure of the Polymer

Another very recent exciting development involves the use of lightly-doped polyaniline film membranes for the separation of gases,[37]. Separations far exceed the selectivity of all known gas separation membranes for many simple gas mixtures, e.g. O_2/N_2. The process by which this occurs is not yet clear although it appears that protonic acid doping of emeraldine base films, followed by complete undoping and then followed by light doping may open up molecular-sized channels previously occupied by dopant anions, within the films. This type of use of conducting polymers opens up new unexpected avenues for research and development.

CONCLUSIONS

Undoubtedly, many important potential applications of conducting polymers have been omitted in this brief summary. The next decade will certainly bring forth new basic science and applications in the field that are logical extensions of what is already known, as well as completely new unexpected fundamental science and technology.

ACKNOWLEDGEMENTS

The authors are particularly indebted to Mr. Michael A. Mancini and Dr. Rakesh K. Kohli for their untiring efforts in assisting with the preparation of this manuscript. The recent work described in this report on polyaniline was supported primarily by the Defense Advanced Research Projects Agency through a grant administered by the Office of Naval Research and in part by N.S.F. Grant No. DMR-88-19885.

REFERENCES

1. M. G. Kanatzidis, Chem. & Eng. News, P. 36, (Dec. 3, 1990); "Handbook of Conducting Polymers", T. A. Skotheim, Ed., Marcel Dekker, New York, 1 & 2, (1986).
2. A. G. MacDiarmid, C. K. Chiang, C. R. Fincher, Jr., Y. W. Park, A. J. Heeger, H. Shirakawa, E. J. Louis and S. C. Gau, Phys. Rev. Lett., 39, 1098 (1977); A. G. MacDiarmid, C. K. Chiang, M. A. Druy, S. C. Gau, A. J. Heeger, E. J. Louis, Y. W. Park and H. Shirakawa, J. Am. Chem. Soc., 100, 1013 (1978).
3. A. G. MacDiarmid and A. J. Epstein, Faraday Discuss. Chem. Soc., 88, 317 (1989).
4. C. C. Han and R. L. Elsenbaumer, Synth. Met., 30, 123 (1989).
5. Based on search of STN International On Line Computer File, Chem. Abs., June 3, 1991.
6. T. Enomoto and D.P. Allen, Bridgestone News Release, (Sept. 9, 1987), Chem. Week, p. 40, (Oct. 14, 1987); T. Kita, in: "Practical Lithium Batteries," Y. Matsuda and C. Schlaikjer, ed., JEC Press, Cleveland, p. 124 (1988).
7. European Chemical News, p. 38, (April 22, 1991).
8. Chem. and Eng. News, p. 8, (April 29, 1991).
9. Plastics Technology, p. 15, (June, 1991); V. G. Kulkarni, W. R. Mathew, J. C. Campbell, C. J. Dinkins and P. J. Durbin, 49th ANTEC Conference Proceedings, Society of Plastic Engineers and Plastic Engineering, Montreal, Canada, (May 5-9, 1991), p. 663; L. W. Shacklette, N. F. Colaneri, V. G. Kulkarni and B. Wessling, Ibid., p. 665.
10. J-C. Chiang and A. G. MacDiarmid, Synth. Met., 13,193 (1986).
11. A. G. MacDiarmid, J-C. Chiang, A. F. Richter and A. J. Epstein, Synth. Met., 18, 285 (1987).
12. A. G. MacDiarmid, J-C. Chiang, A. F. Richter, N. L. D. Somasiri and A. J. Epstein, "Conducting Polymers," L. Alcacér, ed., Reidel Publications, Dordrecht, p. 105 (1987).
13. G. E. Asturias, R. P. McCall, A. G. MacDiarmid and A. J. Epstein, Synth. Met., 29, E157 (1989).
14. S. Kaplan, E. M. Conwell, A. F. Richter and A. G. MacDiarmid, Synth. Met., 29, E235 (1989).
15. A. F. Richter, A. Ray, K. V. Ramanathan, S. K. Manohar, G. T. Furst, S. J. Opella, A. G. MacDiarmid and A. J. Epstein, Synth. Met., 29, E243 (1989).
16. A. J. Epstein, J. M. Ginder, F. Zuo, H-S. Woo, D. B. Tanner, A. F. Richter, M. Angelopoulos, W-S. Huang and A. G. MacDiarmid, Synth. Met., 21, 63, (1987).
17. P. M. McManus, S. C. Yang and R. J. Cushman, J. Chem. Soc., Chem. Commun., 1556 (1985).
18. G. E. Wnek, Synth. Met., 15, 213 (1985).
19. A. J. Epstein, J. M. Ginder, F. Zuo, R. W. Bigelow, H-S. Woo, D. B. Tanner, A. F. Richter, W-S. Huang and A. G. MacDiarmid, Synth. Met., 18, 303 (1987); J. M. Ginder, A. F. Richter, A. G. MacDiarmid and A. J. Epstein, Solid State Commun., 63, 97 (1987).
20. A. G. MacDiarmid, G. E. Asturias, D. L. Kershner, S. K. Manohar, A. Ray, E. M. Scherr, Y. Sun, X. Tang and A. J. Epstein, Polym. Prepr., 30-(1),147 (1989).
21. X. Tang, A. G. MacDiarmid and A. J. Epstein, Unpublished Observations (1990).
22. M. Angelopoulos, G. E. Asturias, S. P. Ermer, A. Ray, E. M. Scherr, A. G. MacDiarmid, M. Akhtar, Z. Kiss and A. J. Epstein, Mol. Cryst. Liq. Cryst., 160, 151 (1988).
23. A. G. MacDiarmid and A. J. Epstein, "Science and Applications of Conducting Polymers," W. R. Salaneck, D. T. Clark and E. J. Samuelsen, eds., IOP Publishing Ltd., Bristol, UK, p. 117 (1990).
24. M. Angelopoulos, A. Ray, A. G. MacDiarmid and A. J. Epstein, Synth. Met., 21, 21 (1987).
25. K. R. Cromack, M. E. Jozefowicz, J. M. Ginder, R. P. McCall, A. J. Epstein, E. M. Scherr and A. G. MacDiarmid, Bull. Am. Phys. Soc., 34, 583 (1989).
26. E. M. Scherr, A. G. MacDiarmid, Z. Wang, A. J. Epstein, M. A. Druy, and P. J. Glatkowski, Synth. Met., 41, 735 (1991).

27. J. Brandrup and E. H. Immergut, "Polymer Handbook," John Wiley and Sons, New York, Ch. VIII, p. 2 (1975).
28. X. Tang, E. M. Scherr, A. G. MacDiarmid and A. J. Epstein, Bull. Am. Phys. Soc., 34, 583 (1989).
29. "Textile World Man-Made Fiber Chart," McGraw-Hill (1988).
30. A. J. Heeger, "Science and Applications of Conducting Polymers," W. R. Salaneck, D. T. Clark and E. J. Samuelsen, eds., IOP Publishing Ltd., Bristol, UK, p. 1 (1990).
31. R. Baughman, "Science and Applications of Conducting Polymers," W. R. Salaneck, D. T. Clark and E. J. Samuelsen, eds., IOP Publishing Ltd., Bristol, UK, p. 47 (1990).
32. H. Naarmann, "Science and Applications of Conducting Polymers," W. R. Salaneck, D. T. Clark and E. J. Samuelsen, eds., IOP Publishing Ltd., Bristol, UK, p. 81 (1990).
33. Varta, A G., advertisement; "A Decisive Step Towards the Future of Rechargeable Batteries," (1989).
34. T. R. Jow and L. W. Shacklette, J. Electrochem. Soc., 136, 1 (1989).
35. D. Bloor, "Science and Applications of Conducting Polymers," W. R. Salaneck, D. T. Clark and E. J. Samuelsen, eds., IOP Publishing Ltd., Bristol, UK, p. 23 (1990), (and references within).
36. J. H. Burroughs, D. D. C. Bradley, A. R. Brown, R. N. Marks, K. Mackay, R. H. Friend, P. L. Burns and A.B. Holmes, Nature, 347, 539 (1990).
37. H. Reiss, R. B. Kaner, M. R. Anderson, B. R. Matts, Science, 252, 1412 (1991).

SYNTHETIC ROUTES TO INTRINSICALLY CONDUCTING ORGANIC MATERIALS ICOMs

H. Naarmann

Plastics Research Laboratory
BASF Corp.
6700 Ludwigshafen FRG

Introduction

Between Goppelsroeder's "electrochemically charging and discharging of polyaniline" in 1891 and highly oriented polyacetylene with metallike characteristics in 1987 lies a century.

What has happened during such a long period? and reading the pioneer-contribution of Goppelsroeder we tend to agree with Goethe "Who can find a wise and original idea which hasn't been thought before?"

Background

In the area of ICOMs we are confronted with several severe problems:
o What do we know concerning the real structure of all the different insoluble "poly...s"

o Who understands and explains the high conductivity and the electron transport mechanismus

o Who synthesizes a soluble and processable electrical conducting polymers

o Who solves the problem of the instable ICOMs

But in spite of all these difficulties the ICOMs represent a class of compounds with fascinating applications and growing interest.

ICOMs are characterized by an extended -C=C-conjugation in combination with electron donor or acceptor compounds forming n-or p-doped systems, reaching conductivities up to 10^5 S/cm.

Frontiers of Polymer Research, Edited by P.N. Prasad and
J.K. Nigam, Plenum Press, New York, 1991

The basic structures are

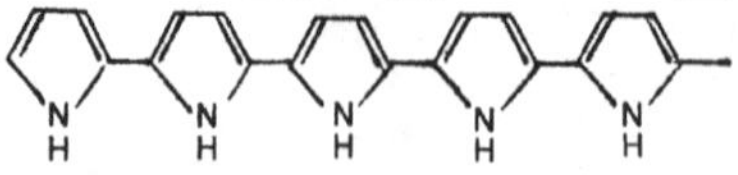

NH, S, O

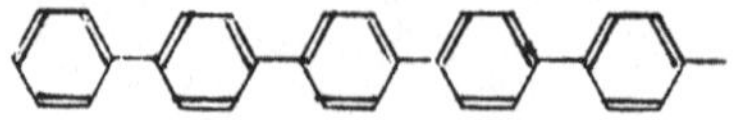

or combinations or condensed aromatic (hetero-) systems.

Polyaniline:

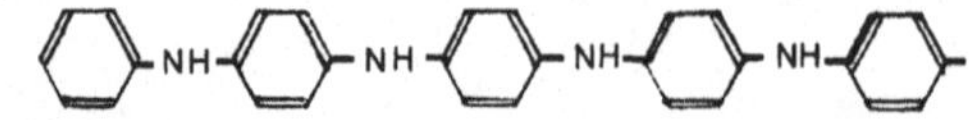

in his different oxidized or reduced forms represents a special type due
to the doping possibility by protonation.

The book "organic semiconducting polymers" by J. E. Katon (1) encompasses
the state of the art up to 1968, the author classifies the polymers

in o covalent organic polymers
    o charge transfer complexes
    o metal organic polymers
    o H-bonded polymers and
    o mixed polymers

and reviews the pioneer work concerning the correlation of the conductivity and complexation with e. g. I_2, $SbCl_3$ including the huge amount
of various types from soluble linear oligoenes up to insoluble fused and
heteroatoms containing aromatic ring systems.

Polyacetylene

     In a retrospective view it was a steady up and down between stimu-
lating enthusiasm and discouragement, but in a summary there are main
trends and progressive developments. A typical example is polyacetylene,
it was first mentioned 1904 (2) a disproportinately long period of time
passed before Reppe 1948 (3) prepared the filmic Cuprene with metallic
lustre and Hatano (1961) reported on the polymerization of acetylene by
$AlEt_3/Ti(OBu)_4$ to polymers with conductivities up to 10^{-3} S/cm (4).
Since then intensive work has been carried out into the various polymer
types e. g. by Shirakawa (11 a); reviews are given in "Polymerisation von
Alkinen" (5). Some important initiator systems are listed chronologically
in the following table 1.

The Feast method (9) for producing "Durham PAC" proceeds according to the
following (scheme 1).

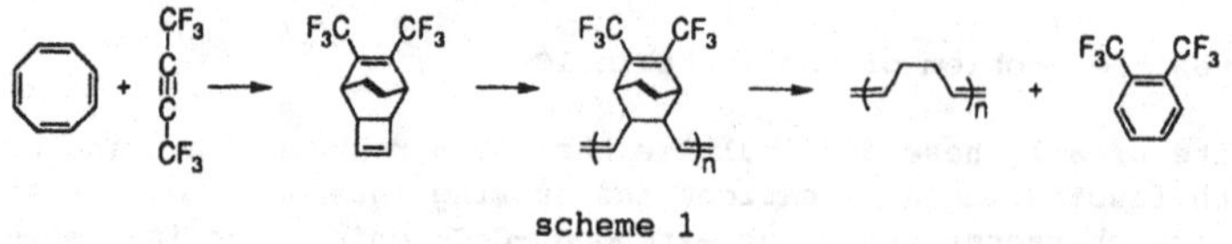

scheme 1

7,3-Bis(trifluoromethyl)tricyclo(4,2,2,0)deca 3,7,9triene polymerizes by
undergoing ring opening and yields polyacetylene through elimination of
1,2-bis(trifluoromethyl)benzene.

Table 1. Preparation and properties of polyacetylenes

Name of method	Catalyst	Polymeri- zation temperture °C	cis/trans ratio	Defects sp³ fraction (rel. %)	Conductivity S/cm (iodine doped)	References
1. Reppe	Ni(acac)$_2$	100	5/95	15	10^{-3}	(5 b) Table 1, No. 2
2. Luttinger-Green	NaBH$_4$/Co^{II}/Ni^{II}	-76	50/50	8	10^{-1}	(5 b) Table 1, No. 6
3. Hatano	AlEt$_3$/Ti/(OBu)$_4$	100	20/80	10	10^{-5}	(5 b) Table 2, No. 8
4. Shirakawa	AlEt$_3$/Ti(OBu)$_4$ in toluene	-76	90/10	3	520	(5 b) Table 1, No. 16
5. Naarmann	a) AlEt$_3$/Ti(OBu)$_4$ in silicone oil	20	85/15	0	500	7)
	b) as a), but with an aged > 100 °C cata- lyst	20	85/15	0	2 000	7)
	c) as b), but oriented		85/15	0	> 100 000	7) 8)
6. Feast	Thermal clearage of tetracyclo bis(trifluo- romethyl)decatriene		60/40	-	100	9)
7. Grubbs	Isomerization of poly- benzvalene		40/60	-	1	10)

In the Grubbs method (10) polybenzvalene is isomerized in the presence of $HgCl_2$ to polyacetylene (scheme 2).

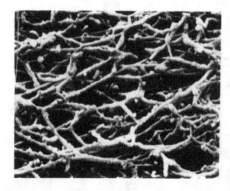

scheme 2

Both of these methods start off with certain monomers that are converted to soluble pre-polymers that then yield insoluble perconjugated polymers after thermal treatment.

Table 2 shows the methods to reach higher conductivities.

All polyacetylenes are insoluble. They can be rendered surprisingly electrically conductive by incorporation of n-type or p-type dopants. They are used as semi-conductors electrodes in polymer batteries, sensors etc. (14).

The polyacetylenes grow primarily in the form of fibrils and have a large internal surface area (up to 200 m^2/g) (15).
   The density lies between 0.4 and 0.8 g/cm^3. Work performed by Haberkorn et al. (6) clearly shows a correlation between conductivity (S/cm) as a function of crystallinity and defects (sp^3 fractions).
   The polyacetylenes are quite unstable (forming e. g. epoxy; carbonyl-carboxyl groups), but the greater stability of the $N-(CH=CH)_x$ samples is probably due to the absence of defect sites at high crystallinity.
   A convincing phenomenon of high anisotropy (1:100) in the stretched samples is the construction of a polarizer. When strips of such a stretched polymer are laid across each other, polarized light (sun light) is extinquished in the region of overlapping (16).
   The following diagramms illustrate the major differences between two polyacetylenes. The Shirakawa type, which is shown in Figure 1, is cross-linked and contains approx. 2 % sp^3 fractions. The polyacetylene shown in Figure 2 has been prepared by the new BASF technique.

It is linear (no sp^3 fractions), is thus highly orientable (up to 660 %) and has a conductivity of more than 100 000 S/cm.

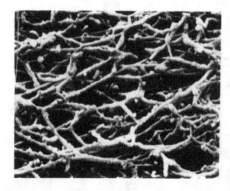

├───────┤
1 um

Fig. 1   Fig. 2

Shirakawatype – Polyacetylene   BASF type

Table 2. Comparative conductivites of several metals and different types of polyacetylenes

Metal	Conductivity (volume)	Density	Conductivity (weight)
Hg	10 365 S/cm	13.546 g/cm^3	767 S/cm2 g^{-1}
Sb	25 575	6.618	3 864
Pt	101 522	21.45	4 733
As	30 030	5.72	5 250
Ir	195 694	22.65	8 639
Cr	77 519	7.19	10 781
Fe	102 986	7.87	13 085
Ni	146 113	8.908	16 402
Co	160 256	8.92	17 965
Au	470 588	19.3	24 382
Ag	671 140	10.491	63 972
Cu	645 000	8.94	72 147
S-(CH)$_x$	897**	before after doping	813
(11 a) (11 b)	30*	0.40 1.23	22
(S-CH)$_x$ oriented in LC matrix (11, 12)	1 600 - 1800***	0.50 1.10	1 750
N-(CH)$_x$ oriented (13)	18 000 - 28 000* doped	0.85 1.12	16 071 - 25 000
ARA method (N-CH)$_x$ oriented (7)	120 000*	0.90 1.15	104 347

* doped with a saturated solution of I_2 in CCl_4

** doped with $FeCl_3$ (11c)

*** doped with I_2 vapor

In addition, the linear polyacetylenes of thicknesses up to 5 um can be stretched, whereby the films become transparent.

Another way of obtaining oriented and thus more highly conductive polymers consists in polymerizing the acetylene at -76 °C in liquid crystal phases that are oriented by strong magnetic fields (12). Conductivities of up to 14 000 S/cm can be achieved by this method.

Substituted polyacetylenes

The synthesis of these polymers is directed towards the preparation of conjugated but substituted chains which ameliorate the negative properties of polyacetylenes such as air-sensitivity, insolubility and infusibility which maintaining the desired electrical characteristics of acetylene's conjugated backbone (17). Result: Soluble and processable but low conductivities (< 10^{-1} S/cm) compared to the unsubst. polyacetylene.

Soluble precursors

The lack of processing scope and of perfect characterization were the reason behind the search that proceed from characterized precursors and soluble and processable prepolymers to the polyenes.
o Pioneer work was performed by Stille (18) e. g. the 1.4 dipolar addition of a dilactone to diethynylbenzene affords quantitative yields of phenylene oligomers.
o Koßmehl (19) made systematical studies with oligothiophenevinylenes, by extended Wittig-condensation, investigating the correlation of the chain length n and the properties of the oligoenes.

—CH=CH—

o Hörhold (20) describes the polycondensation of soluble polymers of the type (-Ar-CR=CR-) were R=H or various aromatic components. A general route to polyphenylenevinylene in the elimination reaction starting from soluble disulfonium xylidenes.

H. H. Hörhold's article entitled "Development of an electroactive polymer material from unmeltable powder to transparent film" describes the progress that has been made in this sector.

Reviews concerning the various synthetic methods are given by Feast (9) and Naarmann (5 b, 11).

New Developments are the "Ring-Opening Metathesis polymerization of Cyclooktatetraenes" (21).

Other soluble systems but with heteroatoms are e. g. oligothienyls (22) and indophenins (23). The indophenins, represent a new class of ICOMs containing a electron donor and acceptor group/molecule leading directly to electrical conducting materials without any additional (external) doping.

A new trend are the investigations and attempts to yield well defined two-dimensional structures (24) e. g. by condensation or by repetitive Diels Alder-addition.

Soluble or solubilized (subst.) polyalkylthiophenes (25) represent a new class of conducting polymers. The molecular weight of these polymers can be determined by gpc, mass spectroscopy and UV/VIS spectroscopy. Transparent and enviromentally stable coatings can be prepared.

The electrochemical polymerization, is an elegant process (26) but a fairly long period elapsed before this method of heterocyclic polymerization attracted general interest. The electrochemical polymerization dates back to Lund in 1957 (27), but interest heightened after Diaz's publication in 1979 (27 a), "Electrochemical polymerization of pyrrole".

In contrast to polyacetylene, polypyrrole is a material with astonishing stability and can be continuously prepared by electrochemical techniques (28).

The quality of the polymers is greatly influenced by the reaction conditions.

If the monomers are of the same purity, the properties of the polymers are reproducibly influenced by the synthesis conditions.

The decisive factors are the electrolyte/conductive salt and the current density but particularly the conductive salt because it is incorporated in the polymer. Thus flexible and smooth films having conductivities of up to 200 S/cm can be produced by judicious selection of the conductive salt, e. g. of $\emptyset-SO_3-N^+Bu_3H$ instead of the BF_4^- salt.

These counterions allow specific effects to be introduced into the polypyrrole. It can thus be understood that it is precisely the anion - or its size, geometry, charge, etc. - that governs the properties of the polymers. In general, one anion is incorporated into 3 pyrrole units.

Some general considerations apply to the choice of anion. If it is organic, flexible smooth films that can be readily detached from the anode are generally obtained.

Changing the synthesis conditions allows different types of product with different surface morphology, e. g. an open porous structure, to be prepared.

Pyrrolesulphonic and thiophenesulphonic acids (29) are interesting variants of conductive salts, they are functionalized counterions acting as monomers.

$$\left(\underset{S}{\boxed{}}^{-SO_3^-}\right)_n$$ Thiophenesulphonic acid and oligothiphene-sulphonic acid

$$\underset{\overset{|}{H}}{\boxed{}}^{-SO_3^-}$$ Pyrrolesulphonic acid

In this instance, the counterion is coupled direct to the monomer. The result is that, under ideal polymerization conditions, there is one counterion (x^-) for every monomer unit. The counterion is not "externally" incorporated but is attached, i. e. chemically bound, to the monomer.

This exclusion of counterions can be performed specifically, and offers, for instance, the possibility of defined releasing optically active counterions or active ingredients of medical interest, such as heparin and penicilline (monobactam), which are incorporated in specific quantities as counterions into polypyrrole.

The extenion of these functionalized counterions lead to a new class of subst. thiophenes namely of

 L.C.—SO_3^-

self dopant and orientable thiophenes (30).

New Monomers

The work carried out in this field concerns the investigations of the effects exerted by substituents, fused rings etc. on polymerization, charge fixation, lowering the band gap, stability, application and processibility. Acetylene with heterocyclic side groups, pyrrole variants, thienyl-pyrrole, azabicyclodienes, benzopyrrolines, thianaphthene, thienobenzenes, heterobridged annellated thiophenes (31 a) oligothiophenes (31 b), arene-methylidenes (31 c), heteropentalenes (31 d, 32)

Acknowledgements

Cited BASF-references are part of the project No 03 C134 - 0, sponsored by BMFT. The author thanks the project partners and colleagues for their contribution and the Bundesministerium für Forschung und Technologie for support.

Considerations

There has been a veritable flood of recent literature and meetings on electrically conductive polymers and it has been difficult to present a complete picture of all that is significant. Many unfilled expectations, many unsolved problems but also a lot of encouraging progress are typical for this area.

Literature

1. Theoretical and Experimental Aspects of the Electronic Behavior or Organic Macromolecular Solid, H. A. Pohl, S. Kandu in "Organic semi-conducting polymers" edited by J. E. Katon, Marcel Dekker, Inc. New York [1968]

2. Vogel, J. H. "Handbuch für Acetylen" (1904) Verlag F. Vieweg und Sohn, Braunschweig p. 194

3. Reppe W., et. al. Anm. Chem. $\underline{560}$, 1 (1948) 104

4. Hatano M., et. al. J. Polym. Sci. $\underline{51}$ (1961) 26

5. a) Kröhnke, Chr. und Wegner, G. in Houben-Weyl (1986) Georg Thieme Verlag Stuttgart Vol. $\underline{20}$/2 12 12

    b) Naarmann, H. (1982) Angew. Makromol. Chemie $\underline{109/110}$ 295

6. Haberkorn, H. Naarmann, H. Penzien, K. Schlag, J. and Simak P. (1982) Synthetic Metals $\underline{5}$ 51

7. Naarmann, H. and Theophilou, N. (1988) Synthetic Metals $\underline{22}$ 1

8. Schimmel, Th., Gläser, D., Schwoerer, M., Schoepe, W. und Naarmann, H., Bayreuther Polymer Symposium 12. April 1989, Synthetic Metals 37 (1990) 1 - 6

9. Edwards, H. H. and Feast, W. J. (1980) Polymer $\underline{21}$ 595
    Feast, W. J., Synthesis of Conducting Polymers in "Handbook of Conducting Polymers" (Skotheim, T. A. ed.) (1986) Marcel Dekker, New York p. 35

10. Swager, T. M., Dougherty, D. A. and Grubbs, R. H. (1988) J. Am. Chem. Soc. $\underline{110}$ 2973
    Swager, T. M., Dougherty, D. A. and Grubbs, R. H. (1989) J. Am. Chem. Soc. $\underline{111}$ 4413

11 a. Ito, T., Shirakawa, H. and Ikeda, S. (1974) J. Polym. Sci., Polym. Chem. Ed. $\underline{12}$ 11

11 b. Chiang, C. K., Fincher, C. R., Park, Y. W., Heeger, A. J.,
       Shirakawa, H., Louis, E. J., Gau, S. C., MacDiarmid, A. G., Phys.
       Rev. Lett. 39, 1098 (1977)

11 c. Pekker, S., Janossy, A., Chemistry of Doping in Polyacetylene in
       Handbook of Conducting Polymers edited by T. A. Skotheim, P. 45 -
       76, (1989) M. Dekker, N. Y.

12.   Chen, Y.-C., Akagi, K. and Shirakawa, H. (1986) Synthetic Metals 14
       173 Araya, K., Mükoh, A., Narahara, T. and Shirakawa, H. (1986)
       Synthetic Metals 14 199
       Aldissi, M. (1985) J. Polym. Sci., Polym. Lett. Ed. 23 167

13.   Naarmann, H. (1987) Synthetic Metals 17 223

14.   Simon, J. and Andrée, J. J. "Molecular semiconductors" (1984)
       Springer Verlag Berlin, p. 50
       Ellis, J. R., in "Handbook of Conducting Polymers" (1986), edited by
       T. A. Skotheim, Marcel Dekker, N. Y., p. 489

15.   Gläser, D., Schimmel, Th., Schwoerer, M., Naarmann, H., Makromol.
       Chem. 190, 3217 (1989)

16.   Naarmann, H. (1990) in Conjugated Polymeric Materials edited by
       Bredas, J. L. and Chance, R. R., Klurer Acad. Publ. 11 - 51

17.   Naarmann, H., Strohriegel, P., "Conducting and Photoconducting Poly-
       mers" in Handbook of Polymer Synthesis Chap. 3.0 (in press) Marcel
       Dekker, N. Y. edited by H. Kreicheldorf

18.   Stille, J. K., Makromol. Chem. 154, (1972) 49

19.   Koßmehl, G., Makromol. Chem. 131 (1970) 37

20.   Hörhold, H. H., 2. Chem. 24 (1987) Heft 4, p. 126 - 127

21.   Ginsburg, E. J., Gorman, C. B., Grubbs, R. H., Klaretter, F. L.,
       Lewis, N. S., Marder, S. R., Perry, J. W., Sailer, M. J. in
       Conjugated Polymeric Materials ed. by Bredas, J. L. and Chance, R.
       R. (1990) Nato ASI Series Vol. 182, p. 65 - 82
       Knoll, K., Schrock, R. R., J. Am. Chem. Soc. III (1989) 7989

22.   Martinez, F., Voelkel, R., Naegele, D., Naarmann, H., Mol. Cryst.
       Liq. Cryst. (1989) 167 p. 227 - 232
       Garnier, F., Roncali, F., Lemaire, M., Garreau, R., Synth. Metals
       (1986) 4, 45 "Polythiophenes and its Derivatives" Tourillon, G.
       Chapt. 9 in Handbook of Conducting Polymers Vol. ed. by T. A.
       Skotheim, Marcel Dekker, N. Y. (1986)

23.   Martinez, F., Naarmann, H., Die Angew. Makromol. Chem. (1990) 178 p.
       1 - 16 Condensation reactions of thiophene and bithiophene with
       isatin

24.   A soluble polyacene precursor
       Vogel, Th., Blatter, K., Schlüter, A. D., Makromol. Chem. Rapid
       Commun. (1989) 10, 427 - 430
       Schlüter, A. D., Aufbruch in die zweite Dimension, Nachr. Chem.
       Tech. Lab. (1990) 38, 8
       Polymere mit Naphthalineinheiten
       DEOS 39 06 812 (03.03. 1989/06.09.1990)

       Müllen, K., Koch, K. H., Naarmann, H. BASF AG, Ludwigshafen FRG
       Polycycloolefine
       DEOS 39 33 901 (14.10.1988/19.04.1990)

Müllen, K., Wohlfahrt, H., Naarmann, H. BASF AG, Ludwigshafen FRG
BMFT-Report 03M 4019, Geb. 1991, Naarmann, H., Müllen, K., Wegner,
G., Hanack, M., Schwoerer, M., Dormann, E.

25.   Synthetic Metals 28 (1989) C 281 - C 545
      Proceedings of ICSM, Santa Fe, USA (1988)
                        - 11 -
26.   "Electrochemical Synthesis of Conducting Polymers", Diaz, A. F.,
      Burgon, J. (Chapt. 3), "Polypyrrole from powders to Plastics",
      Street, B. G., Chapt. 8, in Handbook of Conducting Polymers, Vol. 1,
      edited by T. A. Skotheim, Marcel Dekker, N. Y. (1986)

27.   Lund, H. (1957) Actachem. Scand. 11 1323

27 a. Diaz, A. F., Kanazawa, K. K., Gardini, G. P., (1979) J. Chem. Soc.
      Chem. Commun. 635

28.   US Pat. 4468291 27.06.83/28.05.84 Naarmann, H., Köhler, G.,
      Schlag, J., BASF AG Ludwigshafen, FRG

29.   BASF, DE-OS 3 425 511, Jul. 11, 1984/Jan. 16, 1968, H. Naarmann and
      F. Wudl: Electrochemistry of Polymer Layers, Internat. Workshop,
      Duisburg, F. R. G., Sept. 15, 1986.
      Patil A. O., Ikenoue Y., Wudl, F., and Heeger, A., J. Am. Chem.
      Soc., 109, 1858 (1987)

30.   Naarmann, H., ICSM Tübingen (1990) Sept. Synth. Metals in press

31 a. Naarmann, H., Theophilou, N., "Synthesis of new electronically con-
      ducting Polymers", p. 12 - 31 (1988) in "Electroresponsive molecular
      and polymeric systems", Vol. 1, edited by T. A. Skotheim, Marcell
      Dekker, N. Y.

31 b. Hanack, M., Hüber, G., Dewald, G., Ritter, H., Röhrig, U., IWEP 91
      Kirchberg 11. März "A new class of low gap polymers"

31 c. Fichou, D., Horowitz, G., Xu, B., Garnier, F., Synth. Metals 39, 243
      (1990),
      Martinez, F., Voelkel, R., Naegele, D., Naarmann, H., Mol. Cryst.
      Liq. Cryst. (1989) 227

31 d. Closs, F., Breimaier, W., Frank, W., Gompper, R., Hohenester, A.,
      Synth. Metals 29 (1989) E 537

32.   Kertesz, M., Youg-Sok, L., Electronic Structure of small gap poly-
      mers", Synth. Metals 28 (1989) C 545

NEWER POLYMERIC APPLICATIONS

IN ELECTRONICS

Jai K. Nigam

Shriram Institute for Industrial Research
19, University Road
Delhi-110007 - India

INTRODUCTION

The development and large scale commercial production
of polymers as a class of materials over the last 50 years is
considered the most significant materials development of this
century and a classic success story. Inspite of such a wide
range of applications in virtually every significant field of
human endeavour, the growth in polymers has not yet plateaued.
The entry of new polymeric materials through composites into
automotive & aero-space applications, till recently a primary
domain of metals, has opened further opportunities for polymer
research. However, a great achievement in applications of
polymers has been in the area of electronics which in turn has
become an index of the performance of national economies.

While, most contributions in this publication relate to
specific polymers or applications, in this thematic review,
an attempt is made to identify some trends & examples of the
uses and applications of newly developed polymers in the
emerging areas of electronics. Some techno-commercial aspects
of the two sectors of polymers & electronics can not be
avoided while such a study is undertaken. A number of these
polymers used in electronics have been studied by the author's
group, particularly for developing manufacturing technologies
for commercial production in India.

RELATIONSHIP WITH DEVELOPMENT IN ELECTRONICS

The number and variety of polymers with inter-changeable
applications based on their characteristics are evidently very
wide now. Starting from the first generation plastics of
commodity, or bulk type reaching a level of saturation, the
2nd generation polymers have attained large manufacturing
capacities and consumption levels and now the third generation
are also getting commercial capacities established (Table 1).
The chemical structures, processes of manufacture, characteristic properties and special applications are too well
documented in various polymer publications to be repeated here.

Frontiers of Polymer Research, Edited by P.N. Prasad and
J.K. Nigam, Plenum Press, New York, 1991

Table 1

Evolution of Engineering & Speciality
Polymers

First Generation(Commodity):

- Low Density Polyethylene(LDPE)
- High Density Polyethylene(HDPE)
- Polyvinyl Chloride(PVC)
- Polystyrene(PS)
- Polypropylene(PP)

Second Generation(Engineering):

- Polytetrafluoroethylene(PTFE)
- Linear Low Density Polyethylene(LLDPE)
- Acrylonitrile-Butadiene-Styrene Ter Polymer(ABS)
- Polycarbonates(PC)
- Polyamides(6/6,6) (PA)
- Polyesters(PET/PBT)
- Polyacetals(POM)
- Polyphenylene sulfide(PPS)
- Polyvinylidene fluoride(PVDF)

Third Generation(Speciality):

- Polyphenylene Oxide (PPO)
- Polyarylate(PAR)
- Polyether-ether ketone(PEEK)
- Polyamide-imide(PAI)
- Polyether Sulphones(PSF)
- Polyether-imide(PEI)
- Liquid Crystal Polymers(LCP)

The electronics industry has practically grown simultaneously in parallel with the evolution of plastics. Through
the phases of continuous sophistication in electronics, since
the introduction of transistor in the 50's to the present day
silicon chip, the progress has been significant and long. The
switch over from electro-mechanical systems and signals to
electronics, based on solid-state materials followed by
subsequent miniaturisation and digitalisation, has been quite
rapid. This revolution of universal application of electronic
devices in almost every type of entertainment, domestic appliances, business & office machines, computers, medical, defence
and forensic equipment, besides Telecommunications has
resulted into uncontrollable growth of business opportunities.
Even those countries where the electronics technology could
not commercialise in early stages, have substantial growth
plans now. For example, in India, the growth of electronics
sector between 1988-1995 is expected to be around 300%
reaching a size of about 10 billion U.S dollars per annum.

A number of polymers have been used in the above range
of equipment primarily performing the functions of encapsulants, insulators and body-housing materials. Commodity as
well as engineering plastics have thus found applications in
electronic appliances in substantial quantities, particularly
in developed countries for which data is available as given
in table-2. However, in Japan, the trend is changing fast,
towards using a greater share of new engineering plastics in
structural and mechanical parts in electronic devices as given
in table-3.

NEW TRENDS OF POLYMER USES IN ELECTRONICS

The performance requirements of materials used in
electronic devices have changed considerably over the past
few years. These requirements may include higher temperature
resistance, broad range of workable temperatures, high impact
strength, mouldable components, improved dielectric &
electrical resistivity characteristics etc. Evidently, the
performance of the equipment is directly related to the
performance of materials or polymers used in the construction
of various components, accessories & devices used in the
equipment. With miniaturisation and introduction of micro-
electronics, the choice for new plastics to replace the
earlier candidates has widened considerably. Elimination of
some steps in the present process sequence has also been one
of the factors in enlarging the range of polymers used in the
electronic devices. The Printed Circuit Board(PCB) which
brings together various individual circuit components in a
modular unit is usually made of glass-reinforced epoxies which
had earlier replaced the metallic structure. Epoxy based
PCBs have been replaced by several other polymers like poly-
arryl sulphones, due to which, vapour-phase soldering is
satisfactorily achieved. The introduction of Mouldable
Circuit Boards(MCB) in a 3-dimensional configuration having
features such as connectors, snap fits, capacitors etc., all
moulded on the board is possible through the use of many new
plastics like PSF, LCP etc. The connectors, sockets, switches,
transformer bobbins, switching cams etc. can all now be made
from a number of polymers such as PET, PES, liquid crystal
polymers etc. The most significant example, however, appears
to be the super high heat resistant material for long term

Table 2

Plastics consumption in Electricals/Electronics
in selective countries

Unit: 1000 Tons

Polymer	W.Europe		Japan		U.S.A	
	1989	1990	1989	1990	1989	1990
A. Thermosets:						
– Epoxy	–	–	55	58	26	25
– Phenolics	–	–	–	–	46	42
– Unsaturated polyesters	–	–	–	–	24	24
B. Thermoplastics						
– HDPE	25	27	–	–	70	75
– LDPE	192	203	77	120	190	187
– Polypropylene	–	–	–	–	22	21
– P.V.C	430	445	382	396	253	268
– Polystyrene	–	–	381	405	216	224
C. Engg.Plastics:						
– Nylons	69	74	20	24	36	37
– ABS	88	92	128	131	53	49
– Polyacetals	–	–	–	–	2	1
– Polycarbonate	–	–	–	–	14	14
– Polyesters	–	–	20	25	21	20
– Polyphenylene sulphides	–	–	–	–	13	13

Source: Modern Plastics International, Jan. 1991.
N.B: Blank – data not available

Table 3

Polymers used in structural/mechanical parts
in Electronic domestic appliances in Japan(1989)

Polyacetals	–	36%
Polyamides	–	27%
Polyesters	–	18%
Polycarbonates	–	13%
Polyphenylene Oxides	–	9%
Polyimide/Fluoro polymers	–	9%

Source: Electronics Industries Assn., Japan.

performance with a temperature range from $-43^{\circ}C$ to $480^{\circ}C$ made from polyimides, particularly for use in micro-electronics. Polyimide tapes have also been used for protecting semi-conductor chip, bonded on to the tape. A number of conventional polymers like polypropylene, HDPE, PVC, polystyrene etc. have found use in the form of alloys and blends, particularly for housing and cabinets of electronics equipment both in consumer electronics as also for computers and office equipment like word-processors and photo-copiers. A number of polymers have also been used for providing anti-static and EMI shielding for protection of electronic data storage in the computer memories. Audio, video and data storage materials have undergone a remarkable transformation from polyester tapes to cassettes, to disc drives and finally to compact discs(CDs), essentially made of polycarbonates with memory storage capacities beyond 600 Mbs.

NEW ROLES FOR POLYMERS IN ELECTRONICS

Some of the emerging areas of polymer applications in electronics besides the established ones are:
- Piezo-electric polymers
- Polymers for imaging
- Conducting polymers
- Polymers as optical fibres

Each one of the above areas has received significant attention from Polymer Scientists as well as application development activities for electronic devices.

Piezo-electric polymers: PVDF has established as pre-eminent polymer for piezo-electric responses and is already being used in a number of applications such as electronic calculators, watches, noise cancelling and ultrasonic micro-phones used in telephone head-sets etc., wherever sensors and transducers based on piezo-electric effect are desired.[1]

Polymers for imaging: By the selective treatment of light sensitive polymers to make even smaller images, the number of transistors that can be crammed into a single tiny piece of semi-conductor has been progressively increased to make very large memory and micro-processing capacity possible. The most recent chip like 386, in a very large-scale integrated circuit(VLSI) can now have a 64K random access memory which means that on a piece of silicon less than 12 mm long, there are more than 64,000 places where a transistor can be located together with all the connecting circuitry. The technology for the production of images of less than 2.5 micron in size and having enough physical strength and chemical resistance to enable fabrication of the electronic circuit is now possible through the use of known polymer films which can make such images. The photo-resist polymers for such applications use cyclised polyisoprene which is used as negative resist. Polybutene sulphones and polymethyl meth-acrylate have been reported to be used as positive resists for making images of sub-micron dimension for masks and special requirements.[2]

Conducting plastics: Getting polymers to conduct like metals is still not totally achievable even today, yet a number of variations in the same have been reported. Electrical conductivity in plastics was till recently limited to materials obtained by filling particles of carbon black or metal flakes, but exciting new possibilities emerged with studies on polyacetylene. Surprisingly, polyacetylene invented much earlier by G. Natta, could be converted to exhibit electrical conductivity by Shirakawa with excessive catalyst dosing and exposing to Iodine vapour by MacDiarmid and Heegar.[3] The subject received significant attention to the possibility of practical applications of the phenomenon. However, the most successful application so far, appears to be in the development of light-weight plastics battery, in which the conductive polymer electrode is made of polyacetylene. After polyacetylene, a number of other polymers such as polyaniline, polyparaphenylene, polypyrole and polythiophenes, have also shown potential uses as conducting materials.

Polymeric optical fibres: Although the optical fibre and telecommunication applications fall under the photonics area rather than electronics, it is only for the sake of comprehensive coverage that a mention of this subject is necessary. Optical fibres are widely used now in data transmission of signals. Silica or glass optical fibres are in wide use at present and offer good transmission over long distance, but do have some disadvantages such as brittleness and difficulty in jointing. The disadvantages mean that for short distance data links, optical fibre with higher loss could be tolerated if the core material is easy to handle. Polymeric optical fibres have opened this possibility for the construction of short distance branched information net-works. The two major polymers used as Polymer Optical Fibre(POF) include polymethyl methacrylate(PMMA) and polystyrene(PS). While the polymer preparation for the two polymers is well-known, the POF production requires special equipment and polymerisation conditions as well as in the pre-form production and subsequent fibre drawing, cladding and assessment. Besides, polystyrene and PMMA, a number of companies are investigating the production of fluorinated polymers for POF - not only for core material but also for cladding. Many of the new materials may possess such low refractive indices that conventional claddings used will not be good enough for light guiding. Polycarbonate POF are also reported to have advantages of higher service temperature conditions. POF have already been successfully used in closed systems like computer net-works and inhouse communication systems.

Besides the above new applications for the polymers, the emerging field of high temperature super-conductors opens a totally new field for the conversion of these materials into wires, tapes and films for utilisation in devices and systems requiring substantial support from polymer applications for achieving desired performance characteristics. A number of polymer compatibility studies are already being followed with the well-known $YBa_2 Cu_3 O_{6+}$ - the most common and successful super conducting material to be used in medical devices and super-computer applications.

CONCLUSION

The interaction between polymers and electronics industry
has been interesting and exciting since both technologies have
grown and matured almost in parallel. Continuous inputs of
new polymers with improved performance characteristics have
enlarged their potential use in Electronics not only in the
conventional application as insulators and encapsulants but
also exhibiting optical and electrical current transmission
capabilities. While significant developments have already
been reported in these areas, the new materials continue to
be in the low volume, high value category products. It is
hoped that in order to achieve the best expected performance
of electronic equipment & devices, wide scale use of newer
polymers shall become almost unavoidable and uneconomic not to
use.

REFERENCES

1. P. Pantelis, "Piezo-electric Polymers"; Shell Polymers,
    Vol.8, No.3, 1984, p.78-81.

2. R.C. Daly & J.L.R. Williams; Polymers & the Silicon
    Chip: Shell Polymers, Vol.7, No.1, 1983, p.25-28.

3. W.J. Feast ; Polyacetylene - A Scientific challenge
    and a Technological opportunity; Shell Polymers,
    Vol.8, No.1, 1984, p.24-28.

4. R.S. Ashpole, S.R. Hall, P.A. Luker; Polymer Optical
    Fibres - A case study; GEC Journal of Research, Vol.8,
    No.1, 1990, p.32-41.

THE C=C STRETCH MODE: A PROBE FOR MOLECULAR AND ELECTRONIC STRUCTURE IN CONJUGATED POLYMERS

H. Kuzmany

Institut für Festkörperphysik der Universität Wien and
Ludwig Boltzmann Institut für Festkörperphysik, 1090 Wien
Strudlhofgasse 4, Austria

INTRODUCTION

Physical properties of conjugated polymers are determined to a great deal by the π-electron system. This concerns in particular the electric and electronic, optical, magnetic and even vibronic properties. The influence on the latter is obvious if the widely and very successfully used description of the vibrational structure by phonon renormalization techniques is considered. The key quantity in the π-electron system is the bond alternation amplitude along the carbon backbone. Thus, it is quite natural that all vibrational modes which have a normal coordinate with a strong bond alternation contribution should be excellent probes for the π-electron system. This modes are in particular the C=C stretch mode and related vibrations.

One of the best known and most general parameters which determines the bond alternation amplitude is the length of the macromolecule. The bond alternation amplitude is in particular a strong function of the number N of unit cells on the chain if N is samller than about 40. A typical example is the simple trans-polyacetylene chain with a finite number of double bonds on the backbone. In this case the π-π* transition energy ε_g and the C=C stretch mode frequency ν are strong functions of N as demonstrated in Fig.1 for experimental results from a set of oligoenes and for calculated results from a Hückel type Hamiltonian including σ-compressibility[1]. Obviously, in each case N determines both, the π-π* transition energy and the vibrational mode frequency.

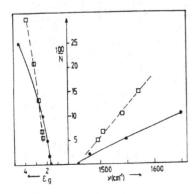

Fig.1

Optical transition energy ε_g and C=C stretch frequency ν for finite conjugated carbon chains as calculated from a Hückel Hamiltonian (squares) and as observed for diphenylpolyenes (points).

In the following paper we want to concentrate on experimental and theoretical results concerning the usefulnes of the vibrational C=C double bond stretch mode to obtain information about the π-electron system. Numerous experiments and calculations have been performed so that only a small amount of carefully selected and most characteristic results can be reported in this reviewing contribution.

THE DISPERSION OF THE C=C STRETCH MODE RAMAN LINES

The most famous example for the connection between the π-electron structure and vibrational modes is the dispersion effect of Raman lines in trans-poly(acetylene). This effect is characterized by a shift of Raman lines from vibrational modes with shifting quantum energy of the exciting laser. Modes with a strong C=C stretch character (amplitude character) exhibit a particular strong lineshift. This effect can be described straight forwardly by a photoselective resonance process in a sense that any distribution of conjugation lenghts in the polymer will lead to a distribution of gap energies and correlated amplitude mode frequencies. Thus, each laser energy will select by resonance excitation a particular part of the polymer and the overall response gives the resulting line shape and line position.

Table 1. Dispersion of Raman lines for various conjugated polymers

polymer	line (cm^{-1})	dispersion (cm^{-1}/eV)	reference
poly(acetylene), cis	1540	7	Ref.2
trans	1500	76	3
poly(diacetylene)	1500	55	4
poly(pyrrole)	1569	6.6	5
poly(3-methylthiophene)	1460	24	6
poly(3-octylthiophene)	1460	40	7
poly(thiophene)	1460	0	8
poly(p-thiophenevinylene)	1410	10	9
poly(p-furylenevinylene)	1432	32	10
poly(p-dihexylsilane)	689	6	11
poly(paraphenylene)	1598	0	12
poly(isothianaphtene)	1470	39	13

The dispersion phenomenon is not restricted to trans-poly(acetylene) but rather observed in nearly all conjugated polymers which have vibrational modes with strong amplitude character. Table 1 compiles a variety of polymers where a dispersion effect of Raman lines has been reported. The amount of dispersion expressed in wavenumber shift of the Raman line peak per energy shift of the laser excitation is both, extrinsic and intrinsic. It is intrinsic to the amount that a change in conjugation length modifies the electronic structure and thus the gap energy and the phonon renormalization but it is extrinsic to the amount that all conjugation lengths from 1 to infinity are really occupied. Thus, the zero dispersion reported for eg. poly(thiophene) may just mean that a reasonable fraction of long segments is not present in these polymers which is in agreement with the well known disorder in this system. The disorder results from the lack of a proper α-α coupling and is in contrast to the higher ordered poly(3-methylthiophene) which, consequently, exhibits a dispersion of 24 cm^{-1}/eV.

The dispersion phenomenon can be described quantitatively in the conjugation length model for resonance Raman excitation of conjugated polymers[14,15]. This model assumes a distribution of conjugation lengths along the polymer chain which originates from an interruption of conjugations by various types of defects. Since a defect may not interrupt the conjugation completely but may just lead to an increase of the gap energy the conjugation length is effective and determined from the observed opening of the gap. Under these assumptions the optical properties and the resonance Raman cross section for the individual segments of the whole chain can be calculated from Albrecht's theory for arbitrary conjugation length N. If the segment length is not very large, eg. smaller than 50 double bonds the A-term in the Hertzberg-Teller expansion of the transition polarizability determines the combined optical and vibronic transitions. Thus, optical absorption and cross sections for first order Raman scattering are, except for thermal averaging, obtained for a zz dominated transition from

$$A(\omega)_{m,n} = C_A \; |(\alpha_{zz})_{m,n}|^2 \tag{1}$$

and

$$(\frac{d^2\sigma}{d\Omega d\omega})_{m,n} = C_R \; |(\alpha_{zz})_{m,n}|^2 \tag{2}$$

respectively. $(\alpha_{zz})_{mn}$ means the zz component of the transition polarizability between the state m and n which can be calculated from

$$(\alpha_{zz})_{m,n} = \sum_e [\frac{\langle m|z|e\rangle\langle e|z|n\rangle}{(\varepsilon_e-\varepsilon_m) - \varepsilon_o} + \frac{\langle m|z|e\rangle\langle e|z|n\rangle}{(\varepsilon_e-\varepsilon_m) - \varepsilon_o}] \tag{3}$$

$$= \sum_e |\langle g^o|z|e^o\rangle|^2 \sum_v [\frac{\langle i|v\rangle\langle v|f\rangle}{(\varepsilon_{ev}-\varepsilon_{g1})-\varepsilon_o+i\Gamma/2} + \frac{\langle i|v\rangle\langle v|f\rangle}{(\varepsilon_{ev}-\varepsilon_{g1})-\varepsilon_o+i\Gamma/2}],$$

where the transition matrix elements must be calculated from the wave functions in the states m,e, and n with the corresponding eigenvalues ε_m, ε_e, and ε_n. ε_o and Γ are the laser energy and the widths of the electronic transitions, respectively. In the limit of the adiabatic and the Franck-Condon approximation the matrix elements can be separated into purely electronic and core contributions where the latter are vibronic overlapp integrals between the initial (i) or final (f) vibrational state in the ground electronic state and the intermediate (v) vibrational state in the excited electronic state as indicated in the second line of Eq.3. For optical absorption i and f are zero whereas for first order Raman scattering i is zero and f is 1. The vibronic overlapp integrals can be evaluated analytically and the electronic transition matrix elements can be determined from a quantum chemical calculation of the electronic structure. Thus, for a comparison with experimental results the only parameter to be fitted is the distribution function of the conjugation length which as a consequence can be evaluated from this type of analysis. The very good and quantitative agreement of calculated results with experiments for a large variety of line shapes with a single distribution function renders this technique as a relyable methode for the determination of the polymer backbone structure. It is very imortant to realize that the electronic transition matrix elements $\rho_{go,eo}$ dependent strongly on the transition energy as it is shown for a Hückel-type calculation in Fig.2 [16]. The symbols represent values for the square of the transition matrix elements for various segment lengths and for various electronic transitions within one segment.

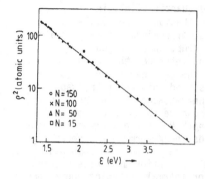

Fig.2

Square of the electronic transition matrix element ρ versus transition
energy on a double logarithmic
scale as calculated from a Hückel approximation for polyene chains of different
lengths N.

They are plotted on a double logarithmic scale. The full drawn line represents
a $1/\varepsilon^4$ law. Thus the resonance is not only determined by the denominator in
Eq.3 but also by the value of the matrix elements. Simplified versions for the
analysis of the resonance Raman spectra like the amplitude mode model[17] or the
effective conjugation coordinate model[18] do not consider the energy dependence
of the matrix elements and can thus not be used to determine distribution
functions of characteristic polymer parameters.

The dispersion effect was studied in many polymers with strongly varying
efford. Interesting results are observed for poly(diacetylene) where both, the
double bond and the triple bond can have amplitude character. Thus, in not
very perfect single crystals both modes show a dispersion effect[4] as
demonstrated in Fig.3. The figure shows the shift of two Raman lines with
shifting laser excitation. The lines are satellites to the main C=C and C≡C
Raman lines from the perfect crystal. The fact that the line shifts merge into
the main line (zero shift) for excitation with the well known exciton
transition energy of 1.9 eV for poly(diacetylene) is a good evidence for the
origin of the two lines from the polymer itself and not from oxygen
impurities.

Other interesting dispersion effects have been observed for
poly(isothionaphthene) (PITN) or poly(octylthiophene) (POT)[7]. In both cases
only one mode with amplitude character is observed. In PITN the Raman line of
this mode, however, shows a fine structure which could be attributed
explicitly to individual perfectly conjugated segments on the backbone. In
POT the amplitude character of the mode changes from one mode to an other with
increasing conjugation length[19].

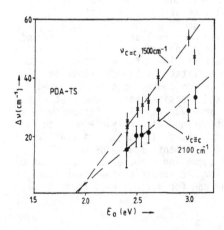

Fig.3

Shift of satellite lines for the C=C
and C≡C stretch modes for various
laser excitation energies ε_0 as
observed for non perfect
poly(diacetylene)-TS crystals.

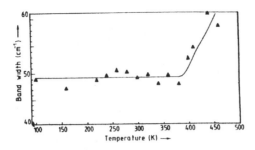

Fig. 4. Line width of the amplitude mode in poly(octylthiophene) versus
temperature below and above the phase transition

THE C=C STRETCH MODE IN ORDER-DISORDER TRANSITIONS OF PROCESSABLE CONJUGATED POLYMERS

Order-disorder transitions in conjugated polymeres are an other target
where the C=C stretch mode can be used as a key probe for polymer structure.
It is to some extend related to the dispersion phenomenon since a transition
into a disordered state certainly reduces the conjugation length. However,
more importantly the transiton may completely quench the Raman intensity if it
is accompanied with a reasonalble shift from resonance condition as it is eg.
observed for the poly(silanes)[11]. In this case the mode with the amplitude
character is not C=C stretching but rather a Si-Si σ-bond stretching which,
evidently, exhibits a very similar nature as the C=C stretching in the carbon
backbone polymers. A very correlated phenomenon was observed recently for the
order-disorder transition in the poly(octylthiophene)[7]. This transition was in
addition characterized by a well expressed discontinuity in the line width of
the amplitude mode at the phase transition as shown in Fig.4. The phase transition for this sample as obtained from the optical spectra ocurred at 400 K.

THE C=C STRETCH MODE IN DOPING INDUCED STRUCTURAL TRANSITIONS

Doping induced structural transition are very familiar in conjugated
polymers. A typical example is the transition of an aromatic type structure of
a backbone into a quinoid structure by doping. Similarly, the reverse will
happen if the neutral polymer is in the quinoid structure as it was observed
recently for PITN[13]. Then the interring stretch mode and the C=C ring stretch
modes are good probes for the structure of the samples.

In an even more dramatic way the transition of an aromatic structure to a
semiquinoide structure can be observed in poly(aniline) (PANI) from a change
of line intensity of the benzene ring mode at 1630 cm⁻¹. This transition
occures in an acidic electrolyte at an electrochemical potential of about 200
mV vs SCE and goes along with an oxidation current as shown in Fig.5., full
drawn line. Simultaneously, polaron states occupied by Curie spins are created
by the oxidation process up to a certain concentration. The polaron state is
characterized by a benzene ring vibration at 1630 cm⁻¹ [20]. As a consequence of
the polaron formation the intensity of this line increases dramatically with
the oxidation current. Finally, for very high polaron concentrations, the
polarons interact to a bipolaron state without spin or to a polaron lattice
state where spins are Pauli type. In both cases the response to the spin
resonance decreases and the Raman line from the polaron species is
simultaneously bleached. More information about the polaron-polaron
interaction process can be obtained from a plot of the Raman line intensity
versus the doping concentration. From this type of analysis an inhomogeneous
doping process was deduced[21].

293

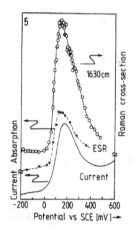
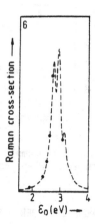

Fig.5. Oxidation current (full line), ESR absorption (full points) and Raman cross section (circles) for the polaron mode at 1630 cm^{-1} in poly(aniline) during an electrochemical oxidation process.

Fig.6. Resonance Raman cross section of the polaron mode at 1630 cm^{-1} as measured (points) and as calculated (dashed line), after Bartonek[21].

Finally, the energetic position of the resonance for the polaron mode allowes to determine very accurately the energy level for the polaron state in the band structure. Experimental results for the Raman cross section and calculated values using Eq.1-3 are show in Fig.6. The structure in the calculated curve originates from the vibronic contribution to the resonance process which is not resolved experimentally. For the electronic transition an energy of 2.75 eV was used for the fit. This value is in very good agreement with results calculated quantumchemically in the valence effective hamiltonian approximation for the polaron state[22].

CONCLUSION

The C=C stretch mode is a fingerprint for physical properties of the conjugated polymers. Its position and intensity in the Raman spectrum allows to analyse the chemical and electronic structure of the systems under consideration. The observed intensity of the modes is always resonance enhanced since it has an amplitude mode character, even in those cases where an amplitude oscillation in the conventional sense is not active.

ACKNOWLEDGEMENT

The support of this work by the Fonds zur Förderung der wissenschaftlichen Forschung in Austria is greatfully acknowledged.

REFERENCES

1. J. Kürti, H. Kuzmany, Phys. Rev. B44, (1991)
2. L.S. Lichtmann, E.A. Imhoff, A. Sarhangi, D.B. Fitchen, J. Chem. Phys. 81:168, (1984)

3. H. Kuzmany, Pure and Applied Chem. 57:235 (1985)
4. J. Kürti, H. Kuzmany, Springer Series in Solid State Sciences 76:43, H. Kuzmany, M. Mehring and S. Roth, ed. Springer Verlag, Berlin (1987)
5. Y. Furukawa, S. Tazawa, Y. Fujii, I. Harada, Synth. Met. 24:329, (1988)
6. E.F. Steigmeier, H. Auderset, W. Kobel, D. Baeriswyl, Synth. Met. 18:219, (1987)
7. T. Danno, J. Kürti, H. Kuzmany, Phys. Rev. B43:4809, (1991)
8. Y. Yacoby, and E. Ehrenfreund, Light Scattering in Solids, M. Cardona ed., Springer Verlag, Berlin, Heidelberg, to be published
9. S. Lefrant, J.Y. Mevelec, J.P. Buisson, E. Perrin, H. Eckhardt, C.C. Han, K.Y. Jen, Springer Series in Solid State Sciences 91:123, H. Kuzmany, M. Mehring and S. Roth, ed. Springer Verlag, Berlin (1989)
10. S. Lefrant, E. Perrin, J.P. Buisson, H. Eckhardt, C.C. Han, Synth. Met. 29:E91, (1989)
11. H. Kuzmany, J.L. Farmer, R.D. Miller, and J.R. Rabold, J. Chem. Phys. 85:7413, (1986)
12. S. Buisson S. Krichene, and S. Lefrant, Synth. Met. 21:229, (1987)
13. W. Wallnöfer, E. Faulques, H. Kuzmany, and K. Eichinger, Synth. Met. 28:C533, (1989)
14. H. Kuzmany, phys. stat sol. B97:521, (1980)
15. H. Kuzmany, Applied Chem. 57:235, (1985)
16. H. Kuzmany, P.R. Surjan, and M. Kertesz, Solid State Commun. 48:243, (1983)
17. E. Ehrenfreund, Z. Vardeny, O. Brafman, and B. Horovitz, Phys. Rev. B36:1535, (1987)
18. J.T. Lopez Navarete, B. Tian, G. Zerbi, Solid State Commun. 74:199, (1990)
19. T. Danno, J. Kürti, and H. Kuzmany, Proceedings of the IWEPP'91, to be published
20. M. Bartonek, and H. Kuzmany, Synth. Met., to be published
21. M. Bartonek, Thesis, University of Vienna 1990
22. S. Stafström, J.L. Bredas, A.J. Epstein, H.S. Woo, D.B. Tanner, W.S. Huang, and A.G. MacDiarmid, Phys. Rev. Lett. 59:1464, (1987)

PHOTOCONDUCTIVE POLYMERS: CHEMICAL STRUCTURES AND TECHNICAL PERFORMANCE

D. Haarer

Physics Institute University of Bayreuth and BIMF
P.O. Box 101251, D-8580 Bayreuth, FRG

Abstract

Photoconductive polymers are presently used in a wide variety of technical applications such as electrophotography (Xerography) and laser printing. From the viewpoint of new materials several improved properties would be very desirable. Whereas the dielectric properties and the charge carrier production yields of polymeric materials are extremely well-suited for the above technical applications, the mobilities of presently used materials are rather low. In this article we will give a short survey over charge carrier production yields and then discuss the mobility issue based on recent experiments and their formal interpretation. As model systems we use carbazole substituted polymers; we also report some recent preliminary data on quasi-conjugated polymers (polysilanes).

I. Introduction

It has been rather surprising that in the late sixties and early seventies polymeric materials were able to substitute and surpass inorganic materials in the area of photoelectric applications, such as electro-photography (Xerography), laser-printing and the production of offset-printing masters. This has been possible since polymers have some unique advantages such as:

- Excellent dielectric properties of thin films (high breakdown strength)
- Inexpensive methods for producing perfect large area films and devices (low pinhole density)

Frontiers of Polymer Research, Edited by P.N. Prasad and
J.K. Nigam, Plenum Press, New York, 1991

- Feasibility of sensitization of the photoelectric response from the ultraviolett to the infrared spectral regime.
- Reasonable mechanical properties combined with excellent adhesion properties.

Besides these qualities there are other properties which need to be improved; they are:

- Low effective charge carrier mobilities
- Charge carrier trapping problems by chemical and structural defects
- low glass transition temperatures

Even though the global figure of light sensitivity as given in Fig. 1 /1/ has many rather different aspects, it shows that the overall performance of photoconductive polymers (electrophotography) is only surpassed by the light induced processes active in silver halides which are known to have a very large gain mechanism. The sensitivity of polymeric photoconductors is several orders of magnitude higher than that of straight forward photochemically active materials.

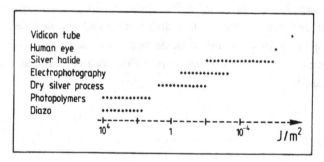

Fig. 1.

Light sensitivity of various technical processes and materials as compared to the sensitivity of the human eye

The basic processes which occur in photoconductive materials can be roughly subdivided into two subsequent steps. First, the electron-hole generation process and second, the carrier transport. The first process will be only briefly discussed. The second process of carrier transport will be discussed in some more detail, since major improvements with new materials can be expected in this area within near future.

II.   The Electron-Hole Separation Process

The light induced electron hole separation process is the primary step of both, the photoconduction process and, more important, the photosynthetic process. This process occurs in the picosecond or sub-picosecond time domain and is not well understood in terms of the involved electronic states (charge transfer states). The macroscopic description of the charge generation efficiency, however, yields a very good agreement between theory and experiment if one uses the wellknown Onsager theory.

The theory describes the diffusion of an electron-hole pair (ion pair) in an external electric field, including the Coulomb attraction of the (geminate) charge carrier pair /2/.

$$\frac{\partial f}{\partial t} = \frac{kT}{e} \mu \ \text{div} \ [\exp(\frac{U}{kT}) \ \text{grad} \ f(\frac{U}{kT})] \tag{1a}$$

$$U = - \frac{e^2}{4\pi\varepsilon\varepsilon_0 r} - eE \ r \cos \Theta \tag{1b}$$

In the above diffusion equation f is the probability for electron hole separation, U is the potential as defined in equ. 1b, E is the electric field, r the electron hole distance and Θ the angle between the field lines and the dipole vector, as defined by the charge carrier pair.

Equation 1 cannot be solved analytically. Numerical solutions have been given for molecular crystals /3,4/; in the case of polymers /5/ the accessible field range was often too limited to verify the validity of the Onsager ansatz.

Fig. 2 shows experimental and numerical data on polyvinylcarbazole /6/ which is often used as model system for polymeric photoconductors (hole conduction). The experimental data agree reasonably well with the numerical data, reflecting a sizeable experimental error bar in the low field regime, in which trapped charge carriers can lead to major field distortions. It is important to note that the S-shaped curve shows very low yields (10^{-4} - 10^{-5}) in the low field regime and a yield of close to unity in the high field regime, which is the regime, which characterizes technical applications like electrophotography and laser printing.

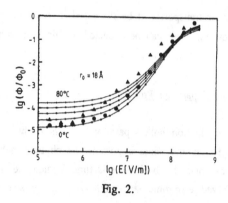

Fig. 2.

Onsager ionization probability Φ as calculated for temperatures between $0°$ C (lower curve) and $80°$ C (upper curve) in increments of $20°$ C. The experimental data are given for $0°$ C (circles) and $80°$ C (triangles).

The Onsager equation can be fitted with one microscopic fit parameter, namely the radius r_0, at which the diffusion mechanism is 'switched on'. This radius is, in the case of polyvinylcarbazole (PVK) 18 Å; it is much larger than a typical excitonic radius which would be on the order of 3 - 4 Å for Frenkel-like excitons and at the most 5 - 7 Å for charge-transfer-like excitations.

Fig. 3 shows a primitive sketch of typical electron-hole separation distances in a carbazole Frenkel-state and in a carbazole charge-transfer state with the acceptor molecule trinitrorfluorenone (TNF). In the latter case the electron-hole wavefunction is in the simplest case delocalized over 3 momoner units (see for instance Ref. /7/).

The present understanding of the process of electron-hole separation is rather limited. Picosecond and sub-picosecond techniques are only emerging recently /8,9/ and the steady state theories give microscopic Onsager radii r_0 on the order of 20 Å, a figure which is too large to be compatible with simple molecular pictures as given in Fig. 3. This may not be surprising since the thermalization process of a geminate electron-hole pair can presently not be discussed in a quantitative fashion and correlated with reliable experimental data. Even more surprising is the rather good agreement between the experimental data and the quasi steady state description of the Onsager model.

From the viewpoint of technical applications of photoconductive polymers the above data show that high quantum yields of close to unity can be obtained at fields above 10^8 V/m i.e. the materials parameters are very close to optimal for electrophotographic applications, which are optimized for the high field regime. Photovoltaic

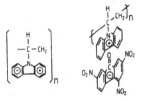

Fig. 3.

Precursor states of photoconductive materials in a schematic view which shows that the 'Onsager radii' of about 20 Å are not in agreement with straight forward pictures of 'molecular excitons'.

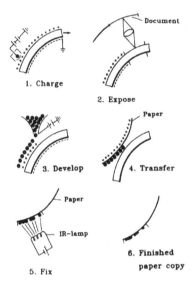

Fig. 4.

Sequence of processes in the electrophotographic imaging technique:
1. Corona charging, 2. Exposure of photoconductor, 3. Development of electrostatic picture, i.e. transfer of charges across the active medium and formation of an electrostatic picture with toner particles, 4. Transfer of toner to paper, 5. Thermal fusion of the toner particles to paper, 6. Finished copy. For the production of offset printing plates, the toner is directly fused to the polymeric photoconductor omitting step 4 /10/.

applications, however, which occur in the low field regime show poor yields. Also the problem of fabrication well defined Schottky-barriers for efficient electron hole separation has, in our opinion, not been solved for organic polymeric materials.

III. Charge Carrier Transport Properties of Polymers

III.1. Aspects of Technical Applications

In most technical applications charge carriers have to be transported over macroscopic distances. In the case of the most commonly discussed process of electrophotography such distances are on the order of 10 μ i.e. the externally created corona charges have to be discharged across a polymer of the above thickness. This process of formation of an electrostatic picture from a 'photonic' picture occurs in steps 2 (expose) and 3 (develop) of the electrophotographic scheme as given in Fig. 4.

The present bottleneck in photoconductive applications is not given by the quantum yield of the process which, as has been shown above, is close to unity. It is given by the transit time of the charge carriers through the photoconductive medium which is presently in the ms time regime. This compares with thin film semiconductors for which this transit timescale can be more than 3 orders of magnitude shorter.

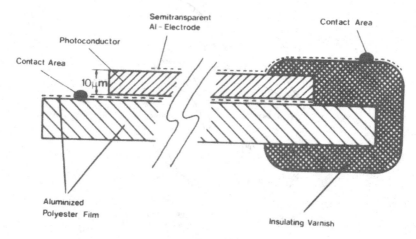

Fig. 5.

Typical sample cross section. The photoconductor layer is on top of an aluminized polyester film. The insulating varnish at the edge of the sample prevents electric breakdown. The sample is illuminated through the top electrode.

III.2.   Charge Carrier Transport: The Mobility Issue

The well established technique for determining charge carrier mobilities is the time of flight technique (TOF) /11,12/. Here the carriers are created with a light pulse (short compared to the transit time τ_t) close to one electrode. In PVK this is achieved by irradiating into the absorption of the carbazole moiety. In this case the penetration depth of the light is less than 0.5 μ i.e. much less than the sample thickness. A typical sample geometry is shown in Fig. 5. The illumination occurs through the top electrode.

An idealized transit current is given in Fig. 6 at a transit time of 100 ms. At τ_t the carriers are discharged through the back electrode and, hence, the photo (displacement) current drops to zero. In this simple scheme the mobility μ is given by

$$\tau_t = \frac{l}{E\,\mu} = \frac{l^2}{V_B\,\mu} \tag{2}$$

where l is the sample thickness and V_B the applied voltage. The total number of light induced charge carriers Q, which is needed for determing the quantum yield,is given by equation 3.

$$Q = \int_0^\infty I\,dt \tag{3}$$

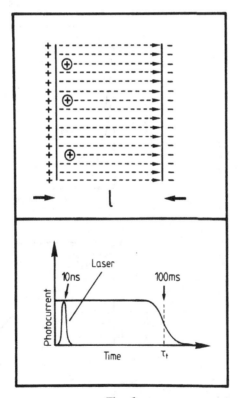

Fig. 6.

Schematic description of the typical TOF experiment (see text)

In crystalline materials the simple TOF scheme can be applied rather successfully. A typical experimental result for a molecular charge transfer crystal is given in Fig. 7a. One can see that the transit time changes from the low field regime (I) to a high field regime (II). The superimposed exponential decay of the photocurrent is due to trapping of charge carriers; it does not interfere with a determination of the mobility at reasonable fields /7/.

Fig. 7.

a) Photocurrent in the molecular CT-crystal anthracene-PMDA for two applied voltages (I = 1000 V; II = 2500 V; the sample thickness was 1,2 mm.

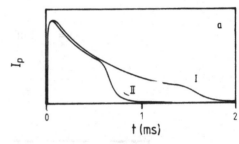

b) Photocurrent in the polymer siloxane at an applied voltage of 200 V and 400 V (Thickness 10 μ).

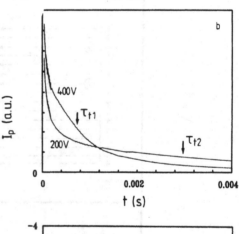

c) Same data as above but plotted on a log-log plot.

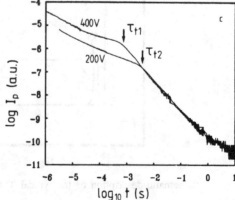

For amorphous polymers straight forward TOF data cannot be described by the simple theory which works well for crystals. Fig. 7b shows two TOF curves for a carbazole substitutes siloxane polymer (see below). Here a double logarithmic plot of the data yields effective transit times τ_{t1} and τ_{t2} at different fields.

The reason why this simple scheme of the TOF experiment breaks down is the dispersive nature of the transport mechanism in polymers.

III.3. Dispersive Transport in Polymers: Energetic und Geometric Disorder

A straight forward description of the mobility does not apply for polymers, since these materials are characterized by disorder. This disorder can be of geometric nature, leading to a hopping transport of carriers for which the time between hopping events varies over a large dynamic range due to the different molecular overlapp between close and distant next neighbors in the polmyeric materials (density fluctuations in the polymer). This situation has been described for the first time by Scher and Montroll /13,14/ in their CTRW approach (continuous time random walk). A one dimensional picture of the CTRW model is given in Fig. 8a. Here the charge carrier is hopping in field direction (downhill) in an equally spaced lattice with a waiting time distribution given by the $\psi(t)$ function which will be defined below.

It was shown later /15/ that an energetic picture of carriers in a narrow band whose transport is determined by traps of various depths leads to results which follow the same mathematical dependency. This model - the multiple trapping model - is symbolically described in Fig. 8b. Here we assume a densitiy of states $g(\varepsilon)$ of traps which falls off exponentially below the conduction band. At a trap depth of energy kT_0 the density of trapping states has, in this model, decayed to its 1/e value.

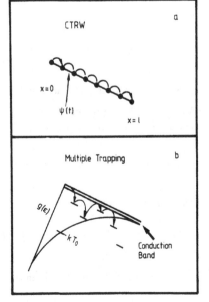

Fig. 8a-b.

Schematic description of the CTRW model and the multiple trapping model (see text)

If one defines the $\psi(t)$ function which describes the hopping mechanism as probability that the charge carrier performs a field induced jump (in field direction) in the time interval between t and $t + \Delta t$, then the most simple ansatz is given by a $\hat{\psi}(t)$ function like

$$\hat{\Psi}(t) = W \exp \{-Wt\} \tag{4}$$

Equ. 4 describes the Gaussian transport mechanism where W is a probability value as given by the molecular overlapp in an ordered, periodic lattice. The $\hat{\psi}(t)$ function is normalized and falls off exponentially.

Clearly the situation in a disordered solid cannot be described by Equ. 4 due to a local variation of the hopping probabilities and the equivalent hopping time distributions. A general ansatz is therefore given by a superposition of $\hat{\psi}(t)$ functions such as

$$\Psi(t) = \sum_{\varepsilon} g(\varepsilon) \, W(\varepsilon) \exp \{-W(\varepsilon)t\} \tag{5}$$

here $g(\varepsilon)$ is a distribution function characterizing the various $\hat{\psi}(t)$ terms. We have chosen an ε-parameter for characterizing the various $\hat{\psi}(t)$ contributions since in the trapping picture the ε-parameter will have a simple interpretation (trap depths; see below).

It was Scher and Montroll and Scher and Lax /13,14/ who showed that the long time behavior of a $\psi(t)$ function as given in Equ. 5 can be approximated by an algebraic law.

$$\Psi(t)\Big|_{t \to \infty} \sim t^{-(1+\alpha)} \tag{6}$$

where α is a number between 0 and 1. Here $\alpha = 0$ would correspond to the largest possible disorder. The algebraic time behaviour of $\psi(t)$ is also reflected in the macroscopic photocurrents which fall off as

$$I = I_0 \; t^{-(1-\alpha)} \tag{7a}$$

before the transit time and

$$I \approx \hat{I}_0 \; t^{-(1+\alpha)} \tag{7b}$$

after the transit time. As has been stated above at the transit time τ_t the charge

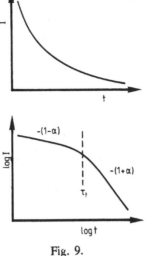

Fig. 9.

Linear and logarithmic plot of a typical dispersive photocurrent (see text)

carriers begin discharging at the back electrode and, hence, the photocurrent shows a steeper slope as given in Fig. 9.

Even though the mathematical description and the results of the theory of dispersive photocurrents are rather straight forward, a microscopic understanding of the α-parameter is not easy. Only in the case of energetic disorder a simple picture can be given within the framework of the multiple trapping model. If one assumes that the detrapping is occuring via Boltzmann activation with

$$W(\varepsilon) = W_0 \exp\{-\varepsilon/kT\} \qquad (8)$$

where ε is the trap depth and W_0 is an 'attempt to jump' probability; and if one further assumes that the density of states of traps falls off exponentially with kT_0 then one can readily show that α is given by the following ratio (see for instance /16/).

$$\alpha = T/T_0 \qquad (9)$$

In this simple picture α is just given by the trap distribution, falling off exponentially with kT_0 and by the absolute temperature T. If one measures photoconduction currents over a large dynamic time window, the above rather simple equation 9 only holds for some polymers. One example is the polymer polysiloxane with pendant

307

carbazole groups. Here the algebraic time dependencies as shown in Fig. 10 are extremely well defined before the transit time. After the transit time the photocurrent deviates slightly from the theoretical curve which is characterized by an α-value of 0.58.

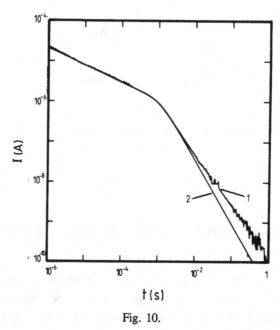

Fig. 10.

Transient photocurrent for a carbazole substitued siloxane polymer /17/. Experimental data (1); theory (2)

III.4. Deviations from the Simple Dispersive Models

The simplest model of dispersive transport as given above can only be tested if a very large dynamic time range is realized experimentally. Among the first data showing such a large dynamic range, the data on PVK /18/ show clearly that the time exponent can change during one current transient over its full range ($\alpha = 0$ at short times; $\alpha = 1$ in the long time regime). Fig. 11 shows the experimental data from 10^{-9} s to 10^{+1} s. The roundoff at the beginning of the transient is given by the 10 ns time response of the experimental setup; the transit time is roughly 10^{-2} s.

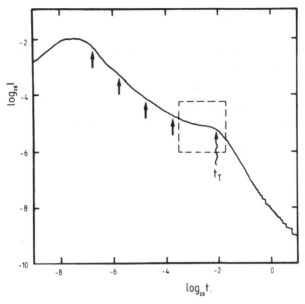

Fig. 11.

Photoconductivity of PVK over a large dynamic range. The arrows mark the onset of different transient digitizers which are optimized for the individual time window. t_T marks the transit time. The dotted line marks the dynamic range of two time decades. Many experiments are, unfortunately, restricted to such a narrow time window and, hence, do not reveal details of the dispersive transport.

The discrepancy between the simplest CTRW model in which α is constant and the experimental data becomes rather obvious in Fig. 11. However, if one replaces the exponential trap distribution which had been assumed so far by a truncated distribution, one can show that a transition between dispersive and non-dispersive transport can occur /16,18,19/. This transition is characterized by the time at which the charge carrier distribution of filled traps reaches the lowest trap level of the truncated exponential distribution /16,20/. In Fig. 11 this change from dispersive to non-dispersive transport occurs about one time decade ahead of the transit time (at 10^{-3} s). More details on the mathematical model /19/ and on calculations with different trap distributions /21/ can be found in the literature. In summary, we can say that only the most simple assumptions lead to simple time dependencies of the photocurrents as given by Equ. 7a,b. In many experimental systems a rigorous mathematical treatment and a set of excellent experimental data is needed to fully characterize dispersive transport in its various aspect.

IV. The Mobility Issue and New Polymers

The mobilities which characterize the 'switching time' of any photoelectric device vary over many orders of magnitude in organic materials. Fig. 12 shows the presently known materials on a scale of about 9 decades. Here organic crystals are characterized by mobilities on the order of 10^{-1} to 10^{0} cm^2/Vs, whereas presently used polymers (like PVK) have mobilities on the order of 10^{-6} cm^2/Vs i.e. they are characterized by values which are about six orders of magnitude below those of amorphous silicon.

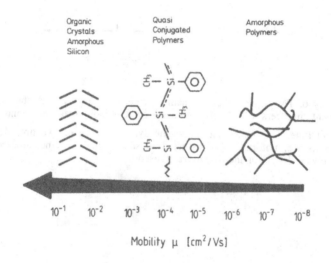

Fig. 12

Mobilities of organic materials (see text)

The question has to be raised, how this situation can be improved in terms of ob-taining higher mobility materials and, hence, faster organic devices. In the presently used materials like PVK or carbazole substituted siloxanes (see Fig. 13) the charge carrier transport mechanism is characterized by a hopping process from one molecular unit to the next (here carbazole). Such a hopping process which is controlled by the molecular overlapp between a adjacent molecules is often well described by a trapping model like the multiple trapping model or by the more general CTRW theory which does not discriminate between energetic and geometric disorder.

The opposite extreme to hopping conduction would be band-like conduction like in semiconductors or conduction through soliton-like mechanisms, as have been claimed for $(CH)_x$-like materials /8,9/. Technically these materials are hard to characterize and are still chemically quite unstable. In this respect polysilanes are a class of polymers which is reasonably stable against oxidation and photochemistry (at least in the

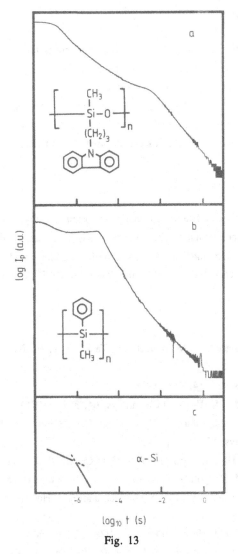

Fig. 13

Photocurrents of various materials on a log-log-plot (see text)

311

solid state). Here first TOF experiments show mobilities which are two orders of magnitude above 'conventional' polymers /22,23/. In polysilanes (see Fig. 13b) the molecular conjugation is believed to go as far as several molecular units (on the order of 6 to 10; quasi conjugation). Correspondingly the transit times are at least two orders of magnitude shorter.

Fig. 13 shows transient photocurrents for several amorphous materials: carbazole substitutes polysiloxane (Fig. 13 a), polysilane (Fig. 13 b; /23/) and amorphous silicon (Fig. 13c; /24,25/). As one can see on the time scale, the transit times are in the ms time regime for siloxane, in the 10 µs time regime for polysilane and in the 0,1 bis 1 µs time regime for amorphous silicon.

For the near future it will be a challenge for chemists to produce materials with delocalized charge carriers which are on one hand chemically stable and which have on the other hand enough delocalized wavefunctions to render band-like mobilities. Efforts with materials like polypyrol and polyphenylenevinylene are underway in various labs (including our lab) and it will be a matter of a few years to prove whether conjugated polymeric materials can be used for technical applications.

Acknowledgement

I would like to thank A. Blumen for many discussions and my coworkers in Bayreuth for their excellent experimental work. This work has been supported by the Deutsche Forschungsgemeinschaft, by the Bundesministerium für Forschung und Technologie and by the Fonds der Chemischen Industrie. We also thank the BASF Corporation for support.

References

1. Methods and Materials in Microelectronic Technology, ed. by J. Bargon, (Plenum Press 1984), pp. 181 ff
2. L. Onsager, Phys.Rev. 54, 554 (1938)
3. R.R. Chance and C.L. Braun, J.Chem.Phys. 64, 3573 (1976)
4. R.H. Batt, C.L. Braun, and J.F. Hornig, Appl.Opt.Suppl. 3, 20 (1969)5. P.J. Melz, J.Chem.Phys. 57, 1694 (1972)
6. H. Kaul and D. Haarer, Ber.Bunsenges. Phys.Chem. 91, 845 (1987)
7. D. Haarer and M. Philpott, in:Spectroscopy and Excitation Dynamics of Condensed Molecular Systems, ed. by V.M. Agranovich and R.M. Hochstrasser, Vol. 4, (North Holland 1983), pp. 27 ff
8. S.D. Phillips and A.J. Heeger, Phys.Rev. B38, 6211 (1988)
9. H. Bleier, S. Roth, J.Q. Shen and D. Schöfter-Siebert, Phys.Rev. B38, 6031 (1988)
10. Patent: K.-W. Klüpfel, M. Tomanck anf F. Endermann, DE-B-11 45 184 (1963)

11. O.H. Le Blanc, J.Chem.Phys. <u>33</u>, 626 (1960)

12. K.G. Kepler, Phys.Rev. <u>199</u>, 1226 (1960)

13. H. Scher and E.W. Montroll, Phys.Rev. <u>B12</u>, 2455 (1975)

14. H. Scher and M. Lax, Phys.Rev. <u>B7</u>, 4491, 4502 (1973)

15. F.W. Schmidlin, Phys.Rev. <u>B16</u>, 2362 (1977)

16. D. Haarer, Festkörperprobleme, Vol. 30, pp. 157-182 (1990) Vieweg Verlag
    D. Haarer, Angewandte Makromol. Chemie <u>183</u>, 197 (1990)

17. H. Domes, R. Fischer, D. Haarer and P. Strohriegl, Makrolmol.Chem. <u>190</u>, 165 (1989)

18. E. Müller-Horsche, D. Haarer and H. Scher, Phys.Rev. <u>B35</u>, 1273 (1987)

19. H. Schnörer, D. Haarer and A. Blumen, Phys.Rev. <u>B38</u>, 8097 (1988)

20. T. Tiedje and A. Rose, Solid St. Commun. <u>37</u>, 49 (1981)

21. H. Schnörer, H. Domes, A. Blumen and D. Haarer, Phil,Mag.Lett. <u>58</u>, 101 (1988)

22. M. Abkowitz and M. Stolka, Polym. for Adv.Technologies <u>1</u>, 225 (1990)

23. H. Kaul and D. Haarer, unpublished results

24. P.B. Kirby and W. Paul, Phys.Rev. <u>B25</u>, 5373 (1982)

25. J.H. Yoon and C. Lee, J.Non-Cryst. Solids <u>105</u>, 258 (1988)

POLYMERS AND DOPED POLYMERS FOR

APPLICATIONS IN ELECTROPHOTOGRAPHY

D. M. Pai

Xerox Corporation
800 Phillips Road
Webster, NY 14580

INTRODUCTION

Copying and duplicating machines are based on xerography[1] in which electrostatic images produced in a photoconductive film are developed into visible images by charged pigment particles. The seven steps of xerography (Fig. 1) are: (1) corona charging the photoconductor film, called a photoreceptor; (2) forming an electrostatic image on the photoreceptor by exposing with light; (3) developing the image by charged pigment particles called toners; (4) transferring the toner image onto the paper; (5) fixing the image onto the paper; (6) cleaning the photoreceptor of residual toner; and (7) erasing the remaining electrostatic image by uniform illumination.

The first step is to charge the photoreceptor, usually by a corotron, which is a wire biased to a potential of 2-8 kV. The high potential produces an intense electric field around the wire that ionizes the air and creates a gaseous discharge. Ions from the discharge drift to the surface of the photoreceptor where they become trapped, thereby charging the photoreceptor to a potential, V.

After charging, the photoreceptor is discharged by light reflected from the original document. The photoreceptor discharges in regions where the light is absorbed and remains charged in the dark regions. The amount of discharge depends on both the light intensity reflected from the document and the discharge properties of the photoreceptor. The resultant electrostatic latent image is usually expressed in terms of a contrast charge density or a contrast potential voltage, which corresponds to the optical contrast of the original image.

The electrostatic image is developed into a visible image by charged powders called toners. Toners are thermoplastic particles that are typically 5 to 25 μm in diameter and are loaded with a dye or pigment to give the desired color. Toners are charged through triboelectric charge exchange by tumbling against other polymeric materials or metals. The charge must be opposite in polarity to the charge on the photoreceptor. Once charged, the toner particles are attracted to the electrostatic image by an electric field above the photoreceptor. This field is directly proportional to the image potential. If the toner is free, the force attracts the toner and deposits it on the charged regions, thereby developing the image.[2]

Frontiers of Polymer Research, Edited by P.N. Prasad and
J.K. Nigam, Plenum Press, New York, 1991

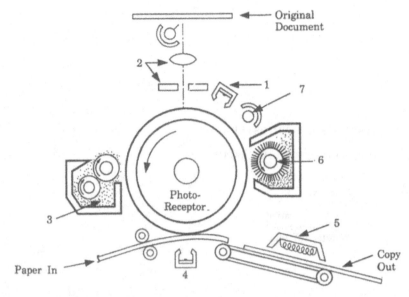

Fig. 1. Schematic of a xerographic reproduction machine indicating the seven important process steps.

In most cases, toners are attached triboelectrically to "carrier beads." Carrier beads serve two functions. They provide a method of transporting the fine toner powders mechanically and a means for charging the toner. Unattached fine powders are difficult to transport and control. Any slight air disturbance would cause such unattached toner to contaminate the xerographic hardware. To minimize this problem, toners are attached triboelectrically to much larger and heavier carrier beads. Carrier beads are magnetic metals that are typically 25 to 300 μm in diameter. These beads can be transported by magnets, which provide an easy method of both bringing toner to the electrostatic image and controlling toner contamination.

After development, the toner is transferred to paper. This is accomplished usually by placing paper over the developed image and corona charging the back of the paper with a polarity opposite to that of the toner charge to attract the toner to the paper.

The final steps in the xerographic process are to fix the image to the paper, clean the photoconductor, and erase the latent image. Fixing is accomplished by heating the toner above its melting point, which causes the melted toner polymer to migrate into the paper. In some cases, toners can be fixed simply with pressure with no external heat applied.

Polymers are employed in xerography to perform a variety of specialized functions. Although some of these polymers were designed for applications other than electrophotography, there are several that are designed specifically for use in electrophotography. This article confines itself to polymers employed in two of the seven steps described earlier. These are the polymers employed for developer/toner applications and for applications in organic photoconductors. The quality of the image depends on how well the electronic activity, such as triboelectrification between carrier coatings and toner, and charge transport through organic materials perform in a predictable way under a variety of temperature and relative humidity conditions.

DEVELOPER MATERIALS

Toner particles are composed mainly of a thermoplastic containing 5-10 percent by weight of colorant.[3,4] Black toners usually consist of a thermoplastic with finely dispersed particles of carbon black less than 1 μm in size. Color toners utilize color pigments, dyes, or mixtures of dyes and pigments. Cyan, magenta, and yellow pigments or dyes are required for full color reproduction. Pigment concentration is generally determined by the desired contrast or color density and toner charging requirements.

The function of the thermoplastic is to fuse and fix the image to the paper physically by heat, pressure, or a combination of both heat and pressure. Although the thermoplastic is chosen primarily for its fusing or melting properties, other characteristics, such as charging, mechanical properties, cost, toner processing, and systems compatibility, must be considered carefully in making the choice for a particular material.

Toners are fabricated usually by melt-mixing pigment and polymer together, followed by impact fracture. In dry powder xerography, the choice of developers is either monocomponent or two-component. With two-component developers, the toner is charged by triboelectrification against a much larger particle called the carrier bead. As discussed earlier, the carrier bead is the vehicle that carries the toner to the latent image. The sign and ultimate magnitude of the toner charge are determined predominantly by the choice of polymer in the toner, the polymer coating on the carrier bead, and mixing conditions. The toner charging process is accomplished through repeated contacts with the carrier surface. The toner acquires a charge opposite to the carrier and remains attached to the carrier by electrostatic forces. During development, the toner particles are stripped from the carrier beads by the latent electrostatic forces, leaving the carrier beads behind to charge and attach new incoming toner. Toners used in single-component development are charged by either induction or triboelectric interaction against a suitable donor roll with or without an applied field.

Magnetic Brush Development

The system employed most commonly to develop the latent charge image across the photoconductor is the magnetic brush development. A schematic diagram of the magnetic brush development is shown in Fig. 2. Magnetic carrier beads are employed to charge the toner and magnets transport the carriers to the latent image. The beads, which range in diameter from 50 to 150 μm in diameter, form bristle-like chains in the magnetic field, hence the name magnetic brush development. Carrier beads are often coated with polymer to provide the desired toner charging and developer conductivity. Once coated, the system is capable of developing toner in proportion to the latent image field in the region between the photoconductor surface and the development roller. Iron, steel, or mixed metal oxides (ferrites), all of which exhibit magnetic properties have been used extensively as carriers in magnetic brush development

Developer Material Design

A new developer design begins by considering the machine hardware and a set of functional requirements. Generally fusing properties are considered first in the design of the toner. Sign of charge (negative or positive) is the next consideration. It can be altered by changing chemical composition of the carrier coating, by choosing the desired pigment concentration and by adding charge control agents. Selection of the basic polymer is based generally upon the work function and the molecular structure of the polymeric material.

With few exceptions, most commercial polymers are not suited for usage in toners. Commercial thermoplastics are generally very high molecular weight materials, designed primarily for durable mechanical properties. These high molecular weight polymers are too tough for efficient micronization to 10 μm toner diameters. In addition, most commercial materials exhibit thermal melting properties that are not compatible with the temperatures required to fuse toner onto paper in practical applications.

Four classes of polymers are used commonly for fabricating dry toners. These include the amorphous random copolymers of styrenes and methacrylates or acrylates; polyesters, epoxies, and the crystalline polyethylenes or copolymers of polyethylene, such as polyethylene/vinyl acetate. The thermal and rheological properties of these polymers (melt viscoelasticity, glass transition temperature, decomposition point, morphology, etc.) are designed specifically to be compatible with a given fuser and are chosen to a lesser extent for their charging characteristics.[5]

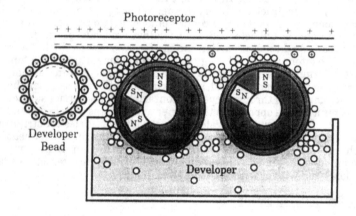

Fig. 2.   Schematic of a magnetic brush development. Carrier beads tend to form chain-like bristles in the presence of magnetic field.

Toner particles are charged most frequently by rubbing in contact with carrier beads coated with a suitable polymer to facilitate rapid charge exchange. Charge exchange between materials brought into contact and then separated is commonly referred to as contact charging, triboelectrification, or contact electrification.[6] The quantity of charge gained through contact charging is expressed in microcoulombs per gram for a given particle size. Charge exchange between insulating materials, such as toner polymers, is not understood well, due to incomplete knowledge of the insulator energy levels. Nevertheless, considerable progress has been made in the identification and characterization of the chemical and electrical properties of polymers. Reviews on contact electrification in polymers have been written by Lowell and Rose-Innes,[7] Seanor,[8] and Harper.[9]

Attempts have been made to correlate the chemical structure of model organic compounds with charging behavior. Gibson[10] reported the first quantitative correlation of solid state triboelectric charging with molecular structure. His results suggest that triboelectric charging of solids is related directly to the molecular structure of the bulk.

Certain polymer classes have been identified empirically as positive charging materials. These include polyamides, polyamines, and polyacrylates. Highly halogenated polymers, such as poly(tetrafluorethylene) and poly(chlorotrifluoroethylene), are found to be negative charging materials. Typically, positive charging polymers have nucleophilic sites such as nitrogen or oxygen atoms attached to the polymer backbone or within the backbone of the polymer. Strongly negative charging atoms, such as fluorine, chlorine, and bromine, are electrophilic in nature. Although highly halogenated polymers are not well suited for toner applications, they are used commonly as carrier coatings when positively charged toners are required.

A selected list of polymers used frequently for toner or carrier fabrication is presented in Table 1.[11] Materials at the top of the list tend to be positive, and negative materials are near the bottom. Therefore, polymers near the top of the list will charge positively with respect to polymers further down the list. It should be noted that a toner is composed of both a polymer and a colorant. The addition of a colorant or any other additives may change the charging outcome from that expected for the pure polymer.

Table 1.   Triboelectric Series

Polyamide (nylon)
Polymethyl methacrylate
Styrene butadiene copolymer
Styrene butylmethacrylate copolymer
Polyethylene terephthalate
Polyacrylonitrile
Polycarbonate
Polystyrene
Polyethylene
Polypropylene
Polyvinylchloride
Polytetrafluoroethylene

Up to this point, the discussion has centered on the polymer design to meet fusing requirements and structural features that tend to drive them in either a positive or negative direction. However, in many cases, additives, such as colorants, plasticizers, flow aids, etc., dominate the electrical properties. In some cases, these effects are desirable. However, in many cases, these additives produce unacceptable charge levels, broad charge distributions, and developer instability. Toners with broad charge distributions or toners that lose charge in time generally produce poor copy quality, high background in the "white" areas of images, and machine dirt. Charge control agents are commonly added to toners to offset these negative effects.

The xerographic process requires the use of developer materials that exhibit a precise balance of physical, chemical, mechanical, and electrical properties matched to the design characteristics of the machine in which they are to be used to develop the latent image. This balance becomes more delicate as machines increase in speed and as the copy quality requirements become more stringent.

POLYMERS IN ORGANIC PHOTOCONDUCTORS (OPC)

The photoreceptor consists of one or more layers of dielectric material on a conductive substrate. The substrate is a rigid cylinder or a flexible belt typically. In some applications plates or flexible sheets are used, but cylinders and belts account for more than 99% of applications. The cylinders are made from aluminum usually, while the belts may consist of electrodeposited nickel or a polymer, usually polyethylene-terephthalate (PET). In the latter case, the PET is made conducting by a thin, vacuum deposited, metal layer and the belt is fabricated from long webs by cutting and welding the two ends, resulting in a seamed belt. The dielectric layer or layers, also referred to as the photoconductor, can range in thickness from a few to hundreds of micrometers. Organic photoreceptors range from 10 to 35 µm in thickness and typically are about 20 µm thick.

The current trend in the industry is to employ two layered structures (Fig. 3) in which the charge generation and transport functions are carried out in two separate layers deposited on a conductive substrate.[11] The charge generation occurs in a thin layer of a photoconductive material called the charge generation layer (CGL). The photogenerated charges are injected subsequently into and transported through a contiguous charge transport layer (CTL). The CGL material is made of an organic dye or pigment. The CTL material is made of either a charge transporting polymer or a dispersion of active organic charge transporting molecules in an appropriate binder. The CTL film is thicker than the CGL layer generally and is transparent in the visible region of the spectrum. Since all photoreceptor systems in use to date employ electron donors to transport the charge, negative polarity is used for charging.

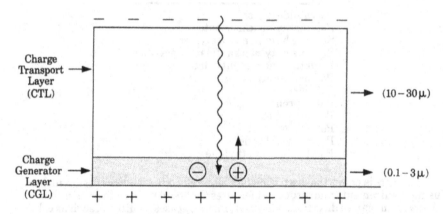

Fig. 3.   Schematic of a two layer organic photoconductor.

Charge Transport Layer Requirements

As discussed in the previous section, the latent image formation requires the corona charging of a photoreceptor followed by an imagewise discharge to the input exposure. The function of the transport layer is to facilitate injection of photogenerated charge from the pigment layer (CGL) and transport the charge to the photoreceptor surface to neutralize the corona deposited charge. For a transport layer to be useful in OPC, the following conditions must be met:

1) The transport layer must be transparent in the visible and IR for the image light to be absorbed in the pigment layer without attenuation in the transport layer.

2) Since the photoreceptor has to hold the corona charge in the dark, the transport layer has to be insulating. This sets a limit on the free carrier concentration in the dark. The free carrier concentration in the dark has to be much less than the corona deposited charge.

Ideally the photoreceptor acts as a capacitor, so the voltage is related linearly to the surface charge, σ, thickness, L, and dielectric permittivity, ε, by,

$$V = \frac{\sigma L}{\varepsilon} \tag{1}$$

Typically a surface charge of 130 nanocoulombs/cm^2, or about 8×10^{11} ions/cm^2, is used. An organic dielectric layer with a thickness, L, of 20 micrometers and dielectric constant, $\varepsilon/\varepsilon_0$, of 3, corresponds to a charging slope, L/ε, of 7.5 V nanocoulomb^{-1} cm^2, or a capacitance, ε/L, of 130 picofarads cm^{-2}. The charge of 130 nanocoulombs cm^{-2} produces a voltage of 980 V on such a photoreceptor. The concentration of free carriers in the dark has to be considerably less than 10^{14}/cm^3($=8\times10^{11}$/L). Additionally it requires that no significant charge be generated thermally on a xerographic time scale of seconds to minutes.

3) The carrier injected from the CGL must traverse the transport layer in a time period that is short compared to the time duration between the exposure and development stations. This sets a lower limit to the charge carrier mobility. The decrease in potential, ΔV, is proportional to the product of the number of photogenerated carriers and the distance they travel. Hence low mobility affects the shape of the photodischarge and reduces the contrast potential. The transit time, t_T, of the injected carrier is related to the charge carrier mobility, μ, by the relation

$$t_T = \frac{L}{\mu E} = \frac{L^2}{\mu V} . \tag{2}$$

Typically the transport layers are 20 μ thick and the devices are discharged from a 900 V initial potential to a background potential of 50 V and the time duration between the exposure and development stations is 0.5 s. The charge carrier mobility has to be larger than 2×10^{-7} cm^2/Vs for a carrier to travel a 20 μ thick film in less than 0.5 s at 50 V. Transport in many organic systems is dispersive. The time-of-flight (TOF) technique employed to determine the charge carrier mobility generally measures the mobility of the fast carriers which are only a fraction of the total number of transiting carriers. The average mobility of the carrier is less than the value measured by the TOF technique by a factor of 2 or 3. Therefore in order not to have any mobility limitation and receive maximum discharge at the development station, the charge carrier mobility as measured by the TOF technique has to be larger than 5×10^{-7} cm^2/V s for a 20 μ thick charge transport layer.

4) The charge carrier range in the CTL, the schubweg, should be larger than the film thickness to minimize bulk trapping. Range is defined as the distance that a photogenerated charge drifts before it is immobilized. The carrier range is determined by the concentration of deep traps and their cross section for trapping. Range limitation leads to residual potentials after the light exposure and erase steps and causes a reduction in contrast potential.

5) The CTL material must be capable of being fabricated into large-area, defect-free films. The materials must be stable to operate in a corona environment containing ozone, oxides of nitrogen, and other effluents, as well as to withstand wear by the development and cleaning actions.

Transfer of charges between localized states is believed to play a dominant role in many organic disordered solid materials employed to fabricate the CTL. Figure 4 illustrates four different classes of charge transporting systems that have been identified thus far. 1) Polymers in which the transport occurs as a result of charge hopping among the pendant-charge transport-active groups. The backbone plays no direct role in charge transport. 2) Molecularly doped polymers that are solid solutions of transport active monomers in essentially non-charge transporting tough polymers. 3) Condensation polymers in which the charge transport groups form building blocks between space linkages such as carbonate, urethane, etc. 4) Silicon and germanium catenated polymers in which the charge transport occurs through the main chain. The pendant groups play no significant role.

Polymers with "Active" Pendant Groups

Photoconductivity has been detected in virtually all known polymers with condensed aromatic groups as pendants to the main chain.[12] These include pendants of napthalenes, anthracenes and pyrenes on vinyl and other backbones as shown in Fig. 5. Photoconductivity involves photon absorption and photo generation of free carriers, followed by their recombination. The observation of photoconductivity *per se* does not provide any information on charge carrier mobility or carrier range, which are the two most critical transport requirements for application of these materials as CTL in OPC. It is found that at best the charge carrier mobility is marginal in these materials. In addition, these polymers have poor mechanical properties, generally, on account of the crowding with large pendant groups.

Marginal charge carrier mobilities are observed also in polymers with aromatic amines (Figure 6). In this group, Poly(N-vinyl carbazole), PVK, was one of the earliest to be studied extensively.[13] Charge transport polymers such as PVK are characterized by saturated polymer chains that ensure low dark conductivity. The photoexcitation in the ultra violet is associated with large pendent planar structures with extended π-electron systems. Weak interactions of the π-electron systems along saturated chains make them analogous to the solid dispersions of similar monomer units in an inert binder. By comparing the hole transport in PVK with that in polycarbonate doped with N(isopropylcarbazole), it has been established[14] that the vinyl backbone of PVK does not contribute to charge transport but merely provides the mechanical stability to the films. The absorption spectrum of PVK in the solid state is also seen to be remarkably similar to that of the free molecule.

Figure 7 compares the hole mobility in PVK with that in poly(1-vinyl-pyrene). Also shown is the line corresponding to requirements for CTL applications in OPC. The mobilities are marginal if the transport layer is to be 20 μ thick.

A variety of polymers have been synthesized without much success to improve the mechanical and charge carrier mobility. Figure 8 shows some of the many structural variants of PVK. Figure 9 shows some structure of polymers with aromatic amines. Figure 10 shows charge carrier mobility in these polymers compared to that in PVK. Once again, the charge carrier mobilities are marginal for CTL applications in OPC.

Molecularly Doped Systems

The concept of molecular doping in which active charge transporting molecules are dispersed into inert polymeric binders is an outgrowth of studies on polymeric systems. The molecularly doped polymers are much more flexible in material design since the concentration and kind of dopant molecule can be selected that allows one to optimize certain transport properties. This flexibility proves to be an essential experimental leverage for unraveling the details of charge generation and transport in these systems.

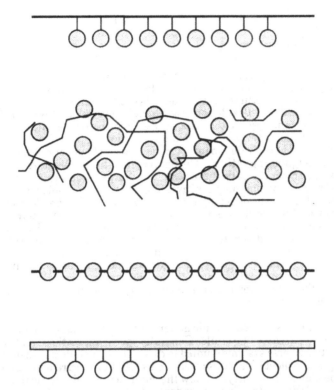

Fig. 4.   Schematic of four classes of CTL materials. In the first three, the charge moves by hopping through states denoted by solid circles. In the fourth case, the charge moves through and between the main chain.

A host of systems has been studied and a large number of interesting features peculiar to the disordered state established. Some of the electron donor molecules employed to date include oxadiazoles, such as 2,5,-bis(4-N,N'-diethyl amino phenyl)-1,3,4 oxadiazole;[15] pyrazolines such as 1-phenyl-3(4'-diethylamino styryl)-5-(4"-diethylamino phenyl) pyrazoline;[16] and hydrazones, such as 4-diethylaminobenzaldehyde-1,1-diphenylhydrazone and N-ethyl-3-carbazole carboxaldehyde-1-phenyl-1-methyl hydrazone[17] and diamines such as N,N'-diphenyl-N,N'-bis(3-methylphenyl)-[1,1'-biphenyl]-4,4'-diamine[18] (TPD). The binder employed is generally a polycarbonate. The choice of binder is based on the solubility of the molecules in it, as well as other properties, such as durability. Polycarbonate is one of the few binders in which most of the molecules employed to date are soluble. Although clear films can be cast with other inert binders, charge trapping is produced as a result of inhomogeneity on a microscopic scale.

There is general agreement that the charge transport is an electric field driven chain of reduction-oxidation processes involving neutral molecules (N) or groups and their charged derivatives: anion radicals ($N_{\bar{\tau}}$) in the case of electron transport and cation radical ($N_{\dot{+}}$) in the case of hole transport.

$$NN \overset{e}{\underset{\cdot}{+}} NN$$

Hole transport →

$$NNN \overset{e}{\underset{\cdot}{-}} N$$

← Electron transport

For hole transport it is essential therefore that the molecules — or groups if attached to a polymer chain — are electron donors in nature, have low ionization potentials so that they form the cation radicals easily, and that the redox processes oxidation/reduction/oxidation is completely reversible. Take hole transport, for example, and assume that as a result of the photogeneration process some dopant molecules are positively charged (radical cation). Under the influence of the applied electric field neutral molecules will transfer repetitively electrons to their neighboring cations. The net result of this process is the motion of a positive charge across the bulk of the film. The charge transport is an electronic process and no mass displacement is involved. Therefore, for hole transport, the dopant molecule has to be donor-like in its neutral state. On the other hand for electron transport where electrons hop from the radical anions to their neighboring neutral molecules, the dopant molecules are acceptor-like. Indeed, for the donor molecules dispersed in polycarbonate only hole transport is observed while for the acceptor molecule dispersed in a binder, only electron transport is observed.[19] Thus the chemical nature of the dopant molecule preserved in the dispersion largely determines the sign of the mobile carrier. Since the molecules are either donor-like or acceptor-like, the molecularly doped films of one type of molecule in a binder is unipolar.

The study of disordered organic materials uncovered many novel features of charge transport in these materials. Generally speaking, all these features are observed in the diverse systems studied to date. These included: 1) dispersion or broadening of the sheet of carriers in the time of flight measurement; 2) concentration dependence of mobility; 3) a low thermally activated electric field dependent mobility; 4) an electric field dependent activation energy; 5) presence of lower ionization potential molecules acting as traps when charge is transported through higher ionization potential molecules; and 6) structural dependence of mobility.

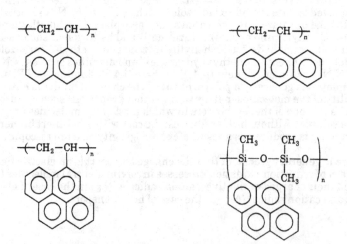

Fig. 5.   Polymers with condensed aromatic rings as pendants.

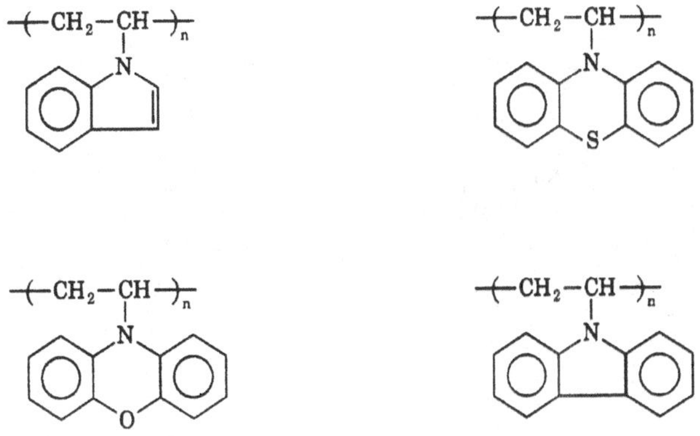

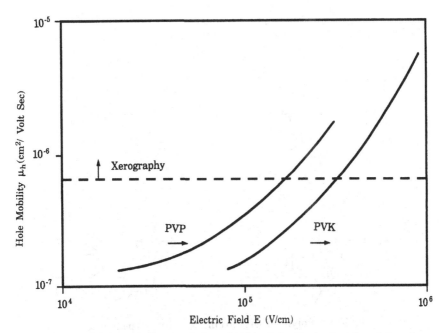

Fig. 6.   Polymers with aromatic amines as pendants.

Fig. 7.   Hole mobility vs electric field in Poly(N-vinyl carbazole) and
           Poly(1-vinylpyrene).

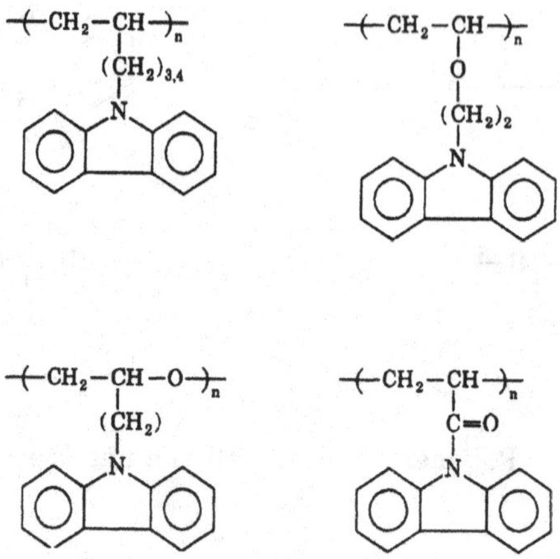

Fig. 8. Other polymers with carbazole pendants.

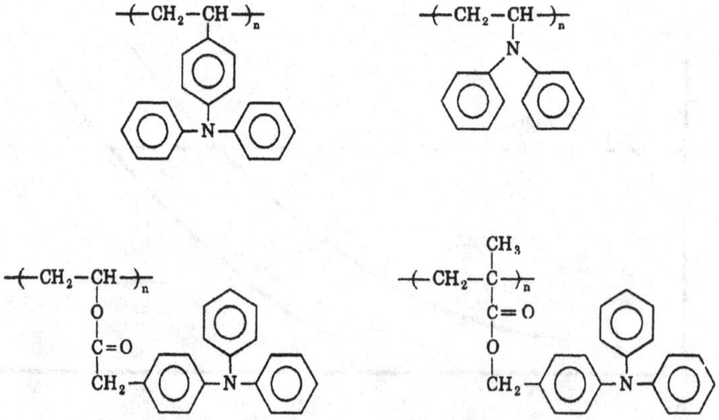

Fig. 9. Polymers with aromatic amine as pendants.

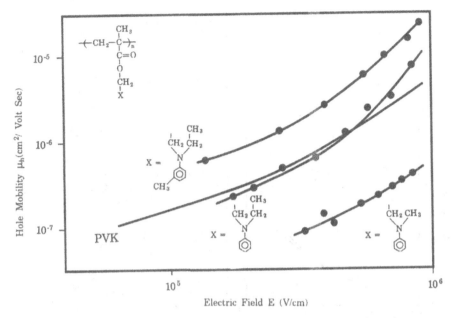

Fig. 10. Charge carrier mobility in some polymers with aromatic amine pendants compared to that of PVK.

The hole mobilities in some of the solid solutions are shown in Fig. 11 for films containing 50 wt.% solution of the molecule in the polycarbonate binder.[20] The data are obtained by the time-of-flight measurements. The mobility is electric field dependent and varies between 3×10^{-5} and 10^{-7} cm^2 V^{-1} s^{-1} for the five electron donor molecules. Also shown is the mobility in PVK.

Since the rate of the electron exchange between electron donor molecules and their cations depends on the separation distance, the mobility will be a strong function of concentration. For most systems, it is found that the empirical relation,

$$\mu_h \sim R^2 \exp\left(-\frac{2R}{R_0} \right) \tag{3}$$

fits the data with R the average intermolecular distance between the molecules, proportional to (concentration)$^{-1/3}$. R_0 is the localization radius of the electron. This exponential dependence on the average intersite distance can be explained in the framework of either non-adiabatic, small polaron hopping or phonon-assisted hopping. The maximum molecular concentration in a practical photoreceptor is dictated by solubility and mechanical considerations.

Molecular structure of the electron donor molecule affects the mobility also. The influence of some substituent groups on the mobility of diethylamino triphenylmethanes (TPMs) in solid solutions of polycarbonate has been studied.[21] It was found that the substituents caused the mobility to vary by four orders of magnitude, Fig. 12, but attempts to correlate the mobility with the Hammett constant for the substituents gave inconsistent results. The aromaticity of the amine as well as its solubility or compatibility with the binder affect mobility also.

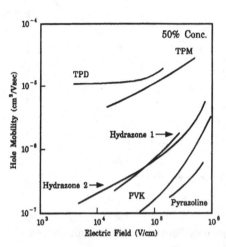
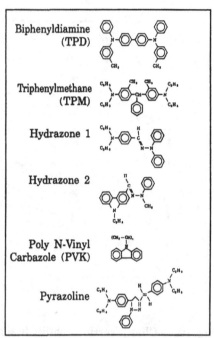

Fig. 11. Hole mobility vs field in several molecularly doped transport layers discussed in the text compared to that of PVK. The molecularly doped layers are fabricated from 50 wt % solid solutions of the molecules in polycarbonate. The six structures on the right are biphenyldiamine (TPD), bisdiethylamino triphenylmethane (TPM), benzaldehyde hydrazone, carabazole hydrazone, PVK, and pryazoline, respectively.

The electric field dependence of the mobility, Fig.11, its temperature dependence, and the dispersion of transport (spread in mobility) are questions of current active research. This research is concerned with constructing a microscopic model to explain the various features of the transport as measured by the time-of-flight experiments. Space does not allow us to discuss these topics adequately, which are less material dependent.

Polymers with Active Groups in the Chain

Unlike the vinyl-type polymers, exemplified by PVK, these polymers possess hole transporting moieties that are built into the macromolecular chain and hence contribute to the mechanical properties of the polymer. The hole transporting units and the linkage part of the structure, joining the transporting units, largely determine the physical properties of the novel polymer and are selected so as to most closely match the polymer properties to the design requirements of the photoreceptor's intended use.

The specific hole transporting polymer (HTP) incorporates a transport group, TPD, which are joined with ether carbonate linkages (Fig. 13).[22]

One can compare the hole transport mobility of the HTP to various compositions of N,N'-diphenyl-N,N'-bis (3-methylphenyl)-[1,1'-biphenyl]-4,4'-diamine in bisphenol-A-polycarbonate (Fig. 14). In the HTP the weight % of the transporting moiety is 72%.

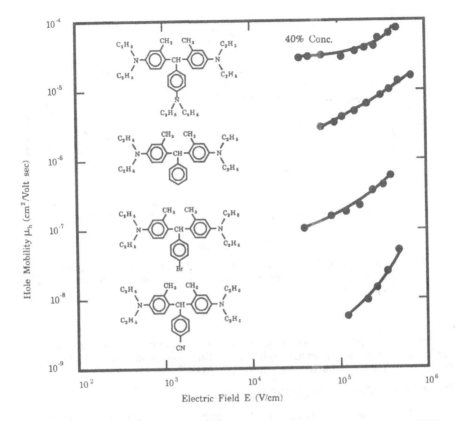

Fig. 12. Variation of hole mobility with substituent at position X for the TPM molecule at 40 wt % loading.

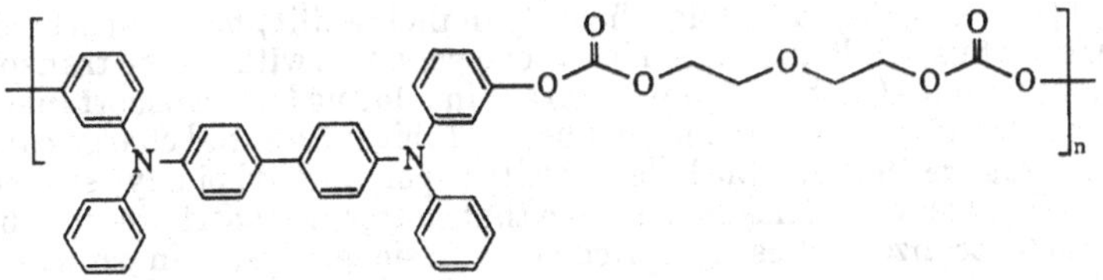

Fig. 13. Structure of the hole transporting polymer incorporating a transport group (TPD) that are joined by either carbonate linkages.

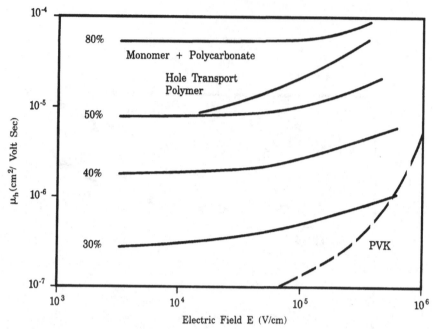

Fig. 14.   Mobility of hole transporting polymer shown in Fig. 13 compared to
that of the various compositions of solid solution of the monomer (TPD)
in polycarbonate.

The hole mobilities are more than adequate for application as transport layers in OPC. These polymers are highly insulating in dark and the charge carrier
range meets the requirements for application in OPC.

Polymers with Charge Transporting Main Chain

Poly(organosilylenes) and Poly(organogermylenes) are high molecular
weight polymers based on a linear σ bonded Si or Ge backbone (Fig. 15). These
polymers are characterized by substantial delocalization of charges over the σ-Si-
Si-chain bonds, as demonstrated by the dependence of ionization potential, electron affinity and peak absorption wavelength on the number of silicon atoms in
the chain. The physical properties of these large class of polymers are dependent
highly on the carbon based groups. Although Si and Ge backbone polymers exhibit dielectric, mechanical, and solubility properties similar to the analogous carbon
chain vinyl polymers, they differ significantly in their ability to transport electronic charge. Thus, while the carbon backbone polymers without the transport
active pendant groups (such as arylamines) are insulating but transport inactive,
many of the substituted Si backbone polymers exhibit substantial charge carrier
mobilities.[23] Figure 16 shows the hole mobility of poly(methyl phenyl silylene)
compared to that of PVK. It has been shown that charge transport proceeds by
hopping among localized states associated not with sidegroups, as in pendant
polymers like PVK, but rather with the main chain. The mobility requirement
for application as charge transport layers in OPC is met easily. The charge carrier range is adequate also for application in OPC.

Although four classes of charge transporting systems have been discussed
and can be employed in principle, most of the organic photoconductors employed
commercially are based on the molecularly doped polymer concept. These are

easy to synthesize and are economical. The operation of the OPC has been described by the two layer schematic shown in Fig. 3, a practical device may have other layers employing speciality polymers. A thin hole blocking layer is required to minimize injection of holes from the substrate during the charging and latent image formation steps. This blocking layer is coated between the conducting substrate and the charge generation layer. To improve adhesion, a thin adhesive layer may be required between the conductive layer and the blocking layer. Both the blocking and adhesive layers employ polymers meeting stringent requirements that are not discussed here.

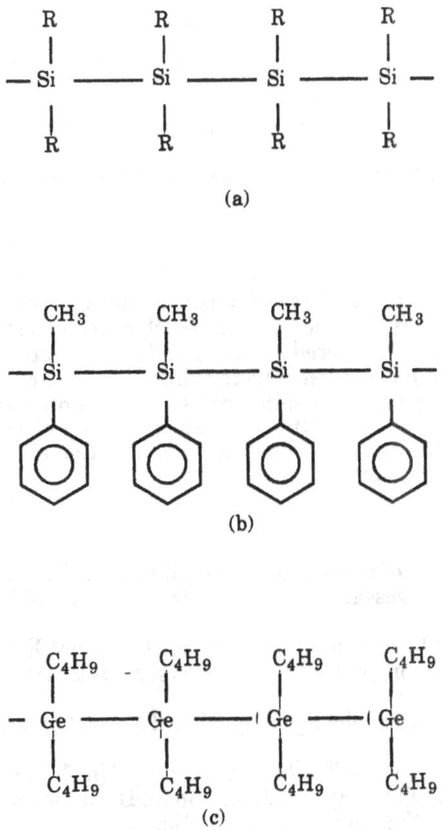

$$- \overset{\displaystyle R}{\underset{\displaystyle R}{Si}} - \overset{\displaystyle R}{\underset{\displaystyle R}{Si}} - \overset{\displaystyle R}{\underset{\displaystyle R}{Si}} - \overset{\displaystyle R}{\underset{\displaystyle R}{Si}} -$$

(a)

$$- \overset{\displaystyle CH_3}{\underset{\displaystyle \bigcirc}{Si}} - \overset{\displaystyle CH_3}{\underset{\displaystyle \bigcirc}{Si}} - \overset{\displaystyle CH_3}{\underset{\displaystyle \bigcirc}{Si}} - \overset{\displaystyle CH_3}{\underset{\displaystyle \bigcirc}{Si}} -$$

(b)

$$- \overset{\displaystyle C_4H_9}{\underset{\displaystyle C_4H_9}{Ge}} - \overset{\displaystyle C_4H_9}{\underset{\displaystyle C_4H_9}{Ge}} - \overset{\displaystyle C_4H_9}{\underset{\displaystyle C_4H_9}{Ge}} - \overset{\displaystyle C_4H_9}{\underset{\displaystyle C_4H_9}{Ge}}$$

(c)

Fig. 15.   Structure of polysilylenes and polygermylenes

(a)   R is either an aliphatic or aromatic group

(b)   Poly(methyl phenyl silylene) (PMPS)

(c)   Poly (dibutyl germylene)

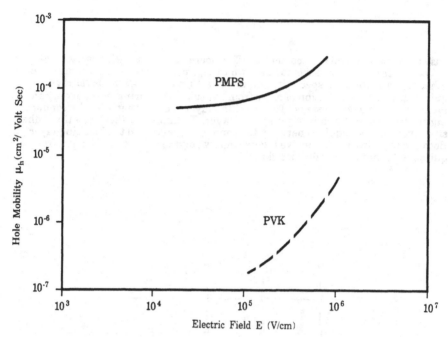

Fig. 16. Hole mobility vs electric field in PMPS compared to that of PVK.

CONCLUSIONS

Polymers are employed in electrophotography to perform a variety of specialized functions. The physical mechanisms (such as triboelectrification and charge transport through disordered systems) underlying application of these polymers are among the most intractable in solid-state physics. Judicious choices are made by combining scientific insight with empirical optimization. Significant progress has been made in the identification and synthesis of new charge transporting systems that meet the requirements for application as charge transport layers in organic photoconductors.

REFERENCES

1. C. Carlson, "History of Electrostatic Recording," in: "Xerography and Related Processes," J. H. Dessauer and H. E. Clark, eds., Focal Press (1965).

2. M. E. Scharfe and F. W. Schmidlin," Charged Pigment Xerography," in: "Advances in Electronics and Electron Physics," Academic Press (1975).

3. R. M. Schaffert, "Electrophotography," Focal Press, London (1975).

4. J. W. Weigl, "Interfaces in Electrophotography," in: "Colloids and Surfaces in Reprographic Technology," M. Hair and M. D. Croucher, eds., American Chemical Society, Washington, D. C. (1982).

5. J. M. O'Reilly and P. F. Erhardt, "Physicl Properties of Toner Polymers," in: "2nd International Conference on Electrophotography," Washington, D. C. (1974).

6. P. C. Julien, "Triboelectrification of Carbon-Polymer Composites," in "Carbon Black Polymer Composites," E. K. Sichel, ed., Marcel Dekker, New York and Basel (1982).

7. J. Lowell and A. C. Rose-Innes, "Contact Electrification," Adv. Phys., 29: 947 (1980).

8. D. A. Seanor, "Electrical Properties of Polymers," K. Frisch and A. V. Patsis, eds., Technomic Press, New York (1972).

9. W. R. Harper, "Contact and Frictional Electrification," Oxford University Press, Oxford (1967).

10. H. W. Gibson, "Linear Free Energy Relationships. V. Triboelectric Charging of Organic Solids," J. Am. Chem. Soc., 97: 3832 (1975).

11. M. E. Scharfe, D. M. Pai and R. J. Gruber, "Electrophotography," in: "Imaging Processes and Materials, Neblette's Eighth Edition," J. Sturge, V. Walworth and A. Shepp, eds., Van Nostrand Reinhold, New York (1989).

12. M. Stolka and D. M. Pai, "Polymers with Photoconductive Properties," in: "Advances in Polymer Science, Vol. 29," H. J. Cantow et al. eds., Springer-Verlag, Berlin (1978).

13. D. M. Pai, "Transient Photoconductivity in Poly(N-vinyl carbazole)," J. Chem. Phys., 52: 2285 (1970).

14. J. Mort, G. Pfister and S. Grammatica, "Charge Transport and Photogeneration in Molecularly-Doped Polymers," Solid State Commun., 18: 693 (1966).

15. G. Schlosser, "A New Organic Double Layer System and its Photoconduction Mechanism," J. Appl. Photo. Eng., 4: 118 (1978).

16. P. J. Melz, et al., "Use of Pyrazoline Based Carrier Transport Layers in Layered Photoconductivity Systems in Electrophotography," Photo. Sci. and Eng., 21: 73 (1977).

17. E. S. Baltazzi, J. Appl. Photogr. Eng., 6: 14 (1980).

18. M. Stolka, J. F. Yanus and D. M. Pai, "Hole Transport in Solid Solutions of a Diamine in Polycarbonate," J. Phys. Chem., 88: 4707 (1984).

19. W. D. Gill, "Electron Mobilities in Disordered and Crystalline Trinitrofluorenone," in: "Proceedings of the 5th International Conference on Amorphous and Liquid Semiconductors," J. Stuke and W. Brenig, eds., Taylor and Francis, London (1974).

20. D. M. Pai and J. Yanus, "Layered Photoconductors in Electrophotography," Photo. Sci. and Eng., 27: 14 (1983).

21. D. M. Pai et al., "Hole Transport in Solid Solutions of Substituted Triarylmethanes in Bisphenol-A-Polycarbonate," Phil. Mag., 48: 505 (1983).

22. W. W. Limburg, D. M. Pai, J. F. Yanus, D. S. Renfer and P. DeFeo, "Designer Polymers for Xerographic Applications," Fifth International Congress on Advances in Non-Import Printing Technologies, SPSE, San Diego, CA, Nov. 12-17 (1989).

23. M. A. Abkowitz, M. J. Rice and M. Stolka, "Electronic Transport in Silicon Backbone Polymers," Phil. Mag., 61: 25 (1990).

CONDUCTING POLYMER ALLOY

J. H. Han, T. Motobe, Y. E. Whang, and S. Miyata

Division of Chemical and Biological Science and Technology
Faculty of Technology, Tokyo University of Agriculture and
Technology, Nakamachi 2-24-16, Koganei, Tokyo 184 (Japan)

INTRODUCTION

Interest in conducting polymers has grown tremendously in the past several years. Especially, heterocyclic polymers such as polythiophene and polypyrrole have received a great deal of attention due to their good environmental stability and high electrical conductivity.[1,2,3,4,5] However, they lack mechanical properties and processibility. These are the major obstacles for practical usage. In recent years, several attempts to overcome these drawbacks have been made.[6,7,8,9] Among them, a method of blending with conventional polymers has been mainly studied. But the mechanical blending in melt state is undesirable because the conducting property of the polymer decreases due to thermal effect during processing. Furthermore a high concentration of conducting polymer is required to make conduction path.

We have developed a novel method to prepare the conducting polymer alloy. The most attractive feature of this method is that polymerization of pyrrole occurrs after fabrication and the construction of conducting networks can be controlled by polymerization conditions.

EXPERIMENTAL

A mixed solution of ferric chloride (oxidant) and poly(vinyl acetate) (PVAc) was first prepared by dissolving them in methanol. The oxidation potential of the mixed solution was measured by using potentiogalvanostat (KIKKO KEISOKU NPGFZ-251B) and controlled whithin optimum range by changing the amount of ferric chloride. After that, pyrrole (Tokyo Kasei Kogyo Co.) was added into the mixed solution with stirring. Pyrrole was purified by distillation prior to use. A conducting polymer composite film was prepared by casting the mixture on a substrate. After drying sufficiently, it was rinsed with methanol to remove remaining ferric chloride and ferrous chloride formed due to the polymerization of pyrrole. The electrical conductivity of film was measured under vaccum by four probe method.

Frontiers of Polymer Research, Edited by P.N. Prasad and
J.K. Nigam, Plenum Press, New York, 1991

RESULTS AND DISCUSSION

When the mixture solution is casted on the substrate. its color changes gradually from light yellow to dark green. It indicates that the polymerization of pyrrole progresses due to the evaporation of solvent causing an increase in the oxidation potential of the casted solution. The mixture of pyrrole and ferric chloride - PVAc solution is homogeneous at initial stage but a phase separation of PVAc and polypyrrole (PPy) occurrs according to the polymerization of pyrrole. The morphology of PPy - PVAc composite film is very different from that of PPy - PVAc composite prepared in solution. PPy aggregates are connected with other in the former case, which results in PPy network over all composite. On the contrary, PPy aggregates are separated from each other in the latter case. It is expected in the former case that a spinodal decomposition occurrs in the course of phase separation process becaues of sudden changes of the concentration of pyrrole and ferric chloride and the viscosity of PVAc. Those sudden changes stem from the evaporation of solvent. High conductivity of the film probably results from the PPy network structure.

The electrical conductivity of solution-cast film depends on the initial oxidation potential of the solution as well as on the holding time of solution state before casting, as shown in Fig. 1. In case that the initial oxidation potential of the solution is either too high or too low, the obtained films do not have good electrical conductivity. It is expected that the pyrrole monomer also evaporates together with solvent when the oxidation potential is too low, while the phase separation has already progressed to a considerable extent before casting in case the oxidation potential is too high. The optimum range of initial oxidation potential of the solution is 480 - 560 mV (vs. SCE). The holding time of solution before casting may be associated with the degree of polymerization of pyrrole and an extent of phase separation. It is noteworthy that the electrical conductivity is as high as 10 S/cm when only 5 wt.% of pyrrole monomer is incorporated, while the incorporate of 40 wt.% of pyrrole leads to 0.01 S/cm in solution polymerization. (Fig. 2)

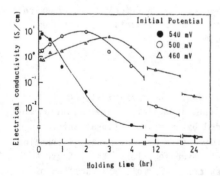

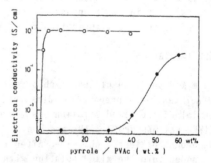

Fig. 1 Electrical conductivity of PPy-PVAc composite film as a function of the initial oxidation potential and the holding time.

Fig. 2 Electrical conductivity versus the content of pyrrole. : ○ PPy-PVAc composite films, ● those prepared by solution polymerization.

336

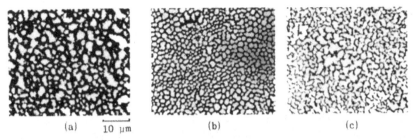

(a) 10 µm   (b)   (c)

Fig. 3 Morphologies of PPy-PVAc composite films of various pyrrole content. All samples are prepared by solvent evaporation method. Weight ratio of pyrrole to PVAc is (a) 0.4, (b) 0.3 and (c) 0.25, respectively.

As shown in Fig. 3, the network structure of PPy is maintained even though the ratio of pyrrole monomer is decreased. Therefore a network structure of PPy may be obtained if only 5 wt.% of pyrrole is incorporated and it brings about the highest conductivity.

Other conditions such as reaction temperature and the presence of water vapor also have an effect on the conductivity of the composite film. The films casted and dried at low temperature show high conductivity. It may be assumed that the high viscosity of PVAc at low temperature prevent a PPy from aggregating, which causes more effective spinodal decomposition.

Furthermore blowing with nitrogen containing water vapor just after casting increases the conductivity of the composite film by about 2 times compared to that dried at ambient condition. This effect may be also associated with the viscosity change of PVAc because water is a nonsolvent to PVAc.

More detailed studies are needed to examine these effects. The PPy-PVAc composite film show a good environmental stability. PPy forms a network by the spinodal decomposition and PVAc is cross-linked by ferric chloride during polymerization process. Thereby the chemical stability of the composite is also excellent. The PPy-PVAc composite are insoluble in any solvent.

The conductivity of composite film is very stable. The initial conductivity persists for more than an year at room temperature, as shown in Fig. 4. In summary, these conductive polymer alloys possess high electrical conductivity and show excellent environmental stability at ambient condition.

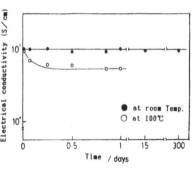

Fig. 4 Time dependence of the electrical conductivity of the PPy-PVAc composite film

REFERENCES

1. G. B. Street, T. C. Clarke, M. Krounbi, K. K. Kanazawa, V. Y. Lee, P. Pfluger, J. C. Scott and G. Weiser, Mol. Cryst. Liq. Cryst., 83 : 253 (1982)

2. R. E. Myers, J. Electron. Mater., 15 : 61 (1986)
3. R. B. Bjorklund, J. Chem. Soc., Faraday Trans. 1, 83 : 1057 (1987)
4. K. Kaneto, Y. Kohno, K. Yoshino and Y. Inuishi, J. Chem. Soc. Chem. Commun., 382 (1983)
5. S. Machida, A. Techagumpuch and S. Miyata, Synth. Met., 31 : 311 (1989)
6. T. Ojio and S. Miyata, Polym. J.,18, No. 1 : 95 (1986)
7. O. Niwa and T. Tamamura, J. Chem. Soc. Chem. Commun., 817 (1984)
8. B. Wessling and H. Volk, Synth. Met., 15 : 183 (1986)
9. T. A. Ezquerra, F. Kremer, M. Mohammadi, J. Ruhe, G. Wegner and B. Wessling, Synth. Met., 28 : C83 (1989)

ELECTRONIC PHENOMENA IN POLYANILINES

Arthur J. Epstein

Department of Physics and Department of Chemistry
The Ohio State University
Columbus, OH 43210-1106 U.S.A.

Alan G. MacDiarmid

Department of Chemistry
University of Pennsylvania
Philadelphia, PA 19104-6323 U.S.A.

ABSTRACT

Polyanilines have been known for over one hundred years. Recent studies of this chemically flexible polymer have demonstrated unusual chemical, electrical, and optical phenomena, both in insulating forms and conducting forms. Polyaniline has three stable insulating forms, leucoemeraldine base (LEB), emeraldine base (EB), and pernigraniline base (PNB). These three forms exhibit differing phenomena ranging from intramonomer luminescing excitons and ring rotation polarons in LEB, to nonluminescing charge transfer excitons and also ring rotations polarons in EB, to a degenerate ground state with bond order parameter solitons and ring order parameter polarons in PNB. The LEB form can be p-doped (oxidatively doped), the EB form can be protonic acid doped and the PNB form can be n-doped (reductively doped) to form conducting systems. Charge conduction studies of oriented films and fibers demonstrate that three-dimensional order between chains is critical for high conductivities. The ability to derivatize polyaniline at ring and nitrogen positions allows one to test concepts for the control of conductivity as well as improved processing.

INTRODUCTION

The polyanilines are among the oldest known synthetic organic polymers, having been first reported in 1862 [1]. Since the mid-1980's there has been a

Frontiers of Polymer Research, Edited by P.N. Prasad and
J.K. Nigam, Plenum Press, New York, 1991

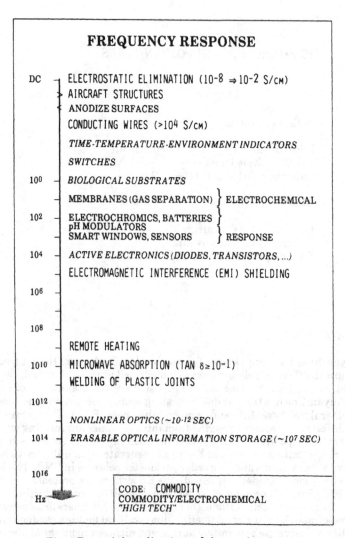

Fig. 1 Potential applications of electronic polymers.

mushrooming interest in this family of polymers [2-8]. Synthesis of polyanilines and their derivatives have been extensively described [7]. The facile chemistry of this system and its ability to be processed into oriented films [9,10] and fibers [11-13] have opened the opportunity to design in properties for function. This has lead to interest in the potential use of polyanilines and related conducting polymers in a variety of commodity, commodity/electrochemical, and high tech type applications [14, 15]. Figure 1 summarizes the range of technologies proposed. It is noted that a wide range of resistivities from insulating through highly conducting are of potential utility.

We review here the essential differences in the physics of the three different polyaniline insulating forms as well as present a summary of the critical aspects controlling conductivity in the emeraldine salt system. The effects of derivatization are briefly indicated. The parent polyaniline can be prepared in a range of oxidation states ranging from the fully reduced form, leuceoemeraldine base (LEB), $\{NH\text{-}C_6H_4\text{-}NH\text{-}C_6H_4\}_x$, (Fig. 2a.), to the fifty percent oxidized emeraldine base (EB), $\{NH\text{-}C_6H_4\text{-}NH\text{-}C_6H_4\text{-}N=C_6H_4=N\text{-}C_6H_4\}$ (Fig. 2b), to the fully oxidized pernigranilne base (PNB), $\{N=C_6H_4=N\text{-}C_6H_4\}_x$ (Fig. 2c) [7,16]. The three base forms are insulators. Protonation of EB or oxidation of LEB leads to formation of the conducting emeraldine salt (ES) shown in Fig. 2d as the hydrochloride salt.

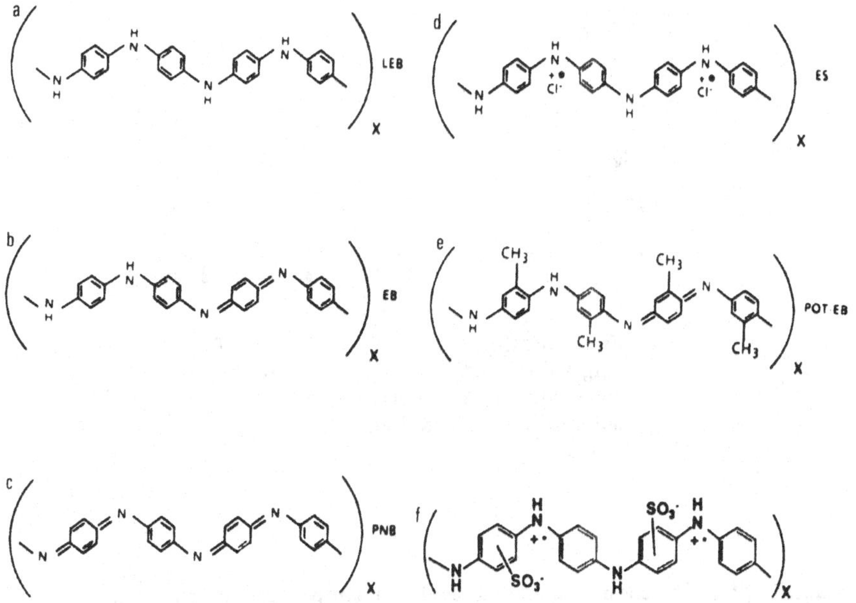

Fig.2. (a) leucoemeraldine base; (b) emeraldine base; (c) pernigraniline base; (d) emeraldine hydrochloride salt polymer; (e) poly(orthotoluidine) (emeraldine base form); (f) sulfonated polyaniline (self-doped form).

DERIVATIZATION

The polyanilines are easily derivatized at the ring [17-19] and nitrogen [20] positions. For example, replacement of one H of each ring with a CH$_3$ group to form poly (o-toluidine), Fig. 2e, enables testing of the roles of interchain separation and crystalline order in determining electronic properties [17,21,22], while replacement of a H on one-half the rings with a sulfonic acid group, -SO$_3$H, Fig. 2f, leads to a self-doped polymer the conductivity of which is independent of pH for pH $\lesssim$ 7.5 [18,19] (Fig. 3) and exhibits greater thermal and environmental

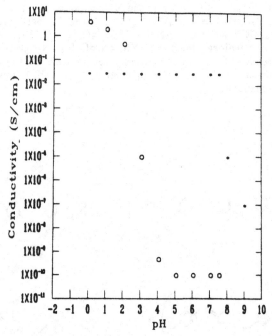

Fig. 3. pH dependence of conductivity of ring-sulfonated polyanilines (*) and the emeraldine hydrochloride form of polyaniline (o) (from Ref. 19).

stability [23]. The change in doping level with temperature for emeraldine hychloride and ring-sulfonated polyaniline is estimated by assuming that the entire weight loss measured by thermogravimetry is due to loss of HCl and SO$_3$H, respectively, Fig. 4. In actuality, loss of polymer fragments may occur as well.

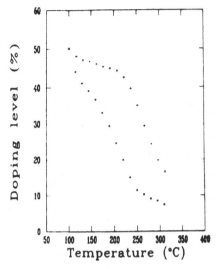

Fig. 4. Doping levels determined by thermogravimetry of emeraldine hydrochloride (*) and sulfonated polyaniline (+) as a function of temperature assuming that the entire weight loss for each polymer is due only to loss of HCl and SO_3H respectively (from Ref. 23).

SPECTROSCOPY AND PHOTOEXCITATION SPECTROSCOPY

The optical absorption spectrum of LEB [24] is shown in Fig. 5 together with the results of a Hückel calculation for the electronic band structure of LEB carried out assuming that the repeat unit is a single (C_6H_4NH) unit [25,26]. This simple calculation reproduces most of the band structure features obtained in more extensive calculations [27,28,29]. The seven energy bands are derived from the six p_z energy levels of benzene together with the p_z energy level of nitrogen. The lower four energy bands are completely filled while the upper three are empty. It is noted that the valence band is relatively broad (its breadth depending upon the ring torsion angle) while the conduction band is dispersionless. Hence, charge conjugation symmetry is absent in leucoemeraldine base. The exact width and separation of each of the bands are sensitive to the torsion angle the C_6 rings make with respect to the plane defined by the nitrogen atoms [25,29].

Excitation of leucoemeraldine solutions in N-methylpyrrolidone (NMP) with bandgap light produces an intense luminescence peak at 3.5 eV [30], Fig. 6. The magnitude of the luminescence intensity decreases linearly with increasing concentration of imine groups within the polymer chain. The luminescence is nearly zero when the emeraldine base oxidation state is achieved. The decay

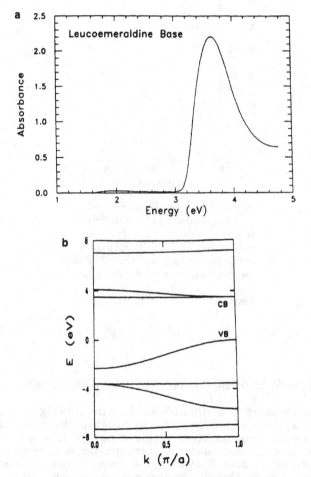

Fig. 5.(a) Absorption
spectrum of leucoemeraldine
base in NMP (from Ref. 33);
(b) Hückel electronic band
structure of leucoemeraldine
(from Ref. 25).

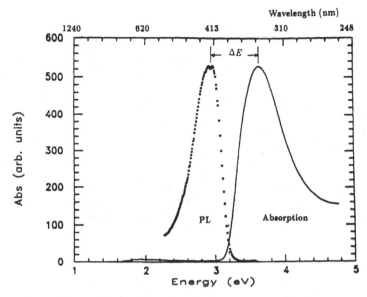

Fig. 6. Absorption and photoluminescence spectra for leucoemeraldine base in NMP, the exciting wavelength for the PL spectrum is ~345 nm (from Ref. 30).

time for the luminescence has been measured and found to be ~ 4ns at room temperature [30]. The presence of a similar luminescence in three-ring oligomers [30] and also in the aniline ($C_6H_5NH_2$) monomer [31] supports that the luminescence derives from radiative decay of an intramonomer polaronic exciton [32].

Photoexcitation of LEB films and powders with superbandgap light leads to new intragap absorptions centered at 0.7 eV and 2.9 eV [24,33]. This photoinduced absorption has been described in terms of formation of a ring rotation polaron [24-26, 33]. Because of the large charge conjugation asymmetry in the system there is only one energy level introduced into the gap by the distortion of the ring torsion angles. This polaron differs substantially from earlier studied polarons in polythiophene and polypyrrole and their derivatives in that it has a very large effective mass (~$50m_e$) due to the effects of a large moment of inertia involved in rotating the polyaniline rings about the N-N axis. Also observed in the photoexcitation spectrum is a bleaching at ~2.0 eV associated with the presence of some residual imine groups in the otherwise pure LEB. At long times a photoinduced absorption at ~1.4 eV is stable. This absorption is associated with a ring rotation polaron trapped adjacent to a quinoid ring. This trapped ring rotation polaron is readily observed in the photoexcitation spectra of emeraldine base [24,33]. Comparison of the magnitude of the photoinduced ir active vibrations and electronic absorption confirms the large mass of the photoexcited defects.

The optical absorbance of pernigranilne base though similar to that of EB is different in origin. In particular, the 2.3 eV absorption in PNB has been assigned to the excitation across a Peierls energy gap. This Peierls gap has been proposed to arise from two independent contributions, variation in bond length order parameter and variation in ring torsion angle order parameter, [25,34,35]. The effects of these two order parameters recently have been shown to vary additively [29]. Steady state photoinduced absorption experiments show the presence of a photoinduced charge defect of effective ~5 m_e and a photoinduced charge defect of effective mass ~300m_e. Because of the enormous difference in their effective mass and their lifetimes, the low energy peak is associated with a solitonic defect in the bond length order parameter while the middle energy peak is associated with a polaronic defect in the ring torsion angle order parameter [36]. Light induced electron spin resonance studies support these assignments [37]. This is the first known material where two order parameters contribute independently to the bandgap and sustain separate excitations.

CHARGE TRANSPORT

The increase in conductivity of emeraldine salt fiber as a function of elongation ratios (l/l_0) demonstrates the importance of orientation and crystallinity (which increases with stretch ratio [13-15]). The conductivity, $\sigma(T) \sim \exp[-(T_0/T)^{\frac{1}{4}}]$ suggests the relevance of two models: quasi-one-dimensional variable range hopping (VRH) [21, 22,38] and hopping of charge carriers between granular metallic regions [39]. The dc conductivity of the hydrochloric salt of poly(orthotoluidine) (POT-ES) has the same temperature dependence but with a six-fold larger T_0 indicative of greater localization of the charge carriers. The thermoelectric power for POT-ES shows only a term for variable range hopping without any temperature range where a metallic-like thermopower dominates [21,22]. Self-doped sulfonated polyaniline (SPAN) [19] provides a similar case study.

The temperature dependence of the microwave frequency (6.5GHz) dielectric constant for unoriented emeraldine hydrochloride (PAN-ES) [40], POT-ES [21,22] and self-doped sulfonated polyaniline [41], Fig. 7, graphically illustrates the importance of interchain interaction in delocalization. Analysis of these data and the data for stretch oriented emeraldine hydrochloric film [30] suggests that the localization length for positive charges in SPAN-ES is ~6A while that for POT-ES is ~8A. Both of these systems have conductivity dominated by quasi-one-dimensional variable range hopping. In contrast, PAN-ES has a parallel localization length in the order of 50A and a perpendicular localization length ~20A even at low temperatures [38]. These dimensions which coincide with the estimated size of the crystalline regions strongly support that the three-dimensional delocalization of charge carriers occurs within the crystalline regions of PAN-ES.

SUMMARY

The varied chemistry and ability of the polyanilines to be processed together with their environmental stability makes the systems suitable for scientific and technological studies. Several important concepts have emerged including the essential roles of ring rotations in electronic phenomena, and three-dimensional order for high conductivity. These materials remain a laboratory for study of chemical, physical, and technological opportunities in conducting polymers.

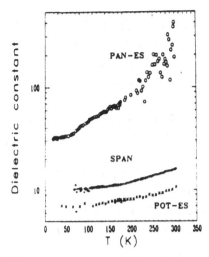

Fig. 7. Microwave frequency dielectric constant vs. temperature for unoriented PAN-ES (from Ref.38) unoriented POT-ES (from Ref. 22) and sulfonated polyaniline (from Ref. 41).

The authors acknowledge the important contribution of their students, postdoctoral fellows and collaborators. This work is supported in part by the Defense Advanced Research Projects Agency through a grant monitored by the U.S. Office of Naval Research.

REFERENCES

1. H. Letheby, J. Chem. Soc. 15:161 (1862).
2. B.D. Humphrey, J-C. Chiang, W-S. Huang, A.F. Richter, N.L.D. Somarisi, A.G. MacDiarmid, X.Q. Yang, A.J. Epstein and R.W. Bigelow, Bull. Phys. Soc. 30:605 (1985).
3. J.P. Travers, J. Chroboczek, F. Devreux, F. Genoud, M. Nechtschein, A. Syed, E.M. Genies and G. Tsintavis, Mol. Cryst. Liq. Cryst. 121:195 (1985).
4. A.G. MacDiarmid, J-C. Chiang, M. Halpern, W-S Huang, S-L. Mu, N.L.D. Somasiri, W. Wu and S.I. Yaniger, Mol. Cryst. Liq. Cryst. 121:173 (1985).
5. E.W. Paul, A.J. Ricco and M.S. Wrighton, J. Phys. Chem. 89:1441 (1985).
6. P.M. McManus, S.C. Yang and R.J. Cushman, J. Chem Soc. Chem. Commun. 1556 (1985).
7. A.G. MacDiarmid and A.J. Epstein, Faraday Discuss. Chem. Soc. 88:317 (1989).
8. For recent progress, see Proc. Int. Conf. on Science and Technology of Synthetic Metals, Kyoto, Japan, June 1986 (Synth. Met. 17-19 (1987));

Sante Fe, NM, June 1988 (<u>Synth Met</u>. <u>27</u>-<u>29</u> (1989)); Tübingen, Germany, September 1990 (<u>Synth. Met</u>. <u>41</u>-<u>43</u> (1991)).

9. K.R. Cromack, M.E. Jozefowicz, J.M. Ginder, R.P. McCall, A.J. Epstein, E. Scherr and A.G. MacDiarmid, <u>Bull. Am. Phys. Soc</u>. 34:583 (1989).

10. K.R. Cromack, M.E. Jozefowicz, J.M. Ginder, R.P. McCall, G. Du, J.M. Leng, K. Kim, C. Li, Z. Wang, A.J. Epstein, M.A. Druy, P.J. Glatkowski, E.M. Scherr and A.G. MacDiarmid, <u>Macromolecules</u> <u>xx</u>, xxx (1991).

11. X. Tang, E. Scherr, A.G. MacDiarmid and A.J. Epstein, <u>Bull. Am. Phys. Soc</u>. 34:583 (1989).

12. A. Andreatta, Y. Cao, J.C.Chiang, A.J. Heeger and P. Smith, <u>Synth. Met</u>. <u>26</u>, 383 (1988); A. Andreatta, S. Tokito, P. Smith and A.J. Heeger, <u>Mat. Res. Soc. Symp. Proc</u>. 173:269 (1990).

13. E.M. Scherr, A.G. MacDiarmid, S.K. Manohar, J.G. Masters, Y. Sun, X. Tang, M.A.Druy, P.J. Glatkowski, K.R. Cromack, M.E. Jozefowicz, J.M. Ginder, R.P. McCall and A.J. Epstein, <u>Synth. Met</u>. (1991).

14. A.G. MacDiarmid and A.J. Epstein, "Science and Application of Conducting Polymers," W.R. Salanek, D.T. Clark and E.J. Samuelsen, eds., Adam Hilger, New York, p.117 (1991).

15. A.J. Epstein and A.G. MacDiarmid, "Science and Application of Conducting Polymers," W.R. Salanek, D.T. Clark and E.J. Samuelsen, eds., Adam Hilger, New York, p.141 (1991).

16. Y. Sun, A.G. MacDiarmid and A.J. Epstein, <u>J. Chem. Soc. Commun</u>. 7:529 (1990).

17. Y. Wei, W.W. Focke, G.E. Wnek, A. Ray and A.G. MacDiarmid, <u>J. Phys. Chem</u>. <u>93</u>, 495 (1989); A.Ray, A.G. MacDiarmid, J.M. Ginder and A.J. Epstein, <u>Mat. Res. Soc. Symp. Proc</u>. 173:353 (1990).

18. J. Yue and A.J. Epstein, <u>J. Am. Chem. Soc</u>. 112:2800 (1990).

19. J. Yue, Z.H. Wang, K.R.Cromack, A.J. Epstein and A.G. MacDiarmid, <u>J. Am. Chem. Soc</u>. 113:2665 (1991).

20. S.K. Manohar, A.G. MacDiarmid, K.R. Cromack, J.M. Ginder and A.J. Epstein, <u>Synth. Met</u>. <u>29</u>,E349 (1989).

21. Z.H. Wang, H.H.S. Javadi, A. Ray, A.G. MacDiarmid and A.J. Epstein, <u>Phys. Rev. B</u> 42:5411 (1990).

22. Z.H. Wang, A.Ray, A.G. MacDiarmid and A.J. Epstein, <u>Phys. Rev. B</u> 43:4373 (1991).

23. J. Yue, A.J. Epstein, Z. Zhong, P. Gallagher and A.G. MacDiarmid, <u>Synth. Met</u>. <u>xx</u>, xxx (1991).

24. R.P. McCall, J.M. Ginder, M.G. Roe, G.E. Asturias, E.M. Scherr, A.G. MacDiarmid and A.J. Epstein, <u>Phys. Rev. B</u> 39:10,174 (1989).

25. J.M. Ginder and A.J. Epstein, <u>Phys. Rev. B</u> 41:10,674 (1990).

26. J.M. Ginder, A.J. Epstein and A.G. MacDiarmid, Solid State Commun. 72:987 (1989).

27. D.S. Boudreaux, R.R. Chance, J.F. Wolf, L.W. Shacklette, J.L. Bredas, B. Themans, J.M. Andre and R. Silbey, J. Chem. Phys. 85:4584 (1986).

28. S. Stafstrom and J.L. Brédas, Synth. Met. 14:297 (1986).

29. J.L. Brédas, C. Quattrocchi, J. Libert, A.G. MacDiarmid, J.M. Ginder and A.J. Epstein, Phys. Rev. B 44:xxx (1991).

30. J.G. Masters, A.G. MacDiarmid, K. Kim, J.M. Ginder and A.J. Epstein, Bull. Am. Phys. Soc. 36:377 (1991).

31. I.B. Berlman, Handbook of Fluorescence Spectra of Aromatic Molecules, 2nd ed., Academic Press (1971).

32. N.F. Colaneri, D.D.C. Bradley, R.H. Friend, P.L. Burn, A.B. Holmes and C.W. Spangler, Phys. Rev. B 42:11,670 (1990).

33. R.P. McCall, J.M. Ginder, J.M. Leng, H.J. Ye, S.K. Manohar, J.G.Masters, G.E. Asturias, A.G. MacDiarmid and A.J. Epstein, Phys. Rev. B 41:5202 (1990).

34. J.M. Ginder and A.J. Epstein, Phys. Rev. Lett. 64:1184 (1990).

35. M.C. dos Santos and J.L. Brédas, Phys. Rev. Lett. 64:1185 (1990).

36. J.M. Leng, J.M. Ginder, R.P. McCall, H.J. Ye, Y. Sun, S.K. Manohar, A.G. MacDiarmid and A.J. Epstein, to be published.

37. K.R. Cromack, A.J. Epstein, J.G.Masters, Y. Sun and A.G. MacDiarmid, Synth. Met. xx, xxx (1991).

38. Z.H. Wang, C. Li, E.M. Scherr, A.G. MacDiarmid and A.J. Epstein, Phys. Rev. Lett., 55:1745 (1991).

39. F. Zuo, M. Angelopoulos, A.G.MacDiarmid and A.J. Epstein, Phys. Rev. B 36:3475 (1987).

40. H.H.S. Javadi, K.R. Cromack, A.G. MacDiarmid and A.J. Epstein, Phys. Rev. B 39:3579 (1989).

41. Z.H. Wang, K.R. Cromack, J. Yue and A.J. Epstein, to be published.

THE PREPARATION OF HIGHLY CONDUCTIVE POLYPYRROLE COMPOSITE FILM BY

CHEMICAL POLYMERIZATION IN SOLUTION

Anuntasin Techagumpuch

Department of Physics, Chulalongkorn University
Bangkok, Thailand

ABSTRACT

Thin films of conducting polypyrrole were prepared by chemical polymerization in solutions : by dipping polyethylene terephthalate film coated
with polymethyl methacrylate (PMMA) into pyrrole monomers and then in $FeCl_3$
aqueous solution. The oxidation potential of this solution, which has
strong influence on the polymerization, was adjusted to appropiate value
by adding suitable amount of $FeCl_2$ to the solution before the reaction. In
the suitable polymerization condition, the polypyrrole films obtained
possess conductivity as high as 110 S/cm. In the case of long polymerization time, the SEM studies of the films indicated that polypyrrole was not
only polymerized on the surface of PMMA, but also penetrated into PMMA to
form a strong composite films. It was found that the polypyrrole films
obtained in this work were very difficult to undope. This suggests that
the prepared films can be used as a conducting surface for electrodeposition of metal on surface of plastic such as PMMA.

INTRODUCTION

In the past highly conductive polypyrrole was prepared only by electrochemical polymerization [1-3]. But recently we found that highly conductive polypyrrole can be prepared by chemical polymerization in solution
if the oxidation potential of the solution is controlled to a suitable
value [4]. In this chemical synthesis the polypyrrole obtained was a
black powder, sheet-type samples can be prepared by pressing; but by this
method a thin film samples are rather difficult to obtain. So in this
report we introduce a new method for preparing a highly conductive polypyrrole films by chemical polymerization in solution.

EXPERIMENTAL

In this preparation we first dissolved PMMA (polymethyl methacrylate)
by solvent such as Benzene and coated the dissolved PMMA on the PET
(polyethylene terephthalate) film. After drying this coated PET film was
dipped in purified pyrrole monomers for about 5s (see Fig. 1), then
transferred this film to $FeCl_3$ solution in distilled water for some period
of time. Polypyrrole film was then synthesized on PMMA surface. It was
found that some part of polypyrrole penetrated into PMMA to form a compo-

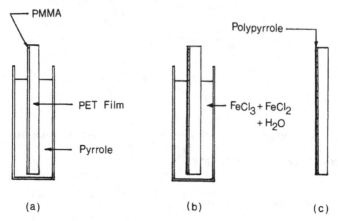

Fig. 1. The preparation of highly conductive polypyrrole composite
film.

site film. The chemical reaction are proposed as [4]

$$n \underset{\overset{|}{N}}{\langle} + 2.33nFeCl_3 \longrightarrow \left[\underset{\overset{|}{N}}{\langle} + 0.33Cl^- \right]_n + 2.33nFeCl_2 + 2nHCl \tag{1}$$

this polypyrrole has Cl^- as a dopant anion.

After this polymerization the film was dried in vacuum at $40^\circ C$ for
24 h. Then we measured the surface resistance ($R_s = \pi V/Iln2$) of the
polypyrrole film by using four-point probe method[5]. After this the
polypyrrole film was separated from the substrate by dissolving off PMMA.
The thickness of the film (d) was then measured and finally we obtained
the conductivity ($\rho = 1/R_s d$) of the synthesized polypyrrole.

In this work $FeCl_3$ solutions in water which had initial oxidation
potential in the range of 500 mV to 700 mV (versus SCE) were used as
oxidant solutions. To prepare these solutions we observe that from
Nernst's equation the oxidation potential of $FeCl_3$ solution is given by
[4, 6]

$$E = E_o + \frac{RT}{nF} \ln \frac{[FeCl_3]}{[FeCl_2]} . \tag{2}$$

From this equation E decreases when $[FeCl_2]$ increases. Thus we can
obtain the solutions of desired initial oxidation potentials by adding
suitable amount of $FeCl_2$ to $FeCl_3$ solutions before the polymerizations,
and from this the dependence of conductivity of synthesized polypyrrole
on oxidation potential of $FeCl_3$ solution can be investigated.

RESULTS AND DISCUSSION

The conductivity of polypyrrole film prepared in this work depended
on polymerization temperature (T_p), polymerization time (t_p), $FeCl_3$
concentration, and oxidation potential of the solution. For $T_p = 0^\circ C$,
$t_p = 5$ min (t_p is a soaking time in Fig. 1b) the dependences of conductivity of polypyrrole on the oxidation potential when $FeCl_3$ concentrations
are 1 M and 3 M are illustrated in Fig. 2. It is clear that 1 M $FeCl_3$
solutions yielded polypyrroles of higher conductivity. But for these
solutions good films were obtained only when their oxidation potentials
were higher than about 600 mV. The highest conductitity obtained from

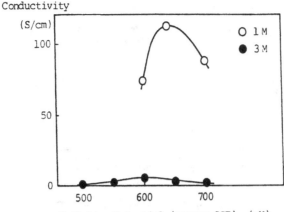

Conductivity
(S/cm)

Oxidation Potential (versus SCE) (mV)

Fig. 2. The conductivity of chemically synthesized polypyrrole film produced by 1 M and 3 M $FeCl_3$ solutions of various oxidation potentials, in these preparations t_p = 5 min and T_p = 0°C.

this solution at 640 mV oxidation potential was about 110 S/cm.

The morphology of the polypyrrole films produced by 1 M and 3 M solutions at the optimum polymerization condition are shown in Fig. 3a and 3b respectively. The 1 M solution yielded film of rather smooth

10 μm

a   b

Fig. 3. The morphology of polypyrrole films prepared by (a) 1 M $FeCl_3$ solution and (b) 3 M $FeCl_3$ solution at optimum polymerization condition.

surface while 3 M solution produced film which surface contained large number of particles aggregated together. We proposed that in 3 M $FeCl_3$ solution polymerization rate was so high such that many polymerized nuclei evolved at the same time, this led to the polypyrrole film of aggregated structure surface which had low conductivity.

For the film prepared by 1 M $FeCl_3$ solution of 640 mV oxidation potential it was found that the thickness of the film increased linearly with polymerization time. But the surface resistance which was large at short polymerization time became smaller and almost constant at longer time. This led to the dependence of conductivity on polymerization time as shown in Fig. 4.

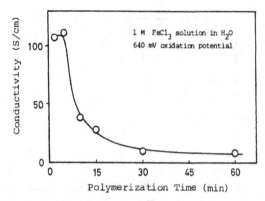

Fig. 4. The dependence of conductivity of prepared polypyrrole on polymerization time when $T_p = 0°C$.

Fig. 5. The morphology of the backside of polypyrrole films when (a) $t_p = 5$ min and (b) $t_p = 1$ h.

The SEM pictures of the backsides of polypyrrole films (the side which attached to PMMA) for $t_p = 5$ min and $t_p = 1$ h' are shown in Fig. 5a and 5b respectively. For $t_p = 5$ min the backside surface of the synthesized film was rather smooth, but for $t_p = 1$ h the backside surface was aggregated structure. We proposed that when the PET film coated with PMMA was dipped into pyrrole monomer (Fig. 1a) polypyrrole did not only wet PMMA but some of them dissolved into PMMA. When the polymerization proceeded for a long time polypyrrole was continuously synthesized and deeply penetrated into PMMA to form an aggregated structure as shown in Fig. 5b. This is a reason why polypyrrole attached very well with PMMA and form a strong composite film. It is clear from Fig. 4 that the aggregated structure at the backside of the film contribute nothing to the conduction process.

The study of the effect of polymerization temperature on the conductivity of the films indicated that about $0°C$ is the optimum temperature for producing conductive polypyrrole. Higher polymerization temperature say $20°C$, produced the film of lower conductivity which had morphology similar to that shown in Fig. 3b [7].

POTENTIAL APPLICATION

Many experiments showed that considerable time is required to electrically undope chemically synthesized polypyrrole. To demonstrate this
phenomena we performed an conventional electrochemical experiment to
undope this polypyrrole [8]. One of platinum electrode of electrochemical
cell was coated with polypyrrole film, this film was chemically synthesized by similar process mentioned in this work, but this time we allowed
some part of polypyrrole to touch platinum in order to have a good electrical contact. An electrolyte used in the cell was 1 M tetrabutylam-
moniumperchlorate in propylenecarbonate, note that the concentration of
this electrolyte was very high compared to those commonly used in this
kind of experiment. First we sweeped the potential down from 0 to -1.5 V
and up to +0.6 V (vs. SCE) with a sweep rate of 40 mV/s. The cyclic
voltammograms of this film is shown in Fig. 6. All of this process was
done in nitrogen atmosphere.

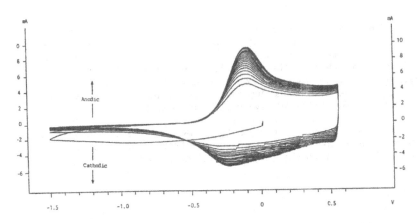

Fig. 6. The cyclic voltammograms of chemically synthesized polypyrrole
measured in 1 M Tetrabutylammonium perchlorate in propylenecarbonate.

For the first cycle of the potential sweep the current was small.
This means that only small percentage of dopant anion was released from
polypyrrole and went into polypyrrole in the cathodic sweep and anodic
sweep respectively, in other word the redox process was very incomplete
at the beginning. However when the sweep was continued the current
gradually increased and finally seem to reach the saturated state, at
this point the dope and undope process was almost complete for each
anodic and cathodic sweep. When the saturated state was reached we
sweeped the potential at different sweep rates and obtained the cyclic
voltammograms of the film as shown in Fig. 7. These cyclic voltammograms
are similar to that obtained by Diaz [8] for the case of electrochemically
synthesized polypyrrole, however in our work the current peaks were more
broader, this may due to the different in thickness of the polypyrrole
film which cannot be controlled by this preparation.

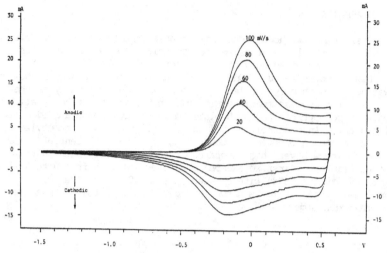

Fig. 7. Cyclic voltammograms of chemically synthesized polypyrrole
at various sweep rates.

The final result of this experiment is the change of dopant anion
from Cl^- to ClO_4^-. Many more analysis was done concerned this electrochemical process and the results are reported elsewhere. In this work
we can only conclude that the chemically synthesized polypyrrole cannot
be undoped in a short time but one this process is done it can be driven
repeatedly between the conducting and nonconducting state the same way
as was done in electrochemically synthesized polypyrrole.

From this fact it can be shown that chemically synthesized polypyrrole
can be used as a conductive surface for electroplating an insulator
surface. Fig. 8a shows the electroplating of copper on surface of

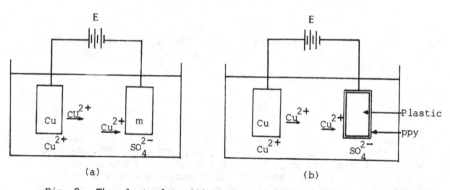

Fig. 8. The electrodeposition of copper on (a) metal surface
(b) polypyrrole surface.

another metal, this can be done because the metal is conductor and can be
used as an electrode. If this metal is replaced by insulator such as
plastic this process is rather impossible. However we can coat the
surface of this plastic by polypyrrole thin film by using process mentioned
in this work or similar process (Fig. 8b). This coated film can work as
a conductive surface for electrodeposition of copper as desired. Since

this polypyrrole is difficult to undope, in using as negative electrode this coated polypyrrole is still conductive material for sometimes which is long enough as required for electroplating. This method is new alternative for electrodeposition of copper on insulator surface, and now is under investigation.

CONCLUSION

The highly conductive polypyrrole films can be prepared by chemical polymerization in $FeCl_3$ solution. The conductivity of the prepared film depends on concentration and oxidation potential of the solution as well as polymerization temperature and polymerization time. When the suitable polymerization condition is chosen the polypyrrole film obtained shows conductivity as high as 110 S/cm. This polypyrrole film has potential application in electrodeposition of metal on surface of insulator such as plastic.

REFERENCES

1) K. Kanazawa, A.F. Diaz, W.D. Gill, P.M. Grant, G.B. Street and G.P. Gardini, Synth. Met., 1(1979/1980)329.
2) G.B. Street, T.C. Clarke, M. Krounbi, K. Kanazawa, V. Lee, P. Pfluger, J.C. Scott and G. Weiser, Mol. Cryst. Liq. Cryst., 83(1982)253.
3) G.B. Street, "Hand Book for Conducting Polymer," T.A. Skotheim, Ed., Marcel Dekker, Inc., New York and Basel, 1986, Chapter 8 and references therein.
4) S. Machida, A. Techagumpuch and S. Miyata, Synth. Met., 31(1989)311.
5) S.M. Sze, "Physics of Semiconductor Devices," Wiley-Interscience, New York, 1981, p. 30.
6) A.J. Bard and L.R. Faulkner, "Electrochemical Methods," John Wileys and Sons, New York, N.Y., 1980, p. 51.
7) T. Yoshigawa, S. Machida, T. Ikegami, A. Techagumpuch and S. Miyata Polym. J., 18(1986)95.
8) A.F. Diaz and J. Bargon, "Hand Book for Conducting Polymer," T.A. Skotheim, Ed., Marcel Dekker, Inc., New York and Basel, 1986, Chapter 3 and references therein.

DIELECTRIC RESPONSE OF POLYPYRROLE AT MICROWAVE AND INFRARED FREQUENCIES

Jayant Kumar, Jeng I. Chen[†], Suresh Ramakoti, Greg Phillips,
Jerry Waldman and Sukant Tripathy[†]

Departments of Physics and [†]Chemistry
University of Lowell
Lowell, Massachusetts 01854

INTRODUCTION

We report measurements on the dielectric properties of Polypyrrole doped with the tosylate anion in the far infrared (FIR) and microwave frequency regimes. DC electrical conductivity measurements were performed on the polypyrrole samples with a four-point probe. The samples were prepared electrochemically providing control of the doping level through reduction of the as grown film. Thin films varying in thickness from 10 to 50 microns were synthesized with dc conductivities in the range of 1 to 100 S/cm.

The heavily doped films exhibit large real and imaginary parts of the complex index of refraction (n+ik) and are highly dispersive at these frequencies[1,2]. The magnitude of these quantities is changed significantly when the conductivity is decreased through the process of reduction. Although large metal-like reflection is observed at the lower frequencies (microwave), the behavior of the complex dielectric constant as a function of frequency[1] does not follow the simple metal (Drude) theory. Characterization of the dielectric constants of doped polymers at microwave and the FIR frequencies is useful in determining the conduction mechanism in these materials. The dielectric relaxation qualitatively follows a Debye-like behavior.

MATERIALS AND METHODS

Sample Preparation

Free standing polypyrrole films were synthesized by inserting a platinum foil as the working electrode in an acetonitrile solution containing 0.1M pyrrole and 0.25M tetraethylammonium tosylate. A carbon plate serves as the counter electrode and a calomel reference electrode is used to monitor the working electrode potential during the galvanostatic synthetic process. By passing a current of 0.4 mA/cm[2] for a period of 4 hrs, oxidation and polymerization occur simultaneously, producing a black conducting film at the

Frontiers of Polymer Research, Edited by P.N. Prasad and
J.K. Nigam, Plenum Press, New York, 1991

electrode ranging in thickness from 5 to 50 microns. The film is immersed in acetonitrile for 15 minutes and dried under vacuum at room temperature[3]. Varying degrees of film reduction were achieved by passing a reverse current at a fixed value of current density (0.4 mA/cm[2]) for different periods of time.

Microwave Measurements

Measurement of the complex permittivity was performed in the frequency range of 8.5 - 40.0 GHz using a HP8510 network analyzer. The sample is held between two waveguide flanges. The network analyzer measures the magnitude and phase of the transmitted and reflected signal from the sample with respect to the incident wave. Calibration is performed using standards such as a short and an offset short, a fixed load and a precise transmission line. The amplitude and phase angles are expressed as complex s-parameters[4] (scattering parameters S_{11} and S_{21}) from which quantities such as standing wave ratios (SWR), complex permittivity and permeability can be computed. Figures 1- 4 are the plots of the real and imaginary parts of the refractive index ($n_c = n + ik$) of a 35μ polypyrrole film under various degrees of reduction in the 8.5-40.0 GHz regime.

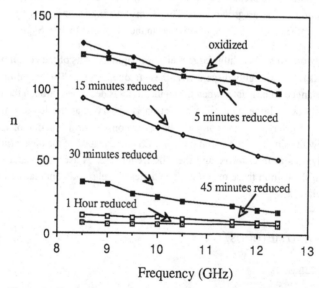

Figure 1. Real part n, of index of refraction of oxidized (as prepared) polypyrrole under various degrees of reduction in the x-band.

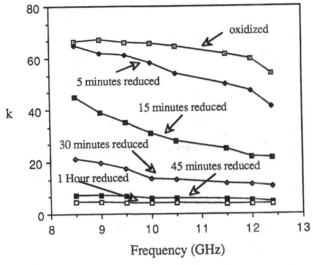

Figure 2.   Complex index of refraction for oxidized and reduced
polypyrrole in the x-band.

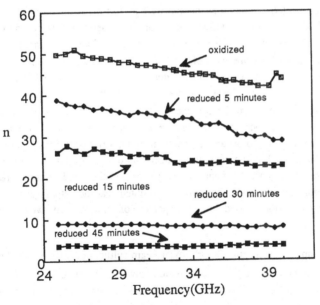

Figure 3.   Real part of the index of refraction of an oxidized and
reduced polypyrrole sample in the k-band.

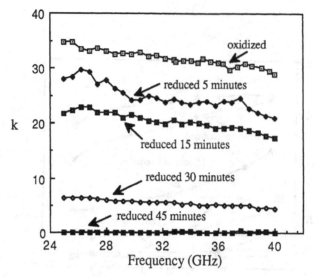

Figure 4.   Imaginary part of the refractive index of oxidized and
reduced polypyrrole in the k-band.

Measurements in the FIR

Measurements of reflection and transmission were performed in the far infrared (FIR)
utilizing an FTIR (Fourier transform infrared) spectrometer to determine the complex index
of refraction.   A Michelson interferometer modified to include an enlarged sample
compartment was used. Measurements of reflection and transmission are performed at
normal incidence with an accuracy of approximately 5 percent. As prepared, the polymer
films are typically 5 cm x 5 cm and when mounted are apertured down to expose an area of
approximately 3 cm^2.   In the reflection measurement, the collimated output of the
interferometer is collected, focused on the sample, and recollected in an approximate f/5
optics system consisting of two off axis parabolic mirrors, a flat gold mirror and a lens.
The sample holder as well as the background mirror is mounted on a translation stage
allowing for consecutive measurements of sample and background without breaking
vacuum. In the transmission mode, the sample is mounted near the output aperture of the
sample box and the radiation focused down to a spot size of approximately two centimeters
in diameter.

A mercury lamp source combined with mylar beam splitters of different thicknesses
provides black body radiation in the spectral region between 10 and 200 cm^{-1}. Since the
radiated power at lower frequencies is only a few microwatts, and the transmittance of the
polymer films may be less than 0.01 percent, a low temperature (1.8 K) silicon bolometer
with a sensitivity of $3x10^{-14}$ Watts/$Hz^{1/2}$ is employed. The entire system is operated under
vacuum to reduce the signal loss associated with water absorption.

To further enhance the reliability of the R and T data, additional measurements are performed with high power (10 mW) HCOOH lasers at discrete frequencies for comparison with the FTIR data. The laser sources have also been used to extend the FIR data to wavenumbers below 8 cm^{-1}.

Determination of the real and imaginary parts of the complex index of refraction is accomplished by direct application of the Fresnel equations to measurements of R and T at normal incidence. The complex reflection and transmission amplitudes are given below:

$$r=\rho(1-e^{i\beta})/(1-\rho^2 e^{i\beta}) \qquad\qquad [1]$$

$$t=(1-\rho^2)e^{i\beta/2}/(1-\rho^2 e^{i\beta}) \qquad\qquad [2]$$

with $\qquad\qquad \rho=(1-n_c)(1+n_c)$

and $\qquad\qquad \beta=4\pi n_c d/\lambda$

where n_c is the complex index of refraction given by: $n_c=n+ik$

Although it is frequently difficult to measure transmission in materials that are highly reflecting or absorbing, we believe this technique to be more accurate than the Kramers-Kronig method and especially useful when collecting data over a narrow frequency range. The Kramers-Kronig method depends on accurate reflection measurements taken over a broad frequency range and extrapolation of data to estimate R_0 (low frequency reflectivity) and R∞ (high frequency reflectivity).

The Fresnel equations in exact form are multiple valued functions of R and T. The transcendental nature of the equations requires a computerized numerical technique for their solution. We have developed software similar to that described by Nestell and Christy[5,2] in which one plots the constant contours of R and T in a plot of n vs k. The intersections of R and T give all the possible paired values of n and k satisfying the Fresnel equations and can be determined by a numerical root searching routine. The selection of the physically correct root from the many possibilities can be accomplished in several ways. Some of the techniques utilized in this study include performing angle scans in reflection utilizing polarized FIR lasers. The fits to the angle scan allow determination of the complex index of refraction unambiguously. It is also possible to determine the complex refractive index unambiguously, from reflection and transmission data of like films of different thicknesses.

The following plots (Figures 5 through 8) report the behavior of n and k as a function of wavenumber (frequency) in the 10 to 200 cm^{-1} region. The large dispersion of n and k is apparent in the doped (as grown) films. Reducing the polypyrrole films produces large changes in the measured levels of dc conductivity as well as the dielectric constants. Figures 5 and 6 are plots of the complex index of refraction for heavily doped films with measured dc conductivities in the 40-100 mho/cm range.

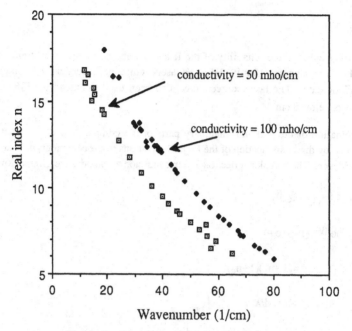

Figure 5.   Real part of the index of refraction.

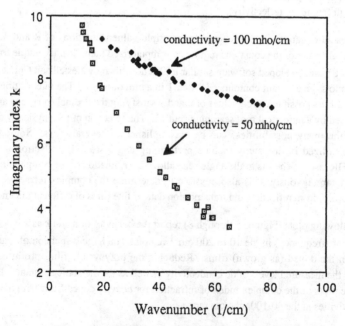

Figure 6.   Imaginary part of the index or refraction.

Figures 7 and 8 are plots of the complex index of refraction for films reduced 1/2 and 1 hour. Typical conductivities for films of this type are in the 5.0 to 0.1 mho/cm range.

The data indicates that the dielectric constants are highly dispersive in the heavily doped films. The values of n and k can be varied over a broad range by controlling the level of doping. In figure 9 the optical (ac) conductivity is plotted as a function of doping for films ranging between heavily doped and reduced 1 hour in which the conductivity of polypyrrole is seen to vary more than two orders of magnitude. The dispersion of complex dielectric constant for a conducting polypyrrole is shown in figure 10.

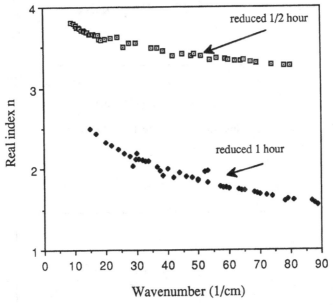

Figure 7.   Real part of the index of refraction.

CONCLUSIONS

The behavior of the optical conductivity versus frequency in conjunction with dispersion of the complex permittivity with frequency obtained for polypyrrole by us do not support a free carrier (Drude type) conduction mechanism[1,2]. The Debye model expression for the complex dielectric constants $\varepsilon_1 = n^2 - k^2$ and $\varepsilon_2 = 2nk$ is given by:

$$\varepsilon = \varepsilon_\infty + \frac{(\varepsilon_0 - \varepsilon_\infty)\omega_\tau}{(\omega_\tau + i\omega)} \qquad\qquad [3]$$

$$\varepsilon_1 = \varepsilon_\infty + \frac{(\varepsilon_0 - \varepsilon_\infty)\omega_\tau^2}{(\omega^2 + \omega_\tau^2)} \qquad\qquad [4]$$

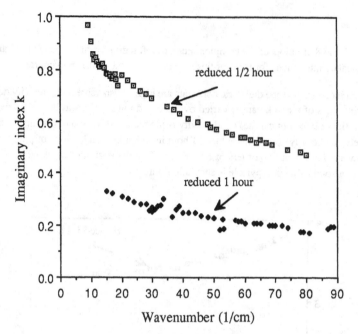

Figure 8.   Imaginary part of the index of refraction.

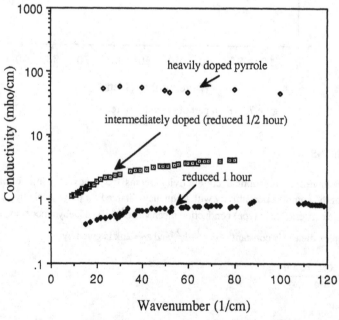

Figure 9.   Conductivity of polypyrrole as a function of doping level.

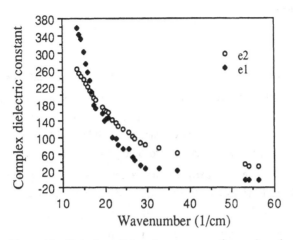

Figure 10. Complex dielectric constants for a doped
polymer with σ=50 mho/cm.

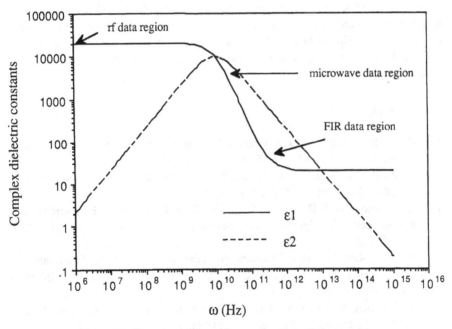

Figure 11. Plot of ε_1 and ε_2 and according to the Debye model.

$$\varepsilon_2 = \frac{(\varepsilon_0 - \varepsilon_\infty)\, \omega\omega_\tau}{(\omega^2 + \omega_\tau^2)}$$ [5]

where $\omega_\tau = \tau^{-1}$ [6]

The behavior of the dielectric constants for a film with $\varepsilon_\infty = 20$, $\varepsilon_0 = 2 \times 10^4$ and $\omega_\tau = 10$ GHz is shown in figure 11. The high and low frequency dielectric constants were estimated from our measurements made in the FIR and radio frequency (1 MHz) regime respectively. Although the data in the frequency range 10 MHz- 1.0 GHz is unavailable at this time, at least qualitatively the data seems to follow a Debye like behavior with ω_τ of the order of 10^{-10}/sec. As the polypyrrole films are reduced the conductivity drops. The magnitudes of the real and imaginary parts of dielectric constants and ω_τ also shift to lower values. Zuo[7] et al. successfully fit dielectric data measured at low frequencies (10^1 - 10^6 Hz) for emeraldine form of the conducting polyaniline polymer ($\sigma = 5$ ohm^{-1} cm^{-1}) to the Debye model. From the Debye model, the relaxation time of their sample was estimated to be of the order of 10^{-10} seconds. This value is of the same order of magnitude as the value of τ inferred for the oxidized polypyrrole films in this work.

The ease with which the electrical and optical properties of polypyrrole can be varied during preparation and the inherent stability of these materials under ambient conditions are useful properties in device fabrication. Of particular interest is the ability to change the real and imaginary part of dielectric constant by almost two orders of magnitude by oxidation or reduction of the films. The large values of n and k provide metal-like reflection in the microwave quite similar to copper. The large changes in the dielectric properties of polypyrrole may make applications such as band pass filters and other microwave and millimeter wave devices feasible.

REFERENCES

1. G. Phillips, R. Suresh, J. I. Chen, J. Waldman, J. Kumar, S. Tripathy, J. C. Huang, J. Appl. Phys. 69 899 (1990).
2. G. Phillips, R. Suresh, J. I. Chen, J. Kumar, J. Waldman, J. C. Huang and S. Tripathy, Mol. Cryst. Liq. Cryst. 190 27 (1990).
3. K. Kanazawa, A. F. Diaz, W. D. Gill, P. M. Grant, G. B. Street, "Polypyrrole: An Electrochemically Synthesized Conducting Polymer", Synth. Met. 1, 329 (1979/1980).
4. A. M. Nicolson, G. F. Ross, "Measurement of the Intrinsic Properties of Materials by Time Domain Techniques", IEEE Transactions, Instrum. Meas., Vol. IM-19, pp. 377-382, November (1970).
5. J. E. Nestell, Jr., and R. W. Christy, "Derivation of Optical Constants of Metals From Thin Film Measurements at Oblique Incidence", Applied Optics, Vol. 3, No. 3, 643 (1972).
6. K. Kanazawa, A. F. Diaz, M. T. Krounbi and G. B. Street, "Electrical Properties of Pyrrole and its Copolymers", Synth. Met., 4 119-130 (1981).

7. F. Zuo, M. Angelopoulos, A. G. MacDiarmid and A. J. Epstein, "AC Conductivity of Emeraldine Polymer", Phys. Rev. B, Volume 39, Number 6, 39 (1989).

MOLECULAR DESIGN OF AMORPHOUS PIEZOELECTRIC POLYMERS WITH THE AID OF NMR

Riichirô Chûjô

Department of Biomolecular Engineering, Tokyo
Institute of Technology, 12-1 Ookayama 2-chome
Meguro-ku, Tokyo 152, Japan

INTRODUCTION

Piezoelectric materials are becoming important as highly functional ones such as ultrasonic (noninvasive) diagnostic instruments, ultrasonic sensors, and electric field sensors.   Polymers from vinylidene cyanide (VDCN) are thought to produce good piezoelectric materials due to large dipole moment in C-CN bonds.   Unfortunately, homopolymers of VDCN are thermally quite unstable and inadequate to any practical application. Instead, copolymers between VDCN and other comonomers are being developed. These copolymers are amorphous, in nature, different from other kinds of piezoelectric polylmers such as poly(vinylidene fluoride).   NMR is, therefore, the most promising charactserization method for these amorphous piezoelectric polymers.   In this paper, NMR characsterization will be discussed for the amorphous piezoelectric polymers and related polymers. Polymers discussed should be Poly(VDCN-co-vinyl acetate), poly(VDCN-co-vinyl benzoate), poly(VDCN-co-vinyl pivalate), poly(VDCN-co-methyl metha-crylate), and poly(VDCN-co-styrene).   Among the results from NMR, copoly-mer composition and stereoregularity were already reviewed elsewhere (Chûjô, 1989), in which these two quantities are not correlated with pie-zoelectric tensors.   In this paper, the other results, i.e., conformation and molecular motion will be reviewed.

EXPERIMENTAL

VDCN was prepared from bi(acetyl cyanide).   All copolymers shown in Table 1 were obtained by radical polymerization.   o,o'-Dichlorobenzoyl peroxide was used as an initiator.   In this paper, only the samples with equimolar feed ratio were used.   Piezoelectric constants, d_{31}'s were also tabulated in Table 1. d_{ij} is an (i,j) component of piezoelectric tensor, $d_{ij} = (\partial p_i / \partial \sigma_j)_{E=0}$, where p is electric polarization, σ is stress, and E is electric field intensity.   Subscripts 1,2, and 3 correspond to the direc-tions of coordinate axes in Cartesian system. For conformational studies ^{1}H NMR spectra were recorded at 500 NHz on JEOL GX-500 superconducting magnet spectrometer.   These measurements were done in solution state. ^{13}C-CP(cross polarization)/MAS(magic angle spinning) NMR spectra were re-corded at 67.8 MHz on JEOL GX-270 superconducting magnet spectrometer.

TABLE 1.   Copolymers used in this paper and their piezoelectric constants.

Copolymer	Comonomer	$d_{31}/pC\ N^{-1}$
P(VDCN/VAc)	vinyl acetate	6.0
P(VDCN/VBe)	vinyl benzoate	4.0
P(VDCN/VPiv)	vinyl pivalate	7.0
P(VDCN/MMA)	methyl methacrylate	0.4
P(VDCN/St)	styrene	0.2

CONFORMATION

From Table 1 we can classify these five copolymers into good and poor piezoelectric ones; the formers are P(VDCN/VAc), P(VDCN/VBe), and P(VDCN/VPiv), and the latters are P(VDCN/MMA) and P(VDCN/St). The other candidate for the classification other than copolymer composition and stereoregularity is the conformation around skeletal bonds in solid state.   Resolution is, however, insufficient to analyze the conformation in solid-state high resolution NMR.   Under the working hypothesis that conformations are identical between solid and solution state in amorphous polymers, we will analyze the conformation with the aids of solution NMR. In Fig. 1 is shown a two-dimensional J-resolved ^{1}H-NMR spectrum of P(VDCN /VAc) in perdeuteraled dimethyl formamide (DMF-d_7) at 86.5°C (Inoue et al., 1990).   From the coupling constants (the ordinate in Fig. 1) the population of each conformer can be evaluated.   With the usage of rotational isomeric state (RIS) model and the conformational vicinal coupling constant values of $^3J_t = 11$ Hz and $^3J_g = 2$ Hz (Bovey, 1972), conformer populations were calculated.   The subscripts t and g stand the conformations between two vicinal protons (not the skeletal conformations).   As typical examples, the conformer populations along $H_AH_BC - CH_X$ are tabulated in Table 2 obtained for ε - meso sequence in DMSO - d_6 solution at 86.5°C.   The symbols, T,G, and $\bar{G}$ stand the skeletal conformations.   In Table 2 there is no data on P(VDCN/MMA).   This is due to the impossibility of the analyses similar to those of the other four copolymers, because of the lack of α - proton (H_X).   Comparing with the conformer populations in Table 2, we can find all of three conformers are realized in good piezoelectric polymers, while G is energetically prohibited in poor piezoelectric polymer.   In order to obtain piezoelectric materials poling process is necessary.   In the poling process electric dipole moments in samples must align parallel to the applied electric field.   It means the change of conformations is necessary in molecular level.   The lack of $\bar{G}$ may prohibit this change.   For P(VDCN/MMA) similar result was obtained by MM2 molecular mechanics calculation.   As a candidate for the classification between good and poor piezoelectric polymers is the allowance and prohibition of the $\bar{G}$ conformer.

As seen in Table 3 (Jo et al., 1987) the magnitude of the calculated resultant dipole moments largely depend on the molecular conformations. If this is the case, the molecular conformations before poling may affect the piezoelectricity.   In order to verify this statement the conformations must be observed.   As already mentioned in the beginning of this section the observation is still unrealistic.   Instead, we will observe the conformations in different solvents.   As typical examples, the conformer population along the $H_AH_BC - CH_X$ in three different solvents are

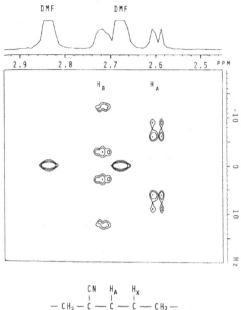

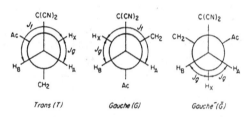

Fig. 1 2D J-resolved NMR spectrum of CH_2 region in P(VDCN/VAc) in DMF-d_7 at 500 MHz and 86.5°C.

tabulated in Table 4 for P(VDCN/VAc) at 85°C. The conformer population depends on the solvent species. The population seems to depend on the polarity of solvent used in NMR measurement. This finding was confirmed by the usage of mixed solvent. This finding implies piezoelectric film cast from solvent has different piezoelectric properties depending of the polarity of cast solvent. Piezoelectric constant, d_{31} were measured for P(VDCN/VAc) films cast from different solvent. Results are tabulated in Table 5. Parallelism is found between the fractions of trans conformer and piezoelectric constants depending on the polarity of cast solvents. In order to obtain piezoelectric materials with larger piezoelectricity from P(VDCN/VAc), the usage of cast solvent with larger polarity is recommended.

TABLE 2. Conformer populations along $H_A H_B C - CH_X$ of ε‑meso sequence in DMSO‑d_6 solution at 86.5°C

Copolymer	T/%	G/%	$\overline{G}$/%
P(VDCN/VAc)	83.3	7.8	8.9
P(VDCN/VBz)	75.6	7.8	16.7
P(VDCN/VPiv)	69.4	22.2	8.4
P(VDCN/St)	80.8	19.2	0.0

TABLE 3. Conformation dependence of the dipole moment in the direction of the vector sum of two CN dipoles bonded to the same skeletal carbon for VDCN ‑ VAc ‑ VDCN ‑ VAc sequence

Conformer	Dipole moment/debye					
	CH	CCN	-CO-	CO	-CH₃	Resultant
TTTTTTT̲	1.39	8.09	1.71	-4.80	-0.34	6.04
TTT̲GTTT̲G̲	0.46	4.04	0.85	-2.14	0.09	3.30
TTT̲GTTTG	0.92	4.04	0.85	-2.44	-0.42	2.95
G̲TTT̲GTTT	0.92	4.04	0.85	-2.44	-0.43	2.96
G̅TTTGTTT	0.46	4.04	0.85	-2.14	0.09	3.31

TABLE 4. Conformer populations along $H_A H_B C - CH_X$ of P(VDCN/VAc) in three different solvent at 85°C

Solvent	DMF ‑ d_7		DMA ‑ d_7		DMSO ‑ d_9	
ε ‑ Tacticity	ε-isot.	ε-synd.	ε-isot.	ε-synd.	ε-isot.	ε-synd
T/%	37.8	44.4	55.6	64.4	83.3	70.0
G/%	5.6	6.7	20.0	20.0	7.8	7.8
$\overline{G}$/%	56.6	48.9	24.4	15.6	8.9	22.2

MOLECULAR MOTION

Not only drawing process, but also poling one is necessary before supply of practical piezoelectric materials. The latter is also innevitable to line up of electric dipoles. This process is accomplished by segmental rearraangement as already mentioned in the previous section. Segmental mobility can be monitored by solid state CP/MAS NMR spectroscopy. Spin-lattice relaxation time ($T_{1\rho}(H)$) in rotating frame is a

TABLE 5. Piezoelectric constant, d_{31} of P(VDCN/VAc) films cast from three kinds of solvent, measured at 10 Hz and at 20°C

Poling Condition	$d_{31}/\text{PC N}^{-1}$	
	20 MV cm^{-1} at 170°C	40 MV cm^{-1} at 170°C
DMF	4.0	6.0
DMA	3.8	6.2
DMSO	5.4	8.0

TABLE 6. Proton $T_{1\rho}$ in ms from proton-carbon cross polarization at H_1 of 44 kHz in P(VDCN/VAc), PVAc, P(VDCN/MMA), and PMMA

Species	P(VDCN/VAc)	PVAc	P(VDCN/MMA)	PMMA
CH$_3$	17.23	21.13		
αCH$_3$			19.5	9.83
OCH$_3$			22.1	15.0
CH$_2$	17.07	20.99	18.25	15.3
CH	18.32	22.52		
quat C (VDCN)	20.43		20.49	
quat C (MMA)			15.1	11.50

TABLE 7. Proton $T_{1\rho}$ in ms from proton-carbon cross polarization at H_1 of 44kHz in three different states of P(VDCN/VAc)

Species	Before Drawing	Before Poling	After Poling
CH$_3$	19.8	19.4	25.8
CH$_2$	15.5	15.3	15.5
CH	17.3	15.9	19.3
quat C	21.4	21.3	19.6

good measure of the low frequency segmental motion.   Values of $T_{1\rho}(H)$ at H_1 of 44kHz were tabulated in Table 6 for P(VDCN/VAc), poly(vinyl acetate) (PVAc), P(VDCN/MMA), and PMMA (Jo et al, 1987a).   All measurements were performed at ambient temperature. Glass transition temperatures of these polymers were determined to 160, 28, 147, and 105°C, respectively, from differential scanning calorimetry.   Experimentally obtained $T_{1\rho}(H)$'s are, therefore, thought to be proportional to correlation times. The proportionality constants are, however, different with each other depending on the species of monomeric units. Comparing the corresponding values of relaxation times between P(VDCN/VAc) and FVAc, we can conclude the larger mobility in this good piezoelectric copolymer than that in the homopolymer. This means poling is easily done in the copolymer. On the other hand, in the comparison between P(VDCN/MMA) and PMMA, we cannot expect the larger mobility in the poor piezoelectric copolymer.

We measured $T_{1\rho}(H)$'s of P(VDCN/VAc) in three different states; i.e., before drawing, before poling and after poling states. Results were summarized in Table 7. After poling, increases of $T_{1\rho}(H)$'s were observed in all of carbon species except for quaternary carbon. This means the segmental mobility in the copolymer decreases after poling. Therefore, relaxation to randomly oriented state is rather constrained.

ACKNOWLEDGEMENT

This work was done under the collaboration with Prof. Y. Inoue, Drs. M. Sakurai, Y. S. Jo, Messrs. Y. Maruyama, A. Kashiwazaki, K. Kawaguchi, Y. Ota (Tokyo Institute of Technology), Prof. S. Miyata, Assoc. Prof. S. Tasaka, Mr. K. Saito (Tokyo University of Agriculture and Technology), Dr. I. Seo, and Mr. M. Kishimoto (Mitsubishi Petrochemical Co. Ltd.).

REFERENCES

Bovey, F.A., 1972, "High Resolution NMR of Macromolecules", Academic Press, New York

Chûjô, R., 1989, NMR Characterization of Amorphous Piezoelectric Polymers, in "Frontiers of Macromolecular Science", T. Saegusa, T. Higashimura, and A. Abe, Ed., Blackwell Scientific Publications, Oxford

Inoue, Y., Maruyama, Y., Sakurai, M., Jo, Y. S., and Chûjô, R., 1990, Conformational Analysis of Piezoelectric Vinylidene Cyanide-Vinyl Acetate Copolymer via Two-dimensional J-resolved [1]H Nuclear Magnetic Resonance Spectroscopy, Polymer, 31:1594

Jo, Y. S., Sakurai, M., Inoue, Y., Chûjô, R., Tasaka, S., and Miyata, S., 1987, Solvent Dependent Conformations and Piezoelectricity of the Copolymer of Vinylidene Cyanide and Vinyl Acetate, Polymer, 28:1583

Jo, Y. S., Maruyama, Y., Inoue, Y., Chûjô, R., Tasaka, S., and Miyata, S., 1987a, Molecular Motions of Amorphous Piezoelectric Palymers Determined by ^{13}C CPMAS NMR Spectroscopy, Polym. J., 19:769

Jo, Y. S., Maruyama, Y., Inoue, Y., Chûjô, R., Tasaka, S., and Miyata, S., 1988, Densification and Orientation on Poling in Copoly(Vinylidene Cyanide/Vinyl Acetate) with the Aids of C-13 CP/MAS NMR Spectroscopy, J. Polym. Sci., Polym. Phys. Ed., 26:463

THERMAL TREATMENT OF EMERALDINE : X.P.S. AND UV-VISIBLE

SPECTROSCOPIC APPROACH

D. Rodrigue, X. Demaret, J. Riga, J.J. Verbist

Facultés Universitaires Notre-Dame de la Paix
Laboratoire Interdisciplinaire de Spectroscopie Electronique
B-5000 Namur, Belgique

ABSTRACT

Polyaniline films are deposited on conducting glass working electrodes. Electropolymerization is performed by the potentiostatic technique (800 mV vs SCE) at room temperature in a 1.0 M HBF_4 or HCl aqueous solution of aniline (0.1 M). Parallely, emeraldine is prepared chemically from oxidation of aniline by ammonium peroxodisulfate.

The thermal analysis of protonated "electrochemical" emeraldine induces the disappearance of the dopant from the surface as well as from the bulk (X.P.S. and UV-visible absorption spectroscopies results). This phenomenom is accompanied by a drastic modification in the chemical structure and in the acid-base properties of the polymer. The X.P.S. N1s and C1s levels spectra reveal cross-linking and degradation reactions. Moreover, the kinetic study shows the deprotonation to be a first order reactions with an activation energy of 8.0 and 6.0 kcal/mol for HBF_4 and HCl doped-polymer respectively. The chemically synthesized emeraldine, submitted to high temperatures (500 K) is also deprotonated. Nevertheless, the counterion is retained inside the structure of the powder particles: the protonation level is never below 25 %. Simultaneously, the electrical conductivity decreases from 6 S/cm to 10^{-7} S/cm.

In addition to the two well-known redox couples, a third one has been identified, corresponding to the quinone/ hydroquinone system.

Frontiers of Polymer Research, Edited by P.N. Prasad and
J.K. Nigam, Plenum Press, New York, 1991

INTRODUCTION

Conducting polymers, especially polyaniline, attract much attention since the two last decades. As is generally observed, their electrical conductivity depends on redox state and moisture content [1]. In the case of polyaniline (PAn), it is also strongly related to the protonation level of the nitrogen atoms [2].

In this paper, X-ray photoemission (X.P.S.) and ultra-violet absorption spectroscopies are used to investigate the influence of the thermal treatment on the chemical and electronic structure of PAn samples obtained by different ways.

EXPERIMENTAL

An one-compartment cell, equipped with conducting glass (ITO) as anode as well as cathode, is used for the electrochemical oxidation of aniline. Polymerization is carried out potentiostatically by applying 0.8V vs SCE in the electrolyte solution containing the monomer (0.1M) and tetrafluoroboric or hydrochloric acid (1M). The application of this potential to the electrode during 30 min for HBF_4 and 20 min for HCl leads to the formation of a homogeneous green polymer film. This is rinsed with distilled water to remove non-adherent oligomers. Then, the polymer is brought to a well-defined redox state by applying a potentiel between - 0.2 et 0.8 V [3]. The basic form (necessary to ease the study of the electronic structure) is obtained by dipping the film in a solution of NaOH 0.1N during 5 min.

Parallely, the emeraldine form has been chemically prepared: oxidation of the monomer by ammonium peroxodisulfate in a sulfuric acid aqueous solution which is stirred during 3.5 hours. The precipitate is recovered by filtration and rinsed several times with distilled water, then treated successively in NaOH 0.1N and HBF_4 1M. In order to reproduce the exact conditions of the electrochemical synthesis, we have repeated the same experiment by using tetrafluoroboric acid instead of sulfuric acid.

RESULTS AND DISCUSSION

Protonation level

The two emeraldine forms (HBF_4 and HCl doped) of polyaniline have been compared after heating in ultra-high vacuum (10^{-7} Torr) or in an oven. As no difference was observed between the two samples, all experiments have been performed at ambient atmosphere, at temperatures varying from 270 to 530 K.

The heating time is fixed to 15 min for each step. The doping level (Figure 1a) and band structure (Figure 1b) evolution are followed respectively by X.P.S. and UV-visible spectroscopies. This doping level is determined by the ratio intensities peaks of N1s/conter-ion.

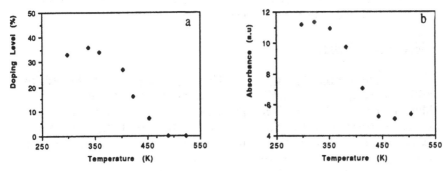

Figure 1. Evolution of (a) the doping level and (b) the band structure ($\lambda = 420$ nm) of emeraldine with temperature.

The similarity of the curves indicates that the removal of the dopant is not limited to the surface, but also occurs in the bulk.

If the heating time is increased, spectacular modifications occur, and a kinetic study has been undertaken. The deprotonation is increasingly fast (Table 1) when temperature raises for HBF4 as well as HCl-doped emeraldine. Nevertheless, the Cl^- dopant seems to leave the bulk faster.

Table 1   Kinetic characteristics of the thermal treated emeraldine.

Temperature (K)	k (min^{-1})	t$_{1/2}$ (min)
HBF4		
348	$2.3 \ 10^{-3}$	301
383	$7.4 \ 10^{-3}$	94
413	$2.6 \ 10^{-2}$	27
473	$5.4 \ 10^{-2}$	13
HCl		
328	$5.1 \ 10^{-3}$	136
353	$1.1 \ 10^{-2}$	63
473	$8.7 \ 10^{-2}$	8
533	$3.1 \ 10^{-2}$	2

The activation energies for deprotonation of the two polymers are 8.0 (HBF4) and 6.0 (HCl) kcal/mol and the reaction are the first ordrer [4]. The limiting speed step seems to be the size of the dopant : to restrict at much as possible the deprotonation at high temperatures, para-sulfonic acid or napthalene sulfonic acid must be used. In fact, deprotonation only occurs beyond 460 K [5] when the PAn is doped with CH_3-φ-SO_3H.

Electronic structure

Doped form treatment. The samples are first heated to different temperatures until the appearance of the characteristic blue color of deprotonated polyaniline. Then, the electronic structure is studied by X.P.S. (Figure 2): the tertiary nitrogen, associated to the phenazine structure, then appears, as well as for HBF4 prepared polymer as for the HCl one.

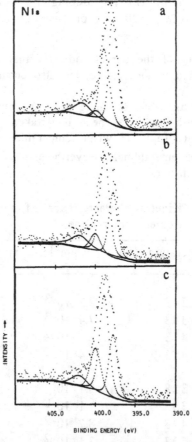

Figure 2. Evolution of the N1s X.P.S. signal of deprotonated emeraldine *vs* temperature (a) 298K , (b) 423K , (c) 513K.

In a second time, after heating 15 min at each selected temperature, the material is dedoped by NaOH 0.1M. A similar evolution can be seen (Table 2).

Table 2   X.P.S. estimation of the tertiary nitrogen content in deprotonated emeraldine as function of temperature.

Temperature (K)	Tertiary Nitrogen (%)	
	HBF_4	HCl
298	5	8
318		9
343	5	
358		14
383	5	
423	10	15
443		15.5
513	22	

Dedoped form treatment.   If samples are heated only after deprotonation , the cyclisation phenomenon is also produced (Table 3).

Table 3   X.P.S. estimation of the tertiary nitrogen content in deprotonated emeraldine synthesized in HBF_4 as function of temperature.

Temperature (K)	Tertiary Nitrogen (%)
298	4
238	5
363	6
453	12
513	15

Consequently, the appearence of the tertiary nitrogen seems not to be due to the leaving of the dopant but rather to a rearrangement of the skeleton with heat.

Electrical conductivity

The four-probe technique is used to determine the electrical conductivity (σ) of the powder. PAn(a) is the shortening for the emeraldine synthesized in H_2SO_4 and PAn(b) in HBF_4. Even if the heating time increases (30 min) compared to the films, the doping level of the powder (table 4) falls less quickly : this level is still at 25 % when σ is already equal to 10^{-7} S/cm.

Table 4 Evolution of the electrical conductivity and the doping level of emeraldine powder as function of temperature.

Temperature (K)	σ (S/cm)	Doping level (%)
PAn(a)		
298	30.0	61
338	1.0	44
353	0.3	48
473	0.2	38
PAn(b)		
298	5.6	45
345	3.5	38
443	2.5	40
503	$4.2\ 10^{-7}$	25

Cyclic voltammetry results

Once upon the polymer films are synthesized, they are dipped in aqueous acid solution 1M, and the potential is swept four times between -200 and 1200 mV vs SCE. The voltammograms are presented in the figure 3.

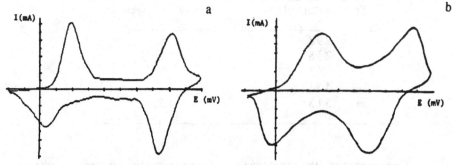

Figure 3. Cyclic voltammograms of polyaniline with temperature (a) 298K and (b) 530K.

The electrochemical behaviour of the polyaniline is not modified by temperature : the third redox couple generally observed, does not appear. Nevertheless, we can determine by X.P.S. the presence of the tertiary nitrogen for the same sample. Consequently, this third redox couple cannot be assigned to the phenazine structure, but rather to the quinone/hydroquinone redox couple.

CONCLUSION

The departure of the dopant is not limited to the surface of the polymer but also extented to the bulk. The activation energies of deprotonation are 8.0 and 6.0 kcal/mol for HBF_4 and HCl doped polymers respectively. The deprotonation can therefore be limited by the dimension of the counter-ion.

Moreover, when heating films, the tertiary nitrogen content increases. A similar evolution is not observed on voltammetric curve. So, the third couple in cyclic voltammetry does not correspond to the phenazine structure but to the quinone/hydroquinone redox couple.

REFERENCES

[1] W.W. Focke, G.E. Wnek, Y. Wei: J. Phys. Chem., **91** (1987) 5813

[2] A.G. MacDiarmid, J.C. Chiang, A.F. Richter, A.J. Esptein: Synth. Metals, **18** (1987) 285

[3] P. Snauwaert, R. Lazzaroni, J. Riga, J.J. Verbist, D. Gonbeau: J. Chem. Phys., **92** (1990) 2187

[4] D. Rodrigue, X. Demaret, P. Snauwaert, J. Riga, J.J. Verbist: Synth. Metals, **41**(3) (1991) 775, Proceedings of ICSM'90, Tübingen, FRG, September 2-7, 1990

[5] to be published

STRUCTURAL FEATURES IN ALKALI-METAL DOPED CONJUGATED POLYMERS

N. S. Murthy, R. H. Baughman and L. W. Shacklette

Allied-Signal Inc., Research and Technology
Morristown, New Jersey 07962

ABSTRACT

Conducting complexes obtained by doping conjugated polymers with alkali-metal ions have channel structures. These structures are discussed using a basis set of simple motifs of alkali-metal ions surrounded by polymer chains, and are illustrated with examples from polyacetylene, poly(p-phenylene) and poly(p-phenylene vinylene). The diameter of the dopant-ion relative to the cross-sectional width of the host polymer determines whether the dopant-ion columns are formed in triangular (three chain) or tetragonal (four chain) channels. Staging observed at low dopant concentrations is shown to be analogous to that in graphite intercalation compounds. The implications of these structural features on the electrochemical curves obtained during doping and dedoping of polymers are discussed.

INTRODUCTION

The packing arrangement of polymer chains and dopant ions is important for determining key properties in conducting polymer charge-transfer complexes. This role of structure is most clearly apparent for transport, since conductivity in all directions is typically limited by interchain hopping. However, various types of disorder severely limit the amount and the quality of obtainable diffraction data, and hence the structural information which can be experimentally derived. The various types of structural disorder in a typical conducting polymer include short coherence lengths, chain misalignment between crystallites, a substantial amorphous component, and nonuniform dopant distribution. The alkali-metal doped conjugated polymers have been most intensely investigated, since less extensive disorder is observed than for many other polymer charge transfer complexes, the counter ions are simple, and high symmetry structures are often obtained.[1-21] Furthermore, detailed electrochemical data available for these polymer complexes[22-25] provide additional information on the structural origins of the phase transformations observed during doping and dedoping. Detailed experimental studies of such comparatively simple systems are useful for developing an understanding of structure-property relationships which is more generally applicable for conducting polymers. We here discuss the influence of the size of the alkali-metal ions, the cross-sectional width of the polymer chain, and the dopant concentration on the structure in alkali-metal doped conjugated polymers: polyacetylene (PA), poly(p-phenylene) (PPP) and poly(p-phenylene vinylene) (PPV). The common features in the structures observed for these polymeric intercalation compounds (PICs), will be described and compared with those of graphite intercalation compounds (GICs).

Frontiers of Polymer Research, Edited by P.N. Prasad and
J.K. Nigam, Plenum Press, New York, 1991

Early work by Baughman et al.[1] on PA heavily doped with Na, K, Rb or Cs demonstrated that alkali-metal ions form columns surrounded by polymer chains. Each alkali-metal column is surrounded by four polymer chains in a tetragonal or a pseudotetragonal arrangement, which corresponds to the four-chain motif shown in Fig. 1a. Additionally, each polymer chain is shared by two alkali-metal columns, so that a two-chain per column structure results. Later work by Winokur and coworkers on sodium-doped PA[14] and sodium-doped PPV[18] and by Murthy et al. on lithium-doped PA,[15] sodium-doped PPP[19] and potassium-doped PPP[19] showed that for ions whose radii are small compared with the chain width, the alkali-metal ion columns are surrounded by three polymer chains (Fig. 1b). Since each polymer chain in such a structure is a nearest neighbor to only one alkali-metal column, a three-chain motif results. For a given polymer, the radius of the channel in the three-chain motif is smaller than in the four-chain motif. The radius of the channel can be decreased for both of these motifs by sequential translation of the polymer chains around the channel axis (Figs. 1c and 1d).

These basic motifs can maximize the coordination of the negatively charged carbon atoms with both the alkali-metal ions and the hydrogens. Structures consistent with the diffraction data are derived by packing these motifs in close-packed, high-symmetry arrangements, which are chosen to provide energetically favorable interactions between chains and between alkali-metal columns in different motifs. The high symmetry motifs shown in Fig. 1 are first approximation models derived using low-resolution data from poorly ordered samples. Various distortions, which include either chain translation or chain rotation, or both chain translation and chain rotation, have been postulated in order to explain the recent, more detailed diffraction data from highly oriented samples.[10-12] These new data require the symmetry to be lower than indicated in Fig. 1. However, the associated structural distortions are not large and the precise nature of these distortions are somewhat ambiguous because of data limitations.

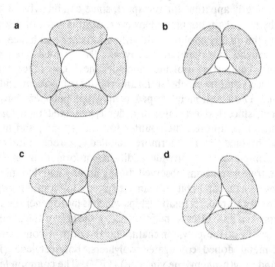

Fig. 1. Basic structural motifs in chain-axis projection: (a) four chain, (b) three chain, (c) four-chain pinwheel, (d) three-chain pinwheel.

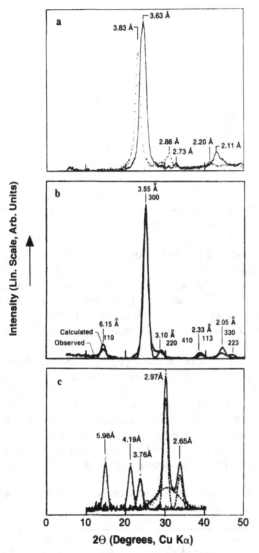

Fig. 2. Equatorial x-ray diffraction scans of (a) undoped, unoriented, highly crystalline cis-PA (dashed line) and trans-PA (solid line), (b) lithium-doped PA (CHLi$_{0.06}$), and (c) potassium-doped PA (CHK$_{0.125}$). Scans (a) and (b) are reproduced with permission from Ref. 17, and scan (c) is reproduced with permission from Ref. 15.

Table I. Comparison of the lattice dimensions of PA heavily doped with various alkali metals. The unit cell dimension provided for the hexagonal lattices is the basal plane axial length, and that for the tetragonal phases is $a' = a/\sqrt{2}$, where a is the basal plane axial length.

		Ion radius[26] (Å)	Unit cell dimension (Å)[1,15,21]
Hexagonal	Lithium	0.60	12.3-12.6
	Sodium	0.95	13.2-14.0
Tetragonal	Sodium	0.95	5.6-5.9
	Potassium	1.33	5.95-6.05
	Rubidium	1.48	6.19
	Cesium	1.69	6.43

XRD data (Fig. 2b) from lithium-doped PA is consistent with a model in which each Li^+ ion column is surrounded by three PA chains in an hexagonal lattice.[15] In such an arrangement, the small radius of the channel formed by three polymer chains is just sufficient to accommodate the smallest of the alkali-metal ions. In sodium-PA complexes, the three-chain channel expands to accommodate the larger Na^+ ion, and this results in a larger unit cell (Table I). Ions larger than sodium can no longer be accommodated in a channel formed by three PA chains. XRD data from potassium-doped PA (Fig. 2c) provides evidence for a tetragonal phase in which each potassium column is surrounded by four PA chains.[1,17] In rubidium-doped and in cesium-doped PA, the channel radius and the cell dimension in the basal plane increase from that in the K-PA complex in order to accommodate these larger ions (Table I).[1]

The above results for alkali-metal doped PA show that the size of the ion determines whether the PA complex crystallizes in a lattice containing the three-chain motif (as for the hexagonal structures) or the four-chain motif (as for the tetragonal structures). But in general, as shown by the recent results from PPP[19] and PPV,[18,20] the motif utilized in the observed structure is determined by the size of the ion relative to the cross-sectional dimension of the host polymer. For example, a tetragonal lattice is observed in PA doped with K,[1-6,8-13] and in some instances even with Na.[1] But the same Na^+ and K^+ ions in PPP yield a hexagonal lattice.[19] Similarly, Cs forms a

Table II. Ratio of the size (w, the longer cross-sectional dimension) of the host polymer and the diameter (d) of the guest ion.

Ion	diameter (Å)[26]	Ratio (w/d) PA	PPP	PPV
Lithium	1.20	4.22 H	5.60	6.46
Sodium	1.80	2.81 H,T	3.73 H	4.31 H
Potassium	2.66	1.90 T	2.53 H	2.91
Rubidium	2.96	1.71 T	2.27	2.62
Cesium	3.38	1.50 T	1.99	2.29 H,T

H: Hexagonal phase (trigonal or hexagonal system) containing the three-chain motif.
T: Tetragonal phase containing the four-chain motif.

388

tetragonal lattice in PA[1,6] and a hexagonal and tetragonal lattice in PPV.[20] The relative sizes of the polymer and dopant ion suggest that Li-doped PA[15] and Na-doped PPP[19] should be equivalent in that both should form hexagonal lattices with three chains per column. These results are summarized in Table II, where the ratio of the maximum Van der Waals chain width (w) to the diameter (d) of the alkali-metal ions is provided for various phases having known structures. According to these results, the three-chain and four-chain motifs are obtained when w/d > 2.8 and w/d < 2.3, respectively; either motif can be observed for 2.3 < w/d < 2.8. If we regard the graphite planes in graphite intercalation compounds as chains having infinite width, then we should expect the guest ions to be surrounded by two chains (i.e., two graphite planes), which corresponds to the observed limit when the ratio of chain-width to the ion-diameter becomes infinite. This analogy with graphite intercalation compounds will be explored further in a following section.

The structural motifs in Fig. 1 provide a means for understanding the dynamics of conversion of the herringbone packing of the polymer chains in the monoclinic lattice of the undoped polymer (which can be regarded as pseudohexagonal) to that of the doped phase (Fig. 3)[15]. The transformation of the chain arrangement in undoped polymer to that found in the four-chain motif structure requires fairly large displacements of chain centers. However, the transformation of the herringbone packing arrangement in the undoped polymer to the hexagonal lattice for the smaller alkali metals (such as lithium in PA) formally requires principally only a chain-axis rotation for two out of every three chains in the undoped phase. For larger ions (such as sodium in PA) significant translations of the polymer chains are also required. Thus, phase boundaries which involve little interfacial strain can exist between undoped and hexagonal doped lattices only when the motif parameter (w/d) is large (>2.8).

INFLUENCE OF DOPANT CONCENTRATION

Based on electrochemical measurements and x-ray diffraction results, Shacklette et al.[22-24] confirmed early conclusions [27] that doping and dedoping of the conjugated polymers proceeds not at random, but via a sequence of crystalline phases. Detailed

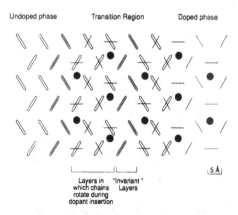

Fig. 3. Coexistence of undoped and hexagonal lattices. Reproduced with permission from Ref. 15.

electrochemical data in the form of electrochemical voltage (E) vs. dopant concentration (y) during very slow doping and dedoping have been obtained with Li⁺, Na⁺ and K⁺ ions for both PA and PPP. The K-doped PA has been studied in most detail since it provides high quality diffraction and electrochemical data (Figs. 2c and 4). The plateaus in voltage vs. composition curves in electrochemical data (Figure 4), which are also observed during the electrochemical intercalation of graphite, are attributed to the coexistence of two phases in the plateau regions. The initial formation of a new phase, or the continuous structural evolution of a single phase, results in the sloping voltage regions at the end of plateau regions. The plateaus correspond to the progressive nucleation and growth of a new phase at the expense of the old. Filling or emptying of channels of a given phase results in successively higher or lower reduction levels for each chain and hence the voltage varies much more rapidly than over the plateaus, where interconversion is occurring between two phases having fixed compositions.

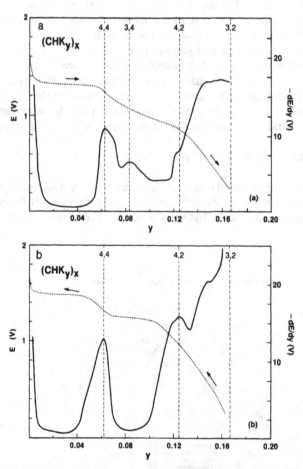

Fig. 4. Electrochemical data for staging, during dopant insertion (a) and dopant removal (b). The dotted lines and the full lines refer to E vs. y and -dE/dy vs. y curves, respectively. Reproduced with permission from Ref. 17.

The sequence of stoichiometric phases observed in electrochemical curves can be described using the notation $(P_n)_m M$, where n is the average number of monomer units (P) of the polymer spanning a length equivalent to the average distance between the alkali-metal ions (M) within a column, and m is the number of polymer chains per alkali-metal ion column in the unit cell.[2,23,24] Variations in dopant level results in changes in either the number of polymer chains per alkali-metal column in the unit cell (m) or the number of ions per monomer unit length within a column (n). As discussed earlier (Table II), m is three for large values of (w/d), and four for smaller ratios. Only one type of structure (m=3) with a three chain motif is presently known for the conducting polymers. In this type of structure, each polymer chain is nearest neighbor to only one alkali-metal column, so that there are three chains per alkali-metal column in the unit cell. However, tetragonal or pseudotetragonal phases containing the four chain motif, but with either four chains per column (m=4) or two chains per column (m=2) in the unit cell, have been reported at low and high dopant levels, respectively.[2,3,8,9,11,12,16,17] The structure at low dopant levels consists of motifs which are non-overlapping, in the sense that each chain is contained in only one motif. In contrast, the structure at high dopant level consists of overlapping motifs in which each chain is nearest neighbor to two alkali-metal columns.

There is an upper limit to the number of ions that can be accommodated in a channel. Once this limit is reached, the structure becomes unstable, and there is a first order transition which provides for a decrease in the number of polymer chains per alkali metal column in the unit cell. Analogous behavior is commonly observed in graphite intercalation compounds, where it is referred to as staging. In K-doped PA, the number of chains per alkali-metal column, m, decrease from 4 to 2 so as to accommodate the increased ion concentration. However, the basic four chain motif is maintained. Such a transformation typically occurs both during doping and dedoping, although substantial hysteresis can give rise to different structures at the same dopant level. The structural model for the stage-1 K-doped PA (m=2),[1] along with the structure for the undoped PA, are shown in Fig. 5.

The density of ions in columns for a given phase, and the relative heights of the ions within the columns of the neighboring motifs, is expected to be dominated by electrostatic energy considerations. Consistent with this view point, a body-centered arrangement is typically observed in that the alkali-metal ions in neighboring columns are displaced relative to each other by one-half of the interion separation. As a consequence of the importance of the electrostatic interactions, the alkali-metal ion and

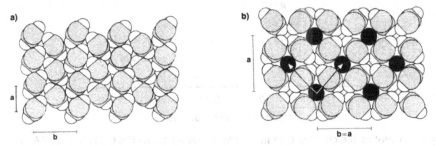

Fig. 5. (a) The structure of the undoped PA. (b) A structural model for stage-1 K-doped PA. The stage-1 structure is tetragonal with a = 8.5 Å; however a smaller cell with a′ = 6 Å can account for the equatorial data which corresponds to the structure in chain-axis projection. Reproduced with permission from Ref. 17.

polymer chain periodicities are generally not commensurate. However, special $(C_4H_4)_m$ phases are stable over substantial ranges in dopant level for PA, because such phases provide commensurability between polymer chain and ion columns, while at the same time minimizing intercolumn electrostatic repulsion by providing a body-centered arrangement of alkali-metal ions in neighboring columns.

COMPARISON WITH GRAPHITE INTERCALATION COMPOUNDS

Changes in the number of chains per alkali-metal ion column (m) with dopant concentration provide a two-dimensional analogue of the one-dimensional stages in GICs. A comparison between the 1-D and 2-D structures provide complementary information on the role of the competing interactions within and between the host lattice elements and the guest ions in determining phase stability. Unlike in the GICs, the dopant-containing layers in the PICs can also include the host polymer. Therefore, staging in PICs is defined with respect to planes containing the maximum dopant concentration. Thus, in a stage-k structure, dopant containing layers (not the dopant layers as in GIC) are separated by k layers of polymer chains.

In GICs, there is typically a progression from stage-1 to higher stages with decreasing dopant concentration. But in PICs, there are no compelling evidence for stages higher than stage-2. The difference is probably because the polymer chains in PICs need to undergo rearrangement (to varying degrees, depending on the size of the dopant ion relative to that of the host polymer), and such rearrangement is not

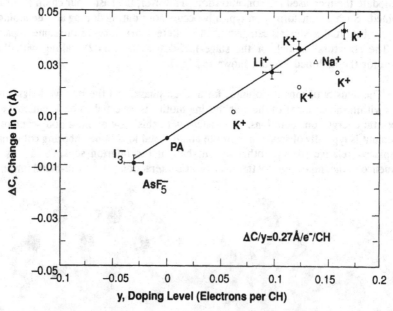

Fig. 6. A plot of the changes (Δc) in the chain-axis length of the C_2H_2 unit in PA vs. electron transfer (e^- per CH). Our data (ref. 30 and unpublished result for AsF_5^-) are shown as filled circles, datum of Winokur et al.[31] is shown as an open triangle, and the data of Saldi et al.[10] are shown as open circles. The full line shown is the least-squares fit to the filled circles.

Table III. Shortest distance (in Å) between alkali-metal and carbon plane in alkali-metal complexes of conjugated polymers (PA) and graphite

Alkali-metal	Polyacetylene[1,15,21]	Graphite[34]
Lithium	2.15	1.85
Sodium	2.62	2.30
Potassium	3.00	2.68
Rubidium	3.10	2.83
Cesium	3.22	2.97

required of the graphite planes in GICs. In higher stages of GICs, the graphite planes separating the dopant layers can maintain the same (or similar) low-energy packing arrangement as that found in unintercalated graphite. On the other hand, if the polymer chains in higher stages of PICs were to have the arrangement similar to that of undoped polymer, then a strained structure would result unless the doped phase provides a compatible interface as in the case of lithium-doped PA. Hence, at concentrations lower than that which supports a stage-2 structure (<6%), segregation of the undoped and the doped phases is energetically favored. Only at very low dopant concentrations is random doping of PA expected, because such random doping results in major strain energy for all alkali-metals except lithium.

Another measurement that is useful in comparing GICs and PICs are the changes in the C-C bond lengths with the extent of electron charge transfer between the dopant and the host lattice. It is known from experimental observations on graphite that C-C bond length depends on charge transfer: the C-C bond length increases with donor intercalation and decreases with acceptor intercalation. Based on electrostatic arguments alone one would expect the C-C bond length to increase due to the repulsion of the charged atoms, irrespective of the sign of the charge transfer. However, based on quantum chemical calculations, Kertesz et al.[28,29] obtained results consistent with observations for graphite, and predicted similar results for PA. Murthy et al.[30] provided the first experimental results in support of these calculations for PA. In heavily doped complexes, the chain-axis repeat per C_2H_2 unit in PA contracts by 0.015 and 0.01 Å for AsF_5^- and iodine, respectively, and expands by 0.026, 0.030 and 0.040 Å for lithium, sodium and potassium, respectively.[30,31] Billaud et al.[10] have obtained similar data for K-doped PA, and Winokur et al.[31] have obtained data for Na-doped PA at varying dopant levels. These results are plotted Fig. 6. In contrast with the case for graphite, the predominant contribution to chain length expansion for PA is due to bond-angle changes.[32,33]

A final comparison between GICs and the PICs is the distance between the alkali-metal ion and the carbon plane in the PA and graphite charge-transfer complexes. Table III shows that these distances are shorter in GICs than in PICs. Because of the 1D and 2D nature of the GICs and PICs, one expects to observe these differences, and the implications of such findings will be discussed elsewhere.

REFERENCES

1. R. H. Baughman, N. S. Murthy and G. G. Miller, J. Chem. Phys. 79, 515 (1983).

2. R. H. Baughman, L. W. Shacklette, N. S. Murthy, G. G. Miller and R. L. Elsenbaumer, Mol. Cryst. Liq. Cryst. **118**, 253 (1985).
3. D. Billaud, J. Ghanbaja, C. Goulon, Synth. Met. **17**, 497 (1987).
4. C. Mathis, R. Weizenhofer, G. Lieser, V. Enkelmann and G. Wegner, Makromol. Chem. **189**, 2617 (1988).
5. O. Leitner, H. Kahlert, G. Leising J. Fink and H. Fritzsche, Synth. Met. **28**, D225 (1989).
6. F. Saldi, J. Ghanbaja, D. Begin, M. Lelaurin and D. Billaud, C. R. Acad. Sci. (Paris) **309**, 671 (1989).
7. J. Ma, D. Djurado, J. E. Fischer, N. Coustel and P. Bernier, Phys. Rev. B **41**, 2971 (1989).
8. N. S. Murthy, L. W. Shacklette and R. H. Baughman, Solid State Commun. **72**, 267 (1989).
9. N. S. Murthy, L. W. Shacklette and R. H. Baughman, Phys. Rev. B **41**, 3708 (1990).
10. F. Saldi, M. Lelaurin and D. Billaud, Solid State. Commun. **76**, 595 (1990).
11. J. E. Fischer, P. A. Heiney and J. Ma, Preprint.
12. P. A. Heiney, J. E. Fischer, D. Djurado, J. Ma, N. Coustel, P. Bernier, D. Chen, M. J. Winokur, and F. E. Karasz, Preprint.
13. N. S. Murthy, L. W. Shacklette and R. H. Baughman, Bull. Am. Phys. Soc. **31**, 232 (1986).
14. M. J. Winokur Y. B. Moon, A. J. Heeger, J. Barker, D. C. Bott and H. Shirakawa, Phys. Rev. Lett. **58**, 2329 (1987).
15. N. S. Murthy, L. W. Shacklette and R. H. Baughman, Phys. Rev. B **40**, 12550 (1989).
16. D. Djurado, J. E. Fischer, P. A. Heiney, J. Ma, N. Coustel and P. Bernier, Synth. Met. **34**, 683 (1989).
17. N. S. Murthy, L. W. Shacklette and R. H. Baughman, Phys. Rev. B **41**, 3708 (1990).
18. D. Chen, M. J. Winokur, M. A. Masse and F. E. Karasz, Phys. Rev. B **41**, 6759 (1990).
19. N. S. Murthy, R. H. Baughman, L. W. Shacklette, H. Fark and J. Fink, Solid State Commun. (1991).
20. D. Chen, M. J. Winokur and F. E. Karasz, Synth. Met. (1991).
21. N. S. Murthy, L. W. Shacklette and R. H. Baughman, Preprint.
22. L. W. Shacklette and J. E. Toth, Phys. Rev. B **32**, 5892 (1985).
23. L. W. Shacklette, J. E. Toth, N. S. Murthy and R. H. Baughman, J. Electrochem. Soc.: Electrochem. Sci. and Tech. **132**, 1529 (1985).
24. L. W. Shacklette, N. S. Murthy, and R. H. Baughman, Mol. Cryst. Liq. Cryst. **121**, 201 (1985).
25. N. Coustel, P. Bernier and J. E. Fischer, Preprint.
26. L. Pauling, The nature of the Chemical Bond (Cornell University Press, 1960) p. 518.
27. R. H. Baughman, N. S. Murthy, G. G. Miller and L. W. Shacklette, J. Chem. Phys. **79**, 1065 (1983).
28. M. Kertesz, Mol. Cryst. Liq. Cryst. **125**, 103 (1985).
29. M. Kertesz, F. Vonderviszt and S. Pekker, Chem. Phys. Lett. **90**, 430 (1982)
30. N. S. Murthy. L. W. Shacklette and R. H. Baughman, J. Chem. Phys. **87**, 2346 (1987).
31. M. J. Winokur, Y. B. Moon, A. J. Heeger, J. Barker, D. C. Bott, Solid State Commun. **68**, 1055 (1988).
32. S. Y. Hong and M. Kertesz, Phys. Rev. Lett. **64**, 3031 (1990).
33. R. H. Baughman, N. S. Murthy and M. Kertesz, Preprint.
34. D. P. DiVincenzo and E. J. Mele, Phys. Rev. B **32**, 2538 (1985).

POLYMER CERAMIC COMPOSITES FOR TEMPERATURE INDEPENDENT

CAPACITORS

V.M.KAMAT, C.S.SHAH, M.J.PATNI AND M.V.PANDYA[*]

Materials Science Centre
Indian Institute of Technology
Powai, Bombay-400076, India

ABSTRACT

The dielectric properties of polymer-ceramic composites
are reported. The polymers used were high density polyethylene
HDPE, a non-polar polymer; polyvinylchloride PVC, a polar
polymer and polyvinylidene fluoride PVDF, a polar polymer
with a high permittivity. Barium titanate was the ceramic used
in particulate form. The polymer and barium titanate were
mixed in different weight proportions to make polymer-ceramic
composites. Dielectric, X-ray diffraction and Scanning
electron microscopy studies were carried out. The dielectric
permittivity of all the polymer-ceramic composites matched
well with the values calculated from Yamada's equation,
assuming polymer as continuous phase. This was confirmed by
SEM studies. Dielectric measurements over a temperature range
of 30° to 180°C indicated that the pure barium titanate transition temperature at 120°C is shifted to higher temperatures
in presence of polymeric phase. The shift of transition temperature was much higher for PVDF than for HDPE.

INTRODUCTION

Polymer-ceramic composites possess a rare combination of
the properties which make them superior to both the constituents i.e. polymer and ceramic.

Ceramic particles can have high dielectric, piezoelectric
and pyroelectric constants. They also exhibit a high electro-
mechanical coupling, but are brittle and cannot withstand mechanical shocks. On the other hand polymers have low dielectric,
piezoelectric and pyroelectric constants but are very flexible
and ductile, which can withstand mechanical shocks. Hence if
polymers and ceramic particles are mixed in a suitable manner,
it can give flexible polymer-ceramic composites with combined
properties. Since polymers can accommodate a very high loading
of ceramic materials while retaining the flexibility, electroactive properties of ceramic particles are also retained in
the polymer-ceramic composites.

---

* : Chemistry Department

Frontiers of Polymer Research, Edited by P.N. Prasad and
J.K. Nigam, Plenum Press, New York, 1991

395

Different methods have ben used till now for the preparation of polymer-ceramic composites, viz., embedding electro-active ceramic fibers in polymers, drilling holes in blocks of ceramics and then filling them with polymers and mixing powdered ceramics with polymers [1-6]. The last method includes solution casting, hot pressing and cold pressing followed by sintering.

In the present work, dielectric properties of high density polyethylene (HDPE) - Barium titanate (BT), Polyvinyl chloride (PVC) - BT and Polyvinylidene fluoride (PVDF) - BT composites in various weight proportions have ben studied. Wide angle X-ray diffraction studies of these composites have been carried out to understand the structural changes taking place in the polymer by incorporation of barium titanate.

EXPERIMENTAL PROCEDURE

Barium titanate powder required for the work was synthesized in the laboratory from $BaCo$ and TiO by solid state reaction. WAX diffraction and infrared analysis were carried out to confirm the formation of barium titanate.

For the preparation of polymer-barium titanate composites, the polymer powder and barium titanate powder were first ground together for 30 minutes. PVC-BT composites were prepared by hot compaction at 120°C. HDPF-BT and PVDF-BT composites were prepared by cold compaction followed by sintering. The sintering temperature for both the composites were 120°C. Circular silver electrodes of 1.5 cm diameter were painted on both sides of pellets. The samples were sandwiched between pressure-contact electrodes for the dielectric measurements, on Hewlett Packard (HP 4192A) LF impedance analyzer. Measurements were carried out in frequency domain at room temperature. Also temperature scans were done at 100 KHz and 1000 KHz between 30°C to 180°C on selected Polymer-BT composites. X-ray diffraction study of polymer-BT compacts were carried out by a Philips PW 1820 diffractometer using Cu target.

RESULTS AND DISCUSSIONS

Following equation due to Yamada et al. [5] gives the permittivity of composites as

$$\epsilon_T = \epsilon_1 \left[1 + \frac{n q (\epsilon_2 - \epsilon_1)}{n \epsilon_1 + (\epsilon_2 - \epsilon_1)(1 - q)} \right] \quad ..(1)$$

where ϵ_1, ϵ_2 and ϵ_T and   are the permittivity of continuous phase, dispersed phase and the composite respectively, q is the volume fraction of dispersed phase and n is the parameter atrributed to the shape of the dispersed ellipsoidal particles. Fig.(1) gives the observed and calculated values of permittivity of HDPE-BT, PVC-BT, PVDF-BT composites respectively based on the equation of Yamada et al. The observed and calculated values of permittivity match well for all the composites for n = 8.

Wide angle X-ray diffraction patterns of HDPE-BT, PVC-BT, PVDF-BT composites (Fig.2a, 2b and 2c) indicate that the

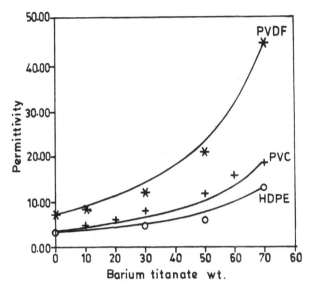

Fig. 1.   Curve Fitting of Permittivity Values
Using Yamada Equation for Polymer
Barium Titanate Composites (n=8).

effect of barium titanate concentration on polymer crystallinity is quite different in PVDF compared to that in HDPE and
PVC which can be explained as follows.

In both HDPE-BT and PVC-BT composites, barium titanate
acts as an inert filler. The changes in crystallinity of PVC
are only due to mechanical shear forces. During compaction of
only polymer under high pressure, the polymer particles initially get restacked, then deformed and ultimately made to cold
flow mostly along the radial direction perpendicular to the
applied pressure. This leads to change in crystallinity. In
presence of filler particles additional factors like shearing
and variation in pressure transfer characteristics also come
into play.

In case of PVDF-BT composites it is seen that even at 10
wt.% barium titanate loading, the PVDF peak at 2(0) = 6.75°C,
8.75°, 10.6°, 20.5°, and 27.0° has nearly vanished. This
indicates that in case of PVDF-barium titanate composites
barium titanate acts as an electroactive filler. The internal
field created by barium titanate creates local strain which
modifies the morphology of continuous phase i.e. PVDF. This
corroborates the fact that loss peak of PVDF is not observed
in PVDF-BT composites.   The anomalous behaviour of PVDF and
its interaction with the electroactive barium titanate particles can be attributed to the highly polar nature of C-F bonds
and also to the small size of F-atoms. Due to this, an internal field created by barium titanate is able to orient the
unit cell dipoles of PVDF and hence change in morphology.

The dielectric measurements of PVDF-BT and HDPE-BT composites over a temperature range (Fig.3,4) indicate that the

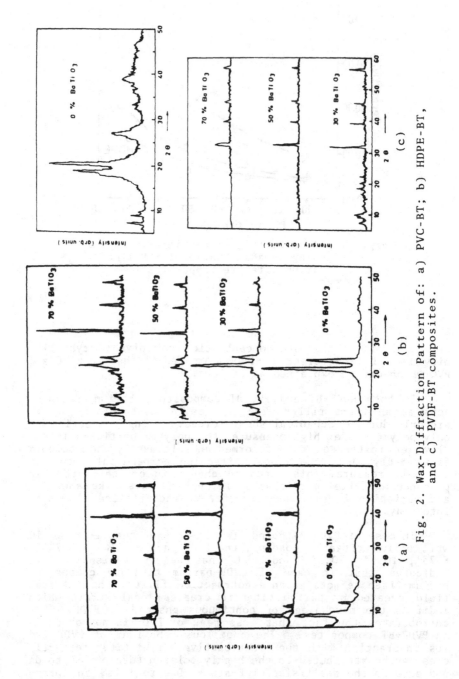

Fig. 2. Wax-Diffraction Pattern of: a) PVC-BT; b) HDPE-BT, and c) PVDF-BT composites.

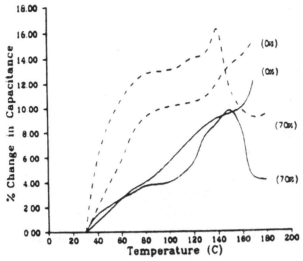

Fig. 3. Effect of Temperature on % Change in
CAPACITANCE of PVDF-Barium Titanate
Composites at 100 KHz (————) and
1000 KHz (- - -).

tetragonal to cubic transition temperature of barium titanate
is shifted towards the right. The effect of transition on
capacitance value is substantially reduced compared to pure
barium titanate. The shift and suppresion are more pronounced
in the case of PVDF-BT composites, where the phase transition
occurs around 155°C. This again indicates that very strong
interactions between the polymer and barium titanate particles
exist. These interactions do not allow the volume increase in
barium titanate which accompanies the transition, at its
actual transition temperature of 120°C. However on increasing
the temperature the supplied energy can overcome the interactions to allow the transition to occur.

    In the case of HDPE-BT composites, phase transition of
barium titanate occurs at 125°C. This indicates that interactions between the dispersed particles and continuous polymer
bulk is not as strong as in the case of PVDF-BT composites.
The accompanying decrease in capacitance during phase transition of barium titanate is not very sharp. Thus one can have a
relatively constant capacitance up to the melting temperature
of HDPE.

CONCLUSIONS

    The experimental values of permittivity fits with the
equation by Yamada et. al. on assumption of polymer being
continuous phase. SEM studies give direct evidence of polymer
as continuous phase in the polymer-barium titanate composites.
Dielectric measurements with varying temperature and wide
angle X-ray diffraction patterns indicate that PVDF has very
strong interactions with the dispersed barium titanate particles where as HDPE and PVC has weak interactions. However, the
capacitance of polymer-barium titanate composites remains
fairly constant with temperature upto melting point of respective polymers.

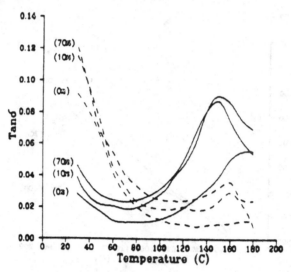

Fig. 4. Effect of Temperature on Tanδ in
PVDF - Barium Titanate Composites
at 100 KHz (────) and
1000 KHz (- - -).

REFERENCES

1. L.J.Bowen and T.S.Gururaja, J.Appl.Phys. 51, 5661, (1980).

2. A.Safari, R.E.Newnham, L.E.Cross and W.A.Schulze, Ferroele-
   ctrics, 41, 197 (1982).

3. D.P.Skinner, R.E.Newnham and L.E.Cross, Mat.Res.Bull., 13,
   599 (1978).

4. T.Furulawa and E.Fukade, Japanese J.App.Phys., 16, 453
   (1977).

5. T.Yamada, T.Ueda and T.Kitayama, J.Appl.Phys.,53,4328
   (1982)

6. C.Muralidhar and P.K.C.Pillai, J.Mat.Sc., 23,1071(1988).

7. V.M.Kamat, M.Tech.Dissertation, Materials Science Centre,
   IIT Bombay, (1991).

METAL/SEMICONDUCTING POLYANILINE HETEROJUNCTIONS

S.C.K. Misra, N.N. Beladakere, S.S. Pandey
B.D. Malhotra and Subhas Chandra

National Physical Laboratory
New Delhi 110012, India

INTRODUCTION

Semiconducting polymers are important for their potential application in electronic device fabrication [1]. The polymers can be prepared in
semiconducting form by manipulating their oxidation, reduction and doping
during polymerization and synthesis [2]. The applicability of a semiconducting polymer is as a metal/polymer electrode or three electrode system
for use in temperature sensor and gas sensor devices. The performance of
a solid state device depends upon the junction characteristics, the work
function of the metal and the polymer, the energy band gap of polymer and
other electronic parameters. The advantages of polymer based electronic
device over the inorganic ones are, the cost effectiveness, dispensing
away with sofisticated crystal growth and its light weight. Metal/poly-
pyrrole junctions have been prepared [3]. The characterization of metal/p-
polyaniline junctions with various metals, is reported here.

EXPERIMENTAL

The polyaniline was chemically synthesized by oxidative polymerization of aniline using ammonium per disulphate as an oxidant [4]. Freshly
distilled aniline (0.2M) was dissolved in 300 ml. of precooled 1M HCl
solution maintained at 0.5°C. 0.05M ammonium perdisulphate dissolved in
200 ml. of distilled water was added slowly for one hr with constant
stirring, resulting into dark green solution. The dark green precipitate
was recovered and washed first with 1M HCl, then with methanol and finally
with di-ethyl ether. Finally it was dried over phosphorous pentaoxide for
48 hrs. Polyaniline pellets of 13 mm dia and thickness 0.5 mm were prepared by hydraulic press. Schottky junctions were prepared by vacuum
evaporation of various metals In, Sn, Al, Ag, Pb etc. on these polyaniline
pellets. Ohmic contact for the configuration metal/polyaniline/ohmic,
were prepared using E+502 as ohmic contact. These polyaniline pellets
had a resistivity 1-5 ohm cm. and the junction area was 0.16 sq.cm.

RESULTS AND DISCUSSION

Polyaniline pellets used in our experiments were p type semiconductors. Polyaniline makes ohmic contact with metals having a work function
greater than that of polyaniline (Fig.1).

Frontiers of Polymer Research, Edited by P.N. Prasad and
I.K. Nigam. Plenum Press, New York, 1991

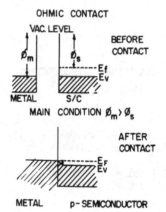

Fig. 1. Energy Band Scheme for a Metal and
p Type Semiconductor Before and
After contact, with $\emptyset_m < \emptyset_s$.

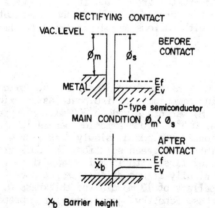

Fig. 2. Energy Band Scheme for a Metal and
p Type Semiconductor Before and
After Contact, with $\emptyset_m < \emptyset_s$.

The metals having a work function lower than that of polyaniline make a rectifying or blocking contact as shown in Fig.2. When the work function of the metal is less than that of polyaniline, a charge buildup takes place at the interface after the contact of metal and semiconductor, and rectification is observed. This is characterized by a barrier height X_b. The V-I characteristics for such a junction with various metal electrodes Al, Sn, In and Pb are shown in Fig.3.

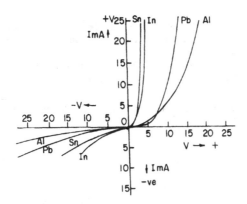

Fig. 3. V-I Characteristics for Metal/Polyaniline with Different Metals.

It is apparant that forward and reverse bias characteristics of Al/poly-aniline junction have little difference indicating incomplete blocking of the charge carriers at the metal/polymer interface. The In/polyaniline junction however appear to be more suitable for fabrication of Schottky barriers as the V-I characteristics resemble that of metal/inorganic semiconductor devices. The current through a Schottky barrier is expressed by the following relation[5]:

$$J = J_0 \exp{(qV/nKT)}$$

where J the reverse saturation current

$$J_0 = \overset{X}{A} T^2 \exp{(-qX_b/KT)}$$

where, q = electronic charge
        n_x = the ideality factor
        $\overset{X}{A}$ = Richardson constant = 120 A/sq.cm degree^2
and   T = temperature

The ideality factor for various metal/polyaniline junctions are shown in Table 1.

Table 1
Electronic Parameters of metal/polyaniline junctions

Metal	Work Function (eV)	Barrier Height(V)	ideality factor
In	4.12	0.4	1.9
Sn	4.11	0.4	4.9
Pb	4.02	0.5	6.9
Al	3.74	0.4	2.8
Ag	4.28	Ohmic contact	

It is observed that In makes a better Schottky contact as is evident from the value of ideality factor [1.9], the value for metal/inorganic junctions is 1.02. The barrier height for various metal/polyaniline junctions has also been estimated, as shown in Table 1. No V-I characteristics could be obtained for high resistivity specimen.

The capacitance voltage characteristics of In/polyaniline/E+502 configuration are shown in Fig.4.

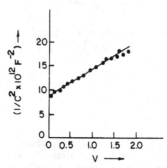

Fig. 4.   C-V Characteristics for Indium Polyaniline Junctions.

The carrier concentration in the metal/polyaniline interface can be estimated from the $1/C^2$ vs V plot and the relation:

$$N = -2A/q\varepsilon\varepsilon_0 \; [d \; (V)/d(1/C^2)]$$

where
ε = Permittivity of polyanine
ε_0 = Permittivity of free space
A = Area of contact
and
N = Carrier concentration in the space charge region.

The carrier concentration in the space charge region as estimated from the above is 10^{17}/cu.cm.

CONCLUSION

It has been possible to prepare Schottky barrier junctions from semiconducting polyaniline using various metals. In/polyaniline junction exhibits excellent rectification characteristics (as indicated by the ideality factor and V-I characteristics). The carrier concentration in the specimen has been estimated to be of the order of 10^{17}/cu.cm. The carrier concentration is the main factor controlling the device performance along with the barrier height and the work function of the semiconductor and the metal. The work function of polyaniline has been estimated to be between 4.12 and 4.28 eV.

ACKNOWLEDGEMENTS

The authors are thankful to Professor S.K. Joshi, Director, National Physical Laboratory, New Delhi for his kind interest and constant encouragement. Thanks are due to Commission of European Economic Community for financial assistance.

REFERENCES

1. J.C. Lacroix, K.K. Kanazawa & A. Diaz; J.Electrochem.Soc., 136(5) 308 (1989).

2. Y. Cao, P. Smith & A.J. Heeger; Synth.Metals, 32, 263, (1989).

3. R. Gupta, S.C.K. Misra, B.D. Malhotra, N.N. Beladakere & S. Chandra Appl.Phys.Lett., 58(1), 51 (1991).

4. A.G. MacDiarmid, J.C. Chiang, M. Helpern, W.S. Huang, S.I. Mu, N.L.D. Somasiri, W. Wu & S.I. Yaniger, Mol.Cryst.Liq.Cryst; 121, 173 (1985).

5. S.M. Sze, Physics of Semiconductor Devices, (Wiley N.Y.) 1981.

CHIRAL CONDUCTING POLY (3-ALKYLTHIOPHENES): SPECTROSCOPIC

AND ELECTROCHEMICAL PROPERTIES

Dilip Kotkar, Pushpito K. Ghosh and Anjan Ray

Alchemie Research Centre, P.O. Box 155, Belapur
Road, Thane-400601, Maharashtra, India

ABSTRACT

Chemical oxidative polymerization of thiophenes
substituted with optically active groups at the 3-position
yields conducting polymers with interesting optical and
electrochemical properties. Carbon-13 NMR, IR and UV-visible
spectral information on these materials are presented,
together with electrochemical and molecular weight data.
Properties of polymers derived from enantiomerically pure
monomers are compared with those obtained from racemic monomer
mixtures and with poly (3-hexylthiophene).

INTRODUCTION

The tremendous worldwide interest in conducting polymers[1]
is due, in large measure, to the "tailorability" of these
materials to specific requirements and desirable combinations
of electronic, optical and mechanical properties. The
introduction of chirality as an added feature is attractive
since it opens up possibilities for enantioselective
electrochemistry[2] and biochemical applications.

Our early work[3] in the area of electrochemically
synthesized chiral polythiophenes is extended in this paper to
oxidative chemical polymerization products of corresponding
monomeric 3-alkylthiophenes. Our preference for thiophene-
based systems is a result of well-established losses in
conductivity of polypyrroles[4] and polyanilines[5] upon alkyl
ring-substitution.

In the present paper, we present spectroscopic and electrochemical characterization of poly[3-(2'- methylbutyl)
thiophene] and poly[3-(3'-methylpentyl) thiophene], comparing
the polymer of each (+)-enantiomeric monomer with that of the
corresponding racemic mixture.

Frontiers of Polymer Research, Edited by P.N. Prasad and
J.K. Nigam, Plenum Press, New York, 1991

EXPERIMENTAL

Synthesis of thiophene monomers

Chiral and racemic 3-substituted alkylthiophene monomers
were synthesized from their corresponding alcohols, as
described in our previous publication[5]. The $NiCl_2$(dppp) -
catalysed coupling reaction[6] between 3-bromothiophene and
alkylmagnesium bromide proceeded in about 70% yield, the
latter being derived in situ from the respective alkyl
bromides, which could be prepared readily in excellent yields
from the alcohols. For comparison with existing literature,
3-hexylthiophene[7,8] was synthesized and characterized under
identical working conditions.

Chemical Polymerization of 3-alkylthiophenes

The appropriate monomer was dissolved in chloroform (1
gm/25ml $CHCl_3$) at room temperature. A 0.5M solution of $FeCl_3$
in MeCN (monomer: $FeCl_3$ = 1:3 mole ratio) was added in one
shot to the stirred solution of $CHCl_3$. A constant air flow
through the solution was maintained for 48 hours while
stirring at room temperature. After removing solvents by
rotary evaporation, the product was suspended in chloroform,
filtered and washed with 4-5 portions of $CHCl_3$ followed by
methanol washes until the filtrate was colourless. The dark
green polymeric solid was compensated ("reduced") by overnight
treatment with aqueous 5% ammonia. The orange red product was
washed copiously with water, then with methanol and finally
dried under vacuum. Yields of the neutral polymer were
typically 20-25%.

Electrochemistry

Cyclic voltammetry was performed using an Ag/Ag^+
reference, a platinum foil counter and a glassy carbon working
electrode on Princeton Applied Research Models 175-173-179
assembly, with the output directed to an X-Y recorder. Thin
films of the reduced forms of the polymers were cast onto the
working electrode from dilute chloroform solutions. The
electrolyte was 0.1M tetraethylammonium tetrafluoroborate in
MeCN. Voltammograms were recorded for a +0.4V ---> +0.0V --->
+0.9V sequence vs the reference at a scan rate of 10 mV/sec.

RESULTS AND DISCUSSION

A summary of elemental analyses of the reduced forms of
the polythiophenes, together with the polymerization yields
and the data from the optical absorption and emission spectra
in the 300-800 nm region is presented in Table 1. The
elemental analytical data is constrained by our inability to
obtain accurate values for C and H in the presence of S,
however, all the materials show similar results. All the
polymers yield UV/visible spectra essentially identical to
poly(3-hexyl-thiophene)[7] with respect to frequency and
oscillator strength, although the lone peak, attributed to the
bandgap absorption, exhibits a small bathochromic shift in
going from the 2'-methylbutyl to the 3'-methylpentyl
derivative, possibly as a result of minor steric differences.

Table 1. Elemental Analyses and Optical Spectral Data for poly (3-alkylthiophenes)

Substituent	Elemental analysis[a]		%Yield	Vis/UV maximum (nm)	
				Absorbance	Emission
Hexyl	F 65.7 C 71.4	7.9 9.5	20	429	560
Chiral 2'- methylbutyl	F 64.1 C 70.0	7.1 9.0	33	423	556
Racemic 2'- methylbutyl	F 65.7 C 70.0	7.3 9.0	22	428	556
Chiral 3'- methylpentyl	F 69.2 C 71.4	8.2 9.5	20	433	570
Racemic 3'- methylpentyl	F 69.3 C 71.4	8.3 9.5	34	432	570

a   We find that both C and H are invariably lower for sulphur-containing compounds. No actual correction for this was made.
b   Excitation at 370 nm.
c   F = found, C = calculated.

Table 2. Gel Permeation Chromatography data[a] for poly(3-alkyl thiophenes)

Substituent	Mn	Mw	polydispersity
Hexyl	16100 (19000)[b]	51900 (40700)	3.22 (2.14)
2'-methylbutyl (chiral)	21700	58100	2.67
2'-methylbutyl (racemic)	20300	50000	2.47
3'-methylpentyl (chiral)	23500	75100	3.20
3'-methylpentyl (racemic)	23900 (19100)[c]	63200 (205000)	2.65 (10.70)

(a)   Polystyrene calibrants, THF solvent.
(b)   Partially oxidised by bromine vapour.
(c)   Electropolymerized[3] at 5V absolute, 2-electrode configuration in MeCN/TBABF$_4$.

There are no surprises in the infrared data, the only point of note being the disappearance of the ~3090 cm^{-1} C-H stretch, corresponding to the α-position of the monomeric thiophene ring. The ß C-H stretch at ~3050 cm^{-1} is predictably retained in the polymer. Thus, the products are indicated to be high molecular weight materials, a conclusion corroborated by the GPC data presented in Table 2. All the polymers are found to have unimodal profiles, weight-average molecular weights of ~50000 and polydispersity (Mw/Mn) indices of 2.5-3. Thus the reproducibility of our polymerization technique appears to be essentially independent of the pendant alkyl group.

Two interesting sidelights arise from the GPC data. The racemic 2'-methylbutyl isomer, when electropolymerized as described previously[3], yields a totally different product with a very high polydispersity and only partial solubility in THF, suggesting possible crosslinking. In another study, oxidation of poly (3-hexylthiophene) with Br$_2$ vapour caused the GPC distribution to change to higher Mn, lower Mw, emphasizing the importance of hydrodynamic volume influences.

Disappointing, but not surprising, is the absence of differences in the NMR spectra of the chiral vs. the corresponding racemic polymers. The aromatic region pattern between 125 and 140 ppm was identical in all five polymers to within 1 ppm. Typical carbon-13 NMR data for racemic poly (3'-methylbutyl) is as follows (ppm vs TMS): 138.9, 133.6, 131.1, 129.6 (aromatic), 36.2, 36.0, 29.5, 19.2, 11.4 (alkyl).

Cyclic voltammograms of the 2'-methylbutyl derivatives show a mundane single redox couple ($E_{1/2}$~0.65 vs Ag/Ag$^+$) like the hexyl-substituted polymer, whereas the wave appears to resolve partially into two distinct oxidations during the anodic scan of the 3'-methylpentyl polymer. This separation between neutral-to-polaron and polaron-to-bipolaron transitions has been noted by other groups working with long-chain substituted conducting polymers[2,9]. It appears likely that beyond a point, kinetic stabilization of the polaron takes place on the CV timescale due to steric factors.

Differential scanning calorimetry of the 3'-methylpentyl polymer from 30°C to 250°C in air shows an onset of decomposition at about 210°C. Given that flexible free-standing films can be cast readily from all the neutral polymers described here, one believes that the glass transition in all these must be well below room temperature.

Although polarimetry of these polymers could not be carried out due to their strong emission near the sodium D line (the hexyl, chiral and racemic polymers all yielded identical rotation!), the chiral 3'-methylpentyl polymer exhibited a strong derivative-type signal in the CD spectrum which crossed the zero-ellipticity baseline at the polymer absorbance maximum. This may indicate some transmission of chirality from the side-chain to the polymer backbone. Detailed studies to probe this effect and to examine chiral behaviour of the relevent polymers are currently under way.

ACKNOWLEDGEMENTS

Assistance from Mr. J.S. Wagh, Mrs. J.M. Tilak, Mr. R. Kelkar and Dr. M. Chakraborty for analytical data, Mrs. C. Pereira and Dr. D. Mukesh in manuscript preparation, Mr. V.V. Joshi for drafting and financial support by ICI India Limited is gratefully acknowledged.

REFERENCES

1. M.G. Kanatzidis, _Chem. Engg. News_, Dec.3 (1990) 36.

2. M. Lemaire, D. Delabouglise, R. Garreau, A. Guy and J. Roncali, _J. Chem. Soc. Chem. Commun._ (1988) 658 and references therein.

3. D. Kotkar, V.Y. Joshi and P.K. Ghosh, _J. Chem. Soc. Chem. Commun._ (1988) 917.

4. G.B. Street, R.H. Geiss, S.E. Lindsay, A. Nazzal and P. Pfluger in _Springer Ser. Solid-State Sci._ 49 (1983), Springer-Verlag (New York).

5. Y. Wei, W.W. Focke, G.E. Wnek, A. Ray and A.G. MacDiarmid, _J. Phys. Chem._ 93(1989) 495.

6. K. Tamao, S. Kodama, I. Nakajima, M. Kumada, A. Minato and K. Suzuki, _Tetrahedron_ 38 (1982) 3347.

7. O. Inganas, W.R. Salaneck, J.E. Osterholm and J. Laakso, _Synth. Met._ 22 (1988) 395.

8. R.L. Elsenbaumer, K.Y. Jen and R. Oboodi, _Synth.Met._ 15 (1986) 169.

9. J.R. Reynolds, J.P. Ruiz, A.D. Child, K. Nayak and D.S. Marynick, _Macromolecules_, 24 (1991) 678.

POLYMER DISPERSED LIQUID CRYSTAL COMPOSITES: ELECTRO-OPTIC RESPONSE

BEHAVIOUR AND THEIR APPLICATIONS

S.C. Jain and D.K. Rout

National Physical Laboratory
K.S. Krishnan Road
New Delhi 110012, India

ABSTRACT

Polymer dispersed liquid crystal composites(PDLCs) are a novel class
of opto-electronic materials which are likely to find many applications
in a wide variety of display areas. The electro-optic studies of various
PDLC films have shown that the threshold voltage, its sharpness and response times depend critically on the frequency and amplitude of the applied signal and on the size of the liquid crystal (LC) droplets. The
threshold voltage is minimum around a few kHz frequency at which the
space charge buildup at the LC-polymer interface does not take place and
the dielectric dissipation is minimum. The rise time increases with increasing droplet size in general and decreases with increasing amplitude
of the voltage. The decay time is, however, a complex function of the
droplet size and is relatively weakly dependent on the driving voltage.
The size of the droplets and the driving voltage primarily governs the
memory effect shown by the PDLCs.

GENERAL DESCRIPTION

Polymer-dispersed liquid crystal composites (PDLCs) are a novel class
of opto-electronic materials which are interesting systems both from a
basic as well as an application point of view. They would be useful not
only in display areas traditionally served by conventional liquid crystal
displays but also for making large and flexible displays, variable transmission windows, room partitions, optical modulating devices and colour
projectors etc. (Doane et al. 1988). In a PDLC film, a LC material with
large positive dielectric anisotropy and large birefringence is embeded
in an isotropic polymer matrix in the form of very fine droplets of
micron to sub micron sizes. The average orientation direction of LC molecules varies from droplet to droplet providing an optical discontinuity
at the LC droplet polymer interface. When white light is incident on such
a film, it is strongly scattered and renders the film a milky white appearance. On application of an appropriate electric field, the LC molecules
realign tehmselves along the field and the film becomes transparent if
the ordinary refractive index of liquid crystal is equal to the refractive index of the polymer matrix. On field removal, the original state is
restored by the elastic and the surface interacting forces of the LC
molecules at the droplet boundary.

Frontiers of Polymer Research, Edited by P.N. Prasad and
J.K. Nigam, Plenum Press, New York, 1991

The electro-optic response bahviour of the PDLC films has been a subject of many investigations (Drzaic, 1988, Erdmann et al. 1989, Wu et al. 1989, Jain, Rout and Chandra, 1990). It has been shown that in a PDLC film, the field seen by a LC droplet is much less than the actual field applied to the film due to a considerable voltage drop across the polymer matrix (Drzaic, 1988). The threshold voltage required to activate a PDLC cell is strongly related to the resistivities of the polymer and the LC and response times can be strongly altered by shaping the droplets (Erdmann et al. 1989).

Our dielectric studies (Rout and Jain, 1990) on the PDLC materials have shown that the frequency of the applied signal should play an important role in controlling the electro-optic response behaviour of the PDLC films. It was argued that the electro-optic response should be best around a few kHz frequency at which space charge build-up at the LC droplet-polymer interface cannot takeplace and the dielectric dissipation factor is minimum. Presented below are the results of our studies on the electro optic response behaviour of PDLCs as a function of applied signal, its frequency and the droplet size.

MATERIALS AND PREPARATION PROCEDURES

PDLC films were prepared using thermoplastic polymer NF-100 of M/s Thick Films Inc., USA and LC mixture E-8 of BDH., UK by solvent induced phase separation techniques (Doane, et al. 1988, Jain, Rout and Chandra 1990). the size of the LC droplets was controlled by cooling the PDLC cell from the isotropic melt at different cooling rates. The electro-optic transmission properties were measured in normal transmission geometry utilizing an Olympus polarizing microscope BH-2, RCA-931A photo-multiplier tube and Kikusui Digital Storage Oscilloscope COM 7201A.

RESULTS AND DISCUSSION

Threshold Behaviour

Figure 1 shows the optical micrographs of three PDLC cells having predominently three different sizes of LC droplets (a) droplet diameter $d < 1\,\mu m$ (b) $d = 2\text{-}5\,\mu m$ and (c) $d = 15\,\mu m$ for which electro-optic properties have been studied. It was observed that in a given PDLC cell irrespective droplet size, transmission at a fixed applied voltage varied significantly

Figure 1.   PDLC cells having three different droplet diameter; (a)<1.0 μm (b) $\approx 2\text{-}5\,\mu m$ (c) $\approx 15\,\mu m$

with increasing frequency. It attains a maximum around a few kHz and then falls off again. Figure 2 shows a typical set of transmission curves

of one such PDLC cell (d = 2-5 μm) as a function of frequency (from 50 Hz
- 50 kHz) at a 10 V r.m.s. signal. The transmission is maximum around
5 kHz and decreases on both the sides of the frequency. These observation
strongly indicate that the optical response is maximum at this frequency
which inturn also implies that the electrical voltage appearing across
the LC droplets is also maximum. This fact was further confirmed by studying the transmission versus applied voltage curves for different PDLC
cells at different frequencies. It was found that the threshold voltage
was minimum and its sharpness, maximum at 50 kHz frequency. These results

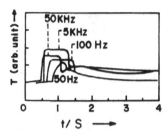

Figure 2.   Electro-optic response of PDLC
(d = 2-5 μm) at different frequencies.

are in accordance with the dielectric studies (Rout and Jain, 1990) according to which best transmission characteristics should be observed at
frequency at which the dielectric loss is minimum and no space charge is
allowed to buildup at LC polymer interface.

Electro-optic transmission studies in three PDLC cells with different
droplet sizes (Fig.1) at 5 kHz also revealed interesting features. It was
seen (Fig.3) that the transmission in the off state increases with increasing droplet size. The threshold voltage and the saturation voltage
increase with decreasing droplet size. These results are in accordance
with the predicted behaviour of PDLCs.

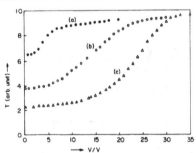

Figure 3. Transmittance vs voltage characteristics
of the PDLCs with (a) d ≈ 15 μm (b) d∼2-5 μm,
(c) d < 1.0 μm.

Response Times

Measurements of rise and decay times of PDLC films as a function of
applied voltage and droplet size have shown many interesting aspects. It
was observed that the rise time τr, (time required to reach 90% of the
initial transmission after the voltage signal is switched on) is a single
step process independent of the droplet size and the amplitude of the
applied signal. However, the rise time increases substantially with increasing voltage signal (Fig.4). But variation in τr, in the PDLC cell
(d < 1 μm) with increasing amplitude is much less.

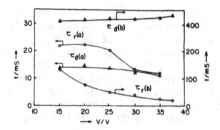

Figure 4. Rise time and Decay time as a function of
applied voltage for PDLCs: (a) d < 1 μm,
(b) d ≈ 15 μm.

The decay time τ_d, (time required to reach 10% of the initial transmission after the signal is switched off) showed complex behaviour as a
function of the LC droplet size. It was observed that in the medium size
LC droplet cell (d= 2-5 μm), the decay was a two step process (Fig.5);
the fast component decay equals to 10 to 20 mS and the slow component
decay equals to 5 S. The two interesting features of figure 5 are; (i)
the slow decay which is about 5 S is present even when the applied voltage is as low as 2 V only and (ii) the fast component of decay becomes
appreciable above 4 V only and reaches saturation at 15 V. But in the
case of PDLC cell (d < 1 μm), only fast decay component was present
(τ_d= 15 mS) and no slow component of the decay was seen evel well beyond
the saturation driving voltage. However, it was found that if the driving
voltage was ≈ 70 V, the cell did show a two component decay. The decay
time for the slow component was about or even greater than a few minutes
giving rise to memory in the PDLC cell. In contrast, in the big size LC
droplet PDLC cell (d > 10 μm), only slow component of the decay was ob-
servable and τ_d was about 300 mS. This complex decay process in different
PDLC cells of different LC droplet sizes is not clearly understood and
is a subject of current investigation.

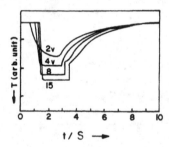

Figure 5. Electro-optic response at various voltages
showing two stage decay process in medium
sized (2-5 μm) PDLC cell.

REFERENCES

Doane, J.W., Golemme, A., West, J.L., Whitehead jr., J.B., and Wu, B.G.,
1988, Polymer Dispersed Liquid Crystals for Display Applications, Mol.
Cryst. Liquid Cryst., 165:511.

Drzaic, P.S., 1988 Reorientation dynamics of polymer dispersed nematic
liquid crystals, Liquid Crystals, 3:1543.

Erdmann, J., Doane, J.W., Žumer, S., and Chidichimo, G., 1989, Electro-optic response of PDLC light shutters, SPIE - Liquid Crystal Chemistry Physics and Applications, 1080:32.

Jain, S.C., Rout, D.K., Chandra, S., 1990, Electro-optic Studies on Polymerd dispersed liquid crystal films prepared by solvent induced phase separation technique, Mol. Cryst. Liquid Cryst. 188:251.

Rout, D.K., and Jain, S.C., 1990, Dielectric properties of a polymer dispersed liquid crystal film, Mol. Cryst. Liquid Cryst. (Communicated)

COMPOSITES OF POLYANILINE FOR ANTISTATIC APPLICATIONS

Dinesh Chandra Trivedi and Sundeep Kumar Dhawan

Central Electrochemical Research Institute
Karaikudi-623 006
India

The preparation of composites of conducting polymers has become a new area of research because of their technological importance. This paper describes the preparation of conductive polyaniline [PAn] composites, whose resistivity lies between $60\,\Omega$ to $1000\,\Omega/\square$. The study indicates that these composites may prove useful materials for dissipation of electrostatic charge (ESD).

INTRODUCTION

The conducting polymers are the new materials having the unique signature like anisotropic electronic conductivity and various other properties like electroactivity, wide range of conductivity has made them the candidate for advance technological application [1-9]. However, the rigid backbone structure of these polymers has made them unprocessible thus limiting their many applications [10-15]. The polyaniline (PAn) is the one of the such conducting polymers, but due to its high environmental stability and the ease with it can be prepared has given PAn an unique status. Though the PAn has a rigid backbone structure like other polymers having phenyl rings with a difference that the chemically flexible group, -NH, is flanked either side by phenyl rings. Thus the presence of -NH group has imparted certain amount of chemical reactivity thereby it is capable of forming a terniary soluble system in presence of reactive dopants [10-14]. Using this behaviour the fabrics have been made conducting by grafting polyaniline [15].

In this paper we report our results on the preparation of flexible polyaniline composites with water soluble polymers like polyvinyl alcohol/carboxymethyl cellulose and their derivatives. Due to the limitation of space we will describe the polyvinyl alcohol composites. The details for other composites have been given elsewhere [11]. These composites may find useful applications as antistatic materials for the dissipation of electrostatic charge in electronic equipments and as zebra connectors for display devices in place of other conductive composites obtained using carbon and metal powders.

EXPERIMENTAL

Aniline was freshly distilled before use. All other chemicals were of high purity grade.

Frontiers of Polymer Research, Edited by P.N. Prasad and
J.K. Nigam, Plenum Press, New York, 1991

Oxidative Polymerization of Aniline

The chemical oxidative polymerization of aniline (0.1 M) in 1.0 M p-toluene sulphonic acid or benzene sulphonic acid was carried out by adding drop by drop aqueous solution of requisite amount of ammonium persulphate at $5 \pm 0.2°C$. The stirring of the reaction mixture was carried out for 2 hours. On completion of the reaction the reaction mixture was filtered and washed repeatedly with distilled water and finally with acetone to remove soluble oligomers and other impurities. The final washing of so obtained polyaniline was carried out with 1.0 M aqueous solution of either p-toluene sulphonic acid (PTSA) or benzene sulphonic acid (BSA).

A 10% aqueous solution of polyvinyl alcohol (PVA) in the presence of varying amount of hydrophilic agents like ethylene glycol and glycerol was prepared as reported earlier[11].

Preparation of Composites/Blends

The composite of PVA:PAn is prepared by homogeneously dispersing the wet polyaniline to PVA solution at room temperature under constant vigorous stirred conditions. The stirring of this solution was continued for 2 hours to ensure the homogeneous dispersion of polyaniline in PVA solution. This viscous homogeneous dispersion on casting gives a rubber like flexible shining composite film. Thus number of composites were prepared by varying the amount loading of polyaniline with PVA. The various results on composite films are tabulated in Table 1. The preparation of composite was carried out in the temperature range of 20-40°C. The lowering of the temperature does not have any adverse effect whereas temperature above 50°C should be avoided. The film can be cast on clean glass plates or on teflon coated conveyor belts, whose speed can be controlled using suitable electric motors. Drying of the composite depends upon the humidity and temperature conditions of drying chamber. In the present case all the composite films cured at 60°C under nitrogen atmosphere for two hours to obtain flexible films.

The cyclic voltammogram of the homogeneous solution of PVA:PAn (1:1 by weight) recorded on a Tacussel bi-pad coupled with PAR programmer (Model 175) and x-y recorder (BBC Model SE - 780). FTIR spectra were recorded on a Nicolet F.T. Infra red spectrometer. Scanning electron micrograms (SEM) were taken with a JEOL SEM - 35 CF. The electrical resistance of the composite film was measured by placing two thin metallic foils of copper, 1 cm. apart. The tensile strength of the composites were determined using Instron equipment.

The static charge of the insulating PVA film and conducting composite sample were measured using Para Electronics Static charge meter SCM 1 mounted on a energy stand with a distance of 2.5 cm. from energy dish whose capacitance at this distance was 23 pf.

RESULTS AND DISCUSSION

The important use of electrostatic dissipating materials is in the packaging of electronic components. Electrostatic shock can often damage these components and therefore, it is necessary to ensure that electrostatic charge does not build up. The levels of conductivity required for this purpose is $10^{-4} - 10^{-8}$ ohm^{-1}□$^{-1}$ and an uniform electronic conductivity.

The polyaniline (PAn) has high environmental stability due to non-degenerate ground state and structurally this polymer differs from various other polymers in the sense that the nitrogen heteroatom present in polyaniline is the part of polymer backbone and hence assume importance from chemical reactivity point of view as can be seen in formula 1.

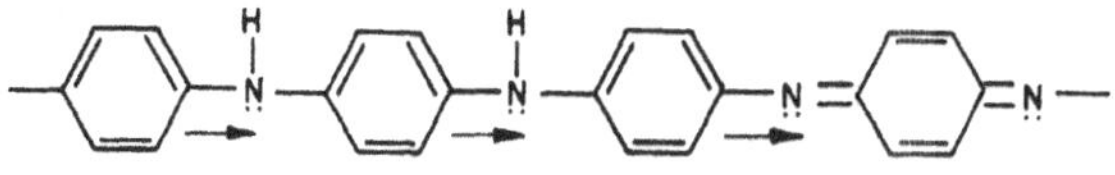

EMERALDINE BASE
Formula-I

The presence of -NH in PAn gives the polymer, a chemical flexibility. This N-H is capable of interacting with other chemical reactive groups like OH group of polyvinyl alcohol as can be seen from the shift in principal iR absorption bands. The complete iR spectra is marked by number of absorption bands which makes difficult to assign the bands. The uniformity of PAn dispersion can also be seen from SEM micrographs. Fig.1 which reveals the uniform globular structure of PAn in PVA matrix which is responsible for good conductivity of the system.

Mechanism of Polymerization

The oxidative polymerization of aniline proceeds via the formation of the radical cation of the monomer which then reacts with the second radical cation to give a dimer as is evident from the cyclic voltammetric studies[12]. Thus after few coupling reactions the filming of anode occurs. In a chemical polymerization the stoichiometric amount of oxidant like ammonium persulphate is used. The conductive polymer obtained by both chemical and electrochemical routes are identical. The possible reaction mechanism can be written as:

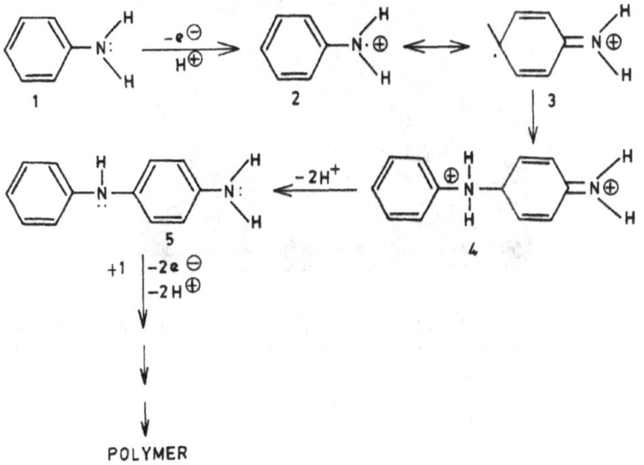

POLYMER

Electrochemical Behaviour of PVA:PAn Solution

Fig. 2 shows the cyclic voltammogram of homogeneous solution of PVA:PAn recorded using Pt electrodes. Two distinct cv peaks at 0.48 and 0.92 volts vs Ag/AgCl are observed. However, the cyclic voltammogram

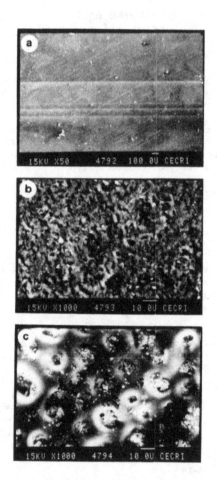

Fig. 1. (a) SEM OF PVA FILM, (b) PVA-PAn COMPOSITE FILM,
          (c) PAn CAST FILM (FROM NMP).

of polyaniline in protonic acid medium shows the characteristic peaks at ≈ 0.1 and 0.7 volts [13] wherein the above peaks are assigned due to the formation of radical cation and diradical dictations.

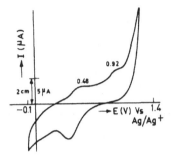

Fig. 2. CYCLIC VOLTAMMOGRAM OF PVA: PAn SOLUTION ON Pt ELECTRODE (0.25 cm^2) Vs Ag/Ag$^+$ AT A SCAN RATE OF 100 mV/s.

The shift in peak potential of the characteristic PAn peaks in case of PVA - PAn composite solution indicates that the possible interaction between PVA and PAn resulting in the shift of the peak potentials. It also indicates that a wide potential window exists for the PVA - PAn homogeneous solution.

Spectral Studies

Principal absorption bands observed in the FTIR spectra of 10% PVA cast film are: 1651, 1239, 920 and 869 cm^{-1}. The characteristic peaks observed in the polyaniline (KBr) are at: 1571, 1483, 1303, 1261, 1144, 1111, 1021 and 803 cm^{-1} [13]. However, the PVA:PAn cast films gives a rather complex absorption spectra showing multiple bands in the region 3600-2800 cm^{-1} & 1600-400 cm^{-1}. From this it can also be inferred that cross-linking might have taken place in the composite. The U.V. visible absorption spectra of PVA:PAn (1:0.1) doped with protonic acid gave a strong absorption band at 835 nm whereas PAn itself gives the absorption band at 820 nm. This indicates better conduction in PVA:PAn composite because this absorption band has been said to be due to delocalization of charge, mainly bipolarons.

The static charge of the insulating PVA film and conducting composite PVA:PAn film were measured using static charge meter. The static charge in coulombs was calculated using the following relationship:

$$Q = C.V$$

Where C is the standard capacitance of the base plate and V is the voltage indicated on the static charge meter. For PVA cast film, the voltage indicated on the static charge meter at 2.5 cm. distance = 400 volts.

∴ Static charge of the PVA film = 92 x 10^{-10} coulombs.

For conducting PVA:PAn composite film having resistance of 0.5 kΩ cm, static charge calculated was zero (as there was no voltage indication in the SCM meter). This shows a complete dissipation of the static charge on employing a PVA:PAn composite film.

The surface resistance tensile strength, extension at break of various composition of composite are tabulated in Table 1. It can be seen from this table that tensile strength progressively decreases on increasing PAn ratio in PVA. Experiments revealed that maximum loading beyond equimolar ratio is impossible.

TABLE 1

Mechanical properties of the conductive composites

Composite Composition	Tensile strength (kg/cm²)	Extension at break (%)	Surface resistance in ohm/□
PVA:PAn 1:0.2	168.9	548.9	1000
PVA:PAn 1:0.5	85.9	415.0	500
PVA:PAn 1:1	22.17	294.3	60

CONCLUSION

We show in this work that PVA:PAn composite of good conductivity and environmental stability associated with good mechanical properties can be prepared by dispersing polyaniline into aqueous solution poly-vinyl-alcohol. These composites can find wide applications as antistatic material for electrostatic charge dissipation and invaries other devices where conducting carbon filled rubbers are in vogue.

REFERENCES

1. A.G.MacDiarmid, S.L.Mu, N.L.D.Somasiri and W.Wu Mol. Cryst. Liq. Cryst., 121 : 187 (1985).
2. D.C.Trivedi, Bull. Electrochem., 4 : 815 (1988).
3. E.W.Paul, A.J.Ricco and M.S.Wrighton, J. Phys. Chem. 89 : 1441 (1985)
4. A.Kitani, J.Yano and K.Sasaki, J. Electroanal. Chem., 209:227 (1986)
5. T.Kobayashi, H.Yoneyana and M.Tamura, J. Electroanal. Chem., 161:419 (1989).
6. D.C.Trivedi, J. Chem. Soc. Chem. Commun. 544 (1989).
7. D.C.Trivedi and S.Srinivasan, J. Chem. Soc. Chem. Commn. 410 (1988).
8. D.C. Trivedi and S.Srinivasan, J. Mater. Sci. Lett. 8:709 (1989).
9. D.C.Trivedi, J. Electrochem. Soc. (India) 35, 243 (1986).
10. S.K.Dhawan and D.C.Trivedi, Bull. Electrochem., 5:208 (1989).
11. D.C.Trivedi and S.K.Dhawan, Indian Patent Filed (193/90).
12. S.K.Dhawan and D.C.Trivedi, British Polymer Journal, 24, 0000 (1991).
13. S.K.Dhawan and D.C.Trivedi, Mater. Chem. & Phys. under print (1991).
14. S.K.Dhawan and D.C.Trivedi, Bull. Mater. Sci, 12 : 153 (1989).
15. D.C. Trivedi and S.K. Dhawan, p.746, Conductive Fibres in Polymer Science Contemporary Themes Vol.II, Ed. S.Sivaram, Tata McGraw-Hill Publishing Comp. New Delhi (1991).

ELECTRONIC STRUCTURE OF POLYACENE: ONE DIMENSIONAL GRAPHITE

A.K. Bakhshi

Department of Chemistry
Panjab University
Chandigarh-160014, India

INTRODUCTION

In the field of synthetic metals, much interest has been taken in the molecular designing of novel intrinsically conducting polymers during the last few years. One exciting possibility in this direction is provided by the hydrocarbon polymers with fused aromatic rings. This class of polymers, frequently referred to as one dimensional graphite family, has been theoretically predicted[1] to have, combined with stability in air, unusual electronic properties: the predicted features range from metallic conductivity to high temperature superconductivity or ferromagnetism. The members of this family include systems such as polyacene(PAc), polyacenacene(PAcA), polyphenanthrene(PPh), polyphenanthro-phenanthrene(PPhP) and polyperinaphthalene(PPN) (Fig.1). Except PPN, the other polymers can be considered to form two series of polymers derived from trans polyacetylene (PA) and cis-PA respectively. PAc and PAcA can be considered as the laddered versions of two and three chains of trans-PA while PPh and PPhP those of the corresponding cis-PA chains. One may also differentiate between these two series on the basis of the mode of growth of condensed aromatic rings. PPN, on the other hand can be considered to be a fused version of planar poly(p-phenylene) and cis-PA structures.

There have been several attempts to actually synthesize members of 1-D graphite family. The synthesis of the conducting films of PPN has been reported by Kaplan et al.[2] by the pyrolysis of 3,4,9,10 perylene tetracarboxylic dianhydride(PTCDA) at temperatures between 700° and $900^\circ C$. Similar films have also been obtained by Schmidt et al.[3] and Forrest et al.[4] by electron and argon ion beam bombardment of PTCDA respectively. No other member of this family has so far been prepared. The synthesis of PAc has been attempted by Ozaki et al.[5] via its prepolymer poly(ethynyl acetylene)

(PEA) which was obtained by the desilylation of poly(trimethyl silyl ethynyl acetylene) (PTMSEA) . Recently Tanaka et al.[6] have studied theoretically the cyclisation reactions of polybutadiyne and polyhexatriyne to PAc and PAcA

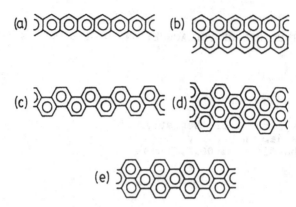

Fig. 1. Structures of (a) Polyacene (PAc), (b) Polyacenacene (PAcA), (c) Polyphenanthrene (PPh), (d) Polyphenanthro-phenanthrene (PPhP) and Polyperinaphthalene (PPN).

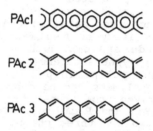

Fig. 2. Structural isomers of Polyacene (PAc)

respectively and have concluded that these reactions are favourable from the energetics point of view. There have also been many theoretical investigations[7-13] of the electronic structure and conduction properties of the members of 1-D graphite family. PAc is the most studied polymer of this family followed by PPN, PPh and PPhP respectively. There are three possible isomers of PAc (Fig.2) and the question of its most stable isomer is still unresolved. In this article we report, for the first time, the results of our investigations of the electronic structure and conduction properties of polyacene on the basis of ab initio Hartree Fock crystal orbital method.

METHODOLOGY

The band structure calculations of the different isomers of PAc were performed using their optimised geometry[8] in the framework of the ab initio Hartree Fock crystal orbital method[14-15]. In this method one solves the pseudo eigenvalue problem

$$\underline{F}(k)\,C_n(k) = \epsilon_n(k)\,\underline{S}(k)\,C_n(k) \tag{1}$$

for different k values in the Brillouin zone. $\underline{F}(k)$ and $\underline{S}(k)$ are the Fock and Overlap matrices in the k representation. The index n denotes the band while N is the number of neighboring cells explicitly taken into account. The Fock and Overlap matrix elements are given by

$$F_{ab}(k) = \sum_{j=-N}^{N} \exp(ikR_j)\, f_{ab}^{oj} \tag{2}$$

$$S_{ab}(k) = \sum_{j=-N}^{N} \exp(ikR_j)\, s_{ab}^{oj} \tag{3}$$

where

$$s_{ab}^{oj} = <\, x_a^o(r) \mid x_b^j(r)\, > \tag{4}$$

$$f_{ab}^{oj} = h_{ab}^{oj} + g_{ab}^{oj} \tag{5}$$

In eq.(5), the one electron term (h_{ab}^{oj}) and the two electron term g_{ab}^{oj} are given by

$$h_{ab}^{oj} = -1/2 <\, x_a^o(r)\mid\triangle\mid x_b^j(r)> + \sum_h \sum_\alpha <\, x_a^o(r)\mid Z\,/\mid r-R_h-R_\alpha\mid\mid x_b^j(r)> \tag{6}$$

and

$$g_{ab}^{oj} = \sum_{h,l=-N}^{N} \sum_{c,d} P_{cd}^{hl}\left[2\begin{pmatrix} o\ j & \mid & h\ l \\ a\ b & \mid & c\ d \end{pmatrix} - \begin{pmatrix} o\ h & \mid & j\ l \\ a\ c & \mid & b\ d \end{pmatrix}\right] \tag{7}$$

respectively with

$$P_{cd}^{hl} = \sum_m \sum_k \exp(-ik(R_h-R_1))\, C_{c,m}^*\cdot\, C_{d,m} \tag{8}$$

and the two electron integral is given by

$$\begin{pmatrix} o\ j & \mid & h\ l \\ a\ b & \mid & c\ d \end{pmatrix} = \int dr_1 \int dr_2 \cdot\, x_a^o(r_1)\, x_b^j(r_1)\,\frac{1}{\mid r_1-r_2\mid}\, x_c^h(r_2)\, x_d^l(r_2) \tag{9}$$

x_b^j is the bth atomic basis function in the unit cell denoted by the upper index j; Z and R are the charge and position of the th nucleus and $C_{d,m}(k)$ are the components of the mth eigenvector of eqn.(1). The first summation in eqn.(8) runs over all doubly filled bands and the second runs over the Brillouin zone. The eigenvalues $\epsilon_n(k)$ give the band structure of the polymer.

All the computations were performed using Clementi's 7s/3p minimal basis set[16] for the heavy atoms and four primitive Gaussian functions contracted to one s function for the hydrogen atoms. All multicentre two electron integrals larger than the threshold value of 10^{-8} a.u. were calculated and the interactions upto fourth neighbours were taken into account.

RESULTS AND DISCUSSION

The most important electronic properties such as the lowest ionisation potential (corresponding to the top of the valence band), maximum electron affinity (corresponding to the bottom of the conduction band), band gap, band widths and

Table 1. Calculated Electronic Properties

	PAc1	PAc2	PAc3	trans-PA
Ionisation Potential	7.650	7.590	7.451	8.482
Electron Affinity	4.921	4.887	4.793	1.968
Valence Band Width	7.961	7.983	8.054	7.826
Conduction Band Width	8.375	8.542	8.533	9.330
Gap	2.728	2.704	2.658	6.514
Total Energy (a.u.)	-152.100992	-152.100868	-152.100843	--

Table 2. Partitioning of the total energy of the different isomers of Polyacene

	PAc1	PAc2	PAc3
Electronic energy	-514.860084	-514.863599	-514.719508
Nuclear repulsion energy	362.759092	362.762731	362.618665
Total Energy (a.u.)	-152.100992	-152.100868	-152.100843

the total energy per unit cell of the three isomers of PAc are given in Table 1. Also given in this table are the corresponding results[17] of the trans polyacetylene (PA) obtained using the same method. The predicted order of stability among the three isomers of PAc is PAc1 > PAc2 > PAc3 implying hereby that the aromatic structure PAc1 is the most stable skeleton of PAc. It, therefore, means that the Peierls transition does not take place in PAc. Baldo et al.[18] also claimed that the ground state PAc is not Peierls distorted but rather antiferromagnetic. Tanaka et al.[8], on the contrary, found that PAc should be distorted with PAc3 to be the most stable and PAc1 the least stable structures. Table 2 shows the partitioning of the total energy per unit cell into the total electronic energy and the nuclear repulsion energy for the isomers of PAc. It can be seen that the total electronic energy stabilises PAc1 the most and Pac3 the least while the nuclear repulsion energy destabilises PAc2 the most and PAc3 the least. Work is in progress to confirm the above predicted order of stability using better basis sets and considering correlation effects

It can be seen from Table 1 that as one goes from trans PA to PAcl, the band gap decreases considerably implying hereby that PAc is expected to be a better intrinsic conductor of electricity than PA. This decrease in the band gap is accompanied by an increase in the electron affinity and a decrease in the ionisation potential of PAcl which means that polyacene is expected to be a much better candidate than trans PA for forming conducting materials through both oxidative (p-) and reductive (n-) doping. It needs to be noted that the calculated values of the band gap for all the systems are far too high (about four times the experimental value). This is the well known overestimation of the band gap in the Hartree Fock method. Using better basis sets and taking into account correlation effects the calculated band gap values are expected to come closer to the experimental values as has been observed earlier[19] in the case of trans-PA.

ACKNOWLEDGEMENT

The author is grateful to the Indian National Science Academy (INSA) for the award of the INSA Research Fellowship - 1990.

REFERENCES

1. S. Kivelson and O. L. Chapman, Phys. Rev.(B) 28:7236 (1983).
2. M. L. Kaplan, P. H. Schmidt, C. H. Chen and W. M. Walsh, Jr. Appl. Phys. Lett., 36:867 (1980).
3. P. H. Schmidt, D. C. Joy, M. L. Kaplan and W. L. Fieldmann, Appl. Phys. Lett., 40:93 (1982).
4. S. R. Forrest, M. L. Kaplan, P. H. Schmidt, T. Venkatesan and A. J. Lovinger, Appl. Phys. Lett., 41:708 (1982).
5. M. Ozaki, Y. Ikeda and I. Nagoya, Synth. Metals, 18:485 (1987).
6. K. Tanaka, S. Yamashita, T. K. Koike and T. Yamabe, Synth. Metals, 31:1 (1989).
7. L. Salem and H. C. Longuet-Higgins, Proc. R. Soc. (Lond.), A255:435 (1960).
8. K. Tanaka, K. Ohzeki, S. Nankai and T. Yamabe, J. Phys. Chem. Solids, 44(11):1069 (1983).
9. M. H. Whangbo, R. Hoffmann and R. B. Woodward, Proc. R. Soc. (Lond.), A366:23 (1979).
10. J. L. Bredas, R. R. Chance, R. H. Baughman and R. Silbey, J. Chem. Phys., 76:3673 (1982).
11. I. Bozovic, Phys. Rev.(B), 32(12):8136 (1985).
12. M. Kertesz, Y. S. Lee and James J. P. Stewart, Int. J. Quant. Chem., 35:305 (1989).
13. A. K. Bakhshi and J. Ladik, Synth. Metals, 30:115 (1989).
14. G. Del Re, J. Ladik and G. Biczo, Phys. Rev., 55:997 (1967).
15. J. M. André, L. Gouverneur and G. Leroy, Int. J. Quant. Chem., 1:427,451 (1967).
16. L. Gianolio, R. Pavani and E. Clementi, Gazz. Chim. Ital., 108:181 (1978).
17. A. K. Bakhshi, M. Seel and J.Ladik, Phys. Rev.(B), 35: 704 (1987).
18. M. Baldo, G. Piccito, R. Pucci and R. Tomasello, Phys. Lett., 95A:201 (1983).
19. S. Suhai, Phys. Rev.B, 27:3506 (1983).

PHTHALOCYANINE BASED ELECTRICALLY

CONDUCTING POLYMERS

S. Venkatachalam and K.V.C. Rao

Polymers and Special Chemicals Division
Vikram Sarabhai Space Centre
Thiruvananthapuram - 695022 (India)

P.T. Manoharan

Department of Chemistry
Indian Institute of Technology
Madras - 600036 (India)

INTRODUCTION

The discovery that electrical conductivity of
polyacetylene could be increased by several orders
of magnitude by doping it with various donors or acceptors to
give p-type or n-type semiconductors has stimulated a lot if
interest on electrically conducting polymers 1,2. Most of
the dopants used are corrosive in nature and majority of the
doped polymers eg. doped polyacetylene, doped polyphenylene,
doped poly pyrrole etc. have very poor stability against heat,
light, moisture and air. Hence there has been a need for new
light weight easily processable electrically conducting
polymers having excellent atmospheric stability against air,
water etc. Metallo phthalocyanines and their polymers have
extremely good thermal and environmental stability. They are
soluble in concentrated acids without decomposition 3,4. They
also possess extended conjugated structures which could reduce
the semiconducting band gap which governs the intrinsic
electrical and optical properties of the system.

Polymers possessing benzenoid and quinonoid segments in
a polythiophene geometry are predicted to have a decrease of
band gap with increase of quinonoid character 5,6. Theoretical
calculations on polymers possessing biphenylene type of structures as shown below

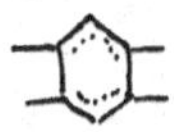

indicate that they have low ionisation potential, small band
gap and wide band width 7 and hence would form environmentally
stable materials of large conductivity. Polymers of metallo
phthalocyanines containing carboxyl terminal groups undergo
decarboxylative polymerisation reactions on thermal treatment
leading to highly conducting conjugated net work structures in

Frontiers of Polymer Research, Edited by P.N. Prasad and
J.K. Nigam, Plenum Press, New York, 1991

Table 1. Elemental analysis for the heated phthalo-
          cyanine polymers, Polymers I B and II B

Polymer with the molecular formulae for the repeat unit	Analytical values (%)			
	C	H	N	metal
Polymer I B $(C_{104}H_{24}N_{32}Cu_4)$ Calcd*	58.05	1.20	20.83	11.7
Obsd.	62.60	0.86	19.44	11.9
Polymer II B $(C_{104}H_{24}N_{32}Co_4)$ Calcd*	58.07	1.11	20.84	10.96
Obsd.	60.01	0.93	21.06	11.10

* all the values are calculated taking into account of 9%
absorbed oxygen in   all the samples.

which phthalocyanine units are joined through biphenylene
moieties[8,9]. In this paper we report some of the aspects of the
physical, chemical and physico chemical investigations on
the electrical and magnetic properties of the above type of
polymers.

EXPERIMENTAL

     The metallo phthalocyanine oligomers of the structure
shown in Fig.1 were prepared by the method reported earlier [8,9]
These polymers are referred to as Polymers I A and II A which
refer to polymeric phthalocyanines of Cu and Co respectively.
They are thermally treated at 753 K in inert atmosphere or in

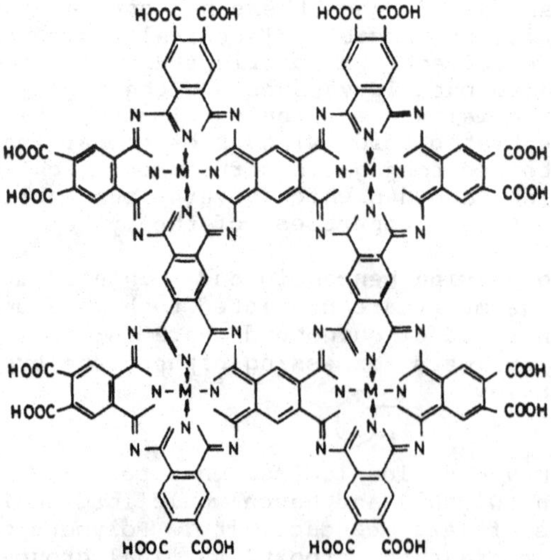

Fig. 1 Structure of metallo phthalocyanine polymers
       (Polymers I-II A) M = Ni or Co

Table 2. Electrical conductivity e.s.r line width and
magnetic susceptibility of Polymers I-II A and
and I-II B at 298 K

Polymer	Electrical Conductivity (S/cm)	Peak to Peak line width (mT)	Magnetic susceptibility (emu/g)
Polymer I A	9.8×10^{-6}	7.0	5.00×10^{-6}
Polymer I B	8.0×10^{-1}	15.0	1.80×10^{-5}
Polymer II A	2.0×10^{-8}	90.0	4.6×10^{-5}
Polymer II B	8.0	No signal	8.82×10^{-5}

vacuum for 1h. by the method reported[8,9]. The corresponding
heated polymers are referred to as Polymers I-II B where I B
and II B contain Cu and Co respectively as the central metal
atom. The heated polymers are characterised by i.r., ESCA and
elemental analyses. The elemental analyses values are in Table 1.
The electrical conductivity, e.s.r and magnetic susceptibility
measurements are carried out as reported [8,9].

RESULTS AND DISCUSSION

The heat treatment of polymeric phthalocyanines
results in formation of large extended conjugated structures
and hence the Polymers I-II B exhibit large conductivity as

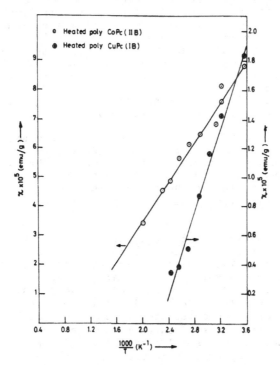

Fig. 2 Plot of Magnetic suseptability vs 1/T of Polymers
I-II B

433

shown in Table 2. The peak to peak line width of the e.s.r
signals and magnetic susceptibility of the heated and unheated
polymers are also given in the Table. There is a large increase
in these properties for the heated polymers which is due to
formation of more number of charge carriers and more delocalised
structures that promote charge carrier mobilities [10] in the
heated polymers than in the unheated ones.

The room temperature electrical conductivity of the
polymers vary systematically from a low value of around 10^{-8}
S/cm to values closer to metallic region 10^{1} S/cm as the
period of heat treatment is increased from about 30 min. to
120 min. The conductivity increase as the weight loss during
heat treatment also increases. The increase of electrical
conductivity with increase of the period of heat treatment
could be explained through the progressive increase
of conjugation length which causes orbital delocalisation
owing to formation of polycondensed ring structures through
the decarboxylative polymerisation reactions. These delocalised π orbitals permit faster transmission of an electrical
disturbance or carriers (electrons or holes) along the backbone
of the molecule, relative to interchain hopping. Orbital
delocalisation, besides increasing the polarizability, mobility
and the thermodynamic stability of the conjugated system, will
also bring about a decrease in the excitation energy required
to promote τ -electrons to triplet and other excited states
(i.e.) decreases the bandgap, eg. of the E_g of the system 11.
As the heating is continued, more and more free radicals are
formed causing an increase in the electrical conductivity.
These free radicals interact with highly extended aromatic ring
systems resulting in formation of dipolar charge carriers of
very high mobility[12-13]. A similar mechanism has been proposed
for the increase in carrier concentration of polymeric copper
phthalocyanines by Epstein and Wildi [14]. The formation of
mobile radical centers which interact with the overlapping
orbitals of the phthalocyanine ring system joined through
biphenylene type of network structures will also be possible
in polymers I-II B of the present study. This has been proved
by e.s.r and magnetic susceptibility measurements.

The static magnetic susceptibility, Χ measurements of
Polymers I-II B indicate large paramagnetic values. These
values decrease with increase of temperature. The plot of
vs 1/T for Polymer I B and II B are given Fig.2. Similar
behaviour was observed for the heated nickelpolymer[8]. The
magnetic susceptibility of all these polymers show Curie-like
behaviour.

The e.s.r signals in these polymers arise due to the
presence of free radical defects or charge carriers in the
polymeric systems. The radical defects probably behave in a
similar manner as the solitons or polarons that are formed in
doped and undoped conjugated polymers such as polyacetylenes
and polyphenylenes[15,16]. In phthalocyanine polymers one could
easily expect formation of soliton type excitations[8,9],because
they possess mesomeric structures resembling the Su, Schreiffer
and Heeger model for trans-polyacetylenes[17]. Therefore,
the electronic transport in both the heated and unheated polymeric phthalocyanines could be explained via hopping of the
soliton bound states from one chain to another. All these

polymers could be treated as lightly doped ones and hence their electrical conductivities could also be explained via bipolarons and polarons[15,18]. From the Table 2, it is observed that the thermally treated polymers show larger linewidths than the unheated ones. In the case of Polymer I A, the sample heated for 30 min. has linewidth of 9 mT and the sample heated for 1 h (Polymer I B) has a linewidth of 15 mT. This behaviour of viz. increase of e.s.r linewidth followed by increase of electrical conductivity resembles the behaviour of thermally treated alkali metal doped polyacetylenes[19,20].

Generally, the linewidths of conjugated polymers are influenced by the presence of three types of spins viz., fixed or localised, mobile and conducting spins[20]. The interactions among these spins control the observed linewidth. The increase in the linewidth could be either due to the increasing influence of conduction electrons through their relaxation rate or due to the contribution to the linewidth from the hyperfine coupling with the nuclei (related to the change in lattice parameters), in this case different metal centres. In metallophthalocyanine polymers the spins centered on the transition metal atoms could cause an increase in the relaxation rate for the interaction among the various types of spins. In the case of Polymer I B, in which the lone pair of electrons on the copper atoms (d^9) is more delocalised through-out the ligand and hence these lone pair of electrons interact with the various spins in the highly extended conjugated system containing the phthalocyanine rings, causing an increase in the relaxation rate and reduces the linewidth.

In the case of Polymer II B, if the central cobalt atom existed as $Co^{+2}(d^7)$, the influence of the lone pair of electrons on the relaxation rate should have reduced the linewidth. But on the other hand, e.s.r signal of Polymer II B is so infinitely broad that no signal could be observed even at 77 K. These observations probably indicate that during heat treatment, could be formation of diamagnetic d^6 Co (III) species, with metal-metal overlap, which causes large increase in conductivity. Martinson et al[22] have observed metal like conductivity through cobalt-cobalt overlaps via formation of diamagnetic d^6 Co (III) species in iodine doped cobalt phthalocyanine complexes

The decrease in intensities of e.s.r signal (i.e.) decrease of e.s.r amplitude with constant linewidth, with increase of temperature from 173 to 473 K for both heated and unheated polymers is indicative of the decrease of unpaired spin concentration occuring due to loss of adsorbed oxygen and also due to mixed valent interactions of radical defects with in the polymer network, leading to spinless dipolar charge carriers[8,9,13].

CONCLUSIONS

Polymeric metallophthalocyanines of Cu and Co containing peripheral carboxyl groups have been prepared. Heat treatment of these polymers results in highly conjugated network structures in which phthalocyanine moieties are joined through biphenylene type structures. These structures give rise to 5 to 8 orders of increase in electrical conductivities, broad e.s.r signals and large paramagnetic susceptibilities. In copper polymer, the linewidth remains constant with temperature. The heated cobalt polymer does not show any e.s.r signal which is due to formation of diamagnetic d^6 Co (III) species. The latter polymer exhibits larger electrical conductivity than the former one. The larger electrical conductivity and e.s.r behaviour of these polymers explained by the hopping of soliton bound states from one chain to another. The Curie-like magnetic susceptibility behaviour is explained by the formation of spinless dipolar charge carriers.

REFERENCES

1. R.H.Baughman, in E.J. Vandenberg (ed), "Contemporary Topic in Polymer Science", Vol.5, Plenum Publishing Corporation, (1984) p.321.
2. T.A. Skotheim (ed) "Hand book of Conducting Polymers", Vols.1&2 Marcel Dekker, New York (1986).
3. J. E. Katon (ed.) "Organic Semiconducting Polymers", Marcel Dekker , New York (1968).
4. D.Wohrle.Adv.Polym.Sci., 50:105 (1983)
5. S.A.Jenekhe, Nature, 322:345 (1986).
6. J.L. Bredas,Mol.Cryst.Fig.Cryst. 118:49 (1985)
7. J.L. Bredas and R.H. Baughman, J.Polym.Sci., Polym.Lett.Ed., 21:475 (1983).
8. S. Venkatachalam, K.V.C. Rao, and P.T. Manoharan, Synth. Met. 26:237 (1988).
9. S. Venkatachalam, K.V.C. Rao and P.T. Manoharan, Synth. Met. (communicated).
10. C.B. Duke and H.W. Gibson in Kirk Oathmer (ed) "Encyclopedia of Chemical Technology", Jown Wiley and sons, (1982)p.767.
11. S. Mrozowaki, Phys.Rev. , 85:609(1952).
12. F. Guttman and L.E. Lyons, "Organic Semiconductors", Wiley New York, (1967).
13. L.Sale, and C.Rowland Angew, Chem.Int.Ed. 11:92 (1972).
14. A. Epstein and B.S. Wildi J.Chem.Phys.32:324 (1964).
15. R.R. Chance, D.S. Boudreaux, J.L. Bredas and R. Silbey in Ref.2 p.825 and references therein.
16. J.L. Bredas, in Ref.2. p.859 and references therein.
17. W.P. Su.J.R.Schrieffer and A.J.Heeger,Phys.Rev.Lett., 42: 1978(1979); Phy.Rev., B 22:2099 (1980).
18. R.R. Chance, D.S.Boudreaux, H.Eckhardt, R.L. Elsebaumer, J.E Frommer, J.L. Bredas and R.Silbey in Ref.7, p.221.
19. P.Dellannoy, G.G. Miller, N.S. Murthy, C.E. Forbes., H. Eckhardt. R.L. Elsenbaumer and R.H. Baughman, Bull.Am. Phys.Soc.,28: 320 (1983).
20. Patric Bernier, in Ref.2 p.1116 and references therein.
21. S.E. Harrison and Assour.J.Chem.Phys., 400:365 (1964) Phy.Rev.A. 136:1368(1964).

DIELECTRIC AND PIEZOELECTRIC PROPERTIES OF A FLEXIBLE POLYMER-CERAMIC COMPOSITE BASED ON POLYVINYLIDENE FLUORIDE AND BARIUM TITANATE

R.P. Tandon

Division of Materials
National Physical Laboratory
New Delhi, India

INTRODUCTION

The traditional lead zirconate titanate (PZT) based ceramics have been used extensively as a piezoelectric transducer material but these suffer from many disadvantages when used as a sensor in hydrostatic applications. The hydrostatic coefficient d_h ($d_h = d_{33} + 2d_{31}$) of the PZT ceramics is small inspite of large d_{33} values. This is due to the fact that d_{33} and d_{31} are of opposite in sign.[1] Similarly, g_{33} ($g_{33} = d_{33}/\epsilon$) and g_h ($g_h = d_h/\epsilon$) are small because of high permittivity values.

In order to improve the values of hydrostatic coefficients and to get rid of the inherent brittleness of the ceramics, a number of composite systems have been fabricated utilizing polymer and ceramics. Such heterostructures have attracted great attention during the recent years because of the performance limitation of the single phase materials. Early attempts to make composites of piezoelectric ceramics and polymers were made by Pauer[2]. Harrison[3] reported the dielectric and piezoelectric constants of the composites based on PZT and silicone rubber. These studies clearly demonstrate that in these composite systems the electric flux pattern, the mechanical stress distribution and the resulting piezoelectric and dielectric properties depend strongly on the manner in which the individual phases are interconnected. By the proper choice of connectivity pattern it has been possible to develop composites in which g_h and d_h coefficients are atleast an order of magnitude higher[4]. These systems also have high mechanical compliance and low density making it easier to achieve good impedancee matching with water.

It has been well established that the connectivity[5] plays a dominant role in designing these composites as it will govern the ultimate properties.

The present investigation deals with the studies carried out on polyvinylidene fluoride barium titanate (PVDF-BaTiO$_3$) composites having 0-3 type of connectivity in which BaTiO$_3$ ceramic is uniformly dispersed

Frontiers of Polymer Research, Edited by P.N. Prasad and
J.K. Nigam, Plenum Press, New York, 1991

in polymer matrix. The permittivity of this system has been explained on the basis of the model proposed by Yamada et al. The effect of conditions and degree of poling on d_{33} parameter have also been examined.

EXPERIMENTAL

The barium titanate ($BaTiO_3$) ceramic was synthesized using conventional solid state reaction by pre-reacting barium carbonate and titanium dioxide at elevated temperatures (1100-1200°C). The X-ray diffraction pattern of the powder confirms the presence of perovskite phase. A few discs of $BaTiO_3$ were pressed and sintered at about 1350°C for 2 hours. These discs were crushed to yield sintered powder of $BaTiO_3$ having particle size ($\sim 10\mu$m) to be used for the fabrication of composites. PVDF granules were dissolved in dimethyl formamide to which sintered $BaTiO_3$ powder was added in various proportions. The resulting thick slurry was homogenized and poured on to a glass substrate and heated in a dust free oven at 115°C for 4 hours to get a smooth and translucent films of PVDF-$BaTiO_3$ composites. The film sample chosen for the present studies had a thickness of about 100 microns.

Silver films were vacuum deposited on both sides of these composites to make an Ag-Composite-Ag structure. The poling was carried out using high d.c. fields at slightly raised temperatures. The dielectric parameters of the omposites were measured by using a HP 4192A LF Impedance Analyzer and d_{33} was evaluated by using a Berlincourt's d_{33} meter.

RESULTS AND DISCUSSION

The variation of measured dielectric constant (ε') as a function of PVDF contents is shown in Fig. 1. The dielectric constant drops with the increase in PVDF contents which is due to the fact that the ε' of $BaTiO_3$ is much higher than the PVDF polymer matrix.

The measured dielectric constant (ε') and loss (ε'') as a function of frequency in the range 10Hz – 10MHz for 60/40 PVDF-$BaTiO_3$ composite are shown in Fig. 2. It is evident from this figure that the dielectric constant has larger values in the low frequency region and tends to saturate in the high frequency region. The dielectric loss (ε'') shows a broad peak around 10kHz and the observed increase in the low frequency region may be due to the presence of space charge polarization.

According to Yamada et al[6] the dielectric constant is given by the following relation :

$$\varepsilon' = \varepsilon_1 \left[1 + \frac{nq (\varepsilon_2 - \varepsilon_1)}{n\varepsilon_1 + (\varepsilon_2 - \varepsilon_1) (1 - q)} \right]$$

where n and g are attributed to the shape and volume fraction of the ellipsoidal particles, ε_1 is the dielectric constant of continuous

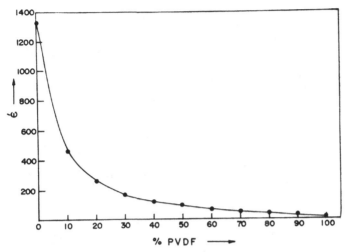

Fig. 1. Effect of PVDF contents on the dielectric constant (ϵ') of the composite measured at 1 kHz.

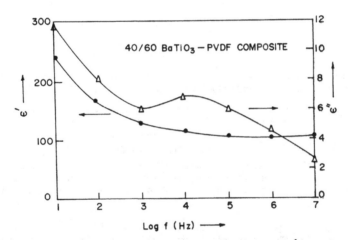

Fig. 2. Variation of dielectric constant (ϵ') and dielectric loss (ϵ'') as a function of frequency.

medium (polymer matrix) and ϵ_2 is the dielectric constant of ellipsoidal particles (BaTiO$_3$ powder). By taking the value of ϵ_1 = 10(PVDF) and ϵ_2 = 1320 (BaTiO$_3$) and assuming the value of n to be 8 to fit the equation, the calculated values of ϵ' comes to be 81.3 (for a 50/50 composite system) which is in good agreement with the

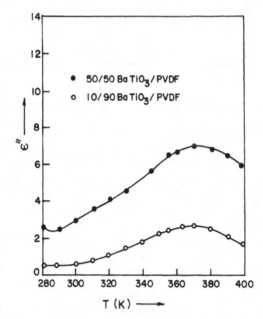

Fig. 3.   Variation of dielectric loss (ε'') as
a function of temperature.

experimentally observed value.   The dielectric loss (ε'') as a function
of temperature in the range 280–400K is shown in Fig. 3 for PVDF/BaTiO$_3$
composites. It may be pointed out here that the results of this figure have
been reported for the measurement done at a frequency   of 1 kHz for this
composite having two compositions 50/50 and 90/10.   It is evident from
this figure that ε'' is higher for higher BaTiO$_3$ content and a
relaxation in the temperature region 345–365°K is clearly visible for
all these compositions which may be associated with the molecular
motions of the main chain in the crystalline region of the polymer
matrix.

The effect of poling was also investigated in terms of variation
of d$_{33}$ (Fig. 4).   The lower most curve corresponds to a poling time of
10 minutes.   It can be seen from the figure that the value of d$_{33}$
increases as the time of poling is increased.   The fact that the d$_{33}$
parameter is still increasing points out that the poling is still
incomplete.   One may have to go in for corona discharge to achieve
better poling results.

CONCLUSIONS

The   dielectric   and   piezoelectric   properties   of   PVDF–BaTiO$_3$
Composites have been evaluated as a function of the relative contents
of the two constituents.   The poling time and poling field are found to

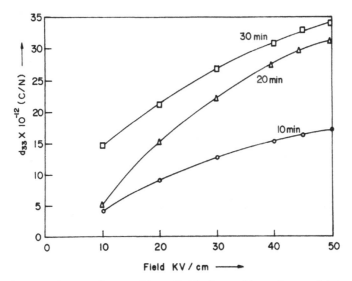

Fig. 4. d_{33} against poling field for poling times of 10, 20 and 30 minutes.

have great influence on the piezoelectric activity.   The conventional poling treatment is found to be inadequate in imparting piezoelectricity to these composites and further refinements are needed.

REFERENCES

1. Robert V. Ting, Evaluation of New Piezoelectric Composite Materials for Hydrophone Applications, Ferroelectric, 67: 143 (1986).
2. L.A. Pauer, Flexible Piezoelectric Materials,   IEEE Intl. Conf. Res., 1 (1973).
3. W.B. Harrison, Flexible Piezoelectric Organic Composites, Proceedings of the Workshop on Sonar Transducer Materials, Office of Naval Research, Arlington, V.A. 257 (1974).
4. R.E. Newnham, A. Safari, J. Giniewicz and B.H. Fox, Composite Piezoelectric Sensors, Ferroelectric, 60:15 (1984).
5. R.E. Newnham, D.P. Skinner and L.E. Cross, Connectivity and Piezoelectric-Pyroelectric Composites, Mat. Res. Bull., 13:525 (1978).
6. T. Yamada, T. Ueda and T. Kitayama, Piezoelectricity of a high-content lead zirconate titanate/polymer composite, J. Appl. Phys., 53:4328 (1982).

ROOM TEMPERATURE MAGNETIC ORDERING IN $FeCl_4^-$ DOPED COPOLYMERS : A NEW OBSERVATION

S.R. Vadera and N. Kumar

Defence Laboratory, Jodhpur - 342 001, India

ABSTRACT

Four new iron chloride doped conducting polymers/copolymers of pyrrole and thiophene have been prepared by electrochemical method. The Mossbauer studies of these samples reveal the presence of iron both in +2 and +3 oxidation states. Two of the samples, viz., iron chloride doped poly (pyrrole+phenylene oxide) and poly (thiophene + 3,methyl thiophene), are the first examples of conducting polymers showing magnetic ordering at room temperature.

INTRODUCTION

Since the first report on iron doped polyacetylene and poly paraphenylene by Pron et al[1], in 1981, there have been consistent efforts to make a variety of metal halide doped conducting polymers.[2-6]

The Mossbauer studies of iron chloride doped polyacetylene by Sichel et al[2] have revealed some very interesting results, i.e., (i) transfer of electron from polyacetylene to dopant results in change of some of the Fe^{3+} ions into Fe^{2+} ions in the lattice, (ii) Fe^{2+} ions form aggregates above certain concentration of the dopant ions and (iii) these Fe^{2+} aggregates give rise to a magnetic ordering below 25K. Subsequent report[6] on Fe^{2+} complexed poly(n-propyl-pyrrole- sulphonate) has shown an anisotropic magnetic behaviour only under high external magnetic field at 4K.

The present work deals with the synthesis and Mossbauer studies of four new iron chloride doped conducting polymers/copolymers derived from

Frontiers of Polymer Research, Edited by P.N. Prasad and J.K. Nigam, Plenum Press, New York, 1991

(i) pyrrole, (ii) pyrrole + phenylene oxide, (iii) thiophene and (iv) thiophene+3,methyl thiophene. The results are very interesting, particularly for iron chloride doped copolymers of pyrrole+phenylene oxide and thiophene+3,methyl thiophene, since they are the first conducting polymers to show magnetic ordering at room temperature.

EXPERIMENTAL

The polymers with simultaneous doping of iron chloride were obtained by a standard electrochemical method.[7]

In a typical case, for example, iron chloride doped conducting polymer of pyrrole and phenylene oxide was prepared according the following procedure :

An electrochemical reaction cell constructed by taking indium tin oxide coated glass plate and a platinum foil (2.5×2.5 cm^2) as the anode and cathode respectively. The solution phase of the cell was consisted of pyrrole and phenol in a predetermined molar ratio, tetraethyl ammonium iron tetrachloride (1g;3 m mole) and acetonitrile (50 ml). The electropolymerization of phenol and pyrrole mixture was carried out at a current density of 10 mA/cm^2, at room temperature. The polymer is obtained as black film on the anode.

The other polymers were prepared similarly by taking appropriate quantities of the respective reactants.

The C,H analyses of the samples were carried at CDRI, Lucknow using Carlo Erba Model 1106 Elemental Analyzer. Perkin-Elmer Model 2380 Atomic Absorption Spectrophotometer was used for estimation of iron present in the samples. The electrical conductivity of the samples was measured using standard four point resistivity probe technique. Mossbauer studies were carried out in a standard transmission geometry using Wiessel driving system and Canberra series 35 plus MCA. The lines were assumed to be Lorentzian and fitted by the method of least squares using computer program.[8]

RESULTS AND DISCUSSION

The electrical conductivity of iron chloride doped polythiophene is the highest 2 ohm^{-1}cm^{-1} whereas it is two orders of magnitude lower in

1. Two of the polymers, viz., iron chloride doped poly pyrrole + phenylene oxide and poly thiophene + 3,methyl thiophene show magnetic ordering even at room temperature (300K) as shown in Fig. 1. The effective magnetic field values are 361 kOe and 309 kOe, respectively.

2. In one of the samples, i.e., iron chloride doped poly pyrrole + phenylene oxide, the value of effective magnetic field increases to 413 kOe at 77K whereas the six line pattern corresponding to magnetic ordering, disappears at 353K.

3. In all the four samples iron is present both in +2 and +3 oxidation states.

4. In two of the polymers, viz., iron chloride doped poly pyrrole and poly pyrrole+phenylene oxide, two different sites for Fe^{3+} are present. At one of the sites showing Fe^{3+} is in regular tetrahedral symmetry while at the other it is in a highly distorted T_d symmetry as evident from the Mossbauer parameters (Table - 1).

There are only few other reports on Mossbauer studies of iron chloride doped conducting polymers with none of them showing magnetic ordering at room temperature. However, Sichel et al[2] and Kucharski et al[4] have reported magnetic ordering in iron chloride doped polyacetylene only below 25K and was correlated with the aggregate formation by Fe^{2+} ions in the lattice in the form of $FeCl_2 \cdot nH_2O$ molecules.[9] The occurence of magnetic ordering in two copolymers at room temperature, in the present report, makes an interesting study as aggregate formation by Fe^{2+} ions in the form $FeCl_2 \cdot nH_2O$ is no longer true. Though at this stage it is rather difficult to explain the true cause of room temperature magnetic ordering in these copolymers some correlation may however be made with the c-1 model[10] which is applicable for magnetic semiconductors.

The c-1 model assumes that the crystal contains magentic ions with non-zero spins of d- or f-shells called "l-spins" and an exchange interaction takes place between the localized l-spins of these ions thus establishing magnetic ordering of some type in crystal as described by

Table. 1

Results of Mossbauer Studies on Iron Chloride Doped Conducting Polymers

Sample	Temp	I.S.* (mm/s)	Q.S. (mm/s)	H_{eff} (kOe)	Remarks
Poly pyrrole	RT	1.08	2.28	–	Fe^{2+}
		0.40	0.77	–	Fe^{3+}
		0.30	0.00	–	Fe^{3+}
Poly pyrrole+	RT	1.11	2.52	–	Fe^{2+}
phenylene		0.39	0.55	–	Fe^{3+}
oxide		0.11	0.00	–	Fe^{3+}
				361	Mag
	77K	1.13	2.00	–	Fe^{2+}
		0.36	0.00	–	Fe^{3+}
				413	Mag.
	353K	0.97	1.98	–	Fe^{2+}
		0.32	0.68	–	Fe^{3+}
Poly thiophene	RT	1.15	2.39	–	Fe^{2+}
		0.35	0.69	–	Fe^{3+}
Poly thiophene+	RT	1.21	2.50	–	Fe^{2+}
3-methyl		0.30	0.70	–	Fe^{3+}
thiophene				309	Mag.

* The isomer shift values are with respect to natural iron.

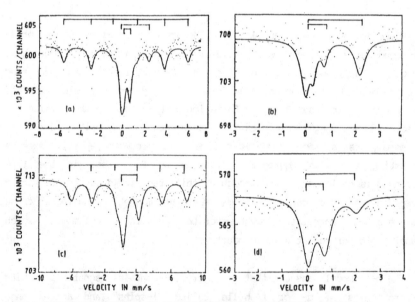

FIG. 1. MOSSBAUER SPECTRUM OF IRONCHLORIDE DOPED CONDUCTING POLYMER OF (a) PYRROLE + PHENYLENEOXIDE AT ROOM TEMPERATURE (b) PYRROLE AT ROOM TEMPERATURE (c) PYRROLE + PHENYLENEOXIDE AT 77K AND (d) PYRROLE + PHENYLENEOXIDE AT 353K.

Heisenberg Hamiltonian. Furthermore, crystal also contains delocalized conduction electron and holes(c-electrons) which in turn interact with l-spins.

Application of the above model, in order to explain the magnetic ordering in two of the polymers may be given in terms of some exchange interaction between the spins of Fe^{2+} and Fe^{3+} ions. It may be recalled that as room temperature magnetic ordering has taken place only in the copolymeric samples, the nature of polymeric backbone is quite important to regulate such a kind of interaction.

REFERENCES

1. A. Pron, I. Kulszewicz, D. Billaud and J. Przyluski, J. Chem. Soc., Chem. Commun., 783 (1981).

2. E.K. Sichel, M.F. Rubner, J. Georger Jr., G.C. Papaefthymoiu, S. Ofer and R.B. Frankel, Phys. Rev. B28, 6589 (1983).

3. A. Pron, M. Zagorska, Z. Kucharski, M. Lukasiak and J. Suwalski, Mater. Res. Bull. 17, 1505 (1982).

4. Z. Kucharski, M. Lukasiak, J. Suwalaski and A. Pron, J de Phys. 44, C3-321 (1983).

5. A. Pron, Z. Kucharski, C. Budrowski, M. Zagorska, S. Krichene, J. Suwalski, G. Dehe and S. Lefrant, J. Chem. Phys. 83,5923 (1985).

6. P. Auric and G. Bidan, J. Polym. Sci.:Part B 25, 2239 (1987).

7. N. Kumar, B.D. Malhotra and S. Chandra, J. Polym. Sci. (Letter Ed.) 23, 57 (1985).

8. E. von Meerwall, Comp. Phys. Commun. 9, 117 (1975).

9. S. Chandra and G.R. Hoy, Phys. Lett. 22, 254 (1966).

10. E.L. Nagaev, Physics of Magnetic Semiconductors, Mir Publishers, Moscow, (1983) p.9.

A MODEL FOR AC CONDUCTION OF CONDUCTING POLYMERS

Ramadhar Singh, R.P. Tandon, V.S. Panwar and Subhas
Chandra

National Physical Laboratory
Dr. K.S. Krishnan Road
New Delhi 110012, India

INTRODUCTION

Conducting polymers have been a subject of extensive investigations[1-3] but little has been done to understand the mechanism of conduction and dielectric relaxation behaviour perhaps due to the fact that the main interest was to enhance the conductivity by doping. Polypyrrole (PPY) has been investigated by several workers[1-3] for its electrical and optical properties. It has been argued that soliton antisoliton pairs in the form of polarons and bipolarons[3,4] can be stable in polymeric conductors with slightly nondegenerate ground states. Kivelson[5] has applied the intersoliton hopping model to explain the electrical conductivity of trans polyacetylene (trans-PA). However, several doubts have been raised[4] on the procedure adopted by Kivelson[5] in formulating his model. It has been concluded[4] that none of the existing models of ac and dc conduction are satisfactory for lightly doped trans-PA. The present investigation is an attempt to provide the exhaustive conductivity data and to propose a model based on polaronic hopping conduction[1,6]. The observed frequency and temperature dependence of ac conductivity of PPY can be explained satisfactorily by the proposed model which considers the contribution from two mechanisms one giving a linear dependence of conductivity on frequency and the other having distribution of relaxation times giving rise to a broad dielectric loss peak.

EXPERIMENTAL

Samples of PPY were prepared by electrochemical polymerization and the details have been described elsewhere[1,2]. The gold electrodes were vacuum deposited on both sides of the samples. The ac conductivity and dielectric constant were measured using an HP 4192A LF Impedance Analyzer in a three terminal cell. The samples of PPY (thickness ~ 15 μm) were electrochemically reduced to produce lightly doped PPY films having room temperature conductivity~$2.62 \times 10^{-4} \Omega^{-1} cm^{-1}$.

The dc conductivity was measured using a Keithley 610C electrometer.

RESULTS AND DISCUSSION

The measured ac and dc conductivities in the temperature range 77–350K as a function of reciprocal temperature are shown in Fig. 1. At low temperatures, a frequency dependent conductivity is observed. However, at higher temperatures, the measured ac conductivity $\sigma(\omega)_m$ is observed to be both temperature and frequency dependent and ultimately it approaches the dc conductivity σ_{dc}. It is observed that a peak in ε'' or $\sigma(\omega)$ defined by[6]

$$\omega \xi_o \, \varepsilon'' = \sigma(\omega)_m - \sigma_{dc} = \sigma(\omega) \tag{1}$$

would be observed at each frequency just below the temperature where $\sigma(\omega)_m$ becomes equal to σ_{dc} and these peaks will shift to higher temperature for higher frequencies. However, these loss peaks cannot be taken as a conclusive evidence of the existence of dielectric loss peaks because the temperature dependence of σ_{dc} is much larger than $\sigma(\omega)$ and so at some temperatures σ_{dc} may become larger than $\sigma(\omega)$ and therefore $\sigma(\omega)_m$ would appear equal to σ_{dc} within the accuracy of measurement.

The frequency dependence of $\sigma(\omega)_m$ is shown in Fig. 2 wherein the low temperature region it is defined by the empirical relation: $\sigma(\omega)=A\omega^s$ where $s = 0.56 < s < 1$ is independent of frequency. At 77K, where $\sigma(\omega) \gg \sigma_{dc}$, the variation of $\sigma(\omega)$ with frequency can be expressed in terms of above equation where parameter $s = 0.56$ and it is observed that this value decreases with the increase in temperature.

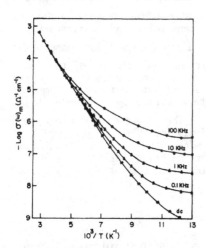

Fig. 1. Variation of total measured ac conductivity $[\sigma(\omega)_m]$ and dc conductivity as a function of reciprocal temperature.

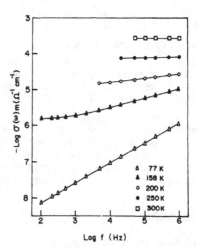

Fig. 2. Variation of total measured ac conductivity $[\sigma(\omega)_m]$ as a function of frequency at different fixed temperatures.

An estimate of the density of states near the Fermi level $N(E_f)$ can be possible in this low temperature range by using the expression derived by Austin and Mott[7]

$$\sigma(\omega) = \pi/3 \ e^2 kT \ [N(E_f)]^2 \ a^{-5} \ \omega \ [\ln(\nu_o/\omega)]^4 \qquad (2)$$

where k is Boltzmann's constant, e is the electronic charge, $\omega = 2\pi f$, ν_o is a frequency factor and a is the radius of the carrier wave function. Assuming $\nu_o = 10^{13}$ Hz and $a = 0.4\overset{o}{A}$ which roughly corresponds to nearest neighbour hopping, the calculated value of density of states (after taking the value of $\sigma(\omega)$ at 77K and 100 kHz) is of the order of $3.2 \times 10^{18} cm^{-3} eV^{-1}$.

The dielectric constant ε' according to Debye relation is given as

$$\frac{(\varepsilon' - \varepsilon_\infty)}{(\varepsilon_o - \varepsilon_\infty)} = \frac{1}{1 + (f/f_o)^2} \qquad (3)$$

where ε_o and ε_∞ are the static and infinite frequency dielectric constants. f and f_o are the measuring and relaxation frequencies. The variation of ε' with temperature at four fixed frequencies is shown in Fig. 3. It is evident from this figure that in the low temperature region the change in ε' with temperature is slow but it has a sharp rise at the temperature at which $f = f_o$. If $f \gg f_o$, the measured value of ε' represents ε_∞ and if $f \ll f_o$, ε' becomes equal to ε_o. It would be observed that strong temperature dependence starts at higher temperatures for higher frequencies indicating thereby that f_o increases with the increase in temperature. It is seen that the region where there is a strong temperature dependence of ε' at a given frequency (Fig. 3) is the same at which $\sigma(\omega)_m$ approaches σ_{dc} (Fig.1).

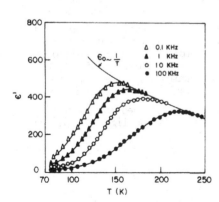

Fig. 3. Variation of dielectric constant (ε') as a function of temperature for four fixed frequencies.

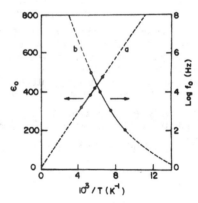

Fig. 4. Plot of (a) ε_o and (b) f_o as a function of reciprocal temperature.

Thus the variation of $\epsilon^\prime$ with temperature confirms the existence of Debye type loss peaks. The ϵ_0 and f_0 estimated from Fig. 3 have been plotted as a function of reciprocal temperature and are shown in Fig.4. The dielectric constant and conductivity in the presence of symmetric distribution of relaxation times can be given by Cole-Cole expression[1]

$$\frac{(\epsilon^\prime - \epsilon_\infty)}{(\epsilon_0 - \epsilon_\infty)} = \frac{1+(f/f_0)^{1-\alpha} \sin(\alpha\pi/2)}{1 + 2(f/f_0)^{1-\alpha} \sin(\alpha\pi/2) + (f/f_0)^{2(1-\alpha)}} \qquad (4)$$

$$\sigma(\omega) = \frac{\omega\xi_0(\epsilon_0 - \epsilon_\infty) [(f/f_0)^{1-\alpha} \cos(\alpha\pi/2)]}{1 + 2(f/f_0)^{1-\alpha} \sin(\alpha\pi/2) + (f/f_0)^{2(1-\alpha)}} \qquad (5)$$

where ξ_0 is a free space permittivity and α is an empirical parameter having the values between 0 and 1.

In the present work the measured ac conductivity arises out of two mechanisms and can be expressed as

$$\sigma(\omega) = \sigma_1(\omega) + \sigma_2(\omega)$$

$$= \int_0^\infty \frac{g^\prime(\tau)\omega^2 \tau\, d\tau}{1 + \omega^2 \tau^2} + \int_0^\infty \frac{g^{\prime\prime}(\tau)\omega^2 \tau\, d\tau}{1 + \omega^2 \tau^2} \qquad (6)$$

The mechanism responsible for $\sigma_1(\omega)$ yields a linear dependence of conductivity on frequency and the distribution function $g^\prime(\tau)$ may be of the form suggested by Austin and Mott[7] as given in Eq. (2). The other mechanism responsible for $\sigma_2(\omega)$ gives rise to dielectric loss peaks with a distribution of relaxation times. If one assumes a Cole-Cole[1] type of distribution function for $g^{\prime\prime}(\tau)$, one arrives at expression (5). The contribution of $\sigma_2(\omega)$ dominates for about 2 to 3 decades above the relaxation frequency and the contribution of $\sigma_1(\omega)$ becomes dominant only at frequencies which are higher than the relaxation frequency. It is evident from Fig. 2 that the conductivity at lower frequencies shows a stronger temperature dependence as compared to that at higher frequencies. This results in a lower value of slope s of $\sigma(\omega)$ to remain unchanged with temperature and any apparent decrease is due to the contribution of $\sigma_2(\omega)$. Then the temperature dependence of ac conductivity at different frequencies can be explained by taking into account the contribution of $\sigma_1(\omega)$ and $\sigma_2(\omega)$. In the light of Eq. (2), $\sigma_1(\omega)$ can be expressed as

$$\sigma_1(\omega) = BT\omega^s \qquad (7)$$

where parameters B and s are independent of temperature. The value of $\sigma_1(\omega)$ at different temperatures and frequencies can be calculated from Eq. (7) and are shown by a dashed line in Fig. 5. The measured values can be fully accounted for by the contribution of $\sigma_1(\omega)$ up to higher temperatures for higher frequencies. However, above a certain

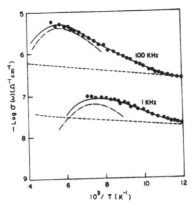

Fig. 5. Comparison between the experimental (filled circles) and the theoretical values of $\sigma_1(\omega)$ $+\sigma_2(\omega)$ predicted by the model (solid line).

temperature range the measured values at different frequencies become substantially higher than the theoretical value of $\sigma_1(\omega)$. This may be due to the fact that in this region the contribution of $\sigma_2(\omega)$ starts to dominate. The value of $\sigma_2(\omega)$ can be calculated from Eq. (5) after taking the value of ε_0, f_0^2 and ε_∞ = 10 (assumed to be temperature independent[1]) from Fig. 4. A temperature independent value of α = 0.5 has been taken for calculation of $\sigma_2(\omega)$ and is shown by dash-dot line in Fig. 5. The combined contributions of $\sigma_1(\omega) + \sigma_2(\omega)$ is shown by solid line in Fig. 5. The good agreement between theoretical and experimental values indicates that the variation of ac conductivity with temperature can be satisfactorily explained by assuming that the two mechanisms contribute to the measured ac conductivity. These two mechanisms of conduction discussed above for PPY film may be associated with hopping near the Fermi level (a linear frequency dependent conductivity) and hopping of the carriers excited to localised band tails (giving rise to dielectric loss peaks). The model discussed here is being extended to other systems such as polythiophene, polyfuran and polyparaphenylene.

REFERENCES

1. R. Singh, R.P. Tandon, V.S. Panwar and S. Chandra, Low frequency ac conduction in lightly doped polypyrrole films, J. Appl. Phys., In Press.
2. ibid, Origin of dc conduction and dielectric relaxation in lightly doped polypyrrole films, Thin Solid Films, 196:L15 (1991).
3. P. Pfluger, G. Weiser, J.C. Scott and G.B. Street, Electronic structure and transport in the organic amorphous semiconductor polypyrrole, in:"Handbook of Conducting Polymers", Marcel Dekker, New York, 2:1369 (1986).
4. K.L. Ngai and R.W. Rendell, Dielectric and conductivity relaxation in conducting polymers, in:"Handbook of Conducting Polymers", Marcel Dekker, New York, 2:967 (1986).
5. S. Kivelson, Electron hopping in a soliton band: Conduction in lightly doped (CH)$_x$, Phys. Rev. B, 25:3798 (1982).
6. N.F. Mott and E.A. Davis, Impurity bands and impurity conduction, in:"Electronic processes in non-cyrstalline mater."Clarendon Press (1979).
7. I.G. Austin and N.F. Mott, Polarons in Crystalline and non crystalline materials, Adv. Phys., 18:41 (1969).

CONDUCTING COMPOSITES

R.P. Singh, A.K. Misra*, A.L. Dutta, S. Deb and S. Sarkar

Materials Science Centre, *Dept.of Metallurgical Eng.
Indian Institute of Technology, Kharagpur
Kharagpur-721 302, India

INTRODUCTION

The conducting polymers have invoked tremendous scientific and industrial research in last decade. The most promising conducting polymers
still suffer from pronounced mechanical and environmental instability.
In many electrical and electronic applications, conducting composites
are performing various functions which are foreseen by conducting
polymers (Bhattacharya, 1986; Delmonte, 1990).

The conducting composites are mainly formed by combining metals in form
of powders, flakes and fibres with plastic powders (Bigg, 1977). The
latest development has been using metal coated glass fibres with plastic
powder to form light weight conducting composites (Crossman, 1985). Many
processing technologies have been applied to develop the conducting
composites such as hot compression, extrusion and in-situ polymerisation
in presence of metallic fillers (Bhattacharya, 1986; Nedorezova et al.,
1987). In authors' laboratory in-situ polymerisation has been adopted
along with mild shearing stimulus which may be categorised as reactive
processing. This is supposed to encourage intimate homogenous mixing and
to enhance the conductivity in the composites.

Various kinds of fillers and matrices have been used in authors' laboratory to develop conducting composites (Dutta 1988, Sarkar,1990). Both
metal powder and metal coated glass fibres have been used as fillers.
PVC, PS, PMMA as well as impact modified PP-EPDM have been used as
matrices. The various aspects of conducting composites developed in
authors' laboratory are reviewed in the present paper.

EXPERIMENTAL

In the present investigation, the conducting composites have been
developed using commercial grade metal powders, and metal coated glass
fibres as filler in various polymeric matrices. These fillers are non
spherical having irregular surfaces. The various polymer/metal composites of PE/Cu, PVC/Cu,PVC/Al, PS/Al, PVC/ Cu-Al, PVC/Aluminium coated
glass fibres (AlCGF), PS/AlCGF, PP-EPDM/AlCGF and PMMA/AlCGF have been
prepared. PVC, PS and PMMA make brittle matrices while PE and PP-EPDM
matrices are ductile. PE/Cu, PMMA/AlCGF and PP-EPDM/AlCGF composites were
prepared by melt compounding and hot processing techniques. Rest of the

composites except PMMA/AlCGF were prepared by dry blending. The dry
polymer and metal powders were tumbled followed by hot compression.

Nedorezova et al, (1987) have observed that by in-situ polymerization in
presence of catalyst coated metallic fillers can result high level of
metallic conductivity at lower percentage of fillers. In authors' laboratory reactive processing has been used alongwith in-situ polymerization
to enhance metallic conductivity in PMMA/AlCGF composites. The details of
the process are as follows.

In order to obtain a better and uniform distribution of fibres,in-situ
polymerization and mixing in presence of filler were carried out
simultaneously in Brabender mixing chamber which consists of a heated
enclosure with two counter rotating screws inside for mixing. The Chamber
is detachable. To prevent the leakage of monomers, the two faces of the
chamber were sealed with Teflon tape. The commercial monomer is first
distilled then catalyst, benzoyl peroxide, was added in 60 gm of monomer
for each batch. The fibres were mixed with slightly prepolymerised
monomer in a beaker and then the mixture was introduced in the chamber.
The rate of rotation of screws was maintained at 5 rpm. The polymerisation was carried out for roughly five minutes. During this time, composites with uniform distribution of fibres were formed. Consequently, the
rotation of the screw also stopped. Composites with varying weight
percentage of filler (viz. 6,8,10,12,15 and 18) were prepared.

DC/AC resistivity, thermal conductivity, elastic modulus, thermal
analysis and morphological studies have been done on these composites
by standard techniques.

RESULTS AND DISCUSSION

In conducting composites, reduced values of electrical resistance is
consequence of metallic particle loading,particle shape and size distribution. The loading of high aspect ratio of metal particles reduces
resistivity more as compared with lower aspect ratio particles (Bigg,1977).
The electrical resistivity of the polymer is not much affected by the
addition of metal until a critical composition (percolation threshold)
is reached where the fillers come into intimate contact with each other
leading to formation of continuous conductive network. The percolation
threshold, limiting metallic conductivity depends on the plastics/metal
particle shape, sizes and electrical properties of metal and polymeric
matrix (Devenport, 1981; Bhattacharya 1986).

In case the particle size of plastics is bigger than that of metal,
segregated network is formed. When the particle sizes of plastics and
metals are of the same order, there is random distribution of metals
particles in the polymeric matrix. In case of segregated network formation,the percolation threshold is reached at lower loading of metallic
particles. Physical mixing and hot compression favour segregated network
formation while melt compounding encourages the random distribution. The
results of electrical conductivity for various composites follow these
generalisations.

The typical results of PVC/Cu, PVC/Al and PVC/AlCGF are shown in figs.1-3.
As the nature of filler changes from particulate to metal coated fibre,
the fall in percolation threshold as well as in limiting resistivity are
obtained at lower loading of the filler. Melt blended PE/Cu composites
show random distribution inspite of large matrix/metal particles size
ratio (51.42) and the fall is not abrupt as well. There is a large
limiting resistivity. An interesting observation has been made in case
of PVC/Al composites, the combination of two metal powders lowers

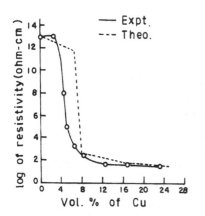

FIG. 1. LOG RESISTIVITY (D.C)
VS COMPOSITION FOR
PVC/Cu COMPOSITES

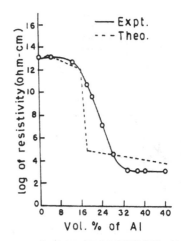

FIG. 2. LOG RESISTIVITY (D.C)
VS COMPOSITION FOR
PVC/Al COMPOSITES

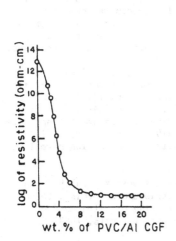

FIG. 3. LOG RESISTIVITY (D.C)
VS COMPOSITION FOR
PVC7Al COATED GF
COMPOSITES

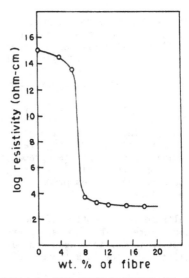

FIG. 4. LOG RESISTIVITY (D.C)
OF COMPOSITES VS
CONCENTRATION OF
AlCGF IN PMMA MATRIX

457

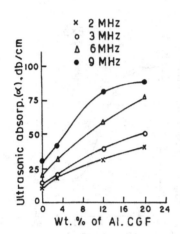

FIG. 5. ULTRASONIC ABSORPTION
VS COMPOSITION FOR
PVC/Al COATED GF
COMPOSITES AT VARIOUS
FREQUENCIES

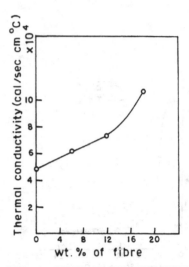

FIG. 6. THERMAL CONDUCTIVITY
OF COMPOSITES VS
CONCENTRATION OF
AlCGF IN PMMA MATRIX

lmiting resistivity more than any of the metals alone. For particulate
composites, various models have been proposed to explain theoretically
the fall of resistivity with increase in filler concentrations.
Bhattacharya's model (1982) gives the nearest fit to our data. Regarding
the mechanism of conduction due to percolation, hopping and tunnelling
(Biggs, 1986), our results also show the manifestation of these processes.

Figure 4 depicts the resistivity versus weight percentage of AlCGF in case
of PMMA/AlCGF composites prepared by in-situ polymerisation and reactive
processing. The resistivities of the composites with 6, 12, 18 wt of fibres
prepared by melt compounding are $3.9x10^{15}$, $5.9x10^3$ and $1.42x10^3$ while
resistivities of composites prepared by reactive processing at respective
compositions are $3.3x10^{13}$, $1.42x10^3$ and $1.38x10^3$ respectively. It is
obvious that composites prepared by the in-situ polymerisation have lower
resistivities.

The ultrasonic absorption of these composites has been typically shown
by the Figure 5. The composites are highly ultrasound absorbing. The
absorption is due to single and multiple scattering of ultrasound by
metallic particles in a typical viscoelastic matrix (Dutta, 1988).

Figure 6 shows the typical results of the thermal conductivity variation
with filler loading. The variation of thermal conductivity with
increasing filler concentration is less dramatic. The thermal conductivity
of the composites is theoretically explained by Nielson's model (1974).

The tensile modulus of PMMA/AlCGF increases with concentration of filler.
The increase is due to reinforcement of the glass fibers. These changes
can be theoretically calculated by using a theoretical equation applicable
for short fibre composites (Phillips and Harris, 1977).

It may be concluded from the above investigation, that in-situ polymerisation under shear field provides composites which have better performance.
Like shape, size and electrical properties of the constituents, the
processing has also significant effect on percolation threshold and
limiting electrical conductivity of the composites. The metal coated
glass fibres also increase thermal conductivity and have reinforcing
effect on composites. These are good fillers to develop light,strong,
sound and vibration absorbing conducting composites for both antistatic
 and EMI shielding in electrical and electronic applications.

REFERENCES

Bhattacharya, S. K. (ed) 1986,"Metal-Filled Polymers Properties and
Applications", Marcel and Dekker, Inc. New York.
Bhattacharya, S.K. and Chaklader A.C.D. 1982,Review on Metal-Filled
Plastics, Polym. Plast Technol. Eng., 19:21.
Bigg, D.M., 1977, "Conductive Polymeric Compositions', Polym. Eng. Sci.,
17:842.
Bigg, D.M.,1986,"Electrical Properties of Metal Filled Polymer Composites,
In "Metal-Filled Polymers, Properties and Applications", p 165,Marcel
Dekker, New York.
Crossman, R.A., 1985, Conducting Composites Past, Present and Future,
Polym. Engg. Sci., 25:507.
Delmonte, J., 1990 "Metal/Polymer Composites",Van Nostrand Reinhold,
New York.
Devenport, D.E., 1981, "Metalloplastics: An Answer to Electromagnetic
Pollution", Polym. Plast. Tecknol. Eng., 17:211.
Dutta, A.L., 1988, "Direct Current Conduction and Ultrasonic Investi-
gations of Polymer/Metal Composites", Ph.D. Thesis, IIT Kharagpur.
Nedorezova, P.M., Shevchenko, B.G., Galashina N.M., Tchmutin, I.A.,
Saratovskikh, S.L.,Tsvetkova, V.I.,Ponomarenko, A.T., Djachkovsky, F.S.,
and Enikolopyan N.S., 1987, Synthesis of Polypropylene-Graphite
Polymerisation Filled Composite, International Conference, Electronics
of Organic Materials, Tashkant USSR.
Nielson, L.E., 1974, The Thermal and Electrical Properties of Two-Phase
System, Ind. Eng. Chem. Fundam., 13:17.
Phillips D.C. and Harris 1977,"The Strength, Toughness and Fatigue Pro-
perties of Polymer Composites", In Polymer Engineering Composites,
Richardson, M.O.W. (Ed)p.97, Applied Science Publishers, London.
Sarkar Supriti,1990), "Investigation of Conducting Polymer Composites
Developed by Reactive Processing", M.Tech Thesis, IIT Kharagpur.

PIEZO – ELECTRIC CHARACTERISTICS OF POLYMER / CERAMIC COMPOSITES IN 1-3 CONNECTIVITY

A.K. Tripathi, P.S. Mal, T.C. Goel and P.K. C.Pillai

Department of Physics, I.I.T Delhi
New Delhi - 1100116 India

INTRODUCTION

Lot of work is being done these days on polymer - ceramic composites (1-3). This is mainly due to the fact that:- (i) composites have higher mechanical strength than the ceramic (ii) high piezo-electric 'd' coefficients due to piezo-electric ceramic phase compared to piezo-electric polymers, (iii) higher piezo-electric voltage coefficients 'g' due to lower dielectric constant of the composite compared to pure ceramic, (iv) higher hydro-static piezo-electric coefficient d_h and g_h and (v) better acoustic impedance matching in water for their use as hydrophones due to their low-density compared to pure ceramic.

Composites being a biphasic system , the design of composite element is an important consideration. Newnham(4) has described the different polymer - ceramic connectivity patterns, depending upon the continuity of the polymeric/ceramic phases in the three dimension.

In the present paper we are reporting the piezo-electric characteristics of PZT - Epoxy composites in 1-3 connectivity. The effect of ceramic volume fraction in the composite and of poling conditions on the piezo-coefficients have been studied. The effect of coating the ceramic element with PMMA or rubber before making the composite was also studied.

EXPERIMENTAL

Lanthanum doped PZT pellets of 1mm thickness, obtained from CEL (India), were used in the present experiments. These pellets were polarized at $100^{\circ}C$ by corona discharge technique. Quick drying silver paint as electrodes. The corona set up is shown in Figure 1. It is a single pin corona discharge unit. The corona is generated at 6kV. A bias voltage of 1.5 kV is used to give a poling current of 50 uA. After poling these pellets were cut into pieces to give squares of 2mm*2mm, 3mm*3mm and 4mm*4mm. These squares

Frontiers of Polymer Research, Edited by P.N. Prasad and
J.K. Nigam, Plenum Press, New York, 1991

were fixed with quickfix on a glass substrate at a regular
spacing of 1mm and Epoxy (Araldite) was poured under vacuum
to give composite samples. These samples were grinded,
polished and electroded when the epoxy was cured. The
curing was done at 60°C for 6h. To make PMMA and rubber
coated samples; PZT squares were dipped once into viscous
PMMA solution in Chloroform/ synthetic rubber adhesive
fevibond respectively. On drying it was observed that a 50
um thick coating of polymer/rubber was formed on the
squares. The epoxy composites were prepared, using the
earlier mentioned technique, from these squares also. Care
has been taken that the same polarity surface of squares
should face the same direction. These samples were used as
such without additional poling. Few composites samples were
also prepared with unpoled PZT. These composite samples were
poled, after grinding, polishing, and electroding at 100°C.
The dielectric constant and the frequencies for maximum and
minimum impedances of these samples were determined on a
L.F.Impedance Analyzer (HP model 4192A). The hydrostatic
piezo - electric coefficient d_h was measured using a
quasistatic technique. The instrument that was designed for
these measurements is shown in Figure 2.

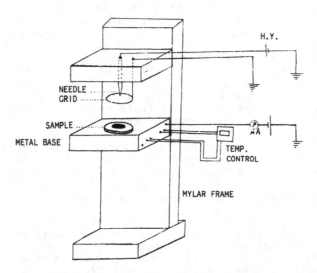

Fig.1   Corona Charging Assembly.

RESULTS AND DISCUSSION

It was observed that the hydrostatic piezo - electric
coefficient d_h of the composites prepared using poled
ceramic is higher than that of the composites prepared with
unpoled ceramic which were poled later on. For example, in
case of composites the composites having 60% volume fraction
of PZT using prepoled ceramic, the d_h value was found to be
98 pC/N, while the composite poled later on it was found to
be about 30 pC/N. This may be due to:

1. PZT discs used for preparing the composites were about 95 to 96 % dense, they still have some micropores which are filled with Epoxy at the time of making the composite. As the Epoxy has very high resistance it will not allow the full electric field to get applied across the ceramic.

2. Moreover epoxy in the pores may exert considerable stress on the ceramic grain while poling, which may act opposite to the ceramic elongation on application of electric field.

(i) the effect of ceramic volume fraction as well as height / width ratio of PZT pillars on the piezo - electric characteristics of the composites are summarized in Table 1.

TABLE I. Piezo-electrics of PZT:Epoxy Composites

S.No.	Ceramic volume fraction	h/w	E	d_h pC/N	$g_h = d_h/EE_o$ (10^{-3}) m.V/N	figure of merit $d_h \cdot g_h$ (10^{-15}) m^2/N
1	100 %	-	1200	40	3.76	150.4
2	70 %	.28	336	65	21.8	1417
3	60 %	.33	326	98	33.34	3322.2
4	50 %	.50	300	100	37.6	3760

It can be seen from this table that as the volume fraction of ceramic is decreased, their dielectric constant goes on decreasing. This is expected as epoxy has low dielectric constant compared to PZT. Another interesting feature is the increase in piezo-electric d_h constant and the resultant increase in g_h and the piezo - electric figure of merit of the composite element.

Now the increase in d_h constant of the composite samples compared to pure PZT is easy to understand. The piezo-coefficient d_h is given by

$$d_h = d_{33} + d_{31} + d_{32}$$

piezo - electric coefficients d_{31} and d_{32} are negative for PZT, therefore d_h coefficient is low. However in a 1-3 composite, the epoxy matrix reduces the stresses that are applied on the ceramic from the perpendicular 1 and 2 direction during the hydrosstatic measurements. Therefore the values of d_{31} and d_{32} lower considerably. As a result of this we get a fairly high d_h constant in the 1-3 composites.

To understand the increase in d_h constant with lowering of volume fraction of PZT, we have to consider the height/width ratio (h/w) ratio of the PZT pillars. It can be seen that although the volume fraction of PZT is decreasing with increasing d_h constant, the height/width ratio is increasing. It seems, therefore, logical to conclude that the increase in d_h constant is mainly due to the increase in h/w ratio of the pillar. By reducing the width of the pillar for the same height allows the polymeric matrix to shield it directions perpendicular to the poling axis.

In our composites we could change the h/w ratio from 0.28 to 0.50 because of the dicing limitations, we feel that any further increase in this will not lead to an increase in the overall performance of the composites as this will considerably lower volume percentage of the piezo active ceramic in the composite. Moreover, our results also indicate that the increase in the d_h value is very small when the h/w ratiois changing from 0.33 to 0.5.

The effect of coating ceramic pillars with PMMA or synthetic rubber on the piezo-electric coefficients is shown in Table-II. Here h/w ratio and volume fraction of ceramic is kept constant.

TABLE II. Piezo-electric Coefficients of Polymer coated PZT : Epoxy Composites.

S.No	Sample Name	E'	d_h pC/N	g_h m.V/N	$d_{h}g_h$ m^2/N
1	Epoxy	336	65	$21.8 * 10^{-3}$	$150.4*10^{-15}$
2	Epoxy + PMMA	395	109	$31.1 * 10^{-3}$	$3389*10^{-15}$
3	Epoxy + Rubber	425	218	$42.1*10^{-3}$	$9177*10^{-15}$

It can be seen from this table that coating PZT pillars with either PMMA or Rubber increases the d_h value and the figure of merit considerably. This is due to the fact that the soft polymer or rubber coating does not allow high stress to develop due to elongation of ceramic element in 1 and 2 direction due to the applied pressure in the perpendicular direction. Moreover, hydrostatic pressure component from 1 and 2 directions also gets reduced considerably in presence of rubber/polymer as they are soft compared to epoxy.
The planar coupling coefficient of these samples are around 0.5 to 0.6.

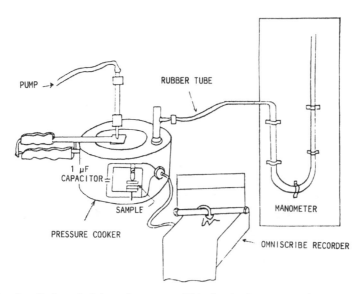

Fig.2 Hydrostatic piezo-coefficient d_h measuring assembly.

REFERENCES

1. H.Banno and S.Saito, Japan J.Appl.Phys. Suppl**22-2**, 67
    (1983).

2. J.Runt and E.C.Galgoci, J.Appl.Polym.Sci.**29**, 611(1984).

3. A.K.Tripathi, R.Sekar and P.K.C.Pillai, Mater.Lett., **9**,
    24(1989).

4. R.E.Newnham, D.P.Skinner and L.E.Cross, Mater.Res. Bull.,
    13,525(1978).

FUTURE AEROSPACE NEEDS

FOR HIGH PERFORMANCE MATERIALS

Ernesto VALLERANI, Paolo MARCHESE, Bruno FORNARI

ALENIA (*)
Alenia Spazio SpA
Turin - Italy

INTRODUCTION

In terms of performance and reliability, the aerospace industry is the most demanding market in which the materials suppliers have to try their strength. As designers define more and more stringent structural requirements on the basis of new space missions and air transportation needs, producers are asked to develop advanced materials to meet demands that cannot be satisfied by existing traditional materials.

Such requirements are becoming even more difficult to meet, because the scenario of aerospace transportation systems is changing very rapidly and the boundaries between areas historically well defined - aircraft and spacecraft - are gradually fading.

The evolution towards fully reusable space vehicles with hypersonic cruise capabilities will see an unusual combination of aviation and space requirements which will pose an exceptional technological challenge.

The premise for accepting this challenge with a definite probability of success has already been laid down: composites and ceramics have now reached such a high degree of development as to threaten the role of metals in airframe and engine applications with their appealing weight savings.

The mass reduction of structural components has always been a main goal of aircraft engineers. It has been estimated that up to 170 liters of fuel can be saved yearly for each kilogram of dead weight reduction in a civil airplane. The mass saving can pay off even more in space missions, as the launch cost of satellites may be as high as 30,000 dollars per kilogram.

However, aerospace manufacturers are unwilling to compromise performance for weight savings; hence the demand for lightweight, high-performance advanced materials.

This group includes qualified materials such as polymer composites, glass matrix and metal matrix composites, carbon-carbon composites, ceramics, light alloys and superalloys. The most exotic of them are still in their infancy and several years will be required before they are sufficiently reliable. When mature, each material will cover a definite field of application, depending on its temperature resistance (Figure 1), but overlaps will not be infrequent.

(*) ALENIA is the new Company resulting from the merger of Aeritalia and Selenia, taking over full responsibility for their functions, responsibilities and activities.

Frontiers of Polymer Research, Edited by P.N. Prasad and
J.K. Nigam, Plenum Press, New York, 1991

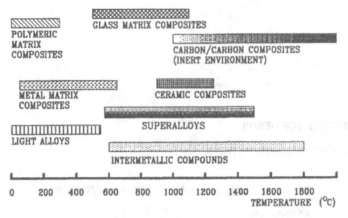

Fig. 1 · Temperature Ranges of Application for High-Performance Materials

The business is highly competitive and polymeric materials - on which the present paper is focused - are gaining a larger share of the market as their mechanical properties improve and the temperature range of application broadens.

Engineering plastic composites are expected to play a major role in the heat management of spacecraft in the near future, achieved through heat-resistant structures and thermal protections. But, as the future has its roots in the present, a glance at today may help to understand tomorrow better.

PRESENT

Polymeric composites were first used in decorative inboard aircraft components some 20 years ago, when they appeared as commercially available materials. Their evolution towards primary structural application developed in parallel with the development and optimization of strong and stiff fibers - such as graphite and aramid fibers - which replaced the initial glass reinforcements.

Today they are widely used in both military and civil aircraft for critical parts such as rudders, elevators, ailerons, fairings, radomes, gear doors, trailing edge panels, tanks, pods, skins. It has been estimated that the latest generation of military fighters employs up to 40% by weight of composites - mainly epoxy resin reinforced with continuous graphite and aramid fibers - and that this figure will increase to 60% in a few years.

In commercial aircraft, the composite replacement of metals has seen a slow but constant growth up to about 20% by weight (present) and is likely to reach 40-50% by the end of the century. The greatest increase is expected in business aircraft, as suggested by the Beechcraft Starship I and the Lear Fan 2100, which are two examples of aircraft with airframes built entirely of composites.

Fiber-reinforced plastics started their journey into space more than 15 years ago; since then satellites and spacecraft have made extensive use of polymeric composites in support truss structures, antenna components and solar arrays. The Giotto spacecraft was successfully protected from dust and meteoroids on its way towards Halley's comet by a shield of polyurethane foam and Kevlar composite. The Magellan interplanetary probe is steering towards Venus with a carbon fiber reinforced epoxy truss structure as a propulsion module support and upper stage assembly. Astronomers rely on the stability of the epoxy composite Metering Truss Structure and Flight Focal Plan Structure of the NASA Hubble Space Telescope to gaze at quasars and galaxies never examined before.

Fig. 2 Primary Structure of EURECA Carrier: Carbon Fiber Struts and Titanium interconnectors

The antenna module of the Intelsat V communications satellite, the European Retrievable Carrier (EURECA, Figure 2) and part of Ariane-4's SPELDA structure are but a few additional examples of how designers have made the best of advanced thermosetting composites to meet stiffness, strength, stability and weight requirements.

Aeritalia has been present in this evolution from the beginning, that is from 1968, when the thermal protection of the ELDO missile was built with glass reinforced phenolic resin (Table 1). The first successful applications to secondary structures and the sound knowledge of materials behavior reached through technological and laboratory tests stimulated the development of specific design tools and manufacturing technologies, leading to sufficient confidence in composites to allow them to replace metals even in primary structures.

Today carbon and aramid fiber reinforced epoxies are employed in the monolithic multispar structure of the AMX aircraft's vertical fin (Figure 3) and graphite fiber reinforced modified bismaleimides in the European Fighter Aircraft's wing box (Figure 4).

Table 1   Aeritalia's Composite Application

APPLICATION	MATERIAL
ELDO Missile Heat Shields	Glass / Phenolic
G222 Air Conditioning System	Glass / Polyester
G222 Air Intake	Glass / Phenolic
Radomes of G222, F104, MRCA B767	Glass/Epoxy
B767 Ailerons and Elevators	Carbon / Epoxy
AMX Vertical Fin, Gear Doors, Horizontal Stabilizers	Carbon, Kevlar / Epoxy
EFA Wing Box, Fuselage Skins	Carbon / Modified Bismaleimide
SAX Satellite Optical Bench	Carbon / Epoxy
SOHO Satellite Optical Bench	Carbon / Epoxy
HELIOS Satellite Adaptation Case	Carbon / Epoxy

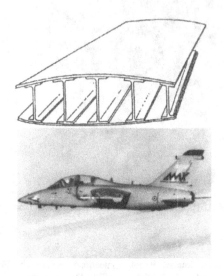

Fig. 3   AMX Aircraft: Monolithic Multispar Composite Structure of the Vertical Fin

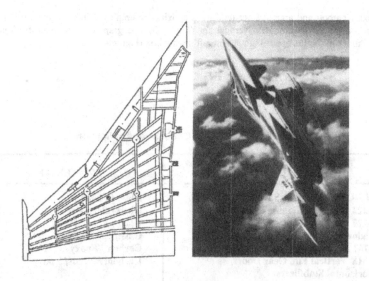

Fig. 4   EFA: Composite Wing Box

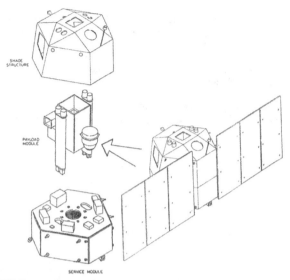

Fig. 5 SAX: Composite Optical Bench and Concentrator Tubes of the Payload Module

Space applications include the optical bench and the concentrator tubes of the Satellite for X-ray Astronomy (SAX, Figure 5) and the optical bench of the UV Coronagraph Spectrometer of SOHO (Solar and Heliospheric Observatory) satellite (Figure 6), all made of high-modulus graphite fiber/epoxy composites.

Most of the R & D studies which prefaced these programs are now resting in the archives, having been superseded by applied research on new advanced materials, although they are the necessary technological heritage required in order to face the challenges of tomorrow and beyond.

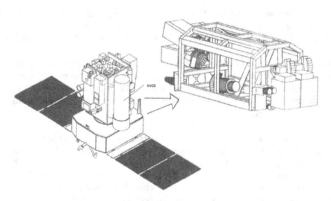

Fig. 6 SOHO: Composite Optical Bench of the Ultra-Violet Coronagraph Spectrometer

FUTURE

In the next decade the advent of spaceplanes will probably overshadow the new generation of satellites for telecommunication/navigation systems and earth observation, the development of probes for deep space exploration and the new types of launch vehicles.

Independently of the relative importance of these programs, however, there is no doubt that they will all require materials with outstanding properties.

Spaceplanes will be winged, fully reusable vehicles meant to provide access to the Space Station with airplane-like operations at reduced launch costs. They are the natural evolution of the

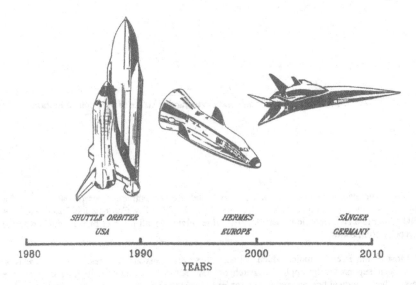

| SHUTTLE ORBITER | HERMES | SÄNGER |
| USA | EUROPE | GERMANY |

1980   1990   2000   2010
YEARS

Fig. 7   Present and Future Space Transportation Systems

American Shuttle Orbiter and European Hermes and will very likely be the ancestors of the hypersonic passenger aircraft (Figure 7). Several vehicle concepts with different mission profiles have been developed: NASP X-30 (USA), HOTOL (UK), SANGER (Germany), HOPE (Japan).

Their common feature is the need for new design solutions for hypersonic flight (> Mach 6) and the aerothermodynamic loads experienced during re-entry operations (temperatures up to 1600°C). The extensive use of advanced materials can easily be predicted, but a great effort in research and development of these systems and related processing and control techniques will be required to bring them to maturity in time.

Fiber-reinforced polymers will continue to be important because of their high specific strength and stiffness: "cold" structures and reusable cryogenic fuel tanks are likely to be the most important applications. In the first case the ability to operate at service temperatures as high as 250° C (with occasional short exposures to 350° C) will be a must. In the second case the ability to withstand manoeuvring aerodynamic and inertial loads at cryogenic fuel temperatures (-250° C for liquid hydrogen) without cracking and crazing will be the choice-leading criterion.

The main requirements to be met by these composites will be an unusual mixture of space and aviation requirements. The most important ones are outlined as follows:

* density lower than 1.8 g/cc
* retention of minimum tensile strength of 800 MPa and modulus of 120 GPa at 250° C (unidirectional composite)
* high vibration damping
* high damage tolerance
* high fracture toughness
* low impact sensitivity
* resistance to thermal cycling and thermal spikes
* long mechanical/acoustic fatigue life
* reduced creep at high temperatures
* good resistance to atomic and molecular oxygen, UV and particle radiations, lightning strikes and electrostatic discharges, chemicals (fuels, lubricants, oils, solvents), hot gases
* no hydrogen embrittlement
* limited moisture absorption
* as low a coefficient of thermal expansion as possible combined with high thermal conductivity
* offgassing lower than 100 µg/g (total non-toxic organics) for manned vehicles
* no inflammability
* no toxicity

A single material possessing all these properties will remain in the dreams of the materials scientists and engineers for a long time. A compromise will be necessary to provide an acceptable balance of properties; otherwise different materials will be tailored and used for different components.

More stringent requirements, though reduced in number, will apply to the new generation of space systems (Space Station, satellites). The present trend towards long mission times (> 10 years) will become more marked in the future (missions on the order of 20 - 30 years) and will add a new question - *How long will it last?* - to the usual ones - *How stiff is it? How strong is it?* - which pestered designers in the past.

Long term durability will become the watchword among spacecraft manufacturers, just as happens today for other producers, with the difference that the space environment is far from being friendly to any object spending its life in orbit.

In Low Earth Orbit application (LEO, 400 - 800 km), atomic oxygen can seriously erode polymeric materials, thus impairing the matrix-dependent properties of composites (Figure 8); in addition, crazing from continuous thermal cycling (thousands of cycles from earth shadow to sun illumination, from -150° C to +150° C) and damage impact from micrometeoroids and man-made debris may reduce the dimensional stability of the structures. In Geosynchronous Orbit (GEO, 36,000 km), ultraviolet solar radiation and ionizing radiation (high-energy electrons and protons) can affect the resin properties during long-term exposures. Finally, in both applications, the high level of vacuum may cause organic materials to outgas and critical surfaces to be contaminated; the acceptance limits of 1.0% total mass loss and 0.10% condensable materials (sometimes reduced to 0.01% in the presence of optical instruments) are currently adopted for standard space applications. A harder test-bench for structural materials could not be imagined; in such environmental conditions they are requested to maintain their performance for many years.

Even small variations in material properties are critical for high-precision systems such as antennas and optical benches, for which pointing accuracy and stability are fundamental. A significant example of stability requirements is the Hubble Space Telescope, in which the composite Metering Truss support of primary and secondary mirrors has to maintain the 5 meter spacing between the optical surfaces stable within ± 0.0015 mm under a temperature variation of ±10° C. Only high-modulus materials, with a near-zero thermal expansion coefficient and no moisture absorption during ground operations (to avoid distortions after outgassing in orbit) can provide such characteristics.

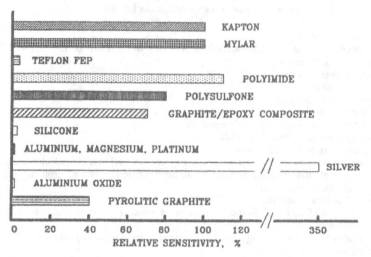

Fig. 8 Relative Sensitivities to Atomic Oxygen Erosion (Kapton = 100%)

Survivability in a harsh environment thus gives the right emphasis to the role of the resin matrix, sometimes overshadowed by the importance attributed to the reinforcement. The matrix is not merely a binder but protects the fibers from abrasion and environmental attack, participates in load transfer and distribution, determines the composite shear strength, compression resistance, toughness and impact/post-impact properties. Of primary importance is its temperature resistance, which affects the thermal and mechanical behavior of the composite; its glass transition temperature (Tg) and/or its melting temperature (Tm) mark the upper limit of application.

As more sophisticated spacecraft are designed, it is gradually becoming evident that epoxy composites, which rule the market today, will hardly be able to cope with the needs of future structures.

Despite recent improvements, brittleness remains their Achilles' heel. Couple this shortcoming with moisture absorption, the tendency to crack at cryogenic temperatures and a maximum service temperature of +175° C, and you have fertile terrain for the cultivation of alternative solutions.

Emerging Structural Materials

The large number of reinforcing fibers commercially available (Figure 9), able to meet almost any structural requirement, indicates the interest and competition existing in this area. Recently a great amount of work has also gone into the development of polymeric matrices with properties superior to epoxies. Some efforts have been directed at the improvement of existing materials, others at the development of innovative resin systems. Even though some of them are still at a stage of pre-commercialization, the replacements for conventional composites in aerospace programs will be found among them.

The game will be played between some thermosets and a crowded group of engeneering thermoplastics which have recently received great attention.

Polyimides (PI) - an especially important family of high-performance resins which includes both thermosets and thermoplastics - are candidates for the leading materials of the future. They are thermally stable, tough, solvent resistant, scarcely affected by moisture and retain their mechanical properties at temperatures unthinkable for other polymers (well above 300° C). Some thermoplastic

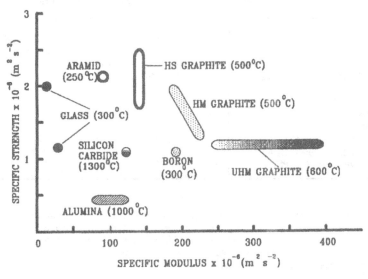

Fig. 9 Mechanical Properties of Commercial Reinforcing Fibers (maximum temperatures of application are indicated)

polyimides, conveniently modified through copolymerization, have shown oxydative stability even up to 370° C. The main restraint on widespread employment of polyimides resides in their very complicated manufacturing process, which requires automatically-controlled, high temperature and pressure autoclaves or molding devices. In some cases, also environmental and safety regulations oppose their introduction into normal production cycles.

Bismaleimides (BMI) are special thermosetting formulations which belong to the PI family but have versatile epoxy-type curing processes. Able to operate at temperatures on the order of 250° C in dry conditions, they may have very high glass transition temperatures (270°-300b C dry, 200°C wet) and are stiff, strong and solvent resistant, but unfortunately they are also very brittle and prone to absorb water (> 1%). Both BMI and PI carbon reinforced composites are being studied for application to control surfaces (flaps) of the next generation of re-entry space vehicles; uses in areas exposed to engine exhausts are also being considered.

An overview of advanced thermosetting resins would not be complete without mentioning polystyrilpyridine (PSP), worthy of attention for its excellent hot/wet properties (well maintained up to 250° C), good fire and chemical resistance, low toxicity, but too brittle to be a serious competitor to PI without further improvements.

Advances in the thermoplastic technology of fiber impregnation have overcome the obstacles to continuous fiber prepreg production and made available high-performance composites which are very promising from the point of view of environmental resistance and processability.

Due to their chemical structure, thermoplastics (both amorphous and semicrystalline) may offer enhanced resistance to aggressive environments, low moisture absorption (< 0.1% for carbon reinforced PEEK), excellent toughness and damage tolerance, lower outgassing than thermosets and good resistance to microcracking under thermal fatigue. The fact that they can be fabricated through rapid thermoforming processes provides additional advantages over the time-consuming curing cycles of thermosets. Since the introduction of polyetheretherketone (PEEK) in 1983, other thermoplastics have been made available on a development or commercial basis with several reinforcements (including graphite and aramid fibers) in tape, fabric or hybrid yarn form.

Table 2   Glass Transition Temperatures and Melting Points of Thermoplastics

	GLASS TRANSITION Tg (˚C)	MELTING POINT Tm (˚C)
Polyimide (Thermoplastic) (TPI)	310-365	-
Polyamideimide (PAI)	290	-
Polyetherimide (PEI)	220	-
Polyethersulfone (PES)	230	-
Polysulfone (PSU)	190	-
Polyarylenesulfide (PAS)	220	-
Polyetheretherketone (PEEK)	143	340
Polyphenylenesulfide (PPS)	90	290

Glass transition temperatures and melting points, which limit the applicability of the composites as structural materials, are listed in Table 2 for the most interesting matrices. Semicrystalline polymers (PEEK and PPS) do have relatively low Tg's but, as their properties do not degrade completely after this temperature, the addition of reinforcement extends their use to temperatures approaching the melting point.

In terms of general aerospace applications, thermoplastic composites seem to possess exceptional potential, but two constraints slow down their widespread diffusion.

The first is the still limited data on their mechanical performance; in spite of successful use in aircraft interiors and flourishing research activities on long term properties, additional studies are needed before these composites are qualified and certified for structural components.

The second constraint is of an economic nature: manufacturers who have infrastructures and equipment for thermoset processing are cautious about making large capital investments in special high-temperature tooling for new fabrication technologies such as those required for thermoplastics. The low-volume aerospace production will derive limited benefit from the high efficiency process of thermoplastic manufacturing, which is more profitable for mass production. Expenditures for developing new technology, therefore, must be compensated by evident improvements in product quality and performance not otherwise achievable.

Materials for Thermal Protection

Important areas for non-structural applications of polymers are the skin insulations of high-speed airplanes and spacecraft and the ablative thermal protections for launchers, space re-entry modules and propulsion units.

In the first case there is a need for non metallic, low-density materials to protect the fuselage skin from overheating during atmospheric flight. Designers' demands for high-insulating, strong and stable systems seem to have found an answer in the Felt Reusable Surface Insulations (FRSI), consisting of non-woven textile fabrics with a 3-dimensional entangled structure. Polymeric fibers (phenolic, acrylic, polyimidic, aramidic and fluorocarbon) offer the advantages of toughness, density and flexibility over their glass and ceramic competitors. FRSI, which can withstand temperatures of 400˚ C, are likely to gain great applicability in the future due to the possibility of tailoring their properties according to the design objectives. Their mechanical characteristics can easily be controlled through the fiber packaging, the use of sizings and the addition of woven layers; surface coatings may be added to improve reflectivity, durability and impact resistance.

The second application concerns launcher fairings, re-entry modules and rocket nozzles and exit cones. These expendable systems necessitate low-cost, simple and light heat shields able to protect the structures below from the intense thermal fluxes originated by the aerodynamic phenomena during atmospheric ascent or re-entry phases (launchers and modules) and by hot gases (rocket motors).

Ablative materials achieve heat dissipation through complex high-temperature chemical and physical transformations such as fusion, sublimation, pyrolysis, carbonization and oxydation; the organic matter decomposes and deposits a surface char layer with insulating and reradiating characteristics. This layer can be eroded away by highly convective fluxes which reduce the heat protection.

Material consumption occurs at a rate inversely proportional to its density and ablation temperature. High-density and high-ablation temperature materials are preferred for re-entry vehicles, which experience severe thermomechanical loads; the reduced ablation rate allows the modules to retain their aerodynamic profile, which in turn results in better control of the re-entry flight path. Launcher fairings, which have to sustain lower temperatures, preferably adopt low-density materials, characterized by good insulation properties but high ablation rates.

Other requirements for ablators include:

* good mechanical properties at high temperature, often obtained with the addition of continuous fibers to the basic polymer
* excellent compatibility and adhesion to metallic supporting structures, in order to avoid disbonding due to the differential thermal expansions
* high oxidation resistance to ozone and atomic oxygen
* good environmental survivability to weathering, corona discharges, ultra-violet radiation, vacuum and thermal cycles

Ablative materials have evolved from the rubbers and resins which were available 40 years ago. The current systems (Table 3) are based on thermosets (phenolic and epoxy resins) or elastomers (ethylene-propylene and silicone rubbers), frequently formulated with various loadings and reinforcements. Cork, cotton, glass, silica, quartz, carbon, silicon carbide, nylon and aramid in the form of powders, discontinuous fibers, filaments, tapes and fabrics are selected to fulfil various performance requirements. The use of continuous, strong refractory fibers and the choice of appropriate fabrication techniques allow thermal composites to retain good mechanical properties up to 2000° C (rocket motors).

CONCLUSIONS

In the next 10 - 15 years the business of advanced materials is destined to grow in conjunction with the aerospace industry's future projects, for which traditional materials seem to be inadequate. This will offer new high-tech materials the opportunity to emerge and dominate those application fields where performance enhancements are cost effective. The arena is crowded with competitors at different stages of development. Because for most materials the birth-growth-maturity cycle takes from 10 to 20 years - a period much longer than the time line from conception to design and fabrication of aerospace vehicles (Figure 10) - the winners, if any, will come out of the material families known today. The surprising discovery of completely new materials seems quite improbable.

Table 3 Thermal Protection Materials

MATERIAL	SERVICE LIMITS T (°C)
FRSI	< 430
Cork Ablators	< 1650
Silicone Elastomers	1650 - 2750
Nylon Phenolics	2750 - 3500
Carbon Phenolics	> 3500

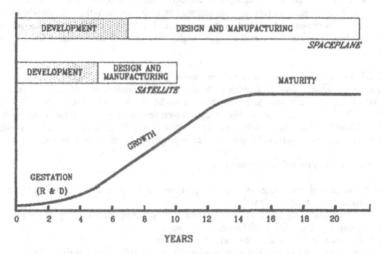

**Fig. 10   Material Life-Cycle compared to Space Programs Timescale
(orders of magnitude only)**

The way ahead for high-performance polymeric composites is well defined by the requirements of future air and spacecraft. Improvements in fiber strength and stiffness will require parallel advances in matrix performance in order to allow composites to exploit their outstanding characteristics to the full. Demands are for tough resins, with negligible moisture absorption, good retention of mechanical properties at temperature extremes and good adhesion to reinforcements.

The success of the enterprise will be achieved only through the coordinated efforts of materials producers and end users. The suppliers will be expected to improve their systems in a tremendously short time, aiming at very specific applications. The users will have to specify their requirements clearly in order to address the material development to the final goals and search for new manufacturing routes and related innovative quality control methods.

This will have to be matched to a continuous effort to better the knowledge of composite behavior and train engineers to take full advantage of the materials' characteristics with proper design criteria.

SOME NOVEL ORGANIC-INORGANIC COMPOSITES

Y. P. Ning,* M. X. Zhao,** and J. E. Mark

Department of Chemistry and the Polymer Research
Center, The University of Cincinnati, Cincinnati, OH 45221
USA

INTRODUCTION

The goal of the present study was to form organic-inorganic
complexes by bonding polymer chains into a ceramic-like material
through reactive groups placed at their ends. For this to occur, the
reactive polymer chains must be present during the generation of
the matrix material, by means of the sol-gel reaction used to prepare
high-performance ceramics.   The reaction used to generate the
ceramic-type matrix was the simple hydrolysis of tetraethoxysilane
(or "tetraethylorthosilicate") (TEOS):

$$Si(OEt)_4 + 2H_2O \longrightarrow SiO_2 + 4EtOH \qquad (1)$$

or of one of the analogous titanates, such as tetra-$\underline{n}$-butyltitanate
(TBOT):

$$Ti(O\underline{n}\text{-Bu})_4 + 2H_2O \longrightarrow TiO_2 + 4\,\underline{n}\text{-BuOH} \qquad (2)$$

Both types of reactions are catalyzed by a variety of acids, bases, and
salts, with reaction (2) generally occurring much more rapidly than
reaction (1)[1-9].   Typically, hydroxyl or alkoxy groups on the poly-

---

* Permanent address: Qingdao College of Chemical Engineering, 52
Zhengzhou Road, Qingdao, People's Republic of China.
** Permanent address: Chengdu Silicone Center for Applications
Research, South Renmin Road, Chengdu, Sichuan Province, People's
Republic of China.

mer chains react with the above silicate or titanate, or their hydrolysis products, thus giving the desired bonding between the polymer chains and the ceramic-like material.

Three aspects of this type of work are described in the present report. They are (i) the effects of increasing the functionality (number of functional groups) at the ends of the polymer chains, (ii) comparison of the reinforcing effects of silica versus titania, and (iii) characterization of the properties of materials having mixed silica-titania components.

EXPERIMENTAL DETAILS

The polymer chosen for the organic phase was poly(dimethylsiloxane) (PDMS), with repeat unit $[-Si(CH_3)_2O-]$, because of its excellent elastomeric properties and thermal stability, and its good miscibility with the metal alkoxides of interest. In order to determine the effect of the molecular weight of the polymer, four samples were employed, with number-average molecular weights $10^{-3} M_n$ of 0.700, 1.70, and 4.20 g mol^{-1}.

Also, an attempt was made to increase the reactivity of the organic component[6,7] by converting the vinyl ($-CH=CH_2$) end groups on vinyl-terminated PDMS chains to triethoxy groups by the reaction

$$CH_2=CH-PDMS-CH=CH_2 + 2HSi(OEt)_3 \longrightarrow$$

$$(EtO)_3CH_2-CH_2-PDMS-CH_2-CH_2(OEt)_3 \qquad (3)$$

In this way, the PDMS chains become hexafunctional, and this increased reactivity should increase the bonding of the polymer chains into the ceramic-like matrix.

The reactions were carried out at 80 °C, for relatively long periods of time (as is done in most fundamental studies in this area). In such reactions, the water required in the hydrolysis reaction can be added directly to the other components[10,11]. Alternatively, it can be generated in-situ by an esterification reaction between acetic acid and ethanol[11,12]:

$$CH_3COOH + C_2H_5OH \longrightarrow CH_3COOC_2H_5 + H_2O \qquad (4)$$

Since silicates generally hydrolyze significantly more slowly than titanates, the silicates to be used in silicate-titanate mixtures were prereacted for either 2 hrs or 24 hrs before the titanate was introduced. The hydrolysis was then continued for the desired period of time, with the reacting mixtures being poured into Petri dishes a few inches in diameter, to a depth of approximately 1 mm. After the reaction was complete, each sheet was removed, and its transparency noted (visually).

The code used for the identification of the samples thus prepared is illustrated in Table I.

Table I

Description of Nomenclature: Example Using TEOS (20) - PDMS (50, 1700) - Ti(OR)4 (30) - 50 - 0.045 - 80

Item	Information
TEOS (20)	Silicate used (wt %)
PDMS (50, 1700)	Polymer used (wt %, Molecular weight)
Ti(OR)4 (30)	Titanate used (wt %)
5 0	% Stoichiometric amt. water used
0.045	Molar ratio of acid/(silicate and titanate)
8 0	Reaction temp. (ºC)

Several strips were cut from each sheet for elongation measurements to characterize its mechanical properties. Of primary interest in this regard are the nominal stress $f^* = f/A^*$, and the reduced stress or modulus

$$[f^*] \equiv f^*/(\alpha - \alpha^{-2}) \tag{5}$$

where f is the equilibrium retractive force, A^* is the cross-sectional area of the undeformed sample, and $\alpha = L/L_i$ is the elongation, or extension ratio relative to the length L_i of the unstretched sample.

In order to obtain estimates of ultimate properties, all measurement are carried out to the rupture points of the samples. These properties are (i) the maximum extensibility (given by the elongation α_r at break), (ii) the ultimate strength (given by either the nominal stress f^*_r at rupture or by the modulus $[f^*]_r$ at rupture), and the energy E_r at rupture (obtained from the area under a plot of f^* vs. α). The last of these three properties is a standard measure of the toughness of a material.

RESULTS AND DISCUSSION

Overview of Properties

The organic-inorganic preparations employed covered a relatively wide range of compositions. In those cases where only one type of metal alkoxide was present, its amount was varied from approximately 30 to 80 wt %. In the case of the mixtures, the relative amounts of silicate and titanate were varied from approximately 10 to 29 wt % at approximately constant total amount of metal alkoxide. The samples generally found to have good transparency, as judged by the simple visual inspections.

Use of Highly-Functionalized PDMS

The effects of replacing some of the usual difunctional PDMS by hexafunctional PDMS of molecular weight 3800 g mol^{-1} in PDMS-silica composites obtained by the direct addition of water are described in Table II and Figure 1. The modulus and ultimate strength are seen to increase by this replacement, which is consistent with better bonding between the PDMS and the silica matrix.

Comparisons Between Silica and Titania Composites

Results obtained upon hydrolysis of either the silicate or the titanate with the usual difunctional PDMS having $M_n = 700$ g mol^{-1} and by the in-situ generation of water are presented in Table III, the last column of which comments on the mechanical characteristic of the samples. The results are also shown graphically in Figures 2 and 3. At constant amounts of ceramic, the titanates give much higher values of the modulus and ultimate strength, but much lower values of the maximum extensibility. These differences could be due, at least in part, to the better coordinating ability of the titanium atoms. Moderate amounts of the metal alkoxide are seen to give tough materials with very attractive mechanical properties.

Combined Silica-Titania Composites

The first set of results on these similarly prepared SiO_2-TiO_2-PDMS composites is summarized in Table IV and Figure 4. Although there is some scatter in these data, replacement of SiO_2 by TiO_2 in these samples seems to increase the modulus and ultimate strength, as does decrease in the molecular weight of the PDMS.

Table II
The Effects of Replacing Some of the Usual Difunctional PDMS by Hexafunctional PDMS[a]

Amount of hexafunctional PDMS, wt %	α_r	f_r^* N mm^{-2}	$[f^*]_r$ N mm^{-2}	$10^3 E_r$ J mm^{-3}
0	1.20	0.303	0.609	0.029
25	1.10	0.386	1.37	0.020
50	1.10	0.552	2.03	0.027

[a] Incorporated in the composite TEOS (50) - Hexafunctional PDMS (x, 3800) - Difunctional PDMS (50 - x, 1700) - 50 - 0.045 - 80

482

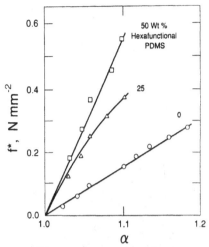

Fig. 1.   The dependence of the stress-strain
isotherms   on   the   amount   of   the   usual
(difunctional) PDMS replaced by hexafunctional
PDMS.   Each curve is labeled with the weight %
of hexafunctional PDMS present in the PDMS
mixture.   (Direct addition of the water required
in the hydrolysis).

Table III

Comparison of Physical Properties of PDMS[a] modified with Titania or Silica

| Metallo-Mechanical Organic | Wt % | Designation | α_r | Ultimate Properties | | Characteristics |
				f_r^* N mm^{-2}	$[f^*]_r$ N mm^{-2}	
Silicate	40	Si-40	1.19	0.080	0.162	Elastic
	50	Si-50	1.11	0.098	0.321	Elastic
	60	Si-60	1.03	0.782	8.16	Tough
	80	Si-80	1.01	0.619	14.6	Glassy
Titanate	30	Ti-30	1.09	0.203	0.829	Tough
	40	Ti-40	1.03	0.552	8.86	Tough
	50	Ti-50	1.02	0.764	7.73	Tough
	55	Ti-55	1.02	2.35	43.1	Glassy

[a] M_n(PDMS) = 700 g mol^{-1}

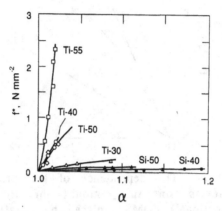

Fig. 2. Typical stress-strain isotherms for composites of PDMS-SiO$_2$ and PDMS-TiO$_2$. (In-situ generation of water)

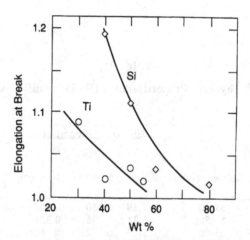

Fig. 3. Values of the elongation at rupture taken from Figure 2 shown as a function of amount of silicate or titanate used in the hydrolysis. (In-situ generation of water).

Table IV

Physical Properties of PDMS Modified by Mixtures of Titania and Silica[a]

PDMS M_n g mol^{-1}	Designation	Ratios of TEOS/TBOT/PDMS	α_r	Ultimate Properties	
				f_r^* N mm^{-2}	$[f^*]_r$ N mm^{-2}
700	1	25.7/14.3/60.0	1.07	1.66	8.74
	2	23.6/21.3/55.0	1.04	1.99	16.0
	3	28.5/21.5/50.0	1.00	1.65	135.
1700	1	25.1/16.3/58.6	1.07	0.465	2.37
	2	23.8/20.6/55.0	1.10	2.22	8.19
	3	22.0/25.0/53.0	1.08	0.889	4.01
	4	21.4/28.6/50.0	1.03	3.88	39.9
4200	1	27.0/10.0/63.0	1.17	0.777	1.75
	2	25.0/15.0/60.0	1.11	0.964	3.29
	3	25.0/16.0/59.0	1.11	1.49	4.99
	4	24.0/21.0/55.0	1.04	2.08	15.3

[a] Silicate pregelation for 24 hrs before addition of titanate

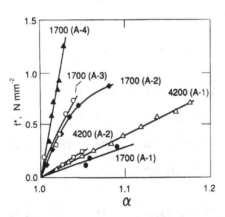

Fig. 4.   Stress-strain   isotherms   for   the
combined   PDMS-SiO_2-TiO_2 composites.   (In-situ
generation of water).   Each curve is labeled with the
molecular weight of the PDMS and the designation
of the corresponding sample in Table V.

The effects of different reaction conditions on the mechanical properties is described in Table V, where the A series of samples were prepared with prereaction of the TEOS with PDMS for 2 hrs, and the B series with prereaction for 24 hrs. Increase in pre-reaction time seems to improve the mechanical properties of the samples.

Table V

Effects of Different Reaction Conditions on the Properties of PDMS with Silica and Titania

PDMS M_n g mol^{-1}	Designation	Ratios of TEOS/TBOT/PDMS	α_r	Ultimate Properties f_r^* N mm^{-2}	$[f^*]_r$ N mm^{-2}	Mech. Char.
700	A-1[a]	25.5/15.5/59.0	- -	--	--	Transparent and glassy
	A-2	23.0/23.0/54.0	- -	--	--	Transparent and glassy
	B-1	25.7/14.3/60.0	1.07	1.66	8.74	Transparent and tough
	B-2	23.6/21.4/55.0	1.04	1.99	16.0	Transparent and tough
1700	A-1	25.7/14.3/60.0	1.09	0.280	1.12	Cloudy and tough
	A-2	23.5/21.5/55.0	1.08	0.879	3.84	Cloudy and tough
	A-3	21.5/28.5/50.0	1.04	0.713	6.27	Transparent and tough
	A-4	20.0/35.0/45.0	1.03	1.29	15.7	Transparent and tough
	B-1	25.0/17.0/58.0	1.07	0.466	2.37	Transparent and tough
	B-2	24.0/21.0/55.0	1.10	2.22	8.19	Transparent and tough
	B-3	22.0/25.0/53.0	1.08	0.889	4.01	Transparent and tough
	B-4	21.4/28.6/50.0	1.03	3.88	39.9	Transparent and tough
4200	A-1	27.6/7.3/65.0	1.18	0.718	1.58	Transparent and tough
	A-2	23.5/21.5/55.0	1.05	0.223	1.59	Cloudy
	B-1	25.0/15.0/63.0	1.17	0.776	1.75	Transparent
	B-2	24.0/21.0/55.0	1.05	2.08	15.3	Transparent

[a] TEOS-PDMS pregelation times were 2 hrs for the A series, and 24 hrs for the B

Summary

The composites described were prepared using techniques very similar to those used in the new sol-gel approach to ceramics. In this case, however, organic-inorganic composites are prepared by having functionally-terminated chains of poly(dimethylsiloxane) (PDMS) present during the hydrolysis of either tetraethylorthosilicate (TEOS) to a silica-like ceramic material, or of tetra-n-butyltitanate (TBOT) to a titania-like material. The end groups on the PDMS were used to obtain bonding between the polymer chains and the silica (SiO_2) or titania (TiO_2). In some cases, the functionality of these ends was increased prior to hydrolysis in a separate chemical reaction to improve the organic-inorganic bonding. The water required for the hydrolysis reaction was either added directly, or generated in-situ by the esterification reaction between ethanol and acetic acid. The resulting composites were characterized by stress-strain measurements in elongation, yielding values of the hardness or modulus, the ultimate strength, and maximum extensibility. The techniques described permit the preparation of hybrid composite materials having a wide range of attractive mechanical properties.

ACKNOWLEDGEMENT

It is a pleasure to acknowledge the financial support provided by the Sandia National Laboratories through Grant #54-4482.

REFERENCES

(1)   L. L. Hench and D. R. Ulrich, Eds., "Science of Ceramic Chemical Processing", John Wiley & Sons, New York, 1986.
(2)   J. D. MacKenzie and D. R. Ulrich, Eds., "Ultrastructure Processing of Advanced Ceramics", Wiley, New York, 1988.
(3)   D. R. Ulrich, CHEMTECH, 18, 242-249 (1988).
(4)   J. E. Mark, in "Ultrastructure Processing of Advanced Ceramics", J. D. MacKenzie and D. R. Ulrich, Eds.,Wiley, New York, 1988. Reprinted in CHEMTECH, 19, 230 (1989).
(5)   D. W. Schaefer and J. E. Mark, Eds., "Polymer-Based Molecular Composites", Materials Research Society Symposium Volume, Pittsburgh, PA, 1990.
(6)   H. Schmidt, in "Polymer-Based Molecular Composites", D. W. Schaefer and J. E. Mark, Eds., Materials Research Society Symposium Volume, Pittsburgh, PA, 1990, pp. 3-13.
(7)   G. L. Wilkes, A. B. Brennan, H.-H. Huang, D. Rodrigues, and B. Wang, in "Polymer-Based Molecular Composites", D. W. Schaefer and J. E. Mark, Eds., Materials Research Society Symposium Volume, Pittsburgh, PA, 1990, pp. 15-29.

(8) J. E. Mark and D. W. Schaefer, in "Polymer-Based Molecular Composites", D. W. Schaefer and J. E. Mark, Eds., Materials Research Society Symposium Volume, Pittsburgh, PA, 1990, pp. 51-56.

(9) B. J. J. Zelinski, C. J. Brinker, D. E. Clark, and D. R. Ulrich, Eds., "Better Ceramics Through Chemistry, IV", ed. by Materials Research Society Volume, Pittsburgh, PA, 1990.

(10) M. X. Zhao, Y. P. Ning, and J. E. Mark, in "Proceedings of the Symposium on Composites: Processing, Microstructure, and Properties", M. D. Sacks, Ed., American Ceramics Society Publications, Westerville, OH, 1991.

(11) J. E. Mark, "Preparation and Characterization of Tagging Materials for Status Verification of Weapons Systems", Contract #54-4482, Sandia National Laboratories, Second Report, May 28, 1990.

(12) Y. P. Ning, M. X. Zhao, and J. E. Mark, in "Chemical Processing of Advanced Materials", ed. by L. L. Hench and J. K. West, John Wiley and Sons, New York, 1991.

DYNAMIC MECHANICAL AND DIELECTRIC PROPERTIES OF

SULFONYLATED POLY(2,6-DIMETHYL-1,4-PHENYLENE OXIDE) COPOLYMERS

Frank E. Karasz, William J. MacKnight and
Heung Sup Kang

Polymer Science and Engineering Department
University of Massachusetts, Amherst, MA 01003

ABSTRACT

   Sulfonylated poly(2,6-dimethyl-1,4-phenylene oxide)
copolymers with a degree of sulfonylation of less than 40 wt %
have been studied by measuring dynamic mechanical and dielectric properties. A very broad mechanical β-relaxation
centered around -10°C at 10Hz was displayed by the dry copolymer. The relaxation disappears upon exposure of the copolymer
to water and can be made to reappear when the copolymer is
fully dried. A broad dielectric secondary β-relaxation is
also observed around 0°C at 50KHz. These findings indicate
that water appears to hinder the localized motion responsible
for the β-relaxation by associating with polar pendant phenyl
sulfone groups. The apparent activation energy for the
mechanical α-relaxation of the copolymer has been estimated
and found to be larger for copolymers with increased degrees
of sulfonylation, indicating that the sulfonylated copolymers
have a much stiffer molecular backbone compared to PPO.

INTRODUCTION

   It is well known that dynamic mechanical properties and
dielectric loss in polar polymers are affected by the presence
of water. There have been several reports[1-3] concerned with
the effect of water content on the β-relaxation of polymers
bearing polar groups on the main chain.
   The secondary β-relaxation of poly(arylene ethers) bearing polar groups on the main chain, for example poly(ether
sulfone), has been found to be enhanced by water absorption.[2]
In previous publications,[4,5] we reported the preparation and
properties of a series of sulfonylated PPO (SPPO) copolymers.
Because these copolymers bear pendant phenylsulfone groups
they may provide a new class of materials for the study of
viscoelastic relaxation behavior.
   In this paper, we report results clarifying the effect of
water on the secondary β-relaxation of SPPO copolymers; this

Frontiers of Polymer Research, Edited by P.N. Prasad and
J.K. Nigam, Plenum Press, New York, 1991

effect is shown most clearly by copolymers with a degree of sulfonylation of 38.8 wt %. The origin of the β-relaxation motion is discussed in conjunction with dielectric measurements.

The activation energy of the mechanical α-relaxation associated with the glass transition of the copolymer has also been estimated from Arrhenius plots, and the results imply that the copolymer chains contain a much stiffer backbone than unmodified PPO.

EXPERIMENTAL

The preparation and properties of SPPO copolymers have been reported elsewhere.[4,5]
Dynamic mechanical measurements were carried out with a Polymer Laboratories DMTA operated in the three point bending mode with a strain amplitude of 60μm. The SPPO copolymer films, which were cast from chlorobenzene solution and dried in a vacuum oven, were annealed above T_g in the DMTA furnace to remove any residual solvent and analyzed over the frequency range 0.1Hz to 30Hz. Measurements were carried out in the temperature range -140°C to 280°C at a heating rate of 1.5°C/min. After completion of dynamic mechanical measurements on a dried sample with a degree of sulfonylation of 38.8 wt%, the sample was immersed in distilled water at room temperature for 7 days. The wet sample was removed, and the excess water was wiped off with filter paper. The water uptake was determined to be 0.25% by weighing.

PPO powders were compression molded into films at 280°C under nitrogen and annealed above T_g in the DMTA furnace before dynamic mechanical measurements.

Dielectric measurements for the SPPO copolymer samples (degree of sulfonylation 38.8 wt%), which were obtained as described above, were carried out with a General Radio model 1689M RLC Digibridge at 50KHz over a temperature range of -155 to 280°C with a nitrogen purge. The temperature was raised at a rate of 3.0°C/min. Samples approximately 0.1mm in thickness were employed with an active electrode cell diameter of 3.3 cm.

RESULTS AND DISCUSSION

The temperature dependencies of the storage moduli (E'), loss moduli (E''), and loss factor (tanδ) at 10Hz for PPO, dry and wet samples of SPPO copolymer (both with a degree of sulfonylation of 38.8 wt%) are shown in Figures 1, 2 and 3, respectively. In contrast to previous results[6-9] which showed that PPO exhibited three mechanical relaxations, in the temperature range employed in the current study, only α-relaxation associated with the glass-rubber transition around 230°C was observed (Fig. 1). The storage moduli for the three samples decrease with increasing temperature and show a marked decrease around T_g. In addition, a slight decrease of the storage modulus for the SPPO copolymer can be observed in the presence of water. The tanδ peak maximum for the SPPO copolymer around 260°C is indicative of the α-relaxation process associated with the glass transition involving the onset of long-range segmental motions of the main chain. This temperature compares well with the T_g (251°C) measured by DSC[4,5] if frequency differences are taken into account.

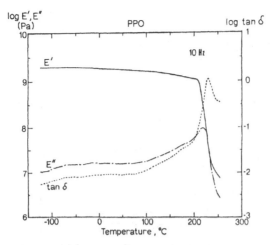

Figure 1. Temperature dependencies of mechanical E′, E″ and tanδ for PPO at 10Hz.

The very broad low-temperature β-relaxation, which is spread over several decades of temperature centered around - 10°C, is readily observed for the dry SPPO copolymer, as shown in Figure 2. However, the β-relaxation for the wet sample, as seen in Figure 3, is not observed. The α-relaxation peak position for the wet sample decreased only slightly (2°C) compared with that of the dry sample owing to gradual redrying during the measurements. However, the removal of water by annealing in the DMTA furnace regenerated a distinct broad β-relaxation.

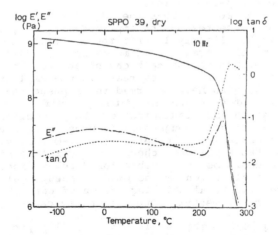

Figure 2. Temperature dependencies of mechanical E′, E″ and tanδ for dry SPPO copolymer (degree of sulfonylation, 38.8 st%) at 10Hz.

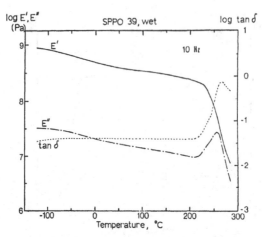

Figure 3. Temperature dependencies of mechanical E', E" and
tanδ for wet SPPO copolymer (degree of sulfonylation, 38.8 st%) at 10Hz.

Motion of the polymer segment in a localized lattice,
absorbed species, impurities or structural irregularities are
known to be associated with low-temperature relaxations.[10] In
the case of the poly(arylene ether sulfones) the amplitude of
β-relaxation was found to increase as the water content increased[2] probably because of the participation of absorbed water
in hydrogen bonds to polar groups.

In contrast to poly(arylene ethers) bearing polar groups
on the main chain, rather peculiar behavior was observed for
the SPPO copolymer. In particular, the secondary β-relaxation
disappeared in the presence of water. The association of
water with the pendant sulfone group may cause either a slow-
ing of the average motion of the phenyl sulfone group by
structure tightening, or a hindrance of the local motion which
gives rise to the β-relaxation.

It has been reported that the addition of water and
methanol to nylon decreased the amplitude of the γ-relaxation
and shifted it to slightly lower temperatures.[11] This effect
was ascribed to the formation of mechanically stable bridges
between amide groups in adjacent chains and to the antiplasti-
cizing effect of water. The decrease in the amplitude of the
β-relaxation has also been observed in bisphenol A-polysulfone
by the addition of 4, 4'-dichlorodiphenyl sulfone,[12-14] which is
known to act as an antiplasticizer of the polysulfone.
However, there are not enough data at present to suggest that
water acts as an antiplasticizer in the SPPO copolymer.

For the SPPO copolymer, there are several possible local
motions which could be responsible for β-relaxation. First,
there may be local motion of the pendant phenyl sulfone
groups. Second, a local twisting motion of the ether group
may occur.[15] Third, methyl group motion may cause a local mode
process.

According to Schaefer et al.[16] a ring flip process is not
permitted for the stiffer PPO main chain. Broad line NMR
studies on PPO[17] revealed that the methyl groups are in motion
at -180°C. Thus, local motions of the phenyl sulfone groups

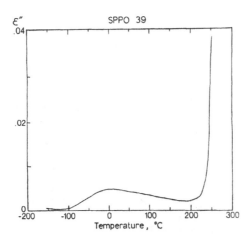

Figure 4. Temperature dependence of dielectric ε'' for dry
SPPO copolymer (degree of sulfonylation, 38.8 wt%)
at 50KHz.

appear to be the source of the very broad β-relaxation at -
10°C. In addition, this motion is considered not to be
entirely intramolecular since it depends on the environment of
the sulfone group.

The temperature dependence of the dielectric ε'' for the
SPPO copolymer with a degree of sulfonylation of 38.8 wt%, is
displayed in Figure 4; DC conductive loss[18] is seen to be
significant at higher temperatures. The broad secondary β-
relaxation at 0°C can be clearly seen. The height of the
dielectric ε'' peak is on the order of 10^{-2}. It is noted that
the amplitude of the β-relaxation for SPPO is quite large
compared to that of PPO, which may be due to the contribution
of the dielectrically active sulfone group.

In the case of PPO only one low-temperature dielectric
relaxation was reported to appear at -100°C, and its molecular
origin was uncertain.[19] The height of the dielectric ε'' for
γ-relaxation is only 3×10^{-3}; the low intensity of the relaxations is ascribed to the relatively nonpolar nature of PPO.

A localized mode of the dipolar sulfone group appears to
be responsible for the β-relaxation of the SPPO copolymer at
0°C. The temperature shift of the dielectric β-relaxation
compared with the dynamic mechanical β-relaxation can be
understood by taking into account the frequency difference or
the differing relative contribution of dipolar and mechanical
motions of the pendant tetrahedral sulfone group, whose exact
nature of motion is unknown.

The activation energies for the mechanical α-relaxation
of the PPO and SPPO copolymer were compared using the relation:

$$\frac{-Ea}{2.3R} = \frac{d\log f}{d(1/T)}$$

In the case of the SPPO copolymer with a degree of sulfonylation of 38.8 wt%, the mechanical tanδ peak temperatures

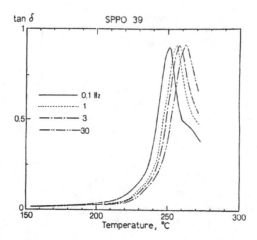

Figure 5. Frequency dependencies of mechanical tanδ curves for
dry SPPO copolymer (degree of sulfonylation, 38.8
wt%).

observed for 0.1, 1, 3 and 30Hz are 251°C, 256°C, 258°C and
263°C, respectively, as shown in Figure 5. Figure 6 shows an
Arrhenius plot of the frequencies as a function of the reciprocal temperature for the peak maxima. The SPPO copolymer of
higher degree of sulfonylation has the larger activation
energy, reflecting the increased intermolecular and steric
barrier to long-range segmental motion.[4,5]

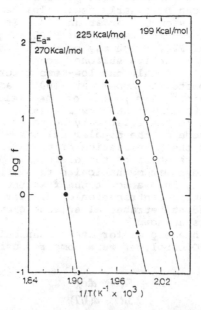

Figure 6. Arrhenius plot of relaxation maxima as a function of
frequency for PPO, SPPO copolymers with the degree
of sulfonylation of 11.8 and 38.8 wt%. O, PPO; Δ,
SPPO, degree of sulfonylation, 11.8 wt%; •, SPPO,
degree of sulfonylation, 38.8 wt%.

494

CONCLUSIONS

Although the SPPO copolymer with a full range of degree of sulfonylation cannot be studied by dynamic mechanical and dielectric relaxation methods, it seems likely that the mechanical β-relaxation for the SPPO copolymer is attributable to the localized motion of the pendant phenyl sulfone group. The presence of a broad dielectric β-relaxation at 0°C, which is known to be absent in the same temperature range for the dry PPO, tends to support this hypothesis.

ACKNOWLEDGEMENTS

This work was supported by DARPA AF49620-85-C-0127.

REFERENCES

1. W. Reddish, <u>Trans. Farad. Soc.</u> 46:459 (1950).
2. G. Allen, J. McAinsh and G. M. Jeffs, <u>Polymer</u> 12:85 (1971).
3. R. M. Ikeda and H. W. Starkweather, <u>Polym. Eng. Sci.</u> 20:321 (1980).
4. H. S. Kang, W. J. MacKnight and F. E. Karasz, <u>ACS, Polymer Preprints</u> 27(2):65 (1986).
5. H. S. Kang, W. J. MacKnight and F. E. Karasz, to be submitted.
6. J. Stoelting, F. E. Karasz and W. J. MacKnight, <u>Polym. Eng. Sci.</u> 10:133 (1970).
7. T. Lim, V. Frosini, V. Zaleckas, D. Morrow and J. A. Sauer, <u>Polym. Eng. Sci.</u> 13:51 (1973).
8. A. F. Yee, <u>Polym. Eng. Sci.</u> 17:213 (1977).
9. J. R. Fried and G. A. Hanna, <u>Polym. Eng. Sci.</u> 22:705 (1982).
10. R. D. McCammon, R. G. Saba and R. N. Work, <u>J. Polym. Sci.</u> Pt.A7:1721 (1969).
11. H. W. Starkweather, <u>ACS Symp. Ser.</u> 127:433 (1980).
12. M. K. Gupta, J. A. Ripmeester, D. J. Carlsson and D. M. Wiles, <u>J. Polym. Sci., Polym. Lett. Ed.</u> 21:211 (1983).
13. S. E. B. Petrie, R. S. Moore and J. R. Flick, <u>J. Appl. Phys.</u> 43:4318 (1972).
14. J. J. Dumais, A. L. Cholli, L. W. Jelinski, J. L. Hedrick and J. E. McGrath, <u>Macromolecules</u> 19:1884 (1986).
15. K. Yamafuji and Y. Ishida, <u>Kolloid Z./Z. Polymer</u> 183:15 (1962).
16. J. Schaefer, E. O. Stejskal, D. Perchak, J. Skolnick and R. Yaris, <u>Macromolecules</u> 18:368 (1985).
17. W. G. Gall and N. G. McCrum, <u>J. Polym. Sci.</u> 50:489 (1961).
18. N. G. McCrum, B. E. Read and G. Williams, "Anelastic and Dielectric Effects in Polymeric Solids," John Wiley and Sons, London (1967).
19. F. E. Karasz, W. J. MacKnight and J. Stoelting, <u>J. Appl. Phys.</u> 41(11):4357 (1970).

MORPHOLOGY OF POLY(VINYLALCOHOL) GELS PREPARED FROM SOLUTION

Masaru Matsuo

Department of Clothing Science
Faculty of Home Economics
Nara Women's University, Nara 630, Japan

INTDODUCTION

Since 1974, the preparation of polymeric fibers and films with high strength and high modulus has been extensively investigated for flexible polymers by gel-state spinning(1), ultradrawing of dried gel films(2), ultradrawing of single crystal mats(3), and two-step drawing of single crystal mats(4). Recently, Matsuo et al. using the method of Smith and Lemstra(2) produced ultradrawn polyethylene(5) and polypropylene(6) whose Young's modulus at 20°C was 216 and 40.4 GPa, respectively. These values were nearly equal to the crystal lattice moduli of polyethylene(5) and polypropylene(7) as measured by the X-ray diffraction technique.

On the other hand, Hyon et al. succeeded to produce high modulus poly(vinylalcohol)(PVA) fibers by drawing of gels prepared by crystallization from semi-dilute solutions containing 70% dimethylsulfoxide(Me2SO) and 30% water(H2O) in volume(8) on the basis of a report by Farrant(9) that solvent mixtures ranging from 50 to 75 vol % Me2SO do not freeze at - 100°C because of hydrating. The resultant tensile and Young's moduli of the drawn fiber reached 2.8 and 60 GPa, respectively. However, the detailed analysis of deformation mechanism remains an unresolved problem.

This paper deals with the sol-gel transition and spinodal decomposition of PVA solutions prepared from a co-solvent of Me2SO and H2O, not only to study drawability of the resultant films in terms of characteristics of molecular chains in solution, but also to study phase separation of non-crystalline polymer solution(10) as well as liquid-liquid phase separation of amorphous polymer blends.(11). The decay rate of the scattered light intensity which can be estimated from autocorrelation function was measured by inelastic light scattering in order to investigate the diffusion of molecular chains in solution based on the correlation function of the light scattered due to density fluctuations(12-14) during spinodal decomposition and gelation. The translational diffusion coefficient of molecules defined on the basis of linearity between the decay rate and the square of scattering vector is discussed in relation to the apparent diffusion coefficient associated with the spinodal decomposition.

EXPERIMENTAL SECTION

The sample used was PVA with a 98% degree of hydrolysis. The weight-average molecular weight, M_w, was estimated to be 88600 by the elastic light scattering technique and M_w/M_n (M_n; the number of average molecular weight) was estimated to be 3.20 by gel permeation chromatography. The solutions were prepared by heating the well-blended polymer/solvent mixture at 105°C for 40 min and the solution was quenched in a water bath at constant temperature. Mixtures of Me2SO and H2O are used as co-solvent and when the content of Me2SO in the mixed solvent is 70 vol %, the co-solvent is designated as the 70/30 composition. In the present experiments, two different co-solvents, with compositions of 70/30 and 50/50 were used. After standing for 5 days in a water bath at a constant temperature, the test tube containing the solutions was tilted. When the meniscus deformed but the specimen did not flow under its own weight, we judged that the solution had gelled. The lowest temperature at which the onset of gelation occured within 5 days was defined as the gelation temperature.

RESULTS AND DISCUSSION

Figures 1 and 2 show the change in the logarithm of the scattered intensity as a function of the time at various q, observed for 1 w/w % PVA solutions with the 50/50 and 70/30 compositions, where q, the magnitude of the scattering vector, is given by $q = (4\pi/\lambda)\sin(\lambda/2)$, λ being the wavelength of light in solution and θ the scattering angle. In the initial stage of phase separation, the logarithm of the scattered intensity increases linearly with time. The change in elastic scattered intensity in Figs. 1 and 2 can be described by a linear theory of spinodal decomposition proposed by Cahn(15), which is given by

$$I(q, t) = I(q, t=0) \exp[2R(q)t] \tag{1}$$

where $I(q, t)$ is the scattered intensity at the time, t, after initiation of the spinodal decomposition, and $R(q)$ is the growth rate of concentration, given as a function of q; $R(q)$ is given by

$$R(q) = -D_c q^2\{ - (\partial^2 f/\partial c^2) + 2\kappa q^2\} \tag{2}$$

where D_c is the translational diffusion coefficient of the molecules in solution, f is the free energy of mixing, c is the concentration of solution, and κ is the concentration-gradient energy coefficient defined by Chan and Hilliard(16). The linear relationship for plots of ln I vs. t at fixed q was also obtained for solutions of other concentrations at various temperatures. According to Eq. (2), a plot of ln I vs. t at fixed q yield a straight line of slop $2R(q)$. With time, the logarithm of scattered intensity tends to deviate from the linear relationship. as has been observed in the later stage od spinodal decomposition.(17-18). The deviation occured for the solution with the 50/50 composition at shorter time scale in comparison with that at the 70/30 composition. Furthermore, although the results are not shown as figures in this paper, it was confirmed that the deviation shifts to shorter time scale with increasing concentration of solution, and this tendency is more pronounced in the case of the 50/50 composition. This indicates that the growth rate of concentration fluctuation is sensitive to the Me2SO/H2O composition.

Figure 3 shows plots of spinodal temperature T_s vs. concentration points, while the curves show the concentration dependence of gelation temperature for data corresponding to the 50/50 composition (dashed curve) and the 70/30 composition (solid curve). The spinodal temperature obtained from the results in Figs. 1 and 2, shifts to higher values as the concentration of solution increases and this tendency is more pronounced in the case of the 50/50 composition. This indicates that the phase separation at the 50/50 composition is more significant than that at the 70/30 composition when both solutions have the same concentration. At temperatures above the curves, the sol-gel transition cannot occur. At temperatures below the curves, the gelation rate became faster as temperature decreased. Here it may be noted that the gelation temperature is higher than the spinodal temperature in the given concentration range. This indicates that both the gelation and the spinodal decomposition

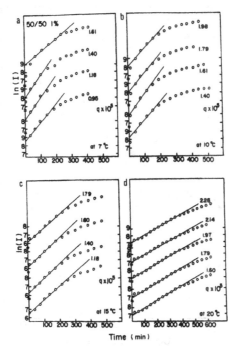

Fig. 1. Change of the logarithm of the scattered intensity against time
at various q, measured for 1 w/w % PVA solution with the 50/50
composition at various temperatures (a) 7°C (b)10°C (c) 15°C (d) 20°C

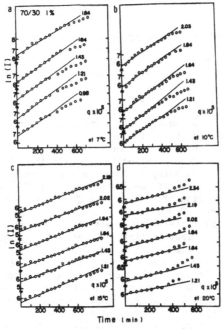

Fig. 2. Change of the logarithm of the scattered intensity against time
at various q, measured for 1 w/w % PVA solution with the 70/30
composition at various temperatures (a) 7°C (b) 10°C (c) 15°C (d)20°C

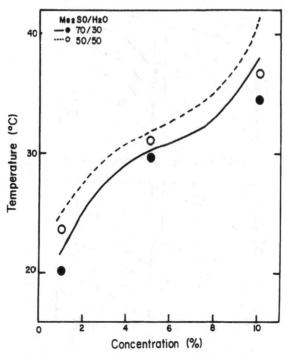

Fig. 3. Plots; Spinodal temperature Ts vs. concentration. Curves;
Gelation temperature vs. concentration for the 50/50 composition
(dashed curve) and for the 70/30 composition (solid curve)

occur simultaneously and this tendency is quite different from the phase
diagram of aqueous solutions of PVA as reported by Komatsu et al.(19)
    In order to obtain more detailed information on the simultaneous prog-
ression of gelation and spinodal decomposition, light scattering patterns
under the Hv polarization condition were observed as a function of time.
Figures 4 and 5 show the results obtained for a 5 w/w % solution at 20°C
and a 10 w/w % solution at 30°C, respectively. The logarithmic plots of
scattered intensity at $\theta = 90°$ as a function of time are presented as references.
In the time scale showing a straight line in the plot of ln I vs. t, the Hv
light scattering patterns cannot be observed except as an indistinct small
spot which is an artifact due to reflection of the incident beam. Furthermore
a very weak broad overlapped X-ray diffraction ring from the (101) and (10$\bar{1}$)
planes, was observed by X-ray diffraction at about 100 min corresponding to
the end of the linear portion of ln I vs. time for 5 w/w % solutions with the
70/30 compositions at about 200 min. The appearance of such a diffraction
ring indicating gelation/crystallization means that the straight line in the
ln I vs. t plot does not reflect the initial stage of pure spinodal decomposition. With increasing time, plots of ln I vs. t deviate from the straight
line. In the present work, we must emphasize that this behavior is independent of the characteristics of the later stage of the spinodal decomposition as pointed out by Langer et al.(17) The X-ray patterns observed
under Hv polarization conditions in the later time scale indicate the
exsistence of rodlike textures, the optical axes being oriented parallel or
perpendicular to the rod axis.(20) The formation of rods is probably associated with the further development of gelation/crystallization in the
polymer-rich phase. Thus, the deviation from the straight line in the ln I
vs. t plots is quite different from the later stage of phase separation as

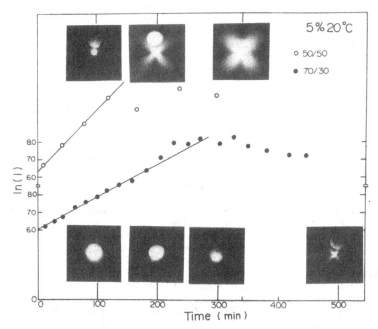

Fig. 4. Change of the logarithm of ln I at θ = 90° and H_V light
scattering patterns with time, measured for 5 w/w % solutions
with the 50/50 and 70/30 compositions at 20°C.

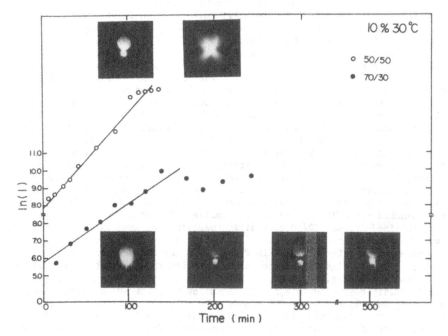

Fig. 5. Change of the logarithm of ln I at θ = 90° and Hv light
scattering patterns with time, measured for 10 w/w % solutions
with the 50/50 and 70/30 compositions at 30°C.

reported for polymer blends.(21) That is, in polymer blend systems, the further growth of concentration fluctuations associated with uphill-diffusion is significant in the later stage of the spinodal decomposition and finally the concentration profile becomes similar to that of nuclear growth. In the present PVA solutions, however, such a further growth of concentration fluctuations is suppressed because of immobilization of the PVA chains owing to crystallization.

To obtain more detailed information on the quasi-spinodal decomposition of the PVA solutions, inelastic light scattering measurements were carried out by assuming that the equilibrium state in solution is maintained during the photon counting period of 5 min. The evaluation of the data was done by the histogram and cumulative methods for determing the distribution of decay rate Γ from the observed correlation function. Figures 6 and 7 show the results at the indicated temperatures, measured for the $1 w/w \%$ solutions with the 50/50 and 70/30 compositions, respectively. The calculations of $\overline{\Gamma}$ were carried out by the cumulative method. Incidentally, the values of $\overline{\Gamma}$ estimated by the histgram method were almost equal to those by the cumulative one. The calculations were followed by the methods reported.(22-24) The value of $\overline{\Gamma}/q^2$ decreases rapidly at the shorter time scale and then tends to level off. Furthermore, the drastic decrease in $\overline{\Gamma}/q^2$ at temperatures > Ts shifts to much longer time scale. Interestingly, in the present analysis, the critical time corresponding to the levelling-off-point of the $\overline{\Gamma}/q^2$ plot is almost equal to the time associated with the derivation from the linear relationship of $\ln I$ vs. t shown in Figs. 4 and 5. In addition to the appearance of the X-ray diffraction ring and the Hv scattering pattern discussed in relation to Figs. 3 - 5, the observed time-dependence of $\overline{\Gamma}$ also supports the interpretation that the linearity in Figs. 1 and 2 is attributable to a quasi-spinodal decomposition corresponding to the simultaneous occurrence of spinodal decomposition and gelation in the PVA solutions. If the linearity was due to the pure spinodal decomposition, the deffusion coefficient estimated by inelastic light scattering would be expected to have a constant value independent of time. Accordingly, it follows that the observed time dependence is not consistent with the linearized spinodal decomposition theory(14) which assures time-independence of $D_{app} = -D_c(\partial^2 f/\partial c^2)$.

Here we must emphasize that maximum draw ratio of the dried PVA gel films could be realized when the gel was prepared by quenching the solution with the 70/30 composition at -50°C. It was found that under these conditions, the crystallinity of the undrawn film is the lowest and no superstructure could be observed by small angle light scattering under Hv polarization. This indicates that spinodal decomposition hampers the drawability of the resultant dried gel films and that the quenching plays an important role in arresting the progression of spinodal decomposition. The occurrence of the lowest crystallinity is thought to be due to the fact that the rapid gelation hampers the promotion of crystallization. To check the validity of this concept, two kinds of gels were prepared from solutions with the 70/30 and 50/50 compositions by the same method as that used for preparation of the specimens for measuring $\ln I$ vs. t in Fig. 4. After 100 min, the two gels were quenched at -50°C. The crystallinities of the resultant dried gel films obtained for the 70/30 composition were in the range 9 - 11%, which is lower than those (15 - 17%) of the films obtained for the 50/50 composition. The maximum draw ratio of the film from the former composition reached 11.5, which is slightly higher than the values obtained for the latter composition.

Returning to Fig. 4, it may be noted that the plots of $\ln I$ vs. t deviate from linearity beyond 100 min for the solution with the 50/50 composition, while they were linear up to 270 min for the solution with the 70/30 composition. This means that 100 min corresponds to the final period of the initial stage of the quasi-spinodal decomposition for the solution with the 50/50 composition, while 100 min is on the way of the progression of the quasi-spinodal decomposition for the latter solution. This is due to the faster progression of the quasi-spinodal decomposition of the former solution than

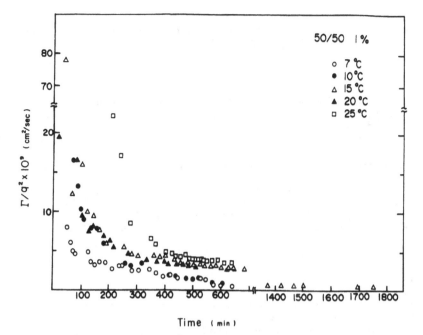

Fig. 6. Time-dependence of Γ/q^2 at $\theta = 90°$ for the 1 w/w % solution with the 50/50 composition, measured at the indicated temperatures.

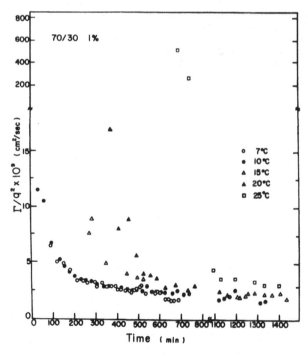

Fig. 7. Time-dependence of Γ/q^2 at $\theta = 90°$ for the 1 w/w % solution with the 70/30 composition, measured at the indicated temperatures.

for the latter solution. Accordingly, it may be concluded that the quasi-spinodal decomposition hampers facile drawability of PVA dried gel films.

REFERENCES

1) P. Smith and P.J. Lemstra, J. Mater. Sci. 15:505 (1980)
2) P. Smith, P.J. Lemstra, J.P.L. Pipper, and A.M. Kiel, Colloid & Polym. Sci. 258:1070 (1980)
3) K. Furuhata, T. Yokokawa and K. Miyasaka, J. Polym. Sci., Polym. Phys. Ed. 22:133 (1984)
4) T. Kanamoto, A. Tsuruta, K. Tanaka and R.S. Porter Polymer J. 15:327 (1985)
5) M. Matsuo and C. Sawatari Macromolecules 19:2036 (1986)
6) M. Matsuo, C. Sawatari and T. Nakano Polymer J. 18:759 (1986)
7) C. Sawatari and M. Matsuo Macromolecules 19:2653 (1986)
8) S.H. Hyon, W.I. Cha and Y. Ikada Rep. of Poval Committee 37:2709 (1989)
9) J. Farrant Nature 205:1284 (1965)
10) J.J. van Aartsen Eur. Polym. J. 6:919 (1970)
11) T. Hashimoto, J. Kumaki and H. Kawai Macromolecules 16:641 (1983)
12) B.J. Berne and R. Pecora "Dynamic light scattering" John Wiley and Sons. Inc. New York N.Y. (1976)
13) B. Chu "Laser light scattering" Academic Press New York N.Y. (1974)
14) D.E. Koppel J. Chem. Phys. 57:4814 (1972)
15) J.W. Cahn J. Chem. Phys. 42:93 (1965)
16) J.W. Cahn and J.E. Hilliard J. Chem. Phys. 28:258 (1958)
17) J.S. Langer, M. Bar-on and H.D. Miller Phys. Rev. A 11:1417 (1975)
18) K. Kawasaki and T. Ohta Progr. Theor. Phys. 59:362 (1969)
19) M. Komatsu, T. Inoue and K. Miyasaka J. Polym. Sci. Polym. Phys. Ed. 24:303 (1986)
20) M.B. Rhodes and R.S. Stein J. Polym. Sci., A-2 7:1539 (1969)
21) T. Hashimoto, M. Itakura and H. Hasegawa J. Chem. Phys. 85:6118 (1986)
22) P. Deby J. Chem. Phys. 31:680 (1959)
23) Y. Tsunashima, N. Nemoto, Y. Makita and M. Kurata Bull. Inst. Chem. Res. Kyoto University 59:293 (1981)
24) Y. Tsunashima, N. Nemoto and M. Kurata Macromolecules 16:584 (1983)

THE ROLE OF MATRIX ON THE MECHANICAL BEHAVIOR OF

GLASS FIBER REINFORCED THERMOPLASTIC COMPOSITES

M.L. Shiao and S.V. Nair

Department of Mechanical Engineering
University of Massachusetts at Amherst
Amherst, MA 01003

ABSTRACT

The role of matrix on the fracture resistance of glass fiber reinforced thermoplastics is investigated. Four broad categories of polymer matrices are considered: intrinsically brittle (SAN), intrinsically brittle but rubber toughened (SAN/ABS), intrinsically tough amorphous (PC) and intrinsically tough semi-crystalline (PBT) matrix. The initiation and propagation toughness of the composites were characterized with and without glass fibers and the relationship of microstructure to toughness was investigated by insitu examinations of deformation and fracture in the crack tip region under the optical microscope. It is found that the properties of the matrix play little role in the intrinsically brittle matrix composite, but play a very predominant role in the other composite systems. The magnitude of the role of the matrix depends on the matrix deformation mechanism and upon the role of glass fibers in modifying the deformation mechanisms.

INTRODUCTION

The use of short glass fiber reinforced thermoplastic composites in engineering applications has been growing rapidly due to their relatively low fabrication cost and many improved properties. It is well known that property improvements through fiber additions, such as increase in strength, stiffness and toughness, are influenced by the intrinsic properties of the added fibers[1,2], fiber/matrix interfaces[3-8] and the base polymers[9-11]. While the enhancement of the composite strength and rigidity is mostly affected by the fiber types, fiber aspect ratio and fiber orientations, the toughness improvement is known to sensitively depend on the properties of fiber/matrix interfaces and moreover on the deformation and fracture behavior of the matrix upon loading of the composites. The effects of fiber/matrix interfaces in the polymer composite materials by various surface treatments or by using thin layers of soft deformable materials around the fibers have been studied by many investigators[3-7]. However, the fundamentals of the deformation and fracture processes in the fiber-matrix systems from the standpoint of the role of the

Frontiers of Polymer Research, Edited by P.N. Prasad and
J.K. Nigam, Plenum Press, New York, 1991

matrix itself is not well understood. For example, the very different trends of toughening by glass fiber additions to various polymer matrices are illustrated in Fig. 1, where the notched Izod impact toughness is summarized for many important engineering thermoplastic composites. As can be seen, the trends of Izod toughness versus glass fiber content varies considerably among various types of polymer matrices. In the case of semi-crystalline polymers, nylon and polypropylene show an increase in Izod toughness as the glass fiber content increases, whereas acetal shows an decrease in the Izod toughness. In the case of amorphous polymers, the intrinsically tough polymers such as polycarbonate exhibit an increase in the Izod toughness, while the very high toughness polyurethane shows the opposite. On the other hand, in the intrinsically brittle polymers such as SAN and polystyrene, an increase in Izod toughness is observed as glass content increases, observed, while in their rubber toughened versions, such as, in ABS, a decrease is found.

Not only very different trends of toughening by glass fibers can be noticed, but the degree of toughness improvement can also vary significantly among different types of polymeric composites. However, a fundamental understanding of these trends cannot be gained directly from the Izod impact toughness, since the Izod toughness is known to depend on the specimen geometry[12,13]. It is therefore required that a more independent fracture parameter be used to systematically study the toughening roles of glass fibers, matrix deformation and possible synergisms between them. Furthermore, micromechanistic understanding of the deformation and fracture processes insitu ahead of a crack tip is also required to unravel the microstructural-macromechanical property relations.

In this paper, we examine the role of matrix on the fracture resistance of glass fiber reinforced polymer composites in an attempt to establish the relationship of micromechanistic fracture behavior to macroscopic toughness properties. The J-integral fracture toughness at crack initiation, J_{ic}, and the fracture resistance curve, J_R versus Δa curve, are used to characterize the fracture properties of the composites. Three systems of glass fiber reinforced polymer composites are investigated involving an intrinsically tough matrix, an intrinsically brittle matrix and finally a rubber toughened matrix. Insitu fracture observations of crack initiation and propagation in pre-cracked specimens are also conducted to reveal the underlying deformation and fracture processes. Special emphasis is given to the toughening roles of various polymeric matrix materials upon loading in order that guidelines for toughness improvement through microstructural modifications may be provided.

EXPERIMENTAL

The materials studied are listed in Table I for the intrinsically tough, intrinsically brittle and rubber toughened polymer composites. In the intrinsically tough materials, amorphous polycarbonate [PC] and semi-crystalline polybutylene terethphalate [PBT], supplied by GE Plastics, were used as matrices. The corresponding composites are made by compression molding with glass fiber mat to 15 wt% and 35 wt% glass fiber containing plates. Tensile specimens (ASTM D638) were then machined from the plates with thickness of 3.18 mm and 6.37 mm for the 15 wt% and 35 wt% glass containing composites respectively.

The polymer matrices used for the other systems are blends of acrylonitrile-butadiene-styrene [ABS] and styrene-maleic-ahydride [SMA] terpolymer, supplied by Monsanto Chemical company. Two levels of butadiene rubber, 9.5 wt% and 2 wt% of the matrix

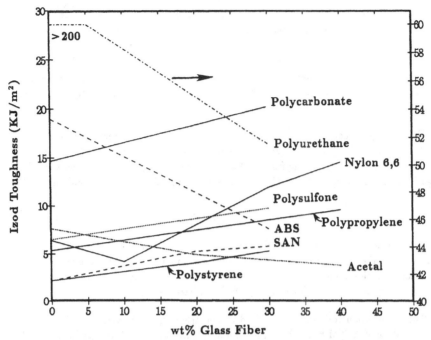

Figure 1 The notched Izod toughness versus glass fiber content for many short glass fiber reinforced thermoplastic composites.

TABLE I

Matrix Type	wt% Glass Fiber	wt% Rubber	Yield Stress MPa	Elongation %
a) Intrinsically Tough Matrix :				
PC	0	–	40.8	125
	15	–	101.9	1.59
	35	–	122.6	1.98
PBT	0	–	45.9	>200
	15	–	106.9	2.37
	35	–	94.3	1.95
b) Intrinsically Brittle Matrix :				
SAN	0	2	44.6	2.12
	5	2	54.0	1.63
	10	2	63.9	1.58
	20	2	72.2	1.41
c) Rubber Toughened Matrix :				
ABS	0	9.5	34.3	24.8
	5	9.5	45.9	3.39
	10	9.5	55.5	2.94
	20	9.5	76.6	2.43

respectively, were studied. The 2 wt% rubber containing material is used as the intrinsically brittle matrix, since it has been shown[14] that such low levels of rubber content do not provide any toughening. The 9.5 wt% rubber containing material is then used for the rubber toughened system in order to compare and contrast the role of rubber particles. Composites of these two systems are injection molded by dry blending of chopped E-glass fibers. Three levels of glass fibers, 5 wt%, 10 wt% and 20 wt%, were investigated in these two systems.

The J-integral fracture toughness at crack initiation, J_{ic}, was determined by the Early and Burns' method[15] using pre-cracked single edge-notched specimens. Controlled pre-cracks with crack length ranging from a/W=0.5 to a/W=0.7 were made by inserting fresh razer blades into the specimens. The initiation toughness J_{ic} was then calculated by[15]

$$J_{ic} = \frac{1}{\Delta A} \oint P dx \tag{1}$$

where Δa is the difference in crack area of specimens with finite differences in crack length, and $\oint P dx$ is the area enclosed by the load-displacement curve and the line connecting the crack initiation points. Initiation toughness data was also obtained from the R-curve measurements on 6.37 mm thick three-point-bend bars, see below.

The propagation toughness is assessed by constructing the J_R versus Δa curves using the procedure of ASTM E813. Pre-cracked 6.37mm thick rectangular bars were loaded in a three-point-bend fixture to measure the fracture resistance curve. The crack advance measurement is carried out by fast fracturing the specimens in liquid nitrogen after being unloaded from various stages of crack growth.

Insitu fracture observations were carried out by loading a pre-cracked specimen in a buckle-plate fixture[16] under an optical microscope as reported previously[17].

RESULTS

Initiation Toughness

Intrinsically Tough Matrix. Fig. 2 shows the initiation toughness J_{ic} as a function of glass fiber content for both the PC and PBT composites. In the unfilled case, PBT is found to have much higher initiation toughness than that of PC, as can be seen in Fig. 2. As the glass fiber content is increased, it is found that in the PBT composites a sharp decrease in the initiation toughness is observed, whereas in the PC composite the initiation toughness is not significantly affected. Above 15 wt% glass fiber content, the initiation toughness of the PBT composite shows a slightly lower value than that of the PC composite.

Insitu fracture observations ahead of the crack tip reveal that in unfilled PC energy is absorbed by shear deformation at the crack tip region (see Fig. 3). On the other hand, in the unfilled PBT a significant region of necking is observed in front of crack tip and the degree of crack blunting appears to be much larger (see Fig. 4). These results suggest that the high value of initiation toughness in PBT matrix material is related to large crack blunting through crack front plasticity in the necking region, whereas in PC matrix· material limited amount of shear deformation in the crack front results in lower initiation toughness. As the glass fibers are added to the materials it is observed insitu that in PBT

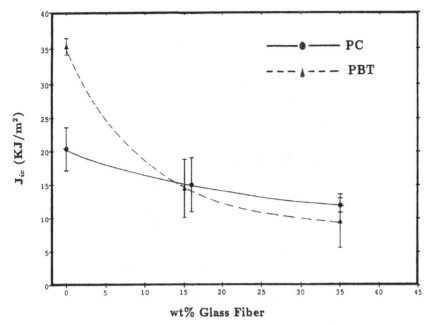

Figure 2 The initiation toughness versus glass fiber content for the intrinsically tough matrix composites.

composite the matrix suffers localized crazing on the crack plane between the crack tip and the nearby glass fibers (see Fig. 5) which facilitates the tendency for crack initiation. In the PC composite however, crack-tip shear deformations are unaffected by glass fiber additions (see Fig. 6). In both cases, debonding at fiber/matrix interfaces prior to crack initiation were also observed. Subsequent crack propagation is achieved by the linkage of the debonded fiber ends. Hence, the sharp drop in initiation toughness of PBT with glass fiber addition is related to the formation of large crack-plane localized craze at the crack tip which promote the crack initiation through its linkage with the debonded fiber ends. In the PC composite the slight decrease in the initiation toughness by the fiber addition is mostly associated with the effect of fiber/matrix interface debonding in the crack front, which increase the tendency of crack initiation and thereby lower the initiation toughness.

Intrinsically Brittle Matrix. The initiation toughness of the composite with an intrinsically brittle matrix is shown in Fig. 7 as a function of glass fiber content. Also shown in Fig. 7 is the initiation toughness of the rubber toughened composite and will be discussed in the next section. As can be seen in Fig. 7, the initiation toughness of the intrinsically brittle composite is relatively low and shows only a marginal increase as the glass fiber content increases. Insitu fracture observations indicate little plastic deformation in the crack tip region prior to crack initiation and the failure appears to be catastrophic. Further analysis, by treating the crack tip region as a hypothetical tensile specimen, show that the slight increase in the initiation toughness by glass fiber addition may be related to fiber strengthening effects within the composite[17]. Significant toughness increases in brittle matrix composites require a sufficiently weak fiber/matrix interface[18]. The interface properties in this system is probably not optimum. Frequently, in brittle-matrix composites the properties of the interface rather than that of the matrix dominate the overall fracture resistance.

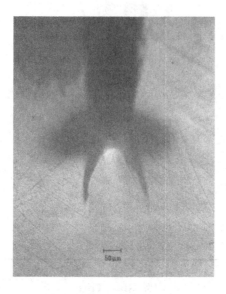

Figure 3 Insitu observation of deformation ahead of a crack tip in the unfilled polycarbonate.

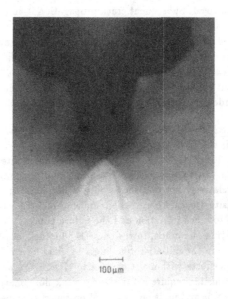

Figure 4 Insitu observation of deformation ahead of a crack tip in the unfilled PBT.

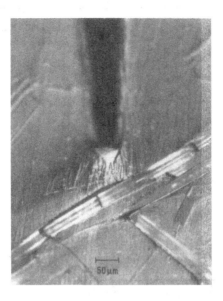

Figure 5 Insitu observation of crack initiation in the PBT composite with 35 wt% glass fibers shows localized crazes formed ahead of the crack tip.

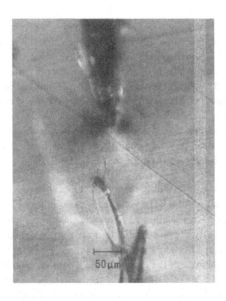

Figure 6 Insitu observation of crack initiation in the PC composite (35 wt% glass fibers) indicate crack-tip shear deformation in the matrix.

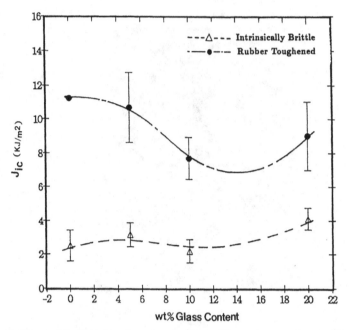

Figure 7 The initiation toughness vs. glass fiber content for the intrinsically brittle and rubber toughened matrix composites.

<u>Rubber Toughened Matrix</u> The initiation toughness for the rubber toughened composite is shown in fig. 7. Note that in this case, the initiation toughness is significantly higher than that of the composite without rubber toughening, i.e., the composite with intrinsically brittle matrix. Although the magnitude of the toughness increase due to rubber addition is reduced at higher glass fiber contents, the increase is still substantial. Note also that glass fiber additions slightly increase the initiation toughness of the untoughened composite as mentioned above, but initially decrease the toughness of the rubber toughened composite. Beyond about 10 wt% glass fibers, however, the initiation toughness of the rubber toughened composite is increased.

The large increase in the initiation toughness of the rubber toughened matrix material is found to relate to formation of a large volume of crazes induced by the rubber particles in the crack tip region[17]. On the other hand, as the glass fibers are added the amount of crazes in the matrix is greatly suppressed due to matrix stress shielding by glass fibers. In addition, void formation at fiber ends were also observed[17], which were found to facilitate crack advance by linking with the crack tip. As a consequence these two mechanisms together would then predict a rapid decrease in the initiation toughness as the glass fiber are introduced. However, the results in Fig. 7 shows a moderate drop of the initiation toughness as the glass fiber content increases. Further insitu examinations indicate that additional energy dissipation mechanism, namely the rubber particle promoted craze formation at fiber/matrix interface, was observed (see Fig. 8). Thus it appears that glass fibers contribute to additional toughness through this process of interaction with rubber particles. Such a toughening mechanism also suggests that in a hybrid composite the interaction between different phases may be important for toughness consideration.

Propagation Toughness

The fracture resistance curves for the rubber toughened composite, both in the unfilled case and the glass fiber filled cases, are shown in Fig. 9. It should be mentioned that such measurements for the composite with an intrinsically brittle matrix were not made, since the matrix material is brittle and failure was catastrophic after crack initiation. For the rubber toughened composite, however, a significant increase in the crack growth resistance is found, as shown in Fig. 9. Note, that in the unfilled case, with rubber toughening the J_R value is found to increase by more than a factor of 2, indicating that additional toughening can be obtained from the crack wake effects. This increase in propagation toughness can result from rubber particle bridging[19] and also from the closure pressure due to unloading effects of the craze zone in the crack wake[20]. Further analysis indicates that the effects of rubber particle bridging are relatively small and the crack wake effects due to unloading of crazes appear to be important in the crack propagation of the rubber toughened polymers.

In the glass fiber containing composites shown in Fig. 9, the propagation toughness is approximately the same for all three fiber contents, with its magnitude being on the order of the initiation toughness component. Note that in the presence of glass fibers craze formation at rubber particles is significantly reduced and hence the contribution to the propagation toughness by unloading of crazes in the crack wake is expected to be smaller. However, unloading effects of crazes formed at fiber/matrix interfaces may still occur. Together with the intrinsic toughening effects of glass fibers by fiber bridging and fiber pull-out, these crack-wake energy dissipation mechanisms result in the observed increase in the propagation toughness of the rubber toughened composite.

The direct assessment of the propagation toughness by constructing the J_R versus Δa curve were not made for the composites with intrinsically tough matrices. However, a simple estimation of the propagation contribution to overall toughness may be provided by using the fracture energy divided by the ligament area for pre-cracked tensile specimens. The results are shown in Fig. 10 for the PC and PBT composites. As can be seen, the estimated propagation component in the PBT composite shows a decrease where as in the PC composite an increase is observed. This result is consistent with the insitu observation that in the PBT composites a major localized craze is formed ahead of crack tip, whereas in PC composite energy are dissipated uniformly throughout the crack tip region by shear deformation during crack propagation. Fiber pull-out and crack bridging by unbroken fibers are also observed in both cases (see Fig. 11), which can also dissipate energy during crack propagation. However, it appears that in PBT composites the toughening mechanisms by fiber pull-out and bridging are limited by the formation of localized crazes, which facilitate crack propagation by linkage between the crack tip and debonded fiber ends and thereby decreasing the contribution of fiber pull-out and bridging effects. Accordingly, the overall toughness of the PBT composite, by combining the initiation and propagation toughness, would then show a significant overall decrease as the glass fiber content is increased. On the other hand, in the PC composites, the overall toughness is expected to show an increase, since the initiation toughness is relatively unaffected but the propagation toughness component may provide a significant increase through fiber toughening and crack tip energy dissipation by shear deformation.

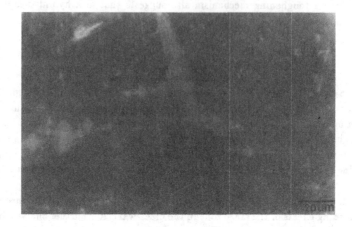

Figure 8 Insitu observation ahead of crack tip region in rubber toughened composite shows craze formation at fiber/matrix interfaces. The craze formation is promoted by the rubber particles.

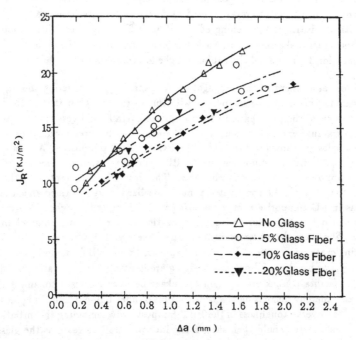

Figure 9 The fracture resistance curves for the rubber toughened matrix composites.

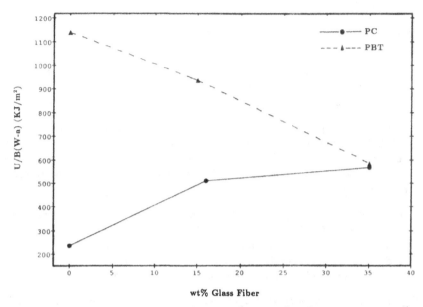

Figure 10 The estimated propagation toughness using fracture energy divided by the specimen ligament area for the intrinsically tough matrix composites.

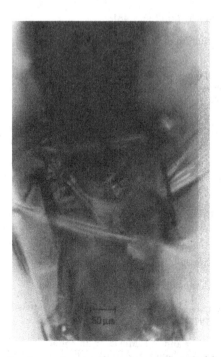

Figure 11 Insitu fracture observation shows fiber bridging and pull-out.

CONCLUSIONS

We draw the following major conclusions :

1. There is no significant role of matrix on the toughening behavior of a brittle-matrix (SAN) polymer composite, consistent with the predominant role of the fiber/matrix interfaces in such systems.

2. In the intrinsically brittle but rubber toughened amorphous matrix composite (SAN/ABS), the substantial additional dilatational toughening contributions from the rubber particles are retained at higher glass levels even though glass fibers limit crazing at rubber particles. Glass fibers and rubber particles can interact synergistically to enhance toughness.

3. In the intrinsically tough amorphous matrix composite (PC) glass fibers do not affect initiation toughness but enhance the propagation toughness. The lack of influence of glass fibers on the initiation toughness is related to the lack of influence of glass fibers on uniform shear deformations in the crack tip region. The increase in propagation toughness is associated with fiber bridging and pull-out.

4. In the intrinsically tough semi-crystalline matrix composite (PBT) glass fibers decrease the initiation toughness and do not appear to provide any propagation toughness component. This appears to be related to the role of glass fibers in craze localization in the crack tip region.

REFERENCE

1. Z. Hashin, J. Appl. Mech. **50** (1983) 481

2. D. Hull, in "An Introduction to Composite Materials," (Cambridge University Press, 1981) p.1

3. C.Y. Yue and W.L. Cheung, in 7th International Conference on Deformation, Yield and Fracture of Polymers, Cambridge, U.K., April, 1988 (The Plastic and Rubber Institute, 1988) p. 84

4. B.J. Briscoe and D.R. Williams, in 1st International Conference on Deformation and Fracture of Composite, Manchester, U.K., March 1991 (The Plastic and Rubber Institute, 1991) p.17

5. S.H. Jao, in "Studies of Interface Properties and Influence of Fiber Coating in Composite Materials," Ph. D. Thesis, MIT (1989)

6. G. Marom and R.G.C. Arridge, Mater. Sci. Eng. **23** (1976) 23

7. D.G. Peiffer, J. Appl. Polym. Sci. **24** (1979) 1451

8. T. Bessell and J.B. Shortfall, J. Mater. Sci. **10** (1975) 2035

9. B.L. Hollingsworth, Composites **1** (1969) 28

10. W.V. Titow and B.J. Lanham, in "Reinforced Thermoplastics," (Applied Science Publishers, London, 1975) p.159

11. J.F. Mandell, D.D. Huang and F.J. McGarry, in "Short Fiber Reinforced Composite Materials," edited by B.A. Sanders, ASTM STP 772, p.3

12. J.G. Williams, in "Fracture Mechanics of Polymers," (Ellis Horwood, Chichester, 1984) p.237

13. W.E. Wolstenholme, S.E. Pregun and C.F. Stark, in "High Speed Testing," Vol. VI (John Wiley & Sons, New York, 1963) p.119

14. C.B. Bucknall, in "Toughened Plastics," (Applied Science Publishers, London, 1977) p.298

15. P.W. Early and S.J. Burns, Inter. J. Fract. **16** (1980) 397

16. P. Chang and J.A. Donovan, J. Mater. Sci. **24** (1989) 816

17. S.V. Nair and M.L. Shiao, in "Fracture Resistance of a Glass Fiber Reinforced Rubber Modified Thermoplastic Hybrid Composite," submitted to the Journal of Material Science.

18. S.V. Nair, J. Am. Ceram. Soc. **73** (1990) 2839

19. S. Kunz-Douglass, P.W.R. Beaumont and M.F. Ashby, J. Mater. Sci. **15** (1980) 1109

20. A.G. Evans, Z.B. Ahmad, D.G. Gilbert and P.W.R. Beaumont, Acta Metall. **34** (1986) 79

STUDIES ON THE THERMAL TRANSITION OF SIDE CHAIN
LIQUID CRYSTALLINE POLY(OXETANES)S

Yusuke Kawakami

Department of Applied Chemistry
School of Engineering
Nagoya University
Chikusa, Nagoya 464, Japan

Abstract

Oxetanes with different mesogenic and spacer groups were synthesized. Some of the polymers obtained by cationic polymerization showed smectic phase. Not only cyano substituted biphenyl but also fluorine substituted biphenyl were found to be good mesogenic groups for these liquid crystalline polymers. Radially substituted mesogen gave nematic phase.

Introduction

Much attention has been paid recently to liquid crystalline polymers, specially to side chain type, because of their potential application in electronics. Polysiloxanes, polyacrylates, and polymethacrylates are usually used as main chain components[1]. There are only few examples in which other main chain like polyolefine, poly(vinyl ether), polyphosphazene, or polyisocyanate are used[2-13]. We have been interested in the effects of the structure of main chain, effects of the length and the structure of the spacers, and effects of the structure of mesogenic group on the liquid crystalline phase exhibited[14-16]. In this article, I would like to report the first example of liquid crystalline polymers which have polyoxetane main chain, and fluorine substituted biphenyl as a new mesogenic group. Liquid crystalline polymers having radially connected mesogenic groups are also reported.

$$X = O(CH_2)_nH \quad : \quad PolyOX\text{-}n\text{-}m$$
$$= CN \quad : \quad PolyOX\text{-}CN\text{-}m$$
$$= F \quad : \quad PolyOX\text{-}F\text{-}m$$

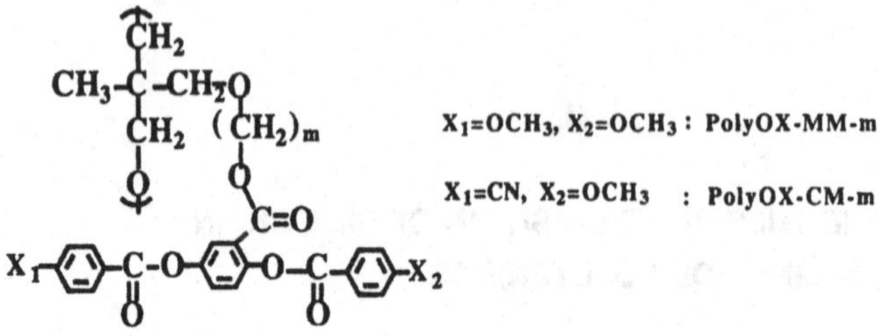

$X_1=OCH_3$, $X_2=OCH_3$: PolyOX-MM-m

$X_1=CN$, $X_2=OCH_3$: PolyOX-CM-m

Structure of Polymers

Experimental

Synthesis of Monomers

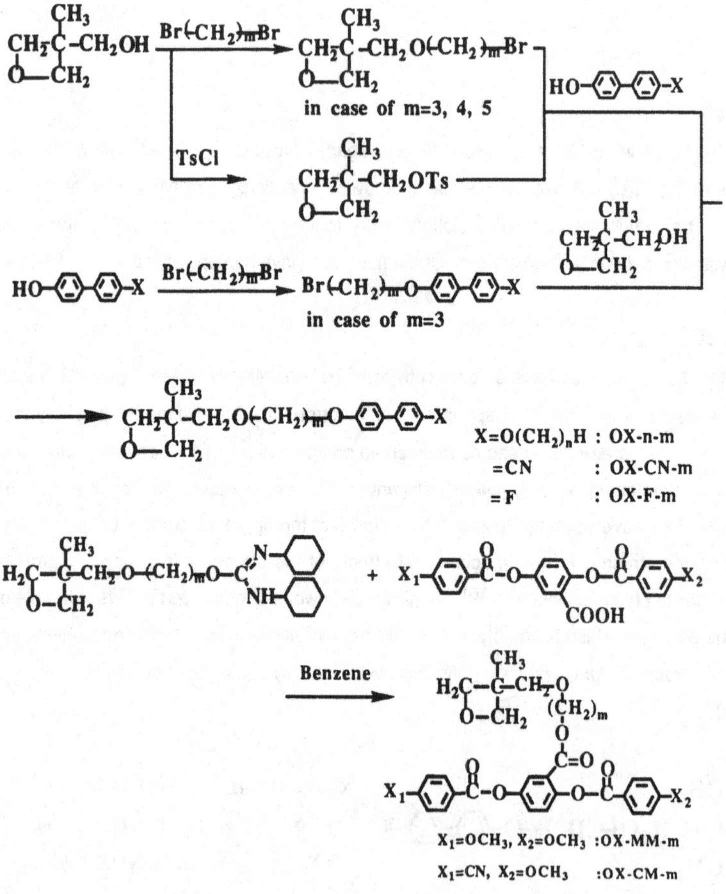

in case of m=3, 4, 5

in case of m=3

$X=O(CH_2)_nH$: OX-n-m

$\quad =CN$: OX-CN-m

$\quad =F$: OX-F-m

Benzene

$X_1=OCH_3$, $X_2=OCH_3$:OX-MM-m

$X_1=CN$, $X_2=OCH_3$:OX-CM-m

The monomers are abbreviated as OX-n-m, OX-CN-m, or OX-F-m in case of having alkoxy, cyano, or fluoro substituted biphenyl mesogen, respectively, and OX-MM-m or OX-CM-m in case of radially attached mesogen, where n indicates the number of carbon atoms in alkoxy group and m the number of carbon atoms of methylene groups in the spacer.

Typical examples of the synthesis are given

3-{4-(4-Cyanobiphenyl-4'-yloxy)butoxymethyl}-3-methyloxetane(OX-CN-4)

To sodium hydride(0.030 g, 1.2 mmole) dispersed in DMF(5 cm^3), 4-cyano-4'-hydroxybiphenyl(0.20 g, 1.0 mmole) was added portionwise. The reaction system was stirred for 30 min at room temperature and for further 10 min at 60°C. To this solution, 3-(4-bromobutoxymethyl)-3-methyloxetane(0.30 g, 1.3 mmole) in DMF(5 cm^3) was added dropwise during 10 min, and the reaction system was stirred for 3 h at room temperature. After 3 h, DMF was removed under vacuum, and chloroform(50 cm^3) and water(30 cm^3) were added to the residual solid. Chloroform layer was separated and washed with aq. sodium carbonate(30 cm^3). The product was isolated by column chromatography after drying and evaporating the solvent. Rf=0.18(eluent; hexane: ether: chloroform=2: 1: 1). Yield 92 %. mp 48.1 $^\circ$C.
Chemical shifts:1.29(s, 3H, C$\underline{H}_3$), 1.80(m, 4H, CH$_2$C$\underline{H}_2$C$\underline{H}_2$CH$_2$), 3.47(s, 2H, C(CH$_3$)-C$\underline{H}_2$), 3.54(t, 2H, J=6.0 Hz, CH$_2$OC$\underline{H}_2$CH$_2$), 4.03(t, 2H, J=6.0 Hz, OC$\underline{H}_2$), 4.34(d, 2H, J=5.8 Hz), 4.50(d, 2H, J=5.8 Hz), 6.97(d with fine coupling, 2H, J=8.8 Hz), 7.51(d with fine coupling, 2H, J=8.8 Hz), 7.61, 7.68(two d with fine coupling, 4H, J=7.8 Hz).

Other monomers were synthesized similarly

3-{4-(4-Fluorobiphenyl-4'-yloxy)butoxymethyl}-3-methyloxetane(OX-F-4)
Rf=0.18(eluent; hexane: ether: chloroform=2: 1: 1). Yield 65 %. mp 42.7 $^\circ$C.
Chemical shifts: 6.93(d with fine coupling, 2H, J=8.8 Hz), 7.07(t with fine coupling, 2H, J=8.8 Hz), 7.44(d with fine coupling, 2H, J=8.8 Hz), 7.47(q with fine coupling, 2H, J$_1$=8.8 Hz, J$_2$=5.4 Hz).
Other protons of this monomer appeared at the corresponding positions with OX-CN-4.

3-{5-(4-Cyanobiphenyl-4'-yloxy)pentyloxymethyl}-3-methyloxetane (OX-CN-5)
Rf=0.18(eluent; hexane: ether: chloroform=2: 1: 1). Yield 68 %. mp 49.4 $^\circ$C.
Chemical shifts:1.29(s, 3H, C$\underline{H}_3$), 1.60, 1.82(m, 4H; quint, 2H: CH$_2$C$\underline{H}_2$C$\underline{H}_2$C$\underline{H}_2$CH$_2$), 3.46(s, 2H, C(CH$_3$)-C$\underline{H}_2$), 3.49(t, 2H, J=6.0 Hz, CH$_2$OC$\underline{H}_2$CH$_2$), 4.00(t, 2H, J=6.0 Hz, OC$\underline{H}_2$), 4.34(d, 2H, J=5.8 Hz), 4.56(d, 2H, J=5.8 Hz), 6.97(d with fine coupling, 2H, J=8.8 Hz), 7.51(d with fine coupling, 2H, J=8.8 Hz), 7.61, 7.68(two d with fine coupling, 4H, J=7.8 Hz).

3-{5-(4-Fluorobiphenyl-4'-yloxy)pentyloxymethyl}-3-methyloxetane (OX-F-5)
Rf=0.18(eluent; hexane: ether: chloroform=2: 1: 1). Yield 72 %. Liquid.
Chemical shifts: 1.58, 1.80(m, 4H; quint, 2H: CH$_2$C$\underline{H}_2$C$\underline{H}_2$C$\underline{H}_2$CH$_2$), 3.98(t, 2H, J=6.0 Hz, OC$\underline{H}_2$), 4.50(d, 2H, J=5.8 Hz), 6.93(d with fine coupling, 2H, J=8.8 Hz), 7.08(t with fine coupling, 2H, J=8.8

Hz), 7.44(d with fine coupling, 2H, J=8.8 Hz), 7.47(q with fine coupling, 2H, J_1=8.8 Hz, J_2=5.4 Hz). Other protons of this monomer appeared at the corresponding positions with OX-CN-5.

3-[5-{2,5-Bis(4-methoxybenzoyloxy)benzoyl}pentoxymethyl)-3-methyloxetane (OX-MM-5)

To benzene(40 cm^3) solution of 2,5-bis(4-methoxybenzoyloxy)benzoic acid(753.9 mg, 1.78 mmole), 2-[5-{3-methyloxetan-3-ylmethoxy}pentyloxy]-1,3-dicyclohexylisourea(1.18 g, 3.0 mmole) was added dropwise and the reaction system was refluxed for 3h. After formed urea was removed by filtration at room temperature, the product was isolated by column chromatography. Rf=0.13(eluent; hexane:ether=1:1). Yield 52 %. mp 92.3 °C.

Chemical shifts:1.18~1.64(m, 6H, CH$_2$CH$_2$CH$_2$CH$_2$CH$_2$),1.25(s, 3H, CH$_3$), 3.35(t, 2H, J=6.4 Hz, OCH$_2$CH$_2$), 3.40(s, 2H, C(CH$_3$)-CH$_2$), 3.88, 3.89(two s, 3H each, OCH$_3$), 4.14(t, 2H, J=6.4 Hz, CO$_2$CH$_2$), 4.30(d, 2H, J=5.8 Hz), 4.45(d, 2H, J=5.8 Hz), 6.98(d with fine coupling, 4H, J=8.8 Hz), 7.24(d, 1H, J=8.8 Hz), 7.45(dd, 1H, J_1=8.8 Hz, J_2=2.8 Hz), 7.87(d, 1H, J=2.8 Hz), 8.15, 8.16(two d with fine coupling, 4H, J=8.8 Hz) .

3-[5-{2-(4-Methoxybenzoyl)5-(4-cyanobenzoyloxy)benzoyloxy]pentoxymethyl)-3-methyloxetane (OX-CM-5)

Rf=0.13(eluent; hexane:ether=1:1). Yield 35 %. mp 94.1 °C.

Chemical shifts:1.19~1.70(m, 6H, CH$_2$CH$_2$CH$_2$CH$_2$CH$_2$)1.25(s, 3H, CH$_3$), 3.35(t, 2H, J=6.4 Hz, OCH$_2$CH$_2$), 3.40(s, 2H, C(CH$_3$)-CH$_2$), 3.89(s, 3H, OCH$_3$), 4.15(t, 2H, J=6.4 Hz, CO$_2$CH$_2$), 4.30(d, 2H, J=5.8 Hz), 6.05(d, 2H, J=5.8 Hz), 6.99(d with fine coupling, 2H, J=8.8 Hz), 7.28(d, 1H, J=8.8 Hz), 7.46(dd, 1H, J_1=8.8 Hz, J_2=2.8 Hz), 7.83(d with fine coupling, 2H, J=8.2 Hz), 7.90(d, 1H, J=2.8 Hz), 8.16, 8.31(two d with fine coupling, 4H, J=8.8 Hz) .

Polymerization

Polymerizations were carried out in dichloromethane at 0°C. Polymers were purified by repeated reprecipitation. The absence of monomer in polymer was checked by thin layer chromatography and GPC.

Analysis

DSC analysis was carried out on a SEIKO thermal analysis system model SSC 5500 equipped with DSC 100 with the heating rate of 3°C/min. The transition temperature is given at the point where transition starts. For the samples which have multiple transitions, peak temperatures are given. The temperature was calibrated by the use of indium and tin metals as the standard(156.6°C for indium and 231.9°C for tin).

Optical polarization micrographs were taken on a Nikon optical polarization micrograph model OPTIPHOTO-POL equipped with Mettler thermal analysis system model FP800 with FP82 hot stage and FP 80 controller. In observing the texture of the samples, they were slowly cooled from isotropic state to a little lower temperature than the peak temperature, and annealed at the temperature. Pictures were taken at appropriate interval.

Table 1. Polymerization of Mesogenic Monomers by Cationic Initiators[a]

Monomer	Conc. (mol/l)	Initiator (mol %)	Yield (%)	Mw[b] x10^3	Mn x10^3	Mw/Mn
OX-1-3	1.4	1.15	68.0	46.8	11.8	3.9
OX-2-3	1.1	1.80	83.3	15.5	6.8	2.3
OX-3-3	1.2	1.41	59.8	14.6	6.6	2.2
OX-4-3	0.89	1.14	70.0	24.4	11.0	2.2
OX-1-4	2.1	1.67	96.0	36.1	12.8	2.8
OX-2-4	0.93	1.01	59.3	22.6	9.5	2.4
OX-3-4	1.2	1.21	68.2	13.4	6.1	2.2
OX-4-4	1.1	0.97	57.2	15.0	7.5	2.0
OX-CN-0	1.0	1.00	43.0	16.6	10.7	1.6
OX-CN-3	0.76	1.00	48.9	39.3	16.6	2.4
OX-CN-4	2.7	1.20	38.4	85.2	38.5	2.2
OX-CN-5	1.0	1.00	33.9	23.0	13.0	1.8
OX-F-0	1.0	1.00	91.2	c)	c)	c)
OX-F-3	0.43	0.91	92.0	16.7	8.3	2.0
OX-F-4	1.0	1.00	99.3	45.6	26.6	1.7
OX-F-5	1.0	1.00	88.2	29.4	13.3	2.2
OX-MM-5[d]	1.2	1.00	58.2	-	-	-
OX-CM-5[d]	1.3	1.00	59.5	-	-	-

a) $BF_3 \cdot OEt_2$ was used as an initiator in CH_2Cl_2, Temp: 0°C, Time: 20~28 h.

b) Estimated by GPC correlating to standard polystyrene.

c) Not soluble in ordinary solvent.

d) Et_3OBF_4 was used as an intiator.

Results and Discussion

The result of polymerization by $BF_3 \cdot OEt_2$ or Et_3OBF_4 as an initiator are shown in Table1. All monomers gave reasonable to good yield in the polymerization. Cyano substituted monomers gave the polymers in rather low yield. Cyano group might act as a weak basic group to suppress cationic polymerization. Fluorine substituted monomer can be easily polymerized under cationic condition. Monomers with radially substituted mesogen having ester function gave moderate yield. Polymers had expected chemical structure formed by ring-opening polymerization of monomer as studied by ^{1}H- and ^{13}C-NMR. Typical example is given for poly[3-{4-(4-cyanobiphenyl-4'-yloxy)butoxymethyl)-3-methyloxetane] (polyOX-CN-4).

Chemical shifts: ^{1}H-NMR; 0.88(broad s, 3H, C$\underline{H}_3$), 1.60-1.89(m, 4H, CH$_2$C$\underline{H}_2$C$\underline{H}_2$CH$_2$), 3.16(broad s, 4H, -C$\underline{H}_2$CC$\underline{H}_2$O- main chain), 3.22(broad s, 2H, C(CH$_3$)-C$\underline{H}_2$), 3.38((m, 2H, CH$_2$OC$\underline{H}_2$CH$_2$), 3.92(t, 2H, J=5.8 Hz, OC$\underline{H}_2$), 6.88(d, 2H, J=8.8 Hz, Hc), 7.42(d, 2H, J=8.8 Hz, Hd), 7.52, 7.59(two d, 4H, J=8.6 Hz, He, Hf). ^{13}C-NMR; 17.66(C1), 26.37(C6, C7), 41.50(C2), 68.09(C3), 71.26, 73.77(C4, C5), 74.55(C8), 110.47(C16), 115.35(C10), 119.34(C17), 127.34(C14), 128.67(C11), 131.67(C12), 132.95(C15), 145.44(C13), 160.13(C9).

Table 2. Thermal Behavior of PolyOX's

Polymer	DSC(Heating)[a]		Texture under Optical
	Tg	Transition	Polarization Microscopy
		(°C)	
PolyOX-1-3	103	115	n.d.[b]
PolyOX-2-3	c)	147	n.d.
PolyOX-3-3	c)	149	n.d.
PolyOX-4-3	c)	148	n.d.
PolyOX-1-4	c)	139	n.d.
PolyOX-2-4	c)	(148,154)	n.d.
PolyOX-3-4	c)	(120,125,	
		136,146)	n.d.
PolyOX-4-4	108	(145,151)	n.d.
PolyOX-CN-0	c)	96	c)
PolyOX-CN-3	33	c)	c)
PolyOX-CN-4	c)	84	fan
PolyOX-CN-5	16	96	fan
PolyOX-F-0	c)	150	c)
PolyOX-F-3	c)	72	n.d.
PolyOX-F-4	c)	103	fan
PolyOX-F-5	c)	90	fan
PolyOX-MM-5	40	83	schlieren
PolyOX-CM-5	48	98	schlieren

a) Values on second heating. Values in parentheses are
    temperatures at peak.

b) n.d. : Phase could not be determined.

c) Did not show.

Thermal behavior is summarized in Table 2.

Only one thermal transition was observed in DSC of polyOX-n-3's with three methylene spacers and polyOX-1-4. Two or more thermal transitions were seen in DSC for other polyOX-n-4's(n=2,3,4). Some organized phase was observed in optical polarization micrograph below the transition temperature for these polymers.

It was found that polyoxetane acts as a possible main chain component for a side chain liquid crystalline polymers with alkoxy functionalized biphenyl as mesogenic group, however, it could not be drawn definite conclusion as if the polymers really show liquid crystalline state or not. Further study is needed to find out a suitable condition to obtain liquid crystalline state.

Although it is generally admitted that polar cyano group on mesogenic group gives better organized phase than alkoxy group[1], polyOX-CN-0 and polyOX-CN-3 did not show any thermal transition other than glass transition temperature in DSC. No liquid crystalline state was observed either, in optical polarization micrograph. Contrary to this, polyOX-CN-4 and polyOX-CN-5 showed discrete transition starting at 84°C and 96°C, respectively and showed discrete mesophase structure in both cooling and heating processes. Typical mesophase structure seen in optical polarization micrograph is shown in Figure1 and 2. Batonett structure formed at 82.5°C or 96.5°C in about 30 min was changed into well-developed fan like focal conic structure under further annealing.

Figure 1.　Optical polarization micrograph of polyOX-CN-4 annealed at 82.5 °C for 2 days.

Figure 2.　Optical polarization micrograph of polyOX-CN-5 annealed at 96.5 °C for 2 days.

Compared with the fact that 6 methylene and ester function are needed to take nematic phase for polyacrylate with cyanobiphenyl or alkoxybiphenyl mesogenic group[1], shorter spacer, namely four methylene and ether function, was enough for polyoxetane with cyano substituted biphenyl as mesogenic group to take more ordered smectic phase. In such polymers, the main chain polyether is considerably flexible compared with polymethacrylate or polyacrylate, which seems to make it easy to organize mesogenic groups.

Structure of mesogenic group is also one of the important factor to determine the mesophase structure. Related to this, fluoro substituted biphenyl was of interest as a possible mesogenic group. Fluorine atom has a sphere like structure and might disturb the requirement for the shape of biphenyl derivative as a mesogenic group.

PolyOX-F-m's showed only one transition. PolyOX-F-3 showed sanded structure. PolyOX-F-4 took discrete smectic fan architecture as shown in Figure 3. PolyOX-F-5 which showed batonnet structure in early stage of annealing, also took well-developed structure as shown in Figure 4 after prolonged annealing. This is the first example in which fluorine atom is used as a tail group for biphenyl mesogen in liquid crystalline polymer.

Polymers with radially attached mesogen showed nematic phase as shown in Figure 5 and 6.

It is of interest to mention that mesogen with electron donating and withdrawing substituents in the molecule showed clearer mesophase. There may be some charge-transfer interaction between mesogenic groups which helps the organization of mesogenic molecules.

525

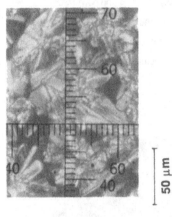

Figure 3. Optical polarization micrograph of polyOX-F-4 annealed at 102.2 °C for 2 days.

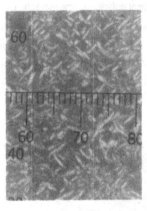

Figure 4. Optical polarization micrograph of polyOX-F-5 annealed at 89.2 °C for 2 days.

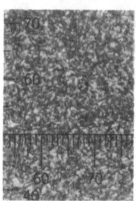

Figure 5. Optical polarization micrograph of polyOX-MM-5 annealed at 84.3 °C for 1.5 days.

Figure 6. Optical polarization micrograph of polyOX-CM-5 annealed at 98.1 °C for 1 day.

Conclusion

It was shown that polyoxetane can act as a main chain of liquid crystalline polymers. Polyoxetanes with cyanobiphenyl as a mesogenic group showed smectic liquid crystalline phase. Fluorobiphenyl was found to be a novel and good mesogenic group for polyoxetane to take smectic phase.   Radially attached mesogenic groups gave nematic mesophase.

Acknowledgments

Financial supports from a Grant-in-Aid for Scientific Research on Priority Areas, New Functionality Materials-Design, Preparation and Control (02205057) and from a Grant-in-Aid for Developmental Scientific Research(63850184) are gratefully acknowledged.

References

1. Plate, N. A.; Shibaev, V. P., Ed.; *Comb-Shaped Polymers and Liquid Crystals* : Plenum: New York, 1987.

2. Magagnini, P. L.; Andruzzi, F. Benetti, G. F. *Macromolecules*, 1980, *13*, 12.

3. Mallon, J. J.; Kantor, S. W. *Macromolecules*, **1989**, *22*, 2070.

4. Mallon, J. J.; Kantor,S. W. *Macromolecules*, **1990**, *23*, 1249.

5. Percec, V. *Makromol. Chem., Macromol. Symp.*, **1988**, *13/14*, 397.

6. Kim, C.; Allcock, H. R. *Macromolecules*, **1987**, *20*, 1727.

7. Singler, R. E.; Willingham, R. A. , Lenz, R. W.; Furukawa, A. *Macromolecules*, **1987**, *20*, 1728.

8. Kojima, M.; Magill, J. H. *Polymer*, **1989**, *30*, 579.

9. Allcock, H. R.; Kim, C. *Macromolecules*, **1990**, *23*, 3881.

10. Moriya, K.; Yano, S.; Kajiwara, M. *Chem. Lett.*, **1990**, 1039.

11. Shiraishi, K.; Sugiyama, K. *Chem. Lett.*, **1990**, 1697.

12. Percec, V.; Hahn, B. *Macromolecules*, **1989**, *22*, 1588.

13. Durairaj, B.; Samulski, E. T.; Shaw, T. M. *Macromolecules*, **1990**, *23*, 1229.

14. Kawakami, Y.; Sakai, Y.; Okada, A. *Polym. J.*, **1990**, *22*, 705.

15. Kawakami, Y.; Takahashi, K.; Hibino, H. *Polym. Bull.*, **1991**, *25*, 439.

16. Kawakami, Y.; Takahashi, K. *Macromolecules*, **1991**, *24*, 4531.

PROCESSABLE RIGID POLYMERS CONTAINING SPIROACETAL UNITS

Kwang-Sup Lee, John M. Rhee and Kil-Yeong Choi

Polymer Lab.2, Korea Research Institute of
Chemical Technology, Taejeon, Korea

Hyun-Min Kim and Soo-Min Lee

Dept. of Chemistry, Han-Nam University, Taejeon
Korea

INTRODUCTION

Recent interest is very high in the practical application of rigid polymers as high performance materials. Some rigid polymers have already been commercialized or are in the development stage. However, one of the glaring drawbacks with rigid polymers has been the lack of processibility because of their insolubility and infusibility resulting from polymer backbone stiffness. To solve this problem, the current molecular designs for processable rigid polymers employ the use of (i) flexible linkages or spacers in rigid polymer main chains,[1,2] (ii) non-linear molecular moieties,[3,4] (iii) copolymerization for breaking the regularity of polymer repeating units[5] and (iv) the derivatization of side chains onto polymer backbone.[6-13] In this work, the molecular design concepts (i) and (iv) were used for improving the processability of the rigid polyspiroacetals.

In general, spiroacetals result when pentaerythritol is reacted with the carbonyl compounds. This synthetic method was employed to make poly-spiroacetals by several researchers.[14-17] Many of polyspiroacetals show excellent transparency, good mechanical strength, good hardness, low birefringence and heat and water resistance. Therefore, some of polyspiroacetals are considered as useful materials for surface coatings, photomemory disks, curing agents of epoxy resins, etc. Especially, polyspiroacetals of rigid rod-like chain structures such as polymers which have spiroacetal units connected directly to phenylene groups or double strand polyspiroacetals, which might be used as high performance materials if they could be made more processible. Thus, we have focused our attention

Frontiers of Polymer Research, Edited by P.N. Prasad and
J.K. Nigam, Plenum Press, New York, 1991

on solving the problems related to the processing of rigid polyspiroacetals. In the following paper we report the synthesis, characterization and physical properties of rigid polyspiroacetals with flexible side chains, and polyspiroacetals having flexible spacers in polymer main chains.

EXPERIMENTAL

Materials

Dowex 50W-X16 ion exchange resin (Dow chemical) was dried by reflux for 24 hrs in 30% sulfuric acid. Dimethylsulfoxide(DMSO)(Aldrich), benzene (Merck) and chloroform(Aldrich) were dried over calcium hydride, calcium sulfate and calcium chloride, respectively, and subsequently distilled. Adipoyl chloride(Fluka) and sebacoyl chloride were purified by distillation under reduced pressure. Pentaerythritol(Aldrich), 4-hydroxybenzalde-hyde and benzyltrimethyl ammonium chloride(BTMAC)(Aldrich) were used without further purification.

Monomer synthesis

2,5-Dialkoxyterephthaloyl chlorides were obtained by saponification and chlorination after Claisen condensation of diethyl 2,5-dihydroxytere-phthalate with alkyl bromide as reported in literature.[6] Dodecandioyl chloride and hexadecandioyl chloride was prepared by chlorination of dodecandioic acid and hexadecandioic acid using excess amount of thionyl chloride at 90-100 °C for 1-2 hrs.

Synthesis of 2,4,8,10-tetraoxaspiro[5,5]undecane-3,9-bishydroxybenzene (TSUH): A mixture with 48g of dried Dowex resin, 266g of DMSO and 330g of benzene was refluxed at 92-94°C for 10hrs. After removing the water which was collected in a Dean-Stark trap, 50g(0.3mol) of pentaerythritol and 91.6g(0.74mol) of 4-hydroxybenzaldehyde were added to the above mixture and refluxed at the same temperature for an additional 24 hrs. Then, after the Dowex resin was filtered off, the residue (filtrant) was slowly poured into benzene, stirring constantly. The precipitated products were separated by filtration, washed with benzene several times, and dried in a vacuum oven. The yield of TSUH was 75g(71%); m.p. 253°C(249-257 °C).[18]

$C_{19}H_{20}O_6$(344.35) Calc. C 66.27 H 5.58
                          Found C 65.63 H 5.58

IR(KBr pellet): 3325(ν(O-H)), 1610, 1520(ν(C=C)), 1260, 1200, 1160, 1120, 1070, 1010(ν(C-O); spiroacetal)

^{1}H NMR(CDCl$_3$): δ=3.5-4.0(8H, m, -CH$_2$-), 4.6-4.8(2H, d, -CH=),
5.47(2H, s, -OH), 6.78-6.87(4H, m, Ar-CH), 7.27-7.35(4H,m,Ar-CH)

Polymer synthesis

A typical polymerization method of PSA-I series of TSUH with
2.5-dialkoxyterephthaloyl chloride is as follows: to a mixture of 1.2g
(2.9×10^{-3}mol) of TSUH, 0.23g(5.8×10^{-3}mol) of sodium hydroxide and 0.2g
(1×10^{-3}mol) of benzyltrimethylammonium chloride(BTMAC) in 10ml of water, a
solution of 1.7×10^{-3}mol of acid chlorides in 10ml of chloroform is slowly
added dropwise and reacted at room temperature for 1hr.   In each case,
polymers are precipitated during the polycondensation. When the reaction
is complete, the polymers are filtered off, washed with water and methanol,
recrystallized from phenol/CCl$_4$(3:2) mixture, and dried in a vacuum oven.

Synthesis of polymer PSA-II series: To a soulution of 6.0×10^{-3}mol of
sodium hydroxide in 10ml distilled water, 0.01g(5×10^{-5}mol) of benzyltrimethyl ammonium chloride(BTMAC) was added while stirring. Then 2.5g
(6.0×10^{-3}mol) of TSUH was added, followed by dropwise addition using the
syringe of a solution of diacid chloride(3.0×10^{-3}mol) in 5ml chloroform.
After stirring for 1hr, the precipitated product filtered off, washed with
warm water and methanol and dried at 50^{0}C in a vacuum oven for 24hrs.   All
stirring speeds during the reaction were kept constant for comparison of
the obtained polymer properties.

Measurements

IR spectra were obtained with a Shimadzu IR-435 IR spectrophotometer.
A Brucker AM 300(300 MHz) NMR spectrometer was employed for the NMR
analyses. A differential scanning calorimeter and thermogravimetric
analyzer(Model 9900) of Dupont were used for the study of thermal behavior
and thermal stability of polymers. Wide angle X-ray patterns were recorded
using nickel-filtered CuK$_\alpha$ (λ=1.5406A) radiation on a RigaKu Geiger Flex
D-Max X-ray diffractometer.

RESULTS AND DISCUSSION

Two polymer series, ie. wholly aromatic polyspiroacetals having
flexible side chains(PSA-I series) and polyspiroacetals containing
flexible spacers in polymer backbone(PSA-II series) were prepared by an
interfacial polymerization reaction of spiroacetal monomer, TSUH, and
disubstituted aromatic acid chlorides or aliphatic diacid chlorides using
the phase transfer catalyst, BTMAC.

Scheme

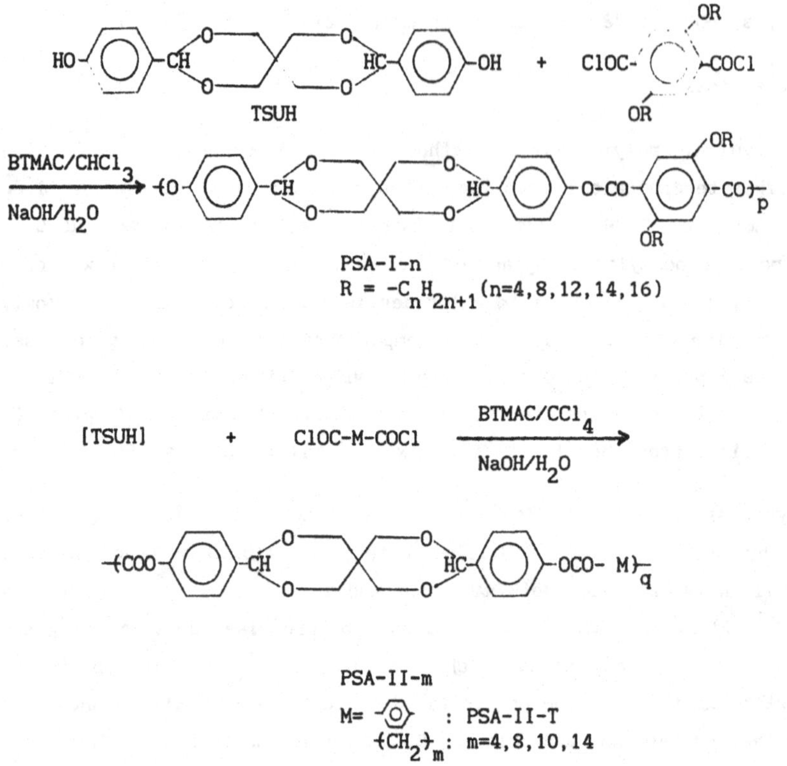

HO~⟨◯⟩-CH(O-⟨◯⟩-O)HC-⟨◯⟩-OH + ClOC-⟨◯⟩(OR)(OR)-COCl

TSUH

$\xrightarrow[\text{NaOH/H}_2\text{O}]{\text{BTMAC/CHCl}_3}$ ⟨O-⟨◯⟩-CH(O-⟨◯⟩-O)HC-⟨◯⟩-O-CO-⟨◯⟩(OR)(OR)-CO⟩_p

PSA-I-n

$R = -C_nH_{2n+1}$ (n=4,8,12,14,16)

[TSUH] + ClOC-M-COCl $\xrightarrow[\text{NaOH/H}_2\text{O}]{\text{BTMAC/CCl}_4}$

⟨COO-⟨◯⟩-CH(O-⟨◯⟩-O)HC-⟨◯⟩-OCO- M⟩_q

PSA-II-m

M= ⟨◯⟩ : PSA-II-T

$\{CH_2\}_m$: m=4,8,10,14

The structures of the obtained polymers were easily identified by IR or ^{1}H-NMR spectroscopy. In the case of polymer PSA-I series, strong absorption bands resulting from spiroacetal groups appeared at 1000-1300cm^{-1}, a carbonyl group of ester linkage at 1750cm^{-1}, and a CH-rocking vibration of alkoxy side chain at 720cm^{-1} in the IR spectra. The peaks which appeared at 3.5-5.5ppm in the ^{1}H-NMR spectra were attributed to the spiroacetal ring protons. Aliphatic protons on the side chains exhibited signals at 0.8-2ppm and aromatic protons showed signals at 7.5-7.8ppm.

The structures of polymer PSA-II series were also substantiated by IR spectroscopy. Strong absorption bands in IR spectra appeared at 1735cm^{-1} (C=O stretching of ester linkage) and 1000-1250cm^{-1} (C-O-C stretching of spiroacetal groups).

The elemental analysis results, yields, molecular weights, and inherent viscosities of PSA-I and PSA-II series are listed in Table.1. Satisfactory C/H ratios by elemental analysis were obtained for all

Table 1. Elemental Analysis Data, Yields, and Molecular Weights or Inherent Viscosities(η inh) of Polyspiroacetals, PSA-I and PSA-II Series.

Polymers	C(%)		H(%)		Yield	MW [a]	η inh
	Cal	Found	Cal	Found	(%)		(dl/g)
PSA-I-4	67.90	67.63	6.14	6.30	74	15000	—
PSA-I-8	70.60	70.85	7.50	7.58	68	13500	—
PSA-I-12	71.16	70.57	8.37	8.81	65	12700	—
PSA-I-14	73.49	72.28	8.68	9.11	57	9700	—
PSA-I-16	74.10	73.53	9.05	9.51	47	7600	—
PSA-II-4	66.17	66.41	5.77	5.69	90	—	0.15
PSA-II-8	68.22	67.42	6.71	6.96	85	—	0.21
PSA-II-10	69.13	67.70	7.11	7.51	82	—	0.29
PSA-II-14	70.73	70.65	7.80	7.88	75	—	0.24
PSA-II-T	68.35	68.42	4.67	4.59	89	—	0.16

a) Determined by GPC using polystylene standards.

polymers. The yields of PSA-I and PSA-II series were 47-74% and 75-90%, respectively. Yield values of both polymer series decreased with increasing of side chain length or spacer length due to reactivity of acid chlorides. Polymer PSA-II series was obtained in good yields, while, in the case of PSA-I series, yields were not very high. The molecular weights($\overline{Mw}$) of PSA-I series ranged 7600-15000 and the inherent viscosities of PSA-II series were 0.15-0.29 dl/g.

Polymer properties

It is known that wholly aromatic polyspiroacetals are insoluble in common organic solvents.[19] However, all of the PSA-I series were soluble in THF, DMF, chloroform and phenol/CCl_4(1:1) mixture. This clearly indicates that the solubility of rigid polymers, PSA series, is improved by the derivatization of flexible side chain onto the rigid polymer backbone. PSA-II series except PSA-II-T was soluble by heating or without heating in polar solvents like m-cresol and phenol/tetrachloroethane(1:1) mixture.

The thermal properties of the polymer were evaluated by differential scanning calorimetry(DSC) and thermal gravimetry analysis(TGA). The results are presented in Table 2. Figure 1 shows DSC curves for PSA-I series. In general, unsubstituted para-linked wholly aromatic polyspiroacetals did not melt, but decomposed around 300°C. In our case, PSA-I-4 bearing short side chains, it did not show any melting point besides the glass transition(T_g) at 117°C. However, the polyspiroacetals which had

Table 2. Thermal Properties of Polyspiroacetals, PSA-I and PSA-II Series.

Polymers	T_1[a)]	T_2[b)]	T_g	T_m	T_{id}[c)]	5% weight loss	Residual weight at 600°C
	(°C)	(°C)	(°C)	(°C)	(°C)	(°C)	(%)
PSA-I-4	—	—	117	—	335	370	39
PSA-I-8	48/78	165	—	228	313	367	21
PSA-I-12	50	128	—	224	304	360	25
PSA-I-14	56	102	—	208	302	359	16
PSA-I-16	62	98	—	182	303	350	14
PSA-II-4	—	—	125	—	298	318	30
PSA-II-8	—	—	115	253	299	331	18
PSA-II-10	—	—	75	220 (173)[d)]	300	339	21
PSA-II-14	—	—	60	175 (151)[d)]	302	337	27
PSA-II-T	—	—	—	—	315	358	19

a) First melting transition of side chains
b) Second melting transition of side chains
c) Initial decomposition temperature
b) Melting transition of less crystalline domain.

alkoxy chains of 8, 12, 14 and 16 carbons, had observed melting points of 228, 224, 208 and 182°C, respectively. As expected, the melting point decreased with increasing the length of the side chains. In the low temperature region there were two additional endothermic peaks. These peaks might be interpreted as enthalpy changes which resulted from side chain crystallization of the PSA-I series. Thus, the rigid polyspiroacetals with flexible side chains had four different phases. Each phase can be postulated as seen Figure 2. Phase A is a regular packing of the hydrocarbon side chains on the rigid polymer backbones. Phase B illustrates the side chain defects which result from the increase in the freedom of thermal motions of side chain segments, namely the rotator phase in long chain alkanes.[20] Phase C shows relaxation of the layered structure of polymers, but rigid polymer backbone ordering is preserved. Finally, phase D is a transition to isotropic state of the rigid polymer backbones.

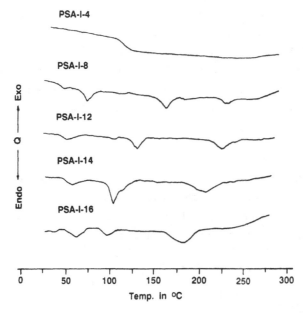

Figure 1. DSC curves of PSA−I series (heating rate 10 °C)

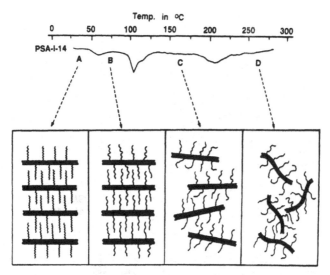

Figure 2. Schematic drawing of tne possible molecular arrangement in four different phases A to D, of PSA−I series

In the case of PSA-II series, the T_g's were detectable at 60-125°C, whose values decreased with increasing aliphatic chain length, but PSA-II-T without flexible chain in the polymer backbone could not find T_g (Figure 3). The PSA-II series shows some different endothermic peaks in comparison with PSA-I series in DSC curves. PSA-II-T polymer lacks the flexible groups. Therefore, there is no melting point. Rather, there is a direct decomposition and no T_g measured. In the case of P-II-4, the only peak was T_g at 125°C. In addition to having a T_g, the polymers PSA-II-8, PSA-II-10 and PSA-II-14 having the higher methylene carbon number show definite melting points at 253, 220, and 175°C, respectivity. For the PSA-II-10, and PSA-II-14, one additional endothermic peak was observed. This transition may be contributed by less crystalline domain of polymer resulting large flexible chains in polymer backbone. Similar thermal behaviors were reported from other polymer which composed of rigid and flexible moieties.

All synthesized polyspiroacetals were relatively stable thermally. The TG curves revealed 5% weight loss at 318-370°C under N_2-gas. The majority of the polymers changed the color from white to brown around 290-300°C, indicating that decomposition starts from this temperature on.

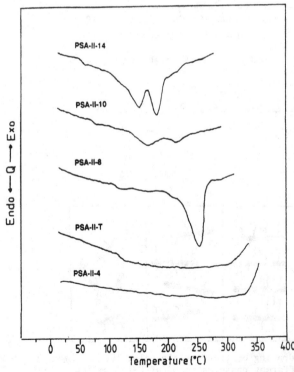

Figure 3. DSC curves of PSA-II series (heating rate 10 °C)

Figure 4 shows X-ray powder patterns of polyspiroacetals at room temperature, which are termed Bragg reflections. They indicate that these polymers have considerable crystallinity. The relatively sharp reflections which appear at low angles suggest that these polymers crystallized in the form of a layered structure, similar to what was found in aromatic polyesters,[6,21] polyimides,[7] polyamides,[8] etc. Figure 5 represents the relationship between the number of side chain carbons, n, and the layer spacing. An almost linear relation between the side chain lengths and layer spacing is observed. However, the diameters of rigid polyspiroacetals obtained from extrapolation were much larger than those of polyesters or polyamides. This suggests that the polyspiroacetals do not crystallize as densely packed as other polymers.

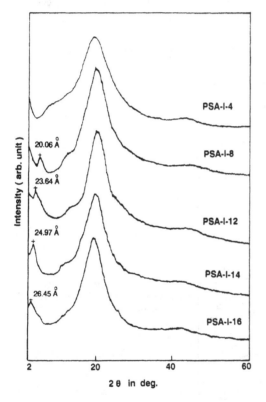

Figure 4. X-ray diffraction patterns of PSA-I series at room temperature

Figure 6 shows X-ray diffraction patterns of PSA-II series. All polymers do exhibit some cystalline properties, especially at low angles which reflects the repeating units of the size of the CH_2 group on the angle. These assumptions are supported by molecular dynamic calculations which use the Chem-X model. In this model(as shown Figure 7), the repeating unit is 19.32 Å long which is consistent with an X-ray diffraction pattern of 19.08 Å.

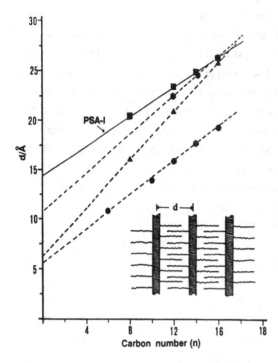

Fig.5. Relationship between layer spacing (d) obtained from WAXS diffractograms and carbon number (n) of alkoxy side chains of several rigid polymers; (●): poly(1,4-phenylene-2,5-dialkoxyterephthalate)s[6], (▲): polyimides prepared from pyromellitic anhydride and 2,5-dialkoxy-1,4-phenylene diisocyanates[9], (●): polyazomethines prepared from 1,4-phenylenediamine and 2,5-dialkoxyterephthalaldehyde,[13] and (■):PSA-I series

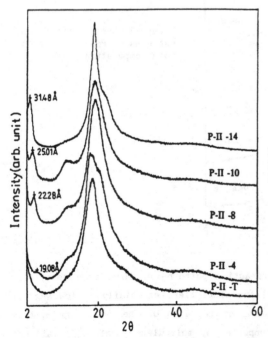

Figure 6. X-ray diffraction patterns of PSA-II series at room temperature

In many cases, the polymers which have both rigid segments and flexible spacers show a thermotropic liquid crystalline phase. These

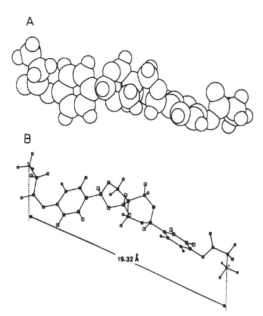

Figure 7. Space-filling model (A) and stick model (B) of PSA-II-4 repeating unit

19.32 Å

polymers containing spiroacetals units cannot be excluded from such phases. To examine this behavior we performed quenching studies. Sample A was crystallized from m-cresol. This sample shows first diffraction peak at 31.48 Å (Figure 8). Sample B was crystallized by quenching from the molten state. No such peaks as for sample A at 31.48 Å are available for B. We also performed polarized microscopic studies on these polymers. Again, we could not find any texture related to liquid crystalline phase. Thus, we can conclude that these polymers are not liquid crystal. The reason can be seen Figure 7. Since the spiroacetal units are not in the same plane, it is difficult to form liquid crystalline phase.

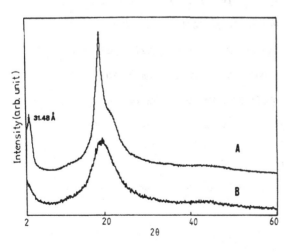

Figure 8. X-ray diffraction patterns of PSA-II-14 with different histories; sample (A) was crystallized from m-cresol and sample (B) was crystallized by quenching rapidly from molten state

ACKNOWLEDGEMENT

This research was supported by the Ministry of Science and Technology of Korea under grant No. KRICT BS N 89-0144.

REFERENCES

1) W.J.Jackson and H.F.Kuhfass, *J. Polym. Sci., Polym. Chem. Ed.* **14**, 2043(1976).

2) S.-M.Lee, K.-S.Kim, K.-S.Lee and S.-K.Lee, *Polymer(Korea)* **13**, 888(1989).

3) B.P.Griffin and M.K.Cox, *Brit. Polym. J.* **12**, 147(1980).

4) Y.Oishi, H.Takado, M.Yoneyama, M.Kakimoto and Y.Imai, *J. Polym. Sci., Polym. Chem. Ed.* **28**, 1763(1990).

5) M.G.Dobb and J.E.McIntyre, *Adv. Polym. Sci.* **60**, 63(1984) and references cited therein.

6) M.Ballauff, *Makromol. Chem., Rapid Commun.* **7**, 407(1986).

7) R.Stern, M.Ballauff and G.Wegner, *Makromol. Chem., Macromol. Symp.* **23**, 373(1989).

8) M.Ballauff and G.F.Schmidt, *Makromol. Chem., Rapid Commun.* **8**, 93(1987).

9) M.Wenzel, M.Ballauff and G.Wegner, *Makromol. Chem.* **188**, 2865(1987).

10) E. Orthmann and G.Wegner, *Angew. Chem., Int. Ed. Engl.* **25**, 1105(1986).

11) Th.Sauer, Dissertation, Mainz 1989.

12) J.Watanabe, H.Ono, I.Uematsu and A.Abe, *Macromolecules* **18**, 2141(1985).

13) K.-S.Lee, J.C.Won and J.C.Jung, *Makromol. Chem.* **190**, 1547(1989).

14) S.M.Cohen and E.Lavin. *J. Appl. Polym. Sci.* **6**, 503(1962).

15) W.J.Bailey and A.A.Volpe, *Polym. Prepr.* **8(1)**, 292(1967).

16) W.J.Baily, C.F.Beam and I.Haddad, *Polym. Prepr.* **12(1)** 169(1971).

17) A.Akar and N.Talinli, *Makromol. Chem., Rapid Commun.* **10**, 127(1989).

18) J.Harding, U.S.Pat. 3,347,871(1967), CA **67**, 11112(1967).

19) S.Hirose, T.Hatakeyama and H.Hatakeyama, *Sen-i Gakkaishi* **38**, 507(1982).

20) K.-S.Lee, Dissertation, Freiburg 1984.

21) K.-S.Lee, B.-W.Lee, J.-C.Jung and S.-M.Lee, *Polymer(Korea)* **13**, 47(1989).

CHARACTERISTICS OF MODIFIED POLYACRYLONITRILE (PAN) FIBERS

R.B. Mathur, O.P. Bahl and J. Mittal

Carbon Technology Unit
National Physical Laboratory
Dr. K.S. Krishnan Road
New Delhi-110012 (India)

INTRODUCTION

Limitations during spinning prevents to obtain desired characteristics in a PAN precursor, especially for producing high performance carbon fibers from it. Such deficiency has lately been overcome by post spinning modifications of the precursor (1,2,3,4,5). Out of the several such attempts the recent trend is towards thinning of the precursor fiber to obtain low diameter carbon fibers, with diameter 4-5 μ as compared to 7-8 μ usually encountered. Such carbon fibres contain lesser number of defects per unit volume and hence should possess very superior mechanical properties in conformity to Weibull "weakest link" principle (6). Another advantage with low diameter PAN fibres is the uniformity of thermal stabilization in very short durations.

In our laboratory we have initiated such studies, wherein the as received 'Courtelle' PAN precursor fibres are first impregnated with certain plasticizers and then stretched, at 100°C to produce low diameter precursor PAN fibers (7,8). In the recent study the same precursor was modified by using carboxylic acid (benzoic acid) and the results were presented in the last Biennial conference (9). Although the benzoic acid results did show remarkable plasticizing effect, it failed to deliver high performance carbon fibres. The reason could be the large size of benzoic acid molecule, which could not impregnate to the core. Therefore during stretching the molecular chains on the surface and that inside the core suffered relative degree of stretch, leaving behind residual stresses in the fibre structure.

In the present study, which falls under the same series of experiments, we have chosen an aliphatic acid (acetic acid), as plasticizer for the smaller size of its molecule than that of benzoic acid. It is believed that the smaller size of the acetic acid molecule should be able to diffuse to the center of the core and should overcome the problems faced with benzoic acid.

In addition to this the acetic acid modified precursor is further impregnated with $KMnO_4$ to have an additional effect of enhanced rate of cyclization. The present study therefore for the first time combines two different routes of modification and may be termed as bi-modification of PAN precursor.

Frontiers of Polymer Research, Edited by P.N. Prasad and
J.K. Nigam, Plenum Press, New York, 1991

The impregnated fibre characteristics are compared with those of the as received fibres by following, mechanical, thermal and x-ray diffraction analysis. Carbon fibres have been prepared with these samples under identical conditions of processing parameters.

EXPERIMENTAL

Courtelle fibrer 10K, 1.2 d'tex, was first treated with acetic acid and then with KMnO$_4$ under different conditions as specified in table-1. The sample "M" which had the maximum uptake of acetic acid (5.8%) was chosen for further studies. Untreated as well as treated fiber samples were then subjected to further treatment with 5% solution of potassium permangnate for three minutes at 85°C, washed and dried. All these samples were oxidized to 235°C in presence of air and then carbonized to 1000°C under the identical conditions and the results are tabulated in table-2. Siemens D-500 X-ray diffractometer with CuK$_\alpha$ radiations as source has been used to record diffractograms of the oxidized samples Mechanical properties of the fibres have been measured on Instron 1122 tensile testing machine. DSC and TGA of the samples has been carried out on a Mettler T-3000 thermal analyzer.

RESULTS AND DISCUSSION

PLASTICIZING EFFECT OF ACETIC ACID

Fig. 1 shows elongation behaviour in which single fibre tows of modified precursor (M) and that of the unmodified precursor (U) have been stretched under identical conditions of load and temperature. The sample 'M' shows overall stretching of 20% which is almost double that of sample 'U'. The real difference in stretching is confined to temperature range 70-130°C. Since the B.P. of acetic acid is only 120°C, it seems that the acid is removed completely upto this temperature only and no residue is remained. It further shows that the acid molecule does not form part of the acrylonitrile chain through chemical bonds but it is linked through the intraction with the -CN groups. Due to the highly polar nature of the nitrile groups, there is partial positive charge at carbon and partial negative charge at the nitrogen as shown in fig.2. The oxygen in the acid group is attracted to the carbon whereas hydrogen is attracted towards the nitrogen and there is the hydrogen bonding between nitrogen and hydrogen. This causes the decrease in the intramolecular intractions among the nitrile groups or in other words decreases the cohesive forces of the molecules, thus facilitating more stretch during heating of PAN fibers without bond breakage. Fig. 3 shows Differential TGA curves for samples 'M' and 'U'. A sharp weight loss between 120 and 130 of sample 'M', correspond to 4% of the total wt. of the sample. Since there is no chemical reaction in the PAN at this temperature the wt. loss can be easily attributed to removal of acetic acid only. It might however be interesting to look for the plasticizer which has boiling point greater than 180-200°C; so that it combines natural plasticizing of PAN and thus contribute to a larger degree of stretch.

EFFECT OF STRETCHING ON THE PHYSICAL PROPERTIES OF THE PRECURSOR

As seen from table-2, the value of Tensile strength of sample M decreases w.r.t sample U. This reduction in strength is due to decrease in the cohesive forces between the molecules of PAN as a result of decrease in the polar bonds between the nitrile groups. Similarly the relaxation of molecular chains during heating due to plasticizing

Table 1
Treatment Conditions of PAN Precursor

COMPOSITION	CONDITION	% OF AcOH
20% AcOH	COLD	NIL
30% AcOH	COLD	NIL
PURE AcOH	COLD	NIL
30% AcOH in Methanol	COLD	NIL
AcOH	70-80'C	4.8%
AcOH	92-95'C	6.8%
50% WATER- AcOH	93'C	3.7%

Table 2
Mechanical Properties of Fibre Samples

Treatment	fibre Type	Dia. microns	T.S MPa	Y.M. GPa	% Elon to break
Untreated (U)	Precursor	12.57	522	6.5	15.6
	Oxidized	11.41	290	7.8	18.6
	Carbonized	7.23	1961	213.8	0.9
Modified with AcOH (M)	As Treated	13.02	426	5.4	19.5
	Oxidized	10.71	303	8.0	19.8
	Carbonized	6.96	2317	210.6	1.1
Modified with $KMnO_4$ (K)	As Treated	12.40	587	6.8	21.1
	Oxidized	10.95	330	6.8	22.3
	Carbonized	7.11	2231	229.8	0.9
Modified with $KMnO_4$+AcOH (MK)	As Treated	11.80	559	7.1	23.8
	Oxidized	10.32	401	8.4	20.6
	Carbonized	6.88	2449	200.7	1.2

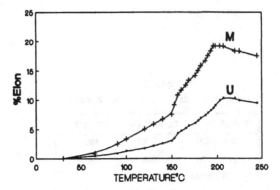

Fig. 1. Elongation Behaviour of the Treated
and Untreated PAN Precursor at
Different Temperatures.
U–Untreated   M–Treated with AcOH

ACETIC ACID (Ac OH)   PAN FIBRE   MODIFIED PAN FIBRE

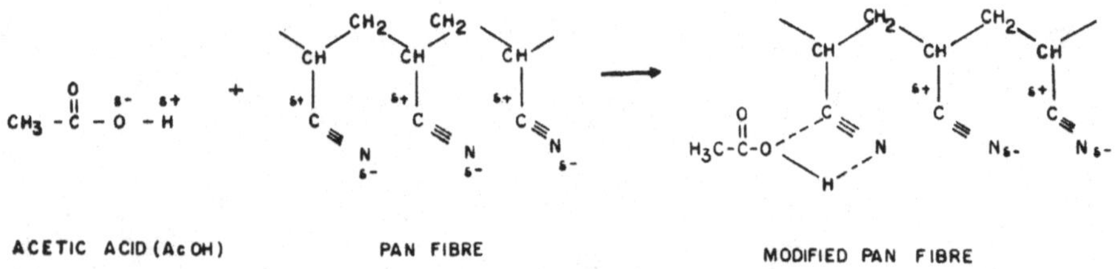

Fig. 2

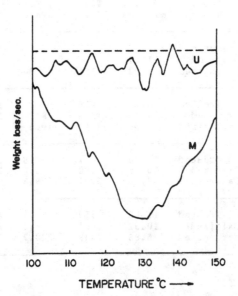

Fig. 3. DTG of the Treated and Untreated
PAN Precursors.
U–Untreated   M–Treated with AcOH

effect of acid causes misorientation of the chain and consequently there is reduction in the value Young's Modulus of the sample M. Obviously this results into higher elongation to brake. The diameter of the modified precursor registers slightly increase because of swelling caused by diffusion of acid molecules. However most importantly the diameter of the stretched modified sample (MS) is reduced to 10.71 µ as compared to 11.41 µ for sample (U), a decrease of 0.7 µ. Further, diameter of the bi-modified and stretched sample (KM) is reduced to 10.32 µ, a reduction of almost 10% of the original value. A reduction of 1 µ in the diameter at this stage should surely produce low diameter carbon fibers.

DSC STUDIES OF THE SAMPLES

DSC of the samples shows that the activation energy of cyclization for the sample M is reduced to 99.68 KJ/Mol only as compared to 119.32 KJ/Mol for sample U. the $KMnO_4$ treated samples "K" shows slight decrease in activation energy i.e. 112 KJ/Mol. The bi-modified sample "MK" also shows the same value i.e. 112 KJ/Mol. This suggests that cyclization reaction is easier in the modified PAN precursors and is carried out to higher degree.

X-RAY DIFFRACTION

X-ray diffraction is an important tool to determine the extent of cyclization (10) in the PAN fibers. However, as shown in fig. 4, a quantitative estimate from the intensities of the peaks at various angles (see table-3) shows that their is no major change in structure of modified precursor as compared to the untreated oxidized sample corresponding to $2\theta = 26$. However, precursor treated with $KMnO_4$ and AcOH shows greater degree of aromatization and intermolecular cross-linkages as evident from the clear diffraction maxima at $2\theta = 13$ (11). The effect is more pronounced for sample "MK" as seen from table-3.

EFFECT OF MODIFICATION ON CARBON FIBRE PROPERTIES

As seen from table-2 there is a marked improvement in the tensile strength of the carbon fibres prepared from modified precursors. Further, maximum improvement is observed for the bi-modified fibres. The improvement in the properties could be attributed to the reduction in the diameter on the one hand and the better stabilization because of the catalytic effect of $KMnO_4$ on the other hand.

CONCLUSION

The studies clearly shows that there is ample scope to improve the carbon fibre properties by following bi-modification technique of PAN fibres. More experimentation is required to derive optimum impregnation conditions to further push up the carbon fibre properties.

ACKNOWLEDGEMENT

Authors wish to thank Dr. S.K. Joshi Director, NPL, for his encouragement and permission to publish the results. One of us (J. Mittal) is greateful to CSIR for financial assistance.

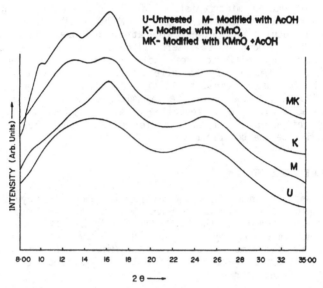

Fig. 4. X-Ray Diffractograms of Oxidized Samples,
(CuKα).

Table 3

Relative Intensity of Various Peaks

Fiber type	Intensity ratio at 2θ		
	$13°$	$17°$	$26°$
Untreated	Unresolved	100	80
Modified with AcOH	80	100	73
with KMnO$_4$	98	100	69
with KMnO$_4$ +AcOH	83	100	62

REFERENCE

1. J.P. Riggs, U.S. Pat. 3,656,882, (1972).
2. J.B. Robin, U.S. Pat. 3,416,874, (1968).
3. N. Popovska, G. Davarsla and I. Mladenov, Carbon 20, 367, (1982).
4. Japan Pat. 59112033 (84,112,031) (1984).
5. A. Aggarwal, H. Yinnon, D.R. Uhlmann, Pepper, T. Rogar, C.R. Desper, J. Mater. Sci. 21 (10), 3455-66 (1986).
6. W. Weibull, 'A stastistical Distribution Function of Wide Applicability', J. Apply Mech. 18, 293 (1953).
7. O.P. B ahl, R.B. Mathur and T.L. Dhami, Mat. Sci. Eng. 73, 105, (1985).
8. R.B. Mathur, O.P. Bahl and V.K. Matta, Carbon, 26, 295, (1988).
9. O.P. Bahl, R.B. Mathur and V.K. Matta, Proc. 19th Bienn Conf. Penn State, USA, p318, (1989).
10. O.P. Bahl and L.M. Manocha, Carbon 12, 417 (1974).
11. R.B. Mathur, O.P. Bahl, J. Mittal and K.C. Nagpal, Proc. International symposium on carbon 1990 Tsukuba, Japan, p594, (Nov.1990).

MECHANICAL PROPERTIES OF SUNHEMP FIBRE -

POLYSTYRENE COMPOSITES

Navin Chand and S.A.R. Hashmi

Regional Research Laboratory (CSIR)
Bhopal-462 026 (M.P.) India

ABSTRACT

Sunhemp fibre-polystyrene composites were developed by using chopped sunhemp organic fibres and polystyrene. Sunhemp fibres impregnated in polystyrene solution were used for making composites under heat and pressure. Different compositions of sunhemp and polystyrene were used for preparing there composites. Fibre reinforcement improved the flexural and impact strengths of composite. Impact strength data showed that maximum impact strength could be achieved by reinforcing 38 weight percent of fibres. Further reinforcement sharply decreased the impact strength. In case of flexural strength behaviour of composite, flexural strength decreased from 1.75 to 1.55 N/mm^2 with the increase of fibre contents from 33.5 to 36 weight percent. Deflection in three point flexural testing decreased with increase of fibre content in the composite. Fracture behaviour of composites has been explained on the basis of fibre-polymer bonding.

INTRODUCTION

Wood is becoming scarce day by day. It has cellulosic structure. Organic fibres such as sisal and sunhemp are also of lignocellulosic nature. Natural fibre reinforced composites can be the alternate of wood. They are cheap, light weight and non-carcinogenic. The presence of fibres in the thermoplastic matrix may lead to a significant improvement of some mechanical properties of composites. Final properties of fibre reinforced thermoplastics are more dependent upon process conditions particularly in case of organic natural fibres. The advantage of reinforcing thermoplastic material is the possibility of combining the tenacity of thermoplastic with strength rigidity of reinforcing agent. It is possible to develop composites of high modulas and impact strength by reinforcing natural fibres (1-4) use of short fibre reinforcement permits a very easy forming of material. But most of processing techniques like injection molding, compression molding, extrusion etc. cause the mechanical and thermal degradation of materials. Two roll milling causes a more severe attrition of fibres than twin screw extrusion. During the injection molding of composites, decrease of fibre length takes place.

Frontiers of Polymer Research, Edited by P.N. Prasad and
J.K. Nigam, Plenum Press, New York, 1991

Arroyo et al. (5) observed the decrease of final fibre length with the increase of percentage of fibre in L D P E after processing. They compared the properties of final composites w.r.t the processing techniques. These results clearly indicate that some new technique is required to over come the mechanical degradation problem caused by the traditional systems of moulding.

In the present investigation a novel mixing technique based on the solution of polymers has been introduced. This technique assures the incorporation of high weight fraction of fibres without causing any mechanical or thermal degradation of fibres. Sunhemp fibres have been used for this study. Polystyrene has been used as a binding material for these fibres. Different weight fraction of chopped sunhemp fibres have been compounded with polystyrene by solution method. Tensile and impact strength of composites was determined. Dependence of impact and tensile strength on V_f was observed and discussed. Possible future studies in this area have been proposed.

EXPERIMENTAL

Polystyrene granules were dissolved in the toluene to prepare a solution. Chopped sunhemp fibres of 5mm length were kept in an oven for one hour at 140°C to remove the moisture and volatiles. Dry, chopped sunhemp fibres were impregnated in the solution of polystyrene. Impregnated fibres were pressed to remove excess solution. Test samples preformed were prepared in the aluminium moulds. Solvent was evaporated at 100°C. After the removal of solvent, the preform was brought to the temperature at 120°C and pressed. 2 psi pressure was maintained till the mixture colled down upto room temperature. Samples were removed from the mould and specimens for tensile and impact strength were prepared.

Flexural strength of the specimens was measured by using a CEAST Italy impact pendulam tester.

RESULTS AND DISCUSSIONS

Figure 1 shows the variation of flexural strength with weight fraction of sunhemp fibre in sunhemp polystyrene composites. It was observed that in the beginning from 33.3% to 35 wt.% of fibre there was an increase in flesural strength from 1.53 N/mm² to 3.04 N/mm². Initially when the fibre content was less, load was mainly taken by polymer. On increase of fibre content load was mainly taken by fibres. Further increase of fibre content did not change the flexural strength which is supported by Figure 2 where at high weight fraction probably weaken the bond and the composite could not tolerate sufficient bonding. This composite had apparent density .89g/cm³ which is lower than the density of polystyrene.

In case of impact strength graph (Fig.3), increase of fibre content increased the impact strength of the composite upto 38 wt%. Further increase in fibre content reduced suddenly the impact strength of the composite which can be explained on the basis that after 38 wt% of fibre matrix could not reach between the two fibres and there was no interface between some fibres which allowed easy crack initiation.

For improving the bonding and for increasing the weight fraction of fibres in this composite modification of fibre surface is necessary. Future work in this direction is in progress.

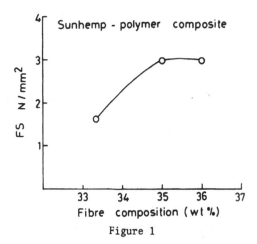

Figure 1

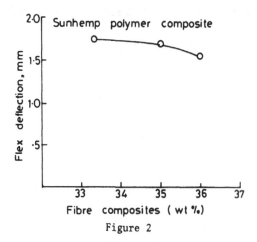

Figure 2

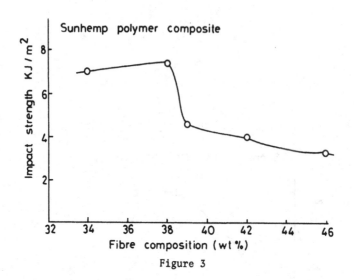

Figure 3

REFERENCES

1. N. Chand, P.K. Rohatgi Europ. Polym. J. (1989) accepted.

2. B.V. Kokta, R. Chen, C. Doneault and J.L. Valade, Polym. Compos. 4, 229 (1983).

3. C. Klason, J. Kubat and H.E. Shomvall, Int. J. Polly. Mater. 10, 159 (1984).

4. R.G. Raj, B.V. Kokta and C. Daneault, J. Appl. Polym. Sci., 37, 1089 (1989).

5. M. Arroyo and F. Avalos Polymer Composites 117, Ap 102 (1989).

RADIATION BASED DEPOLYMERISATION AND DAMAGES

N.K. Gupta, A.K. Aggarwal, J.V. Tyagi and D.A. Dabholkar

Shriram Institute For Industrial Research
19, University Road, Delhi - 110 007, India

INTRODUCTION

Polymeric materials, thermoplastics, elastomers, or thermosets show substantial changes in chemical and physical properties when exposed to ionizing radiation, and the total effect varies with the polymer. In addition to the radiation dose, induced reactions are also influenced by other factors such as temperature, the presence or absence of air or other gases, stresses imposed, and compounding ingredients. The type or radiation source is generally not important.

EFFECT OF HIGH ENERGY RADIATION ON POLYMERS

The primary effects of high energy radiation is to produce excitation and ionization leading to the formation of ionic species andradicals. Three main radical types, terminal (least stable), central, and growing chain radicals (most stable) have been identified[1]. These radicals react via homogeneous combination with radicals elsewhere in the polymer[2,3]; recombination between radicals in pairs[4]; conversion of terminal radicals to central radicals by inter-or intramolecular hydrogen transfer; depolymerisation of growing chain radicals; scission of central radicals; and addition of radical types of unsaturated species, or by hydrogen abstraction from low molecular weight additives containing either unsaturation or labile hydrogen atoms. The result of these reactions is either crosslinking, or cleavage, leading to a high molecular weight or low molecular weight product respectively.

CHEMICAL AND PHYSICAL CHANGES

The changes in properties are reflected in the extent of the above reactions. Chemically, polymers are altered drastically. Aside from crosslinking and cleavage reactions which produce most of the physical changes in the material, a variety of side reactions occur. These may include, gas production; unsaturation; double bond destruction; increased chemical reactivity indicated by increased solubility and production of corrosive by products; increased water absorption; and oxidative degradation, etc. The physical changes may include, increase in density; shrinkage effect; swelling effect by gas leading to a decrease in

Frontiers of Polymer Research, Edited by P.N. Prasad and
J.K. Nigam, Plenum Press, New York, 1991

density; drastic change in electrical properties in the radiation field; discoloration; transparency in some crystalline polymers; changes in mechanical properties, etc. The changes that occur in mechanical properties of irradiated plastics and elastomers depend on the rates of crosslinking and cleavage reactions. Early mechanical property changes in some polymers are often an improvement. Increased tensile strength and increased softening temperature that accompany crosslinking are desirable, but some of the accompanying changes like increased hardness and decreased elongation, may not be desirable. Even though some of the improvements may occur before any serious less desirable changes, for the engineer this is usually a sign of impending radiation damage. Further irradiation will decrease strength and embrittlement may be followed by cracking and powdering.

In the case of vinyl polymers, a general rule[5] states that those polymers in which there are two side chains attached to a single carbon, there is degradation under radiation; while in the polymers where there is only a single side chain or no side chain the result is crosslinking. This rule is not valid for other polymers. Figure 1 depicts the basic chemical structures of some plastics and elastomers listed in order of stability against cleavage. The highest ranking structures are the basic units of polystyrene and aniline formaldehyde polymer which are highly radiation resistant due to the benzene ring side group. In polyethylene, all the early changes in mechanical properties are those attributed to crosslinking. Similar is the case for nylon which shows about the same order of stability. Silicone rubber is ranked near polyethylene as it has the same order of radiation stability as other elastomers and is predominantly crosslinked. In phenol formaldehyde polymer the presence of the benzene ring in the main chain increases cleavage and unfilled phenolics crumble for exposures which do not decrease the strength of polyethylene. Polysulfide rubber is not hardened like most elastomers nor softened, because of a balancing of cleavage against crosslinking. The last five structures are very predominantly cleaved and very marked changes are observed at radiation exposures that would not cause much change in any of the previous structures. Dacron becomes embrittled, polyvinyl chloride softens and produces hydrogen chloride, and celluloics and fluorocarbons such as Teflon show embrittlement at relatively low doses. Butyl rubber is the least stable of elastomers and polymethylmethacrylate the least stable of the rigid polymers. Both are characterized by the quaternary carbon atom attributed as the structure most easily cleaved.

Some general observations of a qualitative nature of the influence of nuclear radiation on some elastomers are shown in Figure 2. The arrow direction indicates increase or decrease and the weight indicates magnitude of change. All elastomers show a loss in tensile strength and decrease in elongation at the same rates except for butyl rubber where the changes are much faster. Compression set of butyl rubber and polysulfide rubber increases due to softening compared to others which are hardened. The magnitutde of change is very large for materials that soften. An acrylate ester elastomer shows very small change in compression set-even less than natural rubber, whereas silicone rubber shows an initial decrease and then an increase for longer irradiation periods. Polysulfide and acrylate ester show a decrease in weight, butyl does not change and other elastomers show an increase. The specific gravity of polysulfide and butyl is decreased, while materials that harden become more dense. Most rapid changes in hardness are observed in butadiene acrylonitrile elastomer, butyl, neoprene, and silicone rubbers. Polysulfide rubber changes at a slow rate due to a balancing of crosslinking and cleavage.

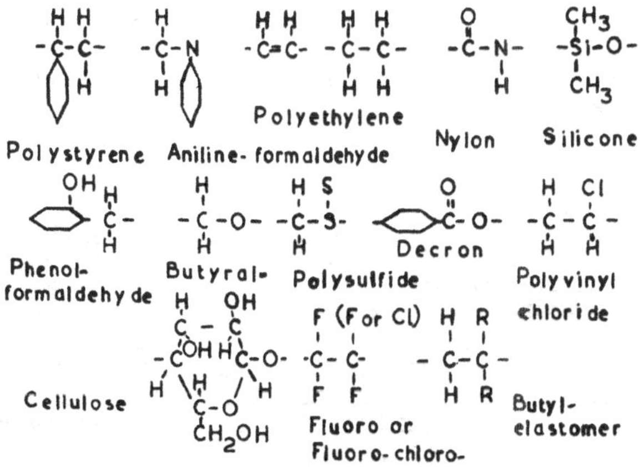

FIGURE 1. BASIC CHEMICAL STRUCTURES.

Polystyrene Aniline-formaldehyde Polyethylene Nylon Silicone
Phenol-formaldehyde Butyral- Polysulfide Decron Poly vinyl chloride
Cellulose Fluoro or Fluoro-chloro- Butyl-elastomer

Elastomer	Ten. Str.	Elong	Comp. Set	Wt.	Sp. Gr.	Hard
Natural Rubber	↓	↓	↓	↑	↑	↑
Polysulfide Rubber	↓	↓	↑	↓	↓	↓
Butadiene Acrylonitrile	↓	↓	↓	↑	↑	↑
Acrylate Ester Rubber	↓	↓	↓	↓	↑	↑
Styrene Rubber	↓	↓	↓	↑	↑	↑
Butyl Rubber	↓	↓	↑	—	↓	↓
Neoprene	↓	↓	↓	↑	↑	↑
Silicone	↓	↓	↻	↑	↑	↑

FIGURE 2. PROPERTIES OF IRRADIATED ELASTOMERS.

OXIDATIVE DEGRADATION

In the presence of oxygen, several types of polymeric chains can form hydroperoxides[6] which are extremely sensitive to radiolysis. In sufficient concentrations, these groups can effect, to some extent, the degradation profile and nature of degradation products in any particular polymer. Due to its strong electron affinity, oxygen may trap electrons ejected by ionization and prevent their return to the positively charged parent molecule.

The relative radiation stability of various thermoplastics, elastomers, and thermosets are depicted in Figure 3 and interprets the degree of dose in terms of a description of the damage as related to the utility of the material after irradiation.

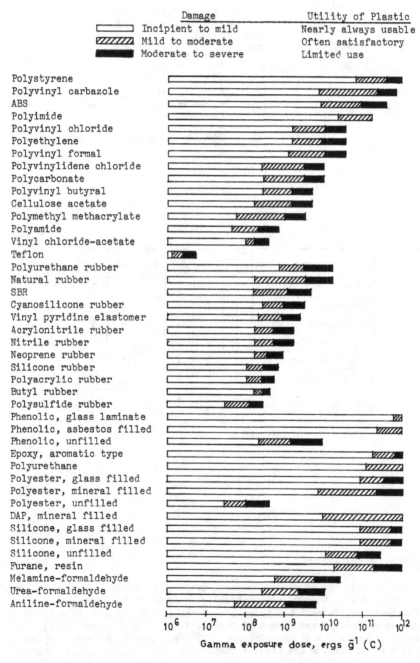

FIGURE 3. RELATIVE RADIATION STABILITY OF THERMOPLASTICS, ELASTOMERS AND THERMOSETTING RESINS.

DEPOLYMERISATION

Depolymerisation has been found to occur in polymethylmethacrylate. No monomer is however formed in irradiated polymer at room temperature, but on heating to 160°-250°C monomer in substantial amounts is liberated. This depolymerisation was shown to be initiated at weak links formed in the chain which could be normal C-C bonds situated in the -position to double bonds produced under irradiation[7], viz.

$$
\begin{array}{cccc}
CH_3 & CH_3 & CH_3 & CH_3 \\
-C-CH_2- & C-CH_2- & \rightarrow -C=CH-C & \text{Weaklink } CH_2- \\
COOCH_3 & COOCH_3 & & COOCH_3
\end{array}
$$

Poly (β-methyl styrene) also undergoes depolymerisation[8] under vacuum at elevated temperatures under the influence of radiation resulting in the formation of monomer and the dimer (1,1,3-trimethyl 3-phenylindane). The yield of dimer is however, reduced when the polymer is irradiated at room temperature or heated to 200°C after irradiating at room temperature. It has been proposed that the chain reaction generating the dimer occurs at high temperature under irradiation. In the presence of cation scavenger, the yield of dimer upon irradiation at 200°C was reduced. This and other results led to the conclusion that the dimer is formed via a cationic process.

APPLICATIONS OF CONTROLLED RADIATIVE DEGRADATION

These include tailored synthesis of polymers and copolymers for controlled photodegradation; photolithography and electron beam lithography for electronics; surface modification of polymers; production of insulating foams, industrial greases and lubricants; fillers; dosimetry; etc. Scope of contemplation also exists for carefully structured plastics which would depolymerize under irradiation so as to facilitate recycling.

REFERENCES

1. D.C. Waterman and M.Dole, J.Phys. Chem., 74 (1970) 1906.
2. T.Waite, J.Chem.Phys., 28 (1958) 103.
3. S. Ya Lebedev, Kinet. Catal. (USSR), 8 (1967) 213.
4. F.Cracco, A.J. Arvia and M.Dole, J.Chem.Phys. 37 (1962)2449.
5. A.A.Miller, E.J. Lawton, and J.S. Balwit, J.Polym.Sci.,14(1954)503.
6. S.W. Shalby, J.Polym.Sci, Macromol. Rev., 14(1979) 419.
7. C.David, D.Fuld, G.Geuskens, and A.Charlesby, Eur. Polym.J., 5(1969) 641.
8. Y.Yamamoto, M.Himei, and K.Hayashi, Macromolecules, 9(1976)874.

A NOVEL POLYIMIDE FOR HIGH PERFORMANCE COMPOSITES

T.M. Vijayan and M.M. Singh Bisht

Polymers and Special Chemicals Division
Propellants, Polymers and Chemicals
Vikram Sarabhai Space Center
Thiruvananthapuram-695022   India

ABSTRACT

Aromatic poly(bis-itaconimide)resin was earlier developed and designated as PIM-750. Application of PIM-750 fibre glass composites in rocket
nozzles and nose cone shield have been carried out. The technology know-how
of the resin synthesis has been transferred for commercial production. Further improvements in molecular structure of PIM-750 are under study. The
process for the synthesis of resin system has been modified to increase the
performance of the composite and to reduce the technological cost. Thermal
and mechanical properties of the modified composite have shown better performance than PIM-750.

INTRODUCTION

Polyimides are the most popular heat-resistant polymers due to its thermal stability, solvent resistance and retention of mechanical properties
over wide temperature range. It has gained acceptance for use in aerospace
composites, electrical, electronic and industrial applications. Different
types of insoluble and soluble condensation linear polyimides based on aromatic dianhydrides and aromatic diamines were developed in many laboratories
and commercialized to suit various applications such as thermal control
films, coatings, sealants and adhesives. Partially crystalline and melt
processable polyimides at a higher temperature, have been reported in the
literature. Many addition type polyimides giving thermoset cross-linked
networks were developed by different laboratories world over mainly for
structural composites. New monomers namely different types of aromatic
diamines and aromatic dianhydrides have been synthesized to develop new
polyimides with various specific properties such as processability and improved stability under thermal, chemical, electrical and radiation environment.[1-8]

Thermosetting polyimides are low molecular weight prepolymers having
imide moieties in their backbone structure and are terminated by reactive
groups which undergo homo and/or copolymerisation by thermal or catalytic
means. The reactive end groups are generally based on maleimides, nadimides,
phenylacetylene or benzocyclobutene. Recently allylnadic-imides are also
reported[9]. In the present paper studies on prepolymer of itaconimide reactive
end groups polyimide matrix resin are presented.

Frontiers of Polymer Research. Edited by P.N. Prasad and
J.K. Nigam. Plenum Press. New York. 1991

BISITACONIMIDE RESINS

Aromatic bisitaconamic acid were synthesized by reacting itaconic anhydride with aromatic amines. A prepolymer of the bisitaconamic acid was prepared with and without the chain extender in a solvent under certain reaction conditions. The reaction product was further converted to imidized form under thermal conditions. The polybisitaconimide prepolymer without chain extender is referred as PIM-700 and with chain extender is referred as PIM-750(Fig.1). These prepolymers were evaluated as resin matrix in glass fibre composites[10]. The PIM-700 glass fibre composites show less flexural properties because of the high crosslink density of the cured polybisitaconimide. In PIM-750 the Michael addition of an aromatic amine to the double bond of the matrix resin backbone has reduced the cross-link density of polyaminobisitaconimide due to chain extension reaction. Hence the flexural properties of PIM-750 glass fibre composites were found better than PIM-700. The glass transition temperature (Tg) of cured PIM-750 is $\sim 203^{\circ}$C.

The chain extended bisitaconimide (PIM-750) was also used for preparing prepregs of high silica cloth and carbon cloth for high temperature applications[11]. It was observed that high silica/polyimide (PIM-750) composites have lower char and erosion rates, high heat of ablation, lower thermal conductivity, higher glass transition temperature and lower coefficient of thermal expansion in comparision to phenolic composites. Temperatures at same depth in the silica/polyimide and silica/phenolic composites were measured in the test rocket motor nozzle and also separately in an arc-jet test. The temperature rise in silica/polyimide (PIM-750) composite is much less than that in silica/phenolic composite, showing that the former is better ablative/insulation liner system for rocket nozzle applications.

Glass fibre composites of PIM-750 have also been used in rocket nose cone shield, radomes, since it has the transparency property for electro-

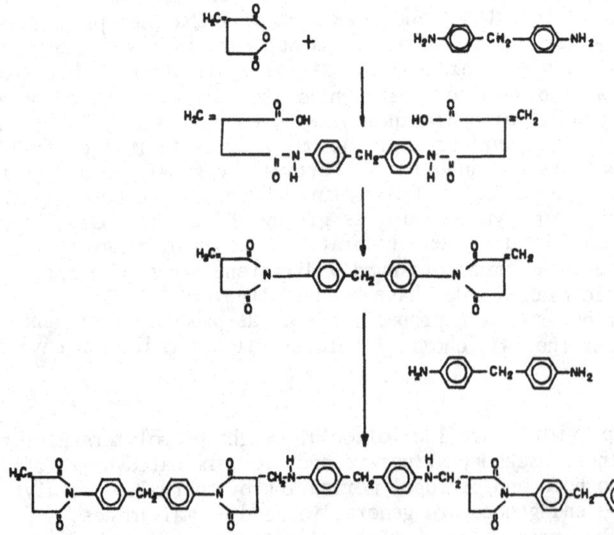

Fig. 1. Synthesis of PIM-750 Polyimide

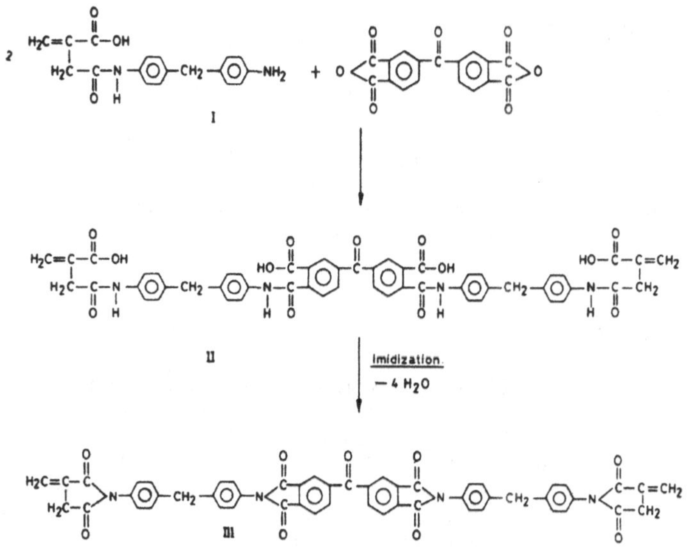

Fig. 2. Synthesis of BTDA-IA Polyimide

magnetic waves. Vikram Sarabhai Space Centre has released the technical know how of PIM-750 and polyamic acid based on pyromellitic dianhydride-oxydianiline system under Technology Transfer to M/s.A B R Organics Ltd., Hyderabad, India for commercial production[12].

MODIFIED BISITACONIMIDES

Thermogravimetric analysis of PIM-700 and PIM-750 glass fibre composites have shown that PIM-700 composite has better high temperature performance. The flexural properties of PIM-750 are better at the cost of its thermal properties. High silica/polyimide (PIM-750) composities have shown lower char yield and lower decomposition termperature compared to high silica/phenolic composites. This may be partly due to the presence of N-H bonds in the structure of PIM-750 matrix resin.

Modified bisitaconimides were prepared by introducing more aromaticity to the back-bone structure and also avoiding N-H bonds in the cured product of the matrix resin.

SYNTHESIS

Itaconic anhydride and 4,4'-diaminodiphenylmethane were reacted at about 5°C in a solvent medium. The product 4-amino-4'-itaconamicacid diphenylmethane (I) was isolated. The compound (I) was reacted with benzo-phenonetetracarboxylic dianhydride at room temperature in a solvent medium. The reaction product (II) was isolated from the solution. The BTDA-Itaconic amic acid (II) was thermally imidized in a solvent medium and a pre-polymer solution (III) of polyimide with reactive end terminal was obtained (Fig.2). Similarly, PMDA-Itaconic pre-polymer solution (IV) of polyimide was also prepared. The pre-polymer solution (III) can be stored at room temperature conditions for nearly six months. Pre-polymer powder can also be isolated for long storage.

COMPOSITE PREPARATION

Prepregs of plain silane treated glass fabric were prepared by impregnating it with prepolymer solution (III) and partially solvent was removed in a controlled way. These prepregs were stacked in a mould and cured in two steps at 150°C and 210°C. Pressure was applied to the stacked prepregs at 210°C for nearly 4 hrs. Post curing of the composite was also done at 200-210°C for 24 hrs. Flexural strength and interlaminar shear strength of the composites were determined. The resin content in the laminates were maintained around 35%.

RESULTS AND DISCUSSION

Amino-itaconamic acid diphenylmethane (I) melts at about 185-186°C and its acid value is 175-180 mg KOH/g.(calculated acid value-181 mg KOH/g.). Elemental analysis of (I) matches with the theoretical values. IR and NMR spectra identify various groups namely - NH_2; - CO-NH- and -C=CH_2 are present in the compound (I). Similarly compound (II) was also analyzed. Its acid value is 228-232 mg KOH/g. (calculated acid value - 237.2 mg KOH/g.). Elemental analysis of (II) matches with the theoretical values. The IR and NMR spectra identify various groups, namely -C-OH of carboxyl group; -CO-NH; - COOH; Ph-CO and C=CH_2 are present in the compound (II). it does not melt.

The DSC curve of the prepolymer (III) does not indicate any melt endotherm but polymerisation exotherm appears at about 197°C. The prepolymer was cured at 210-220°C and the cured polymer has shown thermal stability upto 450°C in N_2 which is 50°C higher than that of PIM – 750.

PMDA-Itaconic prepolymer was also cured at 210-220°C. Its TGA show thermal stability upto 400°C. The char residue at 850°C is 63% in BTDA-Itaconic(BTDA-IA) polyimide. These polyimides show more char residue compared to PIM-750. This is due to the presence of more aromaticity in the modified polyimides. Since BTDA-Itaconic polyimide show better thermal properties than PMDA-Itaconic polyimide hence the former was taken for glass fibre composites studies.

The mechanical properties namely flexural strength and interlaminar shear strength (ILSS) of glass fibre composites were determined at room temperature and at 150°C. Mechanical properties of PIM-750 glass fibre composites were also measured at room temperature and at 150°C. It is observed as shown in Table 1 that in the case of BTDA-IA polyimides glass fibre composites flexural strength and ILSS values are higher than PIM-750 under similar conditions.

Table 1. Mechanical Properties of BTDA-Itaconic Polyimide and PIM-750 Polyimide Composites

Composites	Room Temperature		150°C	
	Flexural Strength	ILSS	Flexural Strength	ILSS
	Kg/cm^2	Kg/cm^2	Kg/cm^2	Kg/cm^2
BTDA-Itaconic	3400	365	2900	317
PIM-750	3090	285	1815	159

The retention of mechanical properties at 150°C is 88% in the case of BTDA-IA polyimide composites whereas for PIM-750 composite it is 60%. The performance of BTDA-IA polyimide composite are better due to more aromaticity and minimum N-H bonding in the molecular backbone structure. Itaconic based polyimides are better in flexibility due to step-ladder type cross-links. Glass transition temperature (T_g) of BTDA-IA polyimide is 256°C which is higher than PIM-750. Processing technique of BTDA-IA polyimide is economic and better compared to PIM-750. The amount of aromatic diamine is also less in the BTDA-IA polyimide process. In the bisitaconimide matrix resin system itaconic anhydride is used which can be obtained from agro-based raw materials.

ACKNOWLEDGEMENTS

The authors gratefully acknowledge the colleagues of Polymers & Special Chemicals Division, Analytical & Spectroscopy Division, Propellants & Special Chemicals Group, Chemical Engineering Complex, Propellants, Polymers and Chemicals entity (PPC) and Composites Group of VSSC for providing excellent technical, analytical and mechanical testing support. Authors are also thankful to Director VSSC for the permission to publish it.

REFERENCES

1. C.E.Sroog in N.M.Bikales, ed, "Ency.Polym.Sci. & Tech.", 1st Ed.Vol.11, John Wiley, New York (1969), P.247.
2. J.Preston in Kirk-othmer "Ency.Chem.Tech."Supp.(1971)p.746.
3. M.I.Bessonov, "Polyimides - Thermally Stable Polymers" Nauka, Lenings Ltd, Leningrad USSR, (1983).
4. K.L.Mittal, "Polyimides: Synthesis, Characterization & Applications" Vols 1 & 2, Plenum Press, New York (1984).
5. Patrick E.Cassidy & Newton C.Fawcett, in Kirk-othmer "Ency.Chem.Tech." Vol 18,(1982) P.704.
6. a)H.D.Stenzenberger et.al, SAMPE Journal 26, No.6 Nov/Dec. p.75,(1990).
   b)J.Richard Pratt & Terry L.St.Clair, ibid p.29.
7. Takekoshi, "Advances in polymer science (94), New Polymer Materials", Springer verlag Berlin Heidelberg (1990).
8. I.K.Varma, G.M.Fohler and J.A.Parker J.Polym.Sci.Polym. Chem Ed.20, (1982) 283.
9. M.A.Chaudhari, Byung Lee, John King & A Renner, "32nd Intl. SAMPE Symp, April 6-9, (1987).
10. T.M.Vijayan, M.M.Singh Bisht, and K.V.C.Rao, J.Polym. Mater.2,81-87 (1985).
11. S.K.Gupta, P.K.Guchhait, S.Alwan and S.Someswara Rao,Chemcon'89 p.735 Indian Chemical Engineers Conf.(1989), Thiruvananthapuram, (Indian Institute of Chemical Engineers, Calcutta, India).
12. "Space Industry news" No.2, Jan (1990) (ISRO Bangalore 560 054, India).

HIGH ENERGY RADIATION RESISTANT ELASTOMERIC COMPOSITES

C.S. Shah, M.J. Patni, and M.V. Pandya*

Materials Science Centre
Indian Institute of Technology
Powai, Bombay-400076, India

ABSTRACT

A scarcity is observed in literature about the work with energy-scavenger additives in radiation environments under oxidizing conditions. In this study, the effect of additives (polyureas adduct and brominated acenaphthene) which were incorporated in chlorosulphonated polyethylene (CSP) rubber and ethylene propylene diene (EPDM) rubber has been evaluated for -irradiation in presence of air, under conditions, which gave rise to oxidation throughout the sample. Both compounds were found to significantly inhibit the changes in tensile properties of the materials. The radiation resistance of the elastomer composites was greatly enhanced when both additives were used together.

INTRODUCTION

There is a great need of radiation stable elastomeric composite materials for their use in nuclear power plants and other high energy radiation environments.

The materials with enhanced resistance to degradation in environments of high energy radiations have been reviewed along with the discussions [1,2]. The efforts in this area range from synthesis of new macromolecules with inherent resistance to high energy radiation to the use of low molecular weight additives.

The interaction of polymer with high energy radiation results in the formation of highly excited electronic states within the materials which subsequently result in bond scission leading to free radicals. These end products are the prime reactive species responsible for the radiation induced chemistry. Most of the additives used for stabilization, generally known as antirads, can be grouped into two categories based on their stabilizing mechanism as follows. (i) energy scavengers, and (ii) radical scavengers.

---

*Chemistry Department.

Frontiers of Polymer Research, Edited by P.N. Prasad and
J.K. Nigam, Plenum Press, New York, 1991

Energy scavengers are primarily aromatic molecules which can act as traps for the excited electronic states generated in the polymer. These energy scavengers after absorbing energy, deactivate back to their ground state with a very high efficiency. The efficiency of absorbing radiant energy by such aromatic compounds is generally related to their size.

Radical scavengers reduce the radiation damage by trapping free radicals which are formed in the elastomer matrix.

Most of the early work on elastomeric composites have been carried out under non-oxidizing conditions. However, for practical applications, the radiation exposure is in the presence of air. The oxidation effects which are induced under these conditions, dominate the degradation chemistry and extent of radiation damage is invariably high [3-5]. Many commercially available antioxidants have been tried and evaluated under radiation exposure in oxidative environment. Most of these compounds are radical scavengers.

There has been surprisingly very little attention given to polycyclic aromatic additives in such environments which act as energy scavengers.

We have carried out a study on effects of an energy scavenger, a radical scavenger and an oxygen scavenger, both individually and jointly, on EPDM and CSP composites for high energy radiation stabilization in oxidative environment. The effect of these additives has been evaluated by monitoring changes in the tensile strength and ultimate elongation over a period of exposure time.

EXPERIMENTAL

Samples of EPDM and CSP composites were prepared in the laboratory as per the compositions given in Table 1 and Table 2. Compounding was carried out on an open roller mill at 80°C. The sheets of dimension 15 cm x 15 cm x 0.2 cm were prepared by compression molding at 165°C for 15 min. Dumbbell shape specimens were cut from the sheets and used for the ageing experiments. These specimens were subjected to ν-irradiation at 45°C in a Co^{60} facility at a dose rate of 6.0×10^2 Gy/hr. Specimens were contained in a cannister of approximate volume of 1.0 litre through which a steady flow of air was maintained during the course of exposure. Tensile testing was carried out on a Instron UTM model-1130 testing machine. Samples were strained at ambient temperature and a speed of 50 cm/min with initial jaw separation of 2.5 cm. At each time interval, three samples were tested and the results were averaged out to obtain the reported values.

RESULTS AND DISCUSSIONS

The three compounds chosen as antirad, were a polymerized amine, [polymeric 1,2 dihydro 2,2,4 trimethyl quinoline (ARD), a polycylic aromatic hydrocarbon [brominated acenaphthene (BrAc)] and polyurea [amine end group adduct of 4, 4' diisocyanate diphenyl methane and 4,4' diamino diphenyl sulphone (MDS)].

ARD is a typical antioxidant used to inhibit oxidative

TABLE 1 RECIPE OF EPDM AND CSP COMPOSITES

INGREDIENTS	EPDM gm	CSP gm
Base Rubber	100.0	100.0
Wax	10.0	10.0
Aluminium Silicate	30.0	30.0
Carbon black (SRF)	10.0	10.0
Vinyl Silane	1.0	1.0
Triallyl Cynurate	–	1.0
Dicumyl Peroxide	3.0	4.0
Sulfur	2.0	–
TMTD	1.0	–
ADDITIVES	*	*

* : As per Table 2.

TABLE 2 ADDITIVE RECIPE IN ELASTOMER COMPOSITES

EPDM or CSP	ARD Resin (phr)	MDS Resin (phr)	BrAc (phr)
A	–	–	–
B	3.0	–	–
C	3.0	5.0	–
D	3.0	–	5.0
E	3.0	5.0	5.0

degradation by capturing the active species. BrAc is a poly-cyclic brominated aromatic compound, ensuring the energy scavenging without any reaction with the singlet oxygen. It also acts as radical quencher as it can stabilize the free radical by resonance. MDS is a polymeric adduct which has amino end groups and repeating urea linkages spaced by aromatic rings. This type of structure helps in scavenging free radicals and singlet oxygen. The aromatic rings, in addition to radical stabilization can also scavenge energy.

The oxygen starvation effect, which can occur due to the high dose rate exposures during the experiments, is minimized by taking sufficiently thin samples (2 mm) and there by ensuring the homogeneous degradation in the samples.

Figure 1 and 2 indicate that CSP composite without any additives (i.e. composite A) has very poor stability in oxidative environment during exposure to high energy radiation. The stability improves on addition of the ARD additive but the degradation takes place at almost same rate after the ARD is consumed completely as one can see from the curves for composites A and B which are almost parallel and close to each other with composite B having a higher induction period. On the addition of either MDS (Composite C) or BrAc (Composite D), a substantial improvement in the stability of the materials is obtained, with BrAc providing a relatively higher

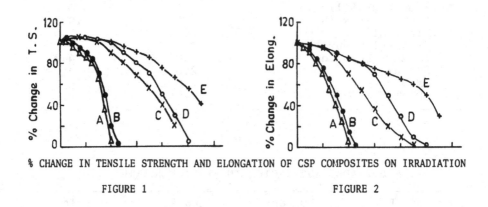

% CHANGE IN TENSILE STRENGTH AND ELONGATION OF CSP COMPOSITES ON IRRADIATION

FIGURE 1   FIGURE 2

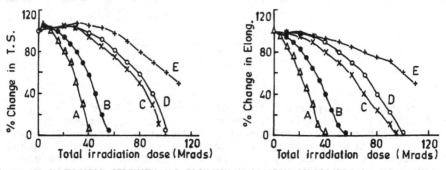

% CHANGE IN TENSILE STRENGTH AND ELONGATION OF EDPM COMPOSITES ON IRRADIATION

FIGURE 3   FIGURE 4

degree of stability. The addition of both MDS and BrAc along with ARD (Composite E) gives the highest degree of stability. The stability comparison of CSP composites can be expressed as follows:

Composite E > D > C > B > A

From Fig.3 and 4, it is observed that a similar kind of stabilization ocurs in EPDM also. The relative stability of EPDM composites can be expressed as follows:

Composite E > D ≈ C > B > A

Thus it is evident that polycyclic aromatic compounds protect the material when used alongwith antioxidants. The stability is seen to be maximum when a combination of stabilizers are added to the polymer due to a synergetic effect.

CONCLUSIONS

Additives like brominated acenaphthene (BrAc) and poly-urea adducts, in addition to the traditionally used ARD anti-oxidant, enhance the serviceability of CSP and EPDM composites in oxidative high energy radiation environment.

The stabilizing mode of each additive is different but effectively they retard the tensile property changes, when used alone or in combination.

The stabilizing effect is maximum when a combination of two or more stabilizers is used because of synergetic action.

REFERENCES

1. Clough R.L. and Gillen K.T., Gamma - radiation induced oxidation and mechanisms of its inhibition, in Oxidation Inhibition in Organic Materials, ed. P.Klemchuk and J. Pospisil, CRC Press, Boca Raton, (1990).

2. Clough R.L., Radiation resistant Polymers, in Encyclopedia of Polymer Science and Engineering, Vol.13, John Wiley, New York, (1988), p. 667.

3. Clouch R.L. and Gillen K.T., J.Polym. Sci., Poly., Chem. Ed., 19, 2041 (1981).

4. Wilski H., Radiat. Phys and Chem. 29, 1 (1987).

5. Clough R.L. and Gillen K.T., Nucl. Techn., 59, 344 (1982).

6. Arakawa K., Seguchi T., Hayakawa N. and Nachi S., J.Polym. Sci., Polym. Chem. Ed., 21, 1173 (1983).

MODIFICATION OF POLYVINYL BUTYRAL PROPERTIES VIA THE

INCORPORATION OF IONOMER GROUPS

Arijit M. DasGupta, Donald J. David and
Ashok Misra*

Monsanto Chemical Company
730 Worcester Street
Springfield, MA 01151, U.S.A.
*Indian Institute of Technology
Hauz Khas, New Delhi 110016, India

ABSTRACT

Polyvinyl butyral ionomers (IPVB) were synthesized in order to modify some of the performance properties of polyvinyl butyral (PVB) used in safety glazings. The ionomers were synthesized via a condensation polymerization process in which ionic groups were permanently incorporated into the polyvinyl butyral backbone by using an ion-containing aldehyde in addition to butyraldehyde during the acetalization of polyvinyl alcohol. Several techniques including 1-H and 13-C NMR spectroscopy, dilute solution viscometry, dynamic mechanical analysis and differential scanning calorimetry were used to characterize the IPVB's. Property modifications such as increased ambient temperature stiffness and high temperature flow comparable to PVB were observed in the IPVB's.

INTRODUCTION

Ionomers are a class of polymers that contain ionic groups along the chain (1,2). These polymers behave as thermally reversible thermoplastics and this unique property has been attributed to the morphological structures formed due to ionic associations, which have been studied in depth by Eisenberg and coworkers (3-5) and MacKnight (6-8). In addition, there is extensive documentation on ionomers in the literature (9-19).

In general, the structure of an ionomer is composed of a hydrocarbon backbone chain containing pendant acid groups which are partially or completely neutralized by an appropriate base to form the ionic component, the

Frontiers of Polymer Research, Edited by P.N. Prasad and
J.K. Nigam, Plenum Press, New York, 1991

concentration of which is usually between one and ten mole percent. The ionic associations which appear as a consequence of the ionic groups, enable the polymer to behave as a crosslinked material at ambient temperatures while at elevated temperatures the ionic associations disappear only to reform upon cooling. This thermally reversible behavior allows easy processibility of the polymer at elevated temperatures and high modulus at ambient temperatures thus making the polymer very versatile. The extent of ionic associations along with the choice of the acid groups and metal (counter) cations effect the final properties of the ionomer. Acid groups such as carboxylic, sulfonic, thioglycolic and phosphoric have been used along with counter ions such as sodium, potassium, zinc, barium and magnesium. The actual incorporation of the ionic groups onto the polymer backbone can be done either by copolymerizing a monomer containing an acid group with another monomer or by reacting acid functionalities onto an existing polymer via a modification reaction . In either case, ionic groups are introduced upon neutralization of the acid groups. An example of the first approach is a copolymer of ethylene and methacrylic acid where the acid groups have been neutralized by magnesium acetate (Trade name: "Surlyn"), while an example of the latter approach is one where sulfonic acid groups are introduced directly into polystyrene at 1-10 mole percent levels via a sulfonation reaction, followed by the neutralization of the acid groups to form the ionomer.

Polyvinyl butyral (PVB) is a polymer that is extensively used in safety glazings. It is prepared via a conventional acetalization reaction between polyvinyl alcohol (PVOH) and butyraldehyde (BA) in the presence of an acid catalyst (20-35). The reaction conditions are normally controlled to provide the proper balance of residual hydroxyl groups and 1,3-dioxane rings to impart desirable end use properties. The resulting polymer is unique in that the residual hydroxyl groups provide both a high tear - resistance and good adhesion to glass while the 1,3-dioxane rings provide chain strength and lack of any detectable crystallinity.These characteristics impart superb optical quality and make it an ideal material for use in preparing laminates suitable for safety glazings. PVB used in safety glazings is plasticized using plasticizers such as dihexyl adipate. The plasticized PVB is then laminated between two sheets of glass to form the final safety glazing.

The aim of this work was to prepare and investigate the properties of ionomeric versions of polyvinyl butyral (PVB). The ionomeric PVB's (IPVB) were prepared by directly reacting ion containing groups onto PVB via the acetalization reaction of an ion containing aldehyde along with BA with PVOH, following the polymer modification approach. The introduction of the ionomeric groups was expected to provide stiffness at ambient temperatures while maintaining flow characteristics at elevated temperatures

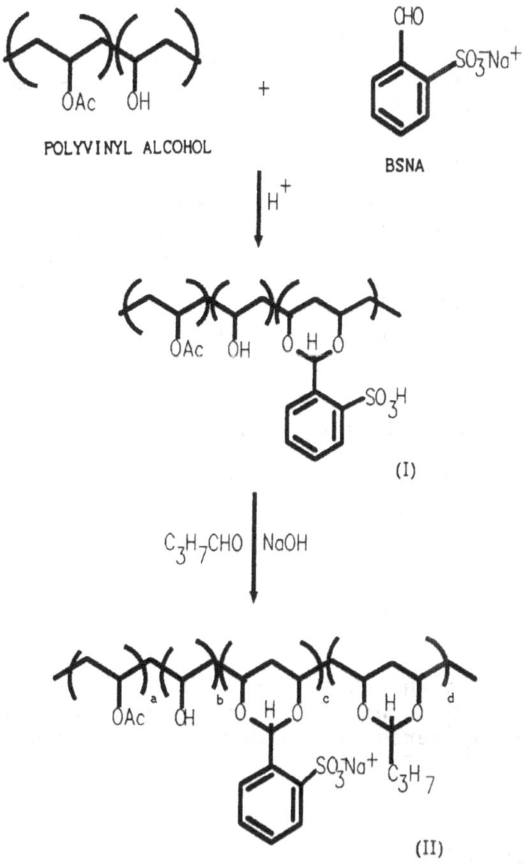

a: 1-2 wt.% b: 18-19 wt.%

c: 1-10 mole% d: 40-50 mole%

Fig. 1. Synthesis route for Ionomeric Polyvinyl Butyral (IPVB).

EXPERIMENTAL

Synthesis of Ionomeric Polyvinyl Butyral (IPVB)

The sodium salt of o-benzaldehyde sulfonic acid (BSNA) obtained from KODAK, was initially reacted with a solution of polyvinyl alcohol (PVOH) in de-ionized (DI) water using nitric acid as the catalyst. The reaction was conducted between 19 to 22°C for periods ranging from 0.5 hour to 2.0 hours depending upon the extent of ionomer substitution required. Next, butyraldehyde (BA) was introduced to the system and after the resulting polymer precipitated out of solution, the reaction mixture was heated to 30°C and held at this temperature for four hours, at the end of which the material was washed, neutralized with either potassium or sodium hydroxide, filtered and dried. A limitation of the IPVB synthesis process is that the BSNA - PVOH reaction is temperature limited, i.e., acetal formation is severly hindered above 60°C., and hence the reaction is conducted below this temperature.

Using this procedure, ionomeric polyvinyl butyral (IPVB) containing 1 - 10 mole percent ionomer groups were prepared as shown in Figure 1. IPVB's of different molecular weights (MW, weight average) were also prepared by using PVOH's of different MW's. The PVOH's were obtained from Nippon Chemical Company under the trade name Goshenol. These were:

a. NL-05 ; DP=550 ; MW = 2.3 x 10^4

b. NM-11 ; DP=1275 ; MW = 5.6 x 10^4

c. NH-18 ; DP=1800 ; MW = 7.7 x 10^4

Minor adjustments in synthesis conditions were made to facilitate ease of recovery and handling of the resulting IPVB's.

A sample of standard Monsanto Chemical Company PVB synthesized from PVOH with a DP of approximately 1800 and containing no ionomeric groups was used as a control for comparison with the IPVB's.

RESULTS AND DISCUSSIONS

Plasticization of IPVB and PVB

IPVB samples were plasticized by mixing with dihexyl adipate The levels of dihexyl adipate used were based on the residual PVOH levels of the polymers. The PVB control was mixed with 25 weight percent plasticizer.

NMR Spectroscopy

Figure 2, is a one dimensional 400 MHz NMR proton spectrum of an IPVB sample containing a targeted ionomer level of 5 mole percent. Aromatic groups resulting from the vinylbenzal sulfonate moiety are clearly seen between 7.0 and 8.0 ppm, thus confirming the above acetal formation. This is further confirmed from the 100 MHz carbon-13 spectrum shown in Figure 3 where in addition to the conventional features of the NMR spectrum of polyvinyl butyral (PVB) reported and discussed previously (36), two carbon resonances at 99.3 and 103.2 ppm are also seen. These signals have been assigned to benzal groups. From the NMR spectra, the efficiency of the BSNA - PVOH acetal reaction has been calculated at approximately fifty percent. Figures 4 and 5, on the other hand, show the proton and carbon NMR spectra respectively, of PVOH acetalized at 75°C. In both spectra there is no evidence of aromatic substitution

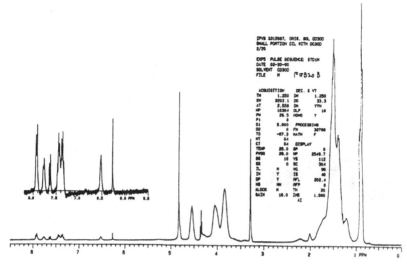

Fig. 2. 400 MHz 1-H NMR Spectrum of IPVB.

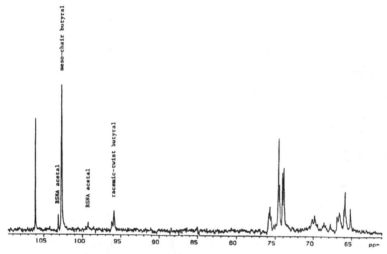

Fig. 3. 100 MHz 13-C NMR Spectrum of IPVB.

resulting from the BSNA - PVOH reaction, and the spectra are essentially those of conventional PVB. This confirms that high acetal temperatures do not favor the formation of ionomer groups via the BSNA - PVOH reaction.

Dilute Solution Viscosity

Evidence of ionomeric behavior in the IPVB's were confirmed from their dilute solution viscosity behaviors. Figures 6 and 7 llustrate the viscosity behaviors of IPVB's and PVB control in tetrahydrofuran (THF) and methanol, respectively. In a non-polar solvent such as THF, the IPVB's

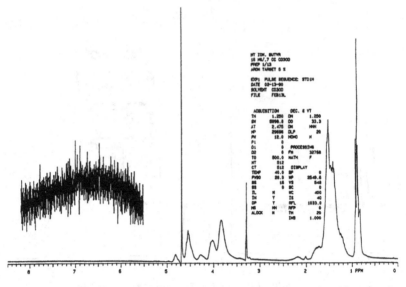

Fig. 4. 400 MHz 1-H NMR Spectrum of Polymer Synthesized at 75 C.

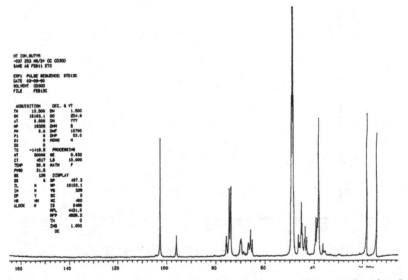

Fig. 5. 100 MHz 13-C NMR Spectrum of Polymer Synthesized at 75°C.

exhibited a sharp rise in reduced viscosity beyond a concentration of 1-2 percent ultimately resulting in gels, while the PVB control exhibited normal viscosity behavior. The dramatic increase in viscosity beyond a particular concentration in a non-polar solvent, is typical of ionomers and is due to the high degree of ionic associations (37,38). On the other hand, in a polar solvent such as methanol, both the IPVB's and PVB exhibited normal viscosity behavior. This is a consequence of the ionic associations breaking apart (solvation) in the polar solvent.

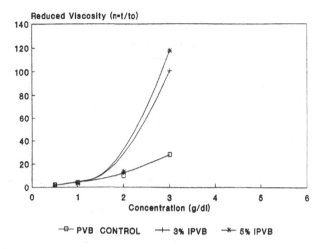

Fig. 6. Reduced Viscosities of IPVB and PVB Measured in Tetrahydrofuran.

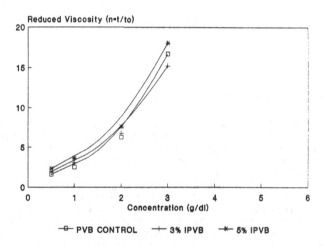

Fig. 7. Reduced Viscosities of IPVB and PVB Measured in Methanol.

Thermal Characterization

The glass transition temperatures (Tg) of the un-plasticized IPVB's and PVB control are presented in Table 1. All the samples have residual PVOH levels between 18 and 20 weight percent. It can be seen that the Tg values of all the IPVB's are higher than the PVB control (Tg=73 C). The Tg's of the IPVB's increase with increases in ionomer levels, which is a direct consequence of the ionic associations, which in turn cause the polymers to increase in stiffness with accompanying increases in ionomer levels.

Table 1.   Effect of Ionomer Groups on the Tg's of IPVB's

SAMPLE	MOLE% IONOMER	Tg($^\circ$C)
PVB CONTROL	0	73
IPVB	3	80
IPVB	5	85
IPVB	10	92
IPVB	15	106

Table 2.   Storage Modulus (G') of IPVB and PVB Control

SAMPLE	%ION	%PVOH	G'(X E-7) (N/$_M^2$)			
			(25C)	(40C)	(60C)	(150C)
PVB-CNTRL	0.0	18.0	3.3	0.57	0.41	0.06
HMW-IPVB	3.0	21.5	24.0	2.4	0.58	0.15
HMW-IPVB	5.0	20.0	45.0	21.0	1.40	0.27
LMW-IPVB	3.0	18.3	42.0	13.0	0.66	0.015
LMW-IPVB	5.0	18.4	49.0	13.0	0.75	0.016

Rheological Properties

Dynamic mechanical properties including storage modulus (G'), loss modulus (G") and tan δ (G"/G') of plasticized IPVB and PVB control samples were measured to provide information pertaining to the stiffness and flow characteristics of the polymers. Table 2, compares the G' values at 25, 40, 60, and 150 C for high molecular weight (HMW) IPVB's which were synthesized using a PVOH with a degree of polymerization (DP) of 1800, low molecular weight IPVB's (LMW) synthesized using a PVOH with a DP of 550, and a PVB control. The table also contains the residual PVOH levels and the theoretical mole percent of ionomer groups in the polymers. Both the HMW and LMW IPVB's had higher G' values at ambient temperatures (25-60 C) than the PVB control, which indicated that the IPVB's possessed greater ambient temperature stiffness than the control. Although higher residual PVOH levels, as in the case of the HMW IPVB's, also contribute towards increased stiffness, the principal source of the stiffness is due to the ionic associations provided by the IPVB's. This is clearly evident in the case of the LMW IPVB's which have similar PVOH levels as the PVB control but yet are considerably stiffer than the latter. It is also seen, as expected, that the G' values increase with increases in the ionomer levels. At 150 C, the G' values of the HMW IPVB's remain higher than the control while the G' values of the LMW IPVB's are lower

Table 3. Storage Modulus (G') and Stress Relaxation Modulus
(SR) of IPVB and PVB Control

$$G'(X \ E-7) \ (N/M^2)$$

SAMPLE	%ION	%PVOH	(25C)	(40C)	(60C)	(150C)	SR150 X E-3
PVB-CTRL	0.0	18.6	6.1	0.51	0.36	0.055	6.9
MMW-IPVB	3.0	17.9	7.1	0.55	0.34	0.036	‹1.0
MMW-IPVB	3.0	18.6	7.5	0.58	0.35	0.004	‹1.0
MMW-IPVB	5.0	16.6	8.4	0.69	0.39	0.007	1.3

than the control and the HMW IPVB's which would indicate
that the LMW IPVB's would have higher flow properties at
these temperatures. Both molecular weight and complete
dissociation of ionic crosslinks are probable factors here.

Dynamic mechanical properties were also measured for
medium molecular weight (MMW) IPVB's synthesized from PVOH
with a DP of 1275. These IPVB's also had higher ambient
temperature G' values than PVB control due to theirincreased
stiffness and like the LMW IPVB's, the MMW IPVB's had lower
G' values at 150°C, than the PVB control. Stress relaxation
measurements at 150°C,(SR150), also showed lower values for
the MMW IPVB's compared to the control (Table 3), which
along with the lower G' values at 150°C indicate that the
elevated temperature flow properties of the MMW IPVB's are
not adversely affected. Both G' and SR150 values increased
with accompanying increases in ionomer concentrations.

CONCLUSIONS

It has been shown that ionomeric polyvinyl butyral
(IPVB) can be prepared by the acetalization of polyvinyl
alcohol (PVOH) using an ionic aldehyde such as the sodium
salt of o-benzaldehyde sulfonic acid (BSNA) in addition to
butyraldehyde (BA), to form side chain groups capable of
conversion to their ionomeric form via the addition of metal
cations. The presence of the vinylbenzal sulfonate groups on
the PVB chain as a result of the above acetalization, were
confirmed by NMR spectroscopy. Dilute solution viscosity and
thermal characterization of the IPVB's showed that the
polymers displayed typical ionomer properties.
Further confirmation of ionomer behavior was obtained from
dynamic mechanical property measurements of the IPVB's. It
was seen that the ionic associations increased the stiffness
of the polymer systems at ambient temperatures. In addition,
modulus at processing temperatures similar to PVB control
prepared using conventional acetalization procedures, could
also be obtained by selecting the appropriate IPVB system.

ACKNOWLEDGEMENTS

We would like to express our sincere gratitude to Mr. Cliff Lin and Mr. Robert Rzeszutek for providing the DMA and DSC data, Dr's Pierre Berger and Edward Remsen for the NMR measurements and interpretation, Ms. Angela Lansberry for conducting the synthesis of the IPVB's, and Mr. Jeff Hurlbut for providing the Rheometrics Solids Analyzer data.

REFERENCES

1. R.W. Rees and D.J. Vaughan, Polym. Prepr. Am. Chem. Soc., Div. Polym. Chem., 6, 287 (1965 a).
2. R.W. Rees and D.J. Vaughan, Polym. Prepr. Am. Chem. Soc., Div. Polym. Chem., 6, 296 (1965 b).
3. A. Eisenberg, Macromolecules, 3, 147 (1970).
4. A. Eisenberg and M. Navratil, J. Polym. Sci., B, 10, 537 (1972).
5. A. Eisenberg and M. Navratil, Macromolecules, 6, 604 (1973); 7, 90 (1974).
6. W.J. Macknight, W.P. Taggert and R.S. Stein, J. Polym. Sci., Polym. Sym., 45, 113 (1974).
7. W.J. Macknight, L.W. McKenna and B.E. Read, J. Appl. Phys., 38, 4208 (1967).
8. P.J. Phillips and W.J. Macknight, J. Polym. Sci., A-2, 8, 727 (1970).
9. A. Eisenberg, Adv. Polym. Sci., 5, 59 (1967).
10. A. Eisenberg and M. King, "Ion-Containing Polymers", Academic Press, New York, 1977.
11. W.J. Macknight and T.R. Earnest, Jr., J. Polym. Sci., Macromol. Rev., 16, 41 (1981).
12. L. Holliday, Ed. "Ionic Polymers", Halstead Press, Wiley,   New York, 1975.
13. A. Eisenberg, Ed. "Ions in Polymers", Adv. Chem. Ser., 187, Amer. Chem. Soc., Washington DC, 1980.
14. C.G. Bauzin and A. Eisenberg, Ind. Eng. Chem. Prod. Res. Dev., 20, 2, 271 (1981).
15. E.P. Otocka, J. Macromol. Sci., C5, 275 (1971).
16. H.P. Brown, Rubber Chem. Technol., 30, 1347 (1957).
17. K.F. Wissburn, Makromol. Chem., 118, 211, (1968).
18. K. Sanui, R.W. Lenz and W.J. Macknight, J. Polym. Sci., Polym. Chem. Ed., 12, 1965 (1974).
19. A. Noshay and L.M. Robeson, J. Appl. Polym. Sci., 20, 1885 (1976).
20. G.O. Morrison, F.W. Skirrow and K.G. Blaikie (to Canadian Electro Products), U.S. Pat. 2,036,092 (Mar. 31, 1936), reissue 20,430 (June 29, 1937).
21. R.W. Hall (to General Electric Co.), U.S. Pat. 2,114,877 (Apr. 19, 1938).
22. H.F. Robertson (to Carbide and Carbon Chemical Corp.), U.S. Pat. 2,167,678 (June 13, 1939).
23. G.O. Morrison and A.F. Price (to Shawinigan Chemicals Ltd.), U.S. Pat. 2,168,827 (Aug. 8, 1939).

24. J. Dahle (to Monsanto Chemical Co.), U.S. Pat. 2,258,410 (Oct. 7, 1941).
25. W.H. Sharkey (to E.I. du Pont de Nemours & Co., Inc.), U.S. Pat. 2,396,209 (Mar. 5, 1946).
26. W.O. Kenyon and W.F. Fowler, Jr. (to Eastman Kodak Co.), U.S. Pat. 2,397,548 (Apr. 2, 1946).
27. E. Lavin, A.T. Marinaro and W.R. Richard (to Shawinigan Resins Corp.), U.S. Pat. 2,496,480 (Feb. 7, 1950).
28. E.H. Jackson and R.W. Hall (to General Electric Co.), U.S. Pat. 2,307,063 (Jan. 5, 1943).
29. W. Haehnel and W.O. Herrmann (to Consortium fur Elektrochemische Industrie), Ger. Pat. 507,962 (1927).
30. H. Hopf (to I.G. Farben Industrie AG), U.S. Pat. 1,955,068 (1934).
31. S. Nomura et. al. (to Sekesui Chemical Co.), U.S. Pat. 4,452,935 (Jun. 5, 1984).
32. H.D. Hermann et. al. (to Hoechst Aktiengesellschaft), U.S. Pat. 4,205,146 (May 27, 1980).
33. I. Tadoki et. al. (to Sekesui Chemical Co.), Jap. Pat. 30706 (Feb. 19, 1982).
34. P. Dauvergne (to Saint-Gobain Industries), U.K. Pat. 2,007,677 (Mar. 10, 1982).
35. G. E. Cartier and P.H. Farmer (to Monsanto Company), U.S. Pat. 4,874,814 (Oct. 17, 1989).
36. P.A. Berger, E.E. Remsen, G.C. Leo and D.J. David, accepted for publication in Macromolecules (May, 1991).
37. R.D. Lundberg, Polym. Prepr. Amer. Chem. Soc., Div. Polym. Chem., 19, 1, 455 (1978).
38. R.D. Lundberg and H.S. Makowski, Adv. Chem. Ser., no. 187, 21 (1980).

STUDY OF PHASE TRANSITIONS IN GAMMA - IRRADIATED POLYVINYLIDENE

FLUORIDE USING X-RAY DIFFRACTION AND ITS CORRELATION TO THERMOLUMINSCENCE

N.V. Bhat and Sunita Pawde

Professor and Lecturer
University Dept. of Chemical Technology
Matunga, Bombay-400019, India

INTRODUCTION

Synthetic polymers are generally regarded as electrically passive materials. However, in the recent past some electroactive polymers have been developed which play vital role in many new technologies. Electrically conductive,pyroelectric and piezoelectric polymers have been developed which can be used for sensors, optical computers, robotics and transducers.

Piezoelectric polymers are of particular interest because of potential application as electro-mechanical transducers. Polymers have advantage that they are easily processible in the form of flexible films having good elastic properties. Polyvinyl Chloride (PVC), Polytrifluroethylene (PTrFE), Polyvinylidene Fluoride (PVDF) and some of its copolymers are of interest.

PVDF have been shown to be ferro & piezo electric with molecular repeat formula $(CH_2-CF_2)n$. The structure and morphology of PVDF was studied in great detail(1). It has been shown that PVDF exists in three crystalline forms namely α, β and γ (form II,I & III). Of these α phase is known to have the chain conformation TG-TG' and crystallizes in such a way that the dipole moments of two chain segments within a unit cell are oppositely oriented perpendicular to the chain axis. The β form on the other hand has a unit cell consisting of two all trans chains packed with their dipoles pointing in the same direction. It has been recently shown that (2) by applying a very strong electric field or corona discharge to a sample in α phase (form II) transforms to β phase (form I). Similarly by drawing the sample at low temperature or by crystallization under pressure produces β transformation (3).

It has also been found that (3) by applying a moderately high electric field (100 MV/meter) to a sample in form II,phase transition leading to a new polar phase called as α_p phase IV (δ) results.

Study of such phase transformations for sample subjected to thermal treatment and electric field have been reported. The effect of γ-irradiation on the ferroelectric properties of PVDF has been reported (4), however, no phase formation have been observed. The present report therefore attempts to investigate the effect of γ-irradiation on the structure and properties of PVDF. Recently we have investigated thermoluminescene of irradiated PVDF and correlated different TL glow peaks with corresponding phase transformations. It was therefore thought interesting to further investigate corresponding phase changes by X-ray diffraction, differential thermal analysis (DTA) and IR spectroscopy.

EXPERIMENTAL

PVDF sample was obtained from Yarsley Technical Institute, England (140 micron thick). Samples were subjected to γ-irradiation using Co-60 Gamma source at a dose rate of 4.45 krads/min for various durations at room temperature. These irradiated samples were subsequently heat treated at temperatures of 350, 370, 440 and 470 K respectively under the nitrogen atmosphere, During heating, samples were properly placed between two glass plates inside the sample cell to avoid wrinkling.

All X-ray diffraction profiles were obtained at room temperature with Philips XRD model PW 1710 using Cu-K$_\alpha$ radiation with Ni-foil filter.

DTA was recorded using Stanton Redcroft 780 model and FTIR was recorded using Bruker's spectrometer.

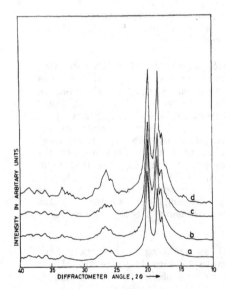

Fig. 1. X-RAY DIFFRACTOMETER SCANS OF PVDF (a) UNIRRADIATED (CONTROL) SAMPLE AND HEATED AT (b) 350 K (d) 440K.

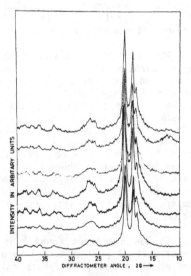

Fig. 2. X-RAY DIFFRACTOMETER SCANS OF PVDF SAMPLES SUBJECTED TO VARIOUS DOSES OF γ-IRRADIATION (a) CONTROL (b) 0.27 MRADS (c) 0.54 MRADS (d) 0.07 MRADS (e) 4.27 MRADS (f) 8.55 MRADS AND (g) 15.5 MRADS.

RESULTS AND DISCUSSION

Typical X-ray diffractogram of the control film is shown in Figure-1 (a). The peaks corresponding to 100, 020 and 110 shows that the film consists of predominently phase with some admixture of γ phase. In order to distinguish clearly between the effects of thermal treatment and irradiation dose the X ray diffraction scan for various samples were analysed. Curves b,c and d of Figure-1 depict X-ray diffractograms corresponding to control sample heated at 350,370 and 440 K respectively. One can note small increase in the intensities of peaks without any additional peaks. The crystallinity index for all these samples were calculated using Manjunath's formula and are given in table I. It may be seen from Table-I that the percentage crystallinity increases initially as heating temperature increases and then saturates.

The samples were also subjected to various doses of γ irradiation. The values calculated for crystallinity index for various doses in Mrads indicate that the crystallinity index suddenly increases initially and then slightly decreases. The X-ray diffractograms corresponding to these are shown in figure 2. It may be noted that an additional peak appears at 17.1° (2θ value).

Figure 3 shows X-ray diffractometer scans for the samples irradiated to dose of 4.27 Mrads and subsequently heated to various temperatures. It can be easily seen from Figure 3 that additional peaks appear at 14.6° and 17.1° (2θ values). The intensity of these peaks have appreciably increased as the temperature of pretreatment increases. The crystallinity values calculated for these samples (Table 1) show increase in the value, the maximum being 52.5 % for the sample heated to 370 K.

PVDF is known to exist in various crystalline modifications such as α,β,γ and δ. Of these α is supposed to be non polar phase and is most common for the films obtained by melt solidification at all temperatures. Transformation of this phase in the most highly polar β phase (all trans) has been reported by several authors. The changes accompanying these transformations on the chain conformation can be investigated by several techniques such as X-ray, IR, NMR and DSC. In the present investigation, the nature of these phase

TABLE 1   EFFECT OF HEAT AND RADIATION ON % CRYSTALLINITY OF PVDF

No.	Sample	Dose (Mrads)	Heating temp (Kelvin)	%Crystallinity
1.	Control	–	–	48.2%
2.	Control	–	350	47.0%
3.	Control	–	370	50.4%
4.	Control	–	440	50.0%
5.	Irradiated	0.27	–	51.4%
6.	Irradiated	0.54	–	50.6%
7.	Irradiated	1.07	–	50.1%
8.	Irradiated	4.27	–	48.8%
9.	Irradiated	8.55	–	50.4%
10.	Irradiated	15.5	–	50.0%
11.	Irradiated & heated	4.27	350	51.5%
12.	Irradiated & heated	4.27	370	52.5%
13.	Irradiated & heated	4.27	440	50.9%
14.	Irradiated & heated	4.27	470	48.3%

TABLE-II INTENSITY RATIOS OF VARIOUS REFLECTIONS WITH (110) PEAK

Sample	Dose Mrads	Heating temp. Kelvin.	Ratios of (100) 2θ=17.7	(020) 18.4	(021) 26.7	(200) 36.3	(210) 37.44
				With (110) Reflection			
Control	-	–	0.479	0.80	0.123	0 049	0.008
Control	-	340	0.496	0.83	0.126	0.049	0.010
Control	-	440	0.486	0.89	0.131	0·053	0·007
Control	-	440	0 500	0·98	0·218	0·059	0·010
Irradiated	0·27	-	0 500	0·83	0·120	0·056	0.018
Irradiated	0.54	-	0 475	0·80	0·135	0·060	0.016
Irradiated	1.07	-	0.520	0.85	0.136	0.063	0.020
Irradiated	4.27	-	0.480	0.81	0.131	0.061	0.008
Irradiated & heated	4.27	340	0.430	0.82	0.133	0.067	0.020
Irradiated & heated	4.27	370	0.500	0.84	0.135	0.064	0.022
Irradiated & heated	4.27	440	0.470	0.83	0.333	0.078	0.007
Irradiated & heated	4.27	470	0.530	0.99	0.34	0.069	0.007

transformations have been analysed using X-ray diffraction. The development of γ or α phase has been detected by the appearance of composite peak (110 + 200) of β at about 20.8° and 021 of γ at 20.7° Such transformation is also accompanied by decrease in the intensity of 021 reflection of α phase at 26.7°.

The ratio of intensities of peak reflections (021 and 110) are given in Table II. It may be seen that the sample just heated as well as the samples just irradiated shows small increase in this ratio. A similar but pronounced increase was observed for samples irradiated and heated.This trend indicates probably perfection in α phase. Similarly the ratio of intensities of 110 reflections and 100 reflections was calculated, the values of which are given in Table-II.

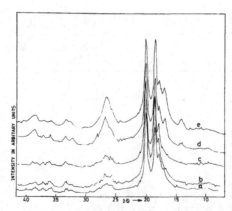

Fig. 3.. XRD SCANS OF (a) PVDF IRRADIATED TO A DOSE OF 4.27
MRADS AND SUBSEQUENTLY HEATED IN PRESENCE OF
NITROGEN TO VARIOUS TEMPERATURES (b) 350K
(c) 370 K (d) 440 K (e) 470 K.

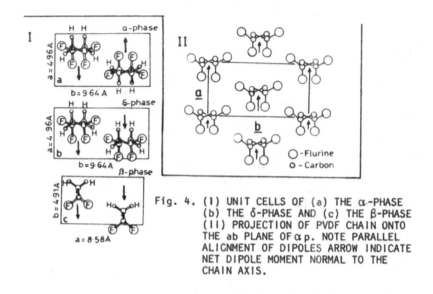

Fig. 4. (1) UNIT CELLS OF (a) THE α-PHASE (b) THE δ-PHASE AND (c) THE β-PHASE (II) PROJECTION OF PVDF CHAIN ONTO THE ab PLANE OF α p. NOTE PARALLEL ALIGNMENT OF DIPOLES ARROW INDICATE NET DIPOLE MOMENT NORMAL TO THE CHAIN AXIS.

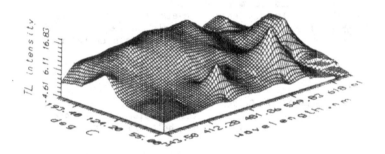

Fig. 5. 3D TL PLOT FOR PVDF.

These values decrease steadily for all these varieties of samples, which shows that apparent content of α phase in the samples increases. The peak intensities of (020) reflection appears to increase and (210) reflection appears to decrease. An increase in the ratio of the intensity of the (100) and (200) reflections leads to the conclusion that new polar crystal phase may be created by aligning the dipoles in form II without altering the chain conformation. However, observation of additional peaks at 14.6^o and 17.1^o leads us to think of some phase transformations. This can be understood on the basis that initially sample consists of a mixture of α & γ phases. During irradition some scission of bonding occurs and chains are some what free to rotate or the energy of radiation can generate a twist wave along the chain. This rearrangement can modify the α phase into δ-phase. It may be pointed out the δ-phase is almost similar in structure with α.phase except for parallel oriented dipoles, Figure 4. It is pertinent to point out that the development of the new phase is slow during radiation and gets enhanced during subsequent heat treatment. The control sample has a reasonable amount of γ-phase is supported by IR spectrum and DTA measurements.

Thermoluminescence glow curve exibits four peaks figure 5 at temperatures of about 350, 370, 440 and 470 K, occuring due to example certain major transitions when sample is gradually heated (5). For example heating of α -PVDF to 350 K shows a maxima in the piezo electric constant, at 440 K it leads to phase transformation (2) and at 470 K it transforms from ferro to para electric state. Such transformation obviously lead to major changes in chain orientation and confirmation thereby altering energy levels within a molecule. Observation of phase transformations when sample is heated to 440 and 470 K, using XRD, in the present study confirms it's correlation to TL.

ACKNOWLEDGEMENTS

We are very grateful to Dr.K.S.V.Nambi of BARC, Bombay for helpful suggestions and discussions. The authors also wish to thank Mr.A.R. Kamat and Miss Rajashree for their help in carrying out experiment.

REFERENCES

[1] A.J.Lovinger, Science Vol.220 No.4602 (1983)
[2] D.K.Das Gupta, K. Doughty, App.Phy Lett 31 No.9 (1977)
[3] G.T.Davis, M.G.Broadhurst, J.Appl.Phy.49, 4998 (1978)
[4] Y.Takase,T.T.Wang, J.Appl.Phy.60,2920 (1986)
[5] Sumita Pawde,K.S.V.Nambi, 'TL Studies on irradiated PvDF above
     RT' Proceeding of National Sym. on Radiation Physics, Ja.90,BARC.

SYNTHESIS, CHARACTERIZATION AND BINDING PROPERTIES OF EPOXY

RESINS BASED ON CARBONOHYDRAZONES AND THIOCARBONOHYDRAZONES

P.M. Thangamathesvaran and Sampat R. Jain

Propellant Chemistry Laboratory
Department of Aerospace Engineering
Indian Institute of Science
Bangalore 560 012, INDIA

INTRODUCTION

Solid fuels capable of igniting spontaneously on coming into contact with liquid oxidizers have been used widely in hybrid propellant motors. Although the hybrid systems combine many advantages of both the solid and the biliquid rocket motors, for smooth operation of the motor the ignition delay, i.e. the elapsed time preceding ignition after the liquid oxidizer comes in contact[1,2] with the solid fuel, must be very short. In our earlier studies[1,2] several hypergolic solid fuel powders igniting with oxidizers, such as white fuming nitric acid with very short ignition delay (WFNA) were developed. In actual use, however, the powder fuels have to be cast in a 'grain' form using polymeric binders. Casting of a highly hypergolic fuel powders using conventional binders like carboxyl or hydroxyl terminated polybutadiene (CTPB or HTPB) although results in imparting good mechanical strength to the fuel grain, the ignition delay becomes intolerably long. The present study was undertaken to develop new high energy resins which could be used as binders for hypergolic fuel powders without affecting their ignition delay significantly, and also providing good mechanical strength to the grain.

Since many of the hydrazine derivatives are known for their hypergolicity with liquid oxidizers it was envisaged that the epoxy resins having N-N bonds in their backbone would have desirable ignition properties. In the literature, there are very few polymers reported with N-N bonds in the backbone or main chain structure. These are, however, unsuitable as binders because of the lack of suitable end groups capable of curing easily without the formation of by-products. Since the resins having epoxy end groups are suitable as binders it is worthwhile to synthesize the epoxy resins based on hybergolic fuels. Of the various such fuels, the bis-thiocarbonohydrazones were found to be extremely hybergolic with WFNA, and also suitable for epoxidation. Herein we report the synthesis and characterization of some new epoxy resins based on bis-carbonohydrazones and thiocarbonohydrazones. The solid loading capacity of the resins and the ignition delays of the cured samples having varied amounts of magnesium powder have been measured.

Frontiers of Polymer Research, Edited by P.N. Prasad and
J.K. Nigam, Plenum Press, New York, 1991

EXPERIMENTAL

Materials: Carbondisulfide, diethylmelonate, hydrazine hydrate, benzaldehyde, o- and m-hydroxybenzaldehydes, m-methoxy-p-hydroxybenzaldehyde (Vanillin), furfuraldehyde and epichlorohydrin.

Preparation of thiocarbonohydrazones and carbonohydrazones:

Thiocarbohydrazide (TCH) was prepared by reacting carbon-disulfide and hydrazine hydrate. The bisthiocarbonohydrazones [3] were prepared by following the method developed by Rajendran and Jain[3] and characterized by comparing their melting points and ^{1}H NMR spectra.   Carbohydrazide and Carbonohydrazones were prepared by following the reported procedure[4].

SYNTHESIS OF EPOXY RESINS[5]

The epoxidation of thiocarbonohydrazones and carbonohydrazones were carried out by refluxing the compound with excess epichlorohydrin for 1-2 hours. The excess epichlorohydrin was vacuum distilled.

The general structure of the difunctional resin :

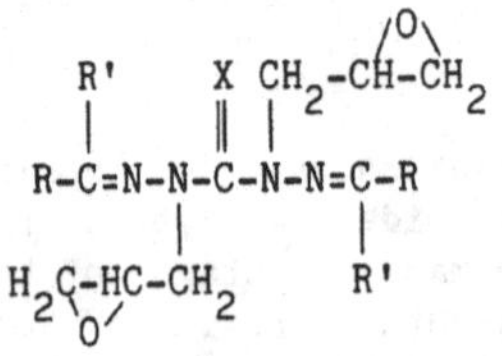

where,

X = S ;	R= (furfuryl ring)	R'= H	Resin 1
	R=-CH$_2$CH$_3$	R'=-CH$_3$	Resin 2
X = 0 ;	R= -CH$_3$	R'=-CH$_3$	Resin 3
	R= (furfuryl ring)	R'= H	Resin 4

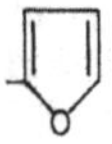

The general structure of the tetrafunctional resin:

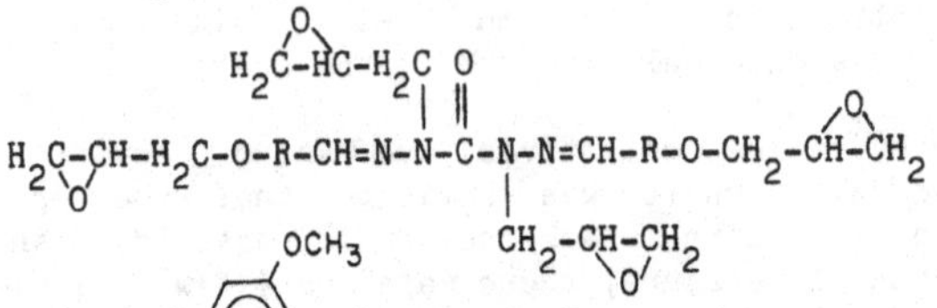

where, R = (benzene ring)   Resin 5

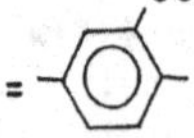

Resin 1 = Diglycidyl ether of furfuralthiocarbonohydrazone
Resin 2 = Diglycidyl ether of butanonethiocarbonohydrazone
Resin 3 = Diglycidyl ether of acetonecarbonohydrazone
Resin 4 = Diglycidyl ether of furfuralcarbonohydrazone
Resin 5 = Tetraglycidyl ether of vanillincarbonohydrazone

CHARACTERIZATION

NMR spectra: The ^{1}H NMR: Instruments:Varian T-60 or BrukerWH-270 FT-NMR
              Solvent:DMSO-d$_6$; Standard: tetramethylsilane (TMS)

Infrared spectra: Instrument: Perkin-Elmer   781

Epoxide equivalent: The epoxide equivalents of the various resins were obtained using the pyridinium chloride method[6].

Viscosity: Instrument: Brookfield viscometer (T-F spindle).

Thermal Analysis: Instrument: Shimadzu DT-40 simultaneous DTA-TG Thermal Analyzer. Heating Rate: 10°C/min; Dynamic N_2 atmosphere.

Curing studies: Curatives: thicarbohydrazide (TCH) ethylenediamine (EDA) and diaminodiphenylmethane (DDM); Ratio:Based on epoxy equivalent.

SEM: Stereoscan 150 Scanning Electron Microscope.

Compression Strength: Universal Testing Machine.

Ignition Delay: Experimental set-up earlier developed in our lab.[7]

RESULTS AND DISCUSSION

The epoxy resins synthesized by the present method are highly viscous liquids at room temperature. The epoxy equivalents of the various resins given in Table 1 confirm the formation of tetra- and di-epoxy resins from the hydroxy and non-hydroxy derivatives respectively.

TABLE 1. Physical characterization of various resins

Resin	Color	State	Epoxy equivalent		Molecular weight	Epoxy content	
			obs.	cal.		obs.	cal.
1.	Brown	viscous	188.90	187.0	374	1.98	2
2.	Light brown	less viscous	167.60	163.0	326	1.95	2
3.	Brown	viscous	141.81	141.0	282	1.94	2
4.	Brown	highly viscous	181.93	179.0	358	1.97	2
5.	Brown	viscous	148.70	145.5	582	3.91	4

obs. = observed ; cal. = calculated

SPECTRAL CHARACTERISTICS

The epoxidation of both the NH and OH protons is further evident from the 1H NMR spectra of the resins. The absence of -NH and -OH & -NH proton resonances in the di- and tetra- glycidyl resins respectively and the appearance of epoxy proton resonance accordingly confirms the formation of epoxy resins. The chemical shift positions of the different protons in the resins are shown in Table 2.

TABLE 2. 1H NMR Chemical Shift,   (ppm)

Resin 1. 3.25-3.8 (m,10H,Epoxy), 6.4-7.7 (m,6H,Ar), 8.2(S,2H,=CH)
Resin 2. 1.0,2.0, (m,16 H, ethyl, methyl), 2.8-3.8 (m, 10 H, epoxy)
Resin 3. 1.8,2.1 (s,12 H,methyl), 3.2-3.7 (m,10 H, epoxy)
Resin 4. 3.2-3.8 (m, 10H, epoxy), 6.45-7.7 (m, 6H, Ar), 8.2 (s, 2H, =CH)
Resin 5. 2.5-3.6(m,20H epoxy), 3.9(s,6H, OCH3), 7.0-7.4 (m,6H, Ar), 8.25 (s,2H, = CH).

m = multiplet, s = singlet

The infrared spectra of the epoxy ring compounds are reported to absorb in the region 1280-1230 cm^{-1} as a result of the ring breathing or symmetric stretching vibration and around at 860-770 cm^{-1}

due to asymmetric stretching vibration of the epoxy ring. The appearance of these two absorption peaks and also the peak around 2900 cm-[1] due to CH and CH2 groups in the ring confirms the presence of epoxy ring in these resins[8]. The infrared spectra of the cured samples have virtually no characteristrics absorption due to epoxy ring which confirms the occurrence of curing reaction.

Viscosity: The variation of viscosity with temperature of the various resins presented in Table.3 shows a steep fall in viscosity with increase in temperature. Generally aromatic resins(Resins 1, 4 & 5) showed higher viscosity than the aliphatic ones (Resins 2 and 3). No significant viscosity difference between the tetra- and di-epoxides was observed.

TABLE 3. Viscosity Data

Temp.	Resin viscosity (in poise)				
(°C)	1	2	3	4	5
26	220	50	100	-	-
35	90	-	32	-	40
40	62	30	-	210	4
50	-	12	-	-	2
54	-	-	-	80	-
60	-	-	-	47.5	-

THERMAL PROPERTIES

The differential thermal analysis data of the various resins (self-cured) are listed in Table 4. In general, the resins and the cured samples decompose exothermally around 450 to 600°C.

TABLE 4. Differential Thermal Analysis Data

Sample (Self-Cured)	DTA peak temperature* (°C)
Resin 1 (DGFTCH)	350 (+) 500 (+)
Resin 2 (DGBuTCH)	230 (+) 450 - 550 (+)
Resin 3 (DGACH)	218 (+) 552 (+)
Resin 4 (DGFCH)	380 (+) 530 (+)
Resin 5 (TGVCH)	268 (+) 600 (+)

* (+) indicates the exotherm.

The DTA of the resins shows an exothermic peak around at 220-380°C and another broad exothermic peak around 510-600°C. The first exothermic peak could be due to isomerisation of the epoxy group into carbonyl group, and also due to the thermal polymerization of the resin, as observed earlier by Anderson[9] in the case of the epoxy resins. The broad exothermic peak in the high temperature region is due to the decomposition of the self-cured resin. However, no exotherm around 300°C was observed in the cured samples.

BINDING PROPERTIES

The resins loaded with upto 70 % solid fuel resulted in a fuel grain with compression strength of ~100Kg/cm[2]. Generally, tetraglycidyl resins have shown better binding properties and loading capabilities

TABLE 5.Ignition delay and compression strength data:Sample weight:200mg

	Weight % Composition Resin : Magnesium : TCH	Amount of WFNA (ml)	Ignition Delay (sec)
Resin 1	100 (self-cured) -	0.5	4.880
Resin 4	40 : 50 : 10	0.5	0.410

Compression strength of all the cured samples were found to be above 15 kg/cm^2.

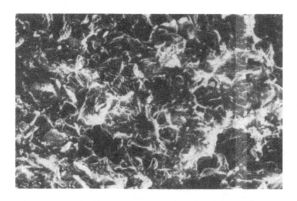

RESIN 2 + 70% TCH($\sim$ 300 microns) [100   micron]

than difunctional resins. The uniformity of the solid fuel distribution in the resin matrix can be observed from the SEM photograph.

IGNITION DELAY

The preliminary ignition delay studies on the fuel grains prepared from   these   novel   hybergolic   resins   showed   their   ignition capabilities. Among the difunctional resins the thiocarbonohydrazone based resin(Resin 1) showed better hybergolic ignition whereas the carbonohydrazone   based resin(Resin 4) showed   better   synergistic hypergolic igntion behaviour.

CONCLUSION

The   synthesized   resins were confirmed by analytical and spectral methods and charaterized by viscosity measurements and thermal analysis. The novel resins had shown not only better binding properties but also shorter ignition delay.

REFERENCES

1. S.R.Jain,P.M.M.Krishna & V.R.Pai Verneker,J.Spacecraft & Rockets
   16:69 (1979)

2. G. Rajendran and S.R. Jain, Fuel, 63:709 (1984)
3. G. Rajendran, S.R. Jain, Indian J. of Chem., 24B:680 (1985)
4. F.Kurzer & M.Wilkinson ,Chem.Rev.,70:111 (1970)
5. P.M.Thangamathesvaran and S.R.Jain, J.Polym.Sci., Polym.Chem.Ed., 29:261(1991)
6. H.Lee, K.Neville, "Hand Book of Epoxy Resins", McGrew Hill Inc., New York, (1967) pp 4-17
7. S.R.Jain,T.Mimani,J.J.Vittal, Combust.Sci.&Tech. 64:29(1989)
8. N.B.Colthup,L.H.Daly and S.E.Wiberley,"Introduction to Infrared and Raman Spectroscopy",Acadamic Press, New York, (1975)
9. H.C. Anderson, Anal. Chem., 32: 12, 1592 (1960)

INFLUENCE OF HYDROXYL CONTENT ON THE FUNCTIONALITY

DISTRIBUTION AND GUMSTOCK PROPERTIES OF POLYBUTADIENE

K.N. Ninan, V.P. Balagangadharan, K. Ambika Devi and
Korah Bina Catherine

Propellants and Special Chemicals Group
Vikram Sarabhai Space Centre, Trivandrum-695 022, India

INTRODUCTION

Hydroxyl terminated polybutadiene (HTPB) is mainly used
as a solid propellant binder in rockets. This fuel binder is
considered superior to other classical binders because of its
better mechanical properties, higher specific impulse and
solid loading capability and the excellent reproducibility of
the HTPB propellant. In Vikram Sarabhai Space Centre (VSSC)
this prepolymer is prepared by free radical polymerisation
using hydrogen peroxide as initiator. During propellant
formulation the prepolymer is mixed with other active
ingredients and cured using a difunctional curator (toluene
diisocyanate - TDI) to obtain a crosslinked urethane network
which imparts the necessary mechanical strength and structural
integrity to the propellant grain.

HTPB is not essentially telechelic as the name implies.
Although the average functionality is close to two, the
polymer molecules differ in the number of hydroxyl groups
present on them even when the hydroxyl value is constant. The
relative amounts of the polymer molecules having various
functionalities affect the network parameters. The
nonfunctional molecules do not partake in the cure reaction.
The monofunctional molecules act as chain terminators,
difunctional molecules as chain extenders and tri- and
multifunctional molecules form crosslinks with the
difunctional curator[1]. In a nutshell, free radical HTPB
exhibits two types of heterogeneities - distribution of
molecular weight and distribution of functionality (Fig.1) -
which are very significant in deciding the ultimate mechanical
properties of the crosslinked binder.

In the present study, four batches of HTPB specially
prepared with varying hydroxyl content are studied for their
functionality distribution and gumstock properties. The
gumstock properties (tensile strength, elongation, modulus,
hardness, crosslink density) are an indication of the resin
contribution towards the properties of the final cured
propellant. The influence of hydroxyl content on the

Frontiers of Polymer Research, Edited by P.N. Prasad and
J.K. Nigam, Plenum Press, New York, 1991

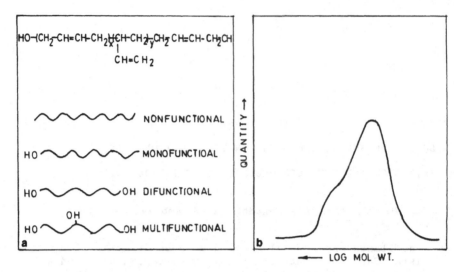

Fig.1   a) Typical molecular weight distribution and b) types of functionalities in HTPB.

functionality distribution and gumstock properties is also discussed.

EXPERIMENTAL

    Four samples of HTPB of different hydroxyl content were prepared at VSSC by varying the initiator concentration. The resins were characterized by chemical and physical methods as described below and the results are given in Table 1. The hydroxyl value was determined by the acetylation procedure. The number average molecular weight (M_n) was measured using a Knauer Vapour Pressure Osmometer, in toluene at 65°C. The molecular weight distribution (MWD) was determined using a Waters ALC/GPC 244 in combination with an R401 differential refractive index (DRI) detector and M730 data station, μ Styragel columns $10^4, 10^3, 500$ & $100\AA$, tetrahydrofuran solvent at 2 mL per minute and the universal calibration method were employed. Viscosity of the resin was measured using a Brookfield viscometer. Determination of functionality distribution involved the following steps: chemical modification of HTPB using 3,5-dinitrobenzoylchloride to form the UV-absorbing ester derivative[2], characterisation of ester (chemical and IR spectrosopy) to ensure complete conversion of OH groups to ester and fractionation of the ester into narrow disperse fractions - Waters Delta Prep 3000, columns 500, 10^3, $10^4\AA$ Styragel, DRI detector, M730 data station. The fractions were analysed with a double detector (DRI & UV) analytical GPC to derive the equivalent weight of each fraction from the ratio of UV signal to RI signal. Functionality is calculated as the ratio of molecular weight to equivalent weight of the individual fractions. The functionality distribution was computed[3] from the plot of functionality vs. cumulative quantity[3] (Fig. 2). Evaluation of gumstock properties involved: curing the HTPB prepolymer with stoichiometric amount of TDI at 70°C for 48 hours and measuring the tensile strength, elongation and modulus (stress at 100% elongation), using a universal testing machine - Instron model 4202.

Cross-link density was determined by equilibrium stress strain method[4]. Dumb bell specimens were subjected to tensile loading in the UTM to different strain levels to obtain the equilibrium modulus from which the crosslink density was computed.

RESULTS AND DISCUSSION

Table 1 lists the properties of the four HTPB resins. The M_n is found to increase with decrease in hydroxyl content. Hydroxyl content is proportional to the initiator concentration in the reaction system. When lesser number of hydroxyl radicals are present in the system fewer chains are initiated which grow to longer lengths, while increased hydroxyl content caused by increased hydrogen peroxide initiator results in more chains of lesser DP. Equivalent weights increase with decreasing hydroxyl content (and increasing molecular weights). Viscosity values are in trend with the molecular weights. Average functionality (M_n(VPO)/Eq.Wt.) and polydispersity values are nearly constant.

Table 1.   Characteristics of the four HTPBs

Props/ B.No.	I	II	III	IV
OH val. (mg KOH/g)	40.0	34.1	27.4	21.1
Mn(VPO)	3070	3400	3510	4480
Mn(GPC)	3500	4650	4800	8380
Mw(GPC)	10440	13130	15700	19700
PD	3.0	2.9	3.3	2.5
F(Mn/eq.wt)	2.2	2.1	1.7	1.7
η(cps) at 30°C	5740	8720	12190	37760

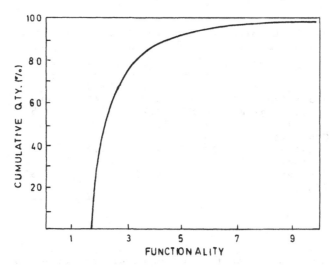

Fig.2   Functionality vs. cumulative quantity in a typical HTPB.

Functionality Distribution

All species with functionality less than 1 do not participate in the cure reaction and are termed here as non-participating species NP^3. Species with functionality between 1 and 3 are termed chain extenders CE and form the major part (60 to 70%) of all four resins (Table 2) The amounts of CE and crosslinking species CR (functionality greater than 3) do not show any major change in these four samples.

Gumstock Properties

The tensile strength and modulus decrease as hydroxyl content decreases (Table 3). Elongation shows a reverse systematic trend. Sol content also shows a steady increase. This may appear contrary to our earlier finding[3] that the gumstock properties are correlatable to the quantities of CE and CR in HTPB. In the earlier work we had used HTPB samples of nearly same hydroxyl value and molecular weight whereas in the present investigation the hydroxyl content steadily decreases from samples 1 to 4. Thus in these samples, even when CE and CR contents are more or less steady, the crosslink density will show a steady decrease since crosslinking takes place through hydroxyl groups. This is clearly indicated in the steadily decreasing crosslink density values and hence decreasing tensile strength and modulus and increasing elongation and sol content for the samples 1 to 4.

Table 2. Functionality distribution in the four HTPBs

Type / B.#	I	II	III	IV
NP(%)	Nil	5	Nil	Nil
CE(%)	71	61	71	65
CR(%)	29	34	29	35

Table 3. Gumstock properties of the four HTPBs

Props/ B.No.	I	II	III	IV
T.S(Kg/cm^2)	7.2	7.1	6.2	5.6
Elong.(%)	200	290	360	600
Modulus(Kg/cm^2)	4.5	3.7	3.4	2.5
Sol(%)	11.3	13.4	17.6	26.6
Crosslink density (moles/cc)	10.7	9.6	8.9	7.5
Hardness (SAH)	30-35	30-35	30	25-30

CONCLUSION

The gumstock properties (mechanical properties and sol content) are directly correlatable to the functionality distribution of HTPB when the hydroxyl content is constant.

However in samples where hydroxyl content is varying the crosslink density and hence gumstock properties depend both on functionality distribution and hydroxyl value. In the present case where functionality distribution is nearly the same, the gumstock properties depend more on the hydroxyl content. Thus in the four samples of HTPB with varying hydroxyl content and molecular weights it is found that the gumstock properties and sol content cannot be simply related to functionality distribution as would have been possible in samples of similar hydroxyl value and molecular weight. However these properties are directly relatable to the crosslink density which reveals the average chain length between croslinks.

ACKNOWLEDGMENT

We thank the Propellant Fuel Complex, VSSC, for supplying the HTPB samples. We also thank DD, PPC and Director, VSSC for their kind permission to publish this work.

REFERENCES

1. Muenker, A.H., Hudson, B.E., J. Macromol. Sci. Chem., 1969, 3, 1465 - 1483.
2. Carver, J.C., Improved Specifications for Composite Propellant Binders for Army Weapon Systems, TR-RK-81-5, June 1981.
3. K.N. Ninan, V.P. Balagangadharan, Korah Bina Catherine, Polymer, 32, 4, 1991, 628.
4. L.R.G. Treloar, "The Physics of Rubber Elasticity", 3rd Edn., Oxford University Press, 1975, p 142.

SYNTHESIS OF AMORPHOUS NYLON 6 USING NEW INITIATOR SYSTEM

M.R. Desai   and   M.V. Pandya

Department of Chemistry, Indian Institute of
Technology, Powai
BOMBAY-400076, INDIA

ABSTRACT

The use of Nylon 6 is restricted to fibre application
because of its high crystallinity. The present work describes
the synthesis of amorphous nylon 6, having potenial
applications in film blowing and molding. Nylon 6 was
synthesized via a two step process using an alkali metal
catalyst and an initiator system. Initiators used were
diisocyanate and polydiisocyanate-caprolactam (CL) block. The
effect of initiator concentration on physical properties like
crystallinity, molecular weight, density and thermal stability
of nylon 6 has been reported.

INTRODUCTION

Commercially nylon 6 is synthesized from ϵ
-caprolactam(CL) by hydrolytic method, which takes place at
265°C and under 15 atm pressure.[1] The nylon 6 formed is highly
crystalline and its use is restricted to fibre applications.
Molding and film blowing applications require nylon 6 of
reduced crystallinity.

Amorphous nylon 6 can be obtained either by chemically
modifying the CL monomer or by melt mixing nylon 6 with other
polymers.[2] Chemical modification of CL leads to a descrease in
its reactivity, resulting in formation of product of low yield
and of low molecular weight. Blending of nylon 6 with other
polymers leads to compatibility problems, hence requires
precision control on the processing parameters.

Introducing a selective organic molecule or a macromer
of low molecular weight in the main pllymer chain should bring
irregularity in the polymer chain, without much affecting the
monomer reactivity, leading to a reduction in the bulk
crystallinity.

The present work describes synthesis of amorphous grade
nylon 6 by a relatively esy and simple anionic polymerization,
using a new initiator system.

Frontiers of Polymer Research, Edited by P.N. Prasad and
J.K. Nigam, Plenum Press, New York, 1991

EXPERIMENTAL

Materials

Commercially available caprolactam (CL) was used without purification. Diisocyanates were used as obtained from Fluka. Meta-cresol, methanol, toluene and carbon tetrachloride were fractionally distilled under nitrogen prior to use. Purity of chemicals was confirmed from their IR spectra.

Synthesis

CL was anionically polymerized via a two step process. The first step involves synthesis of intermediates viz. catalyst and an initiator. Catalyst was prepared by reacting Na-metal with CL at 90°C. Initiator was prepared by reacting diisocyanate / polydiisocyanate (Molecular weight $\sim$ 3000) with CL at 90°C. Both the intermediates formed are quite stable and can be stored under dry and inert atmosphere.

In the second step, the polymerization was carried out by mixing intermediates along with CL in a resin kettle at 140°C in an inert atmosphere. The reaction gets completed within 15 mintues. The polymer obtained was extracted first with hot water and then with hot methanol before characterization.

Two series of nylon 6 were prepared with diisocyanate-CL block and polydiisocyanate-CL block initiators. Effect of initiator concentration on crystallinity, molecular weight, density and thermal stability of nylon 6 was studied.

Characterisation of Polymers

X-ray diffraction: Wide angle x-ray diffraction (WAXD) analysis of extracted polymer samples (1 x 1 x 0.02 cm) was carried out on a Philips PW 1820 diffractometer at a scanning rate 2° min^{-1} from $2\theta = 5°$ to 65° using copper K α target. The percentage crystallinity was calculated from WAXD spectra (Fig.1) by following standard procedure[3].

Viscometric Studies: Intrinsic viscosities of nylon 6 samples were determined on Schott Gerate AVS 400 viscometer at 25 ± 0.1°C using fresh fractionally distilled m-cresol as a solvent. The viscosity average molecular weight was determined using the Mark-Houwink equation[4]

$$[\eta] = (5.26 \times 10^{-4}) \times M_w^{0.74}$$

Density measurements: The density of nylon 6 samples were meausred by floatation method using mixture of carbon tetrachloride and toluene at 30°C. The density of the mixture was determined using a calibrated pycnometer with 0.8 mm bore capillary, giving third decimal accuracy of $\pm$ 1 unit in density measurements.

Thermal analysis: Thermal analysis of nylon samples were carried out on a Dupont 2100 thermal analyzer in nitrogen atmosphere. The temperature was scanned from 25°C to 550°C at

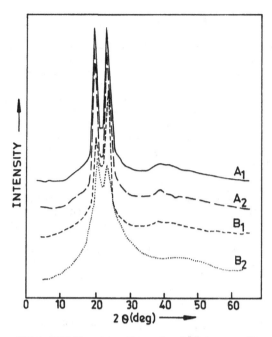

FIG.1. WAXD pattern of series A and series B
polymers.

a heating rate of 5° per minute. The initial decomposition
temperature (idt) and initial procedural decomposition temperature (ipdt) were calculated from thermograms following
standard procedures[5].

TABLE 1 CRYSTALLINITY, MOLECULAR WEIGHT, AND DENSITY OF NYLON 6

SERIES	POLYMER	% CRYST.	INTRINSIC VISCO.	MOL.WT. x 10^{-6}	DENSITY g/cc
A	A1	28	177.9	26.4	1.150
	A2	26	160.4	23.0	1.141
B	B1	15	100.8	12.3	1.125
	B2	8	37.7	3.3	1.092

Series A : Using diiscyanate-CL block initiator
Series B : Using polydiisocyante-CL block initiator.
1 and 2 indicate initiator concentration of 0.1 and 0.2
mol % respectively.

Xray Diffraction Analysis

As seen in Table 1 percentage crystallinity of nylon 6 in
both the series is low compared to that of commercial nylon
which is about 70%[6]. This can be explained as follows.

Commercial nylon 6 is prepared by hydrolytic method which takes place at 285°C. At this temperature polymer formed is in the molten form. This favours the formation of crystal lattice because of strong hydrogen bonding between the amide groups. In the present technique, polymerization takes place at 140°C, which is well below the melting point of the polymer. During most of the polymerization process reaction mixture is in a semi-solid form, restricting the mobility and alignment of polymer chains. Secondly the initiator molecule being present in the main polymer chain, the regular alignment of polyamide chains is disturbed, resulting in a drastic reduction in crystallinity of the bulk.

The decrease in crystallinity is more in case of series B polymers than in series A polymers as the intermediate molecule incorporated in the former series being a macromer, creates more hindrance to the alignment of the polymer chains.

The decreease in crystallinity with respect to initiator concentration is small as addition of new initiator molecules affects only the long range order. The short range order which is mostly unaffected by initiator concentration gives residual crystallinity to the polymer.

As seen in Table 1, very high molecular weights are observed in both series of polymers as the polymerization is anionic, and occurs in solid state.

A decrease in molecular weight is observed in series B polymers, which may be explained as follows.

During initial stage of polymerization, the reaction mixture is in the liquid state. At this stage of polymerization, the more complex nature of macromer initiator in series B hinders the polymer chain growth and hence lower molecular weight in series B polymers than in series A polymers. The decrease in molecular weight with increase in initiator concentration is as expected.

Density

A decrease in density is observed in series B polymers. The same trend is observed in individual series with increase in initiator concentration (Table 1). This can be attributed to decrease in crystalinity, as density of crystalline domains is always more than that of amorphous domains.

Thermal Analysis

The decrease in melting point in series B polymers (Table 2) is attributed to the decrease in bulk crystallinity.

Thermal stability of series B polymer is reduced as compared to series A polymers, which is mainly due to the incorporation of macromer block in the polymer chain as well as the decrease in the bulk molecular weight. The idt values are same for both polymers in same series, indicating similar nature of polymer backbone chain in the polymers. The decrease

TABLE 2 THERMAL DATA OF NYLON 6

SERIES	POLYMER	MELTING POINT °C	idt °C	ipdt °C
A	A1	220	344	380
	A2	215	344	376
B	B1	185	305	339
	B2	150	305	329

in ipdt values for polymers in individual series can be attributed to the reduction in molecular weight.

CONCLUSIONS

The present anionic polymerization process is a fast and convenient method for nylon 6 sunthesis as it requires lower polymerization temperature and it takes place at atmospheric pressure.

Nylon 6 obtained using this methodology are of low crystallinity, high molecular weight and good thermal stability, with the potential application in molding and film blowing.

REFERENCES

1. Hanford, W.E. and Joyce, R.M., J.Polym. Sci., 3, 167 (1948).
2. Kohan, M.I., Nylon Plastics, Wiley-Interscience Pub., New York (1973).
3. Kakudo, M. and Ullman R., J.Polym.Sci., 65, 91 (1960).
4. Mittiusi, A., Gerchele, G.B. and Francesconi, R., J.Polym. Sci.Polym. Phys., 7, 411 (1989).
5. Kilne, G.M., Analytical Chemistry of Polymers, Wiley-Interscience Pub., New York (1959).
6. Gupta, M.C. and Pandey, R.R., J.Polym.Sci. Polym.Chem., 26, 491 (1988).

TRIBOLOGICAL BEHAVIOR OF REDMUD

FILLED POLYESTER COMPOSITES

Navin Chand, S.A.R. Hashmi, and O.P. Modi

Regional Research Laboratory (CSIR)
Bhopal-462 026 (M.P.) India

ABSTRACT

Abrasive wear behaviour of different composition redmud polyester composites was studied by using SUGA abrasion tester. Wear loss was more in case of 5 Phr composite as compared to 15 Phr redmud polyester composites. Effect of particle size on wear loss was also observed and explained on the basis of the microstructures of the worn surfaces.

INTRODUCTION

Redmud is produced as a waste material by aluminium industry during the conversion of bauxite into alumina by Bayer's process. Large quantity of redmud is produced every year in the world as a waste (1). Efforts are being done all over the world for its disposal and utilization. Clay particles are added in the polymers for reducing cost of product, for achieving dimensional stability and for better heat deflection temperature. Apart from this the fillers like NaCl crystals are added for foam production. It has been reported in literature (2-16) that glass beads and NaCl crystals when incorporated in the PMMA, improved the modulus. Improvement in the mechanical and wear properties are also expected in various polymeric materials with the addition of glassy fillers.

Ketomo (11) developed polymer redmud composites and found improvement in mechanical properties of rubber compounded with processed redmud. Red mud was washed with water and then with acids and residue dried and powdered before use. Froges (12) and Camergue sieved red mud through 300 mesh and then acidified with 10% by wt sulphuric acid. Red mud was mixed with rubber gave a strength of 227 kg/cm² as compared to 150 kg/cm².

In this investigation we have mixed red mud particles of different sizes in polyester resin for studying the effect of particle size and loading on tensile, impact, microhardness and abrasion behaviour of composites.

Frontiers of Polymer Research, Edited by P.N. Prasad and
J.K. Nigam, Plenum Press, New York, 1991

MATERIALS AND METHODS

Red mud obtained from BALCO (India) had the size ranging from 10 to 250 um. Chemical constituents of this red mud are mainly iron oxxide, sodium aluminium silicate, titanium oxide. Red mud was classified on the bases of particle size for this study as specified in Table-1. Unsaturated polyester (Isophthalic type), initiator methyl ethyl keton peroxide, and accelerator cobalt naphthalate was provided by M.P. Polymers, India.

Composites were prepared by adding different weight fractions of red mud to polyester resin in which 1.5% initiator and 1.5% accelerator were already mixed. Composition is given in Table-2. After stirring the mixture it was poured into the moulds especially made for impact strength, tensile strength, abrasion resistance and for micro hardness testing. Care was taken to avoid the bubbles formation which causes flaws. The mixture was allowed to set for 24 hours at room temperature. Samples were removed from the moulds after 24 hours and were kept in an oven for 2 hours at 100°C for curing to occur.

Tensile and impact samples were tested on a tensile testing machine "Instron 1182" and on impact pendulam "CEAST" Italy as per ASTMD 638 and D 256 methods respectively. Abrasion test were performed on SUGA Abrasion tester Model NUS-1. These tests were performed on a rectangular piece of composite of the size 60mm x 40mm x 4mm as specified by the equipment manufacturer. An

Table 1. Classification of red mud particles

S.No.	Type of red mud particle	Symbol	Size (dia) in μm
1.	Fine	F	10 – 75
2.	Medium	M	75 – 150
3.	Coarse	C	150 – 250

Table 2. Composition of red mud polyester composites

S.No.	Type of particles of red mud	Composition of composite	
		Red mud (g)	Polyester resin (g)
1.	F	5	100
		10	100
		15	100
2.	M	5	100
		10	100
		15	100
3.	C	5	100
		10	100
		15	100

schematic diagram of Abrasion tester is shown in Fig.1. Water proof silicon carbide paper C-180, HR-4 used for abrasive test was provided by John Akey and Mohan Ltd., India. Microstructure of the samples was evaluated by Leitz Wetzlar 8613 microhardness tester. Surface of the abrased and unabrades samples were observed

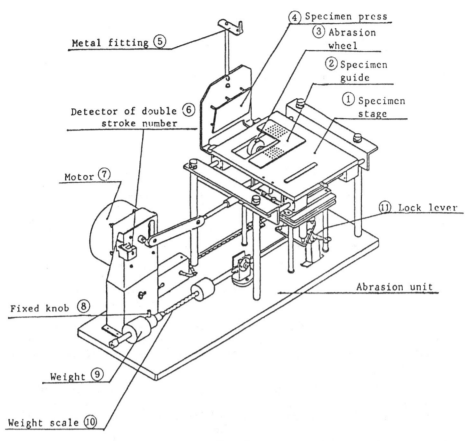

Metal fitting ⑤

④ Specimen press
③ Abrasion wheel
② Specimen guide
① Specimen stage

Detector of double ⑥ stroke number

Motor ⑦

⑪ Lock lever

Abrasion unit

Fixed knob ⑧

Weight ⑨

Weight scale ⑩

Fig. 1

on scanning electron microscope to understand the abrasion mechanism red mud particles in the red mud polyester resin.

RESULTS AND DISCUSSIONS

Effect of red mud particle size and loading of red mud on the tensile strength of red mud polyester composite was determined. Table-3 show the dependency of tensile strength of the red mud in the composite and the influence of red mud particle size. It was found that with the increase of the red mud weight fraction in the composite the decrease in tensile strength was a permanent feature. Effect of particle size was also noticed and it was observed that in the case of medium size particles (75-150 um) there was slight increase in tensile strength.

Incorporation of red mud particles has reduced the tensile strength of polyester from 40 to 10.75 MPa. From 10 to 150 um particle size strength remain same but for coarse particles it reduces further.

Table 3. Impact strength and tensile strength of red mud polyester composites

Partilce size of red mud (μm)	Fraction of redmud in composite (phr)	Impact strength IZOD (kJ/M²)	Tensile strength (MPa)
10 - 75	5	1.52	18.34
10 - 75	10	1.37	13.34
10 - 75	15	1.56	12.63
75 -150	5	0.95	19.10
75 - 150	10	1.20	16.97
75 - 150	15	1.66	12.49
150 - 250	5	0.84	12.04
150 - 250	10	0.87	10.75
150 - 250	15	0.98	9.46

This phenomenon can be explained by the particle to particle contact of red mud that increases with the increase of loading of red mud particles in the composites. Particle-particle contact would create flaws off weak points to initiate the failure of composites.

Impact strength IZOD for these samples was also determined. Table-3 gives the impact strength data for red mud polyester composite. Impact strength decreases with the increase of red mud into composites.

There is an increase in impact strength on increase in loading of red mud which can be explained on the basis of increased particle surface and reduced contacts which would have increased the crack propagation path. This can be further supported by the geometry of the red mud particles, particularly in large size. These particles are of irregular shape as shown in Fig.2 a,b,c.

Table-4 lists the result of microhardness of red mud polymer composite. Microhardness data shows that with the incorporation of red mud particle into polyester, the hardness increases with increase of particle size and red mud content. Microhardness of pure polyester is 19.3 HV whereas in case of fine particle it ranges from 19.3 HV to 57.9 HV. In case of medium particle size microhardness ranges from 21.9 HV to 68.7. For course size particles value of microhardness varies from 19.3 HV to 68.6 HV.

Table 4. Microhardness of red mud polyester composites

Red mud particle type	Red mud content phr	Range of Micro-hardness (Hv)
F	5	19.3 - 21.7
	10	21.7 - 23.7
	15	23.7 - 57.9
M	5	21.9 - 37.1
	10	28.4 - 55.6
	15	28.4 - 68.7
C	5	19.3 - 28.8
	10	21.7 - 28.8
	15	28.4 - 68.6

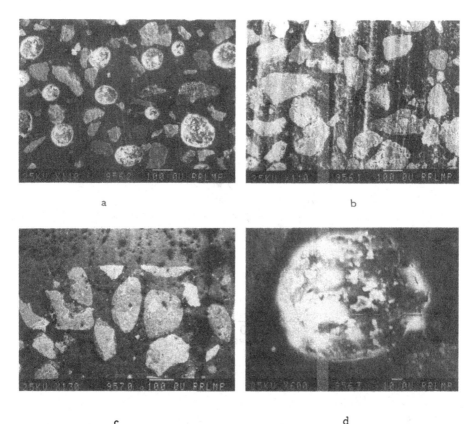

a   b

c   d

Fig. 2

Microhardness results can be explained by considering that composite has basically three regions, the matrix, interface and particles of red mud. At low loading and fine particles the indent point may fall upon the matrix and the interface mostly and on red mud particles may contribute very little. On increase of loading and size of particle, high value of hardness may be due to red mud particles which share the main past in case of hardness.

Fig.2 is the photograph of the fresh polyester red mud composite in which irregular shaped particles and spherical balls are observed. Distribution of particles in the polyester matrix in uniform. Matrix is clear and any cavity, voids or porosity was not observed. In microstructure of abrased composite irregular shaped particles were absent except one to two places. No cavity was observed which showed that binding was very good. Absence of spherical particles indicates that some soft/clay type material was agglomorated. WDX run confirmed that mostly irregular shaped particles contain iron. On further higher resolution, at 1000 times enlarged picture, spherical balls were proved to be made up of powdered material which forms a spherical cluster and possesses porous structure (Fig.2 d).

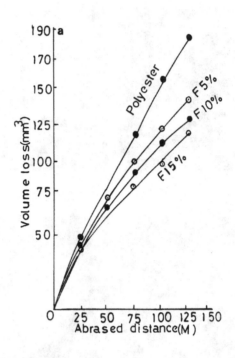

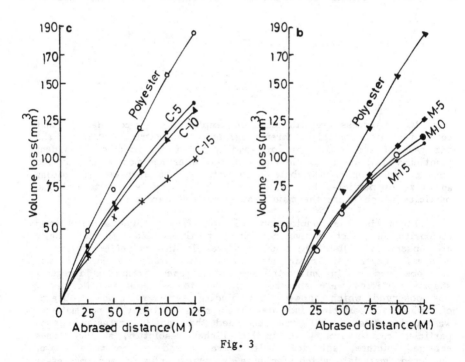

Fig. 3

Fig.3 shows the variation in the wear volume of red mud polymer composite with the applied load at 2.56 m/minutes sliding speed.

Test was conducted on composites having red mud of 5, 10 and 15 phr from each category of red mud i.e. F, M and C type. In each graph wear vvolume loss of composites has been compared with pure polyester. It is evident that there is a significant improvement in the wear volume loss with the increase of red mud particles in the composites. In Fig.3 a volume loss at 100 meter sliding distance is 150 mm^3/m^3 for pure polyester at 15 phr. Similar type of curves were obtained for M&C type composite. Micrograph (Fig.2) shows 10 phr M tyype composite, porous particles were broken during abrasion. Flow lines are observed on matrix and on particles as well.

REFERENCES

1.   Thakur, R.S. and Sant B.R. JSIR 42, Feb. (1983) 87-108.
2.   Navin Chand & Rao T.C. communicated to IMEJ (1990).
3.   Reisner K.H., Chem. Abstr. 74 (1971) 128240 y.
4.   Solymar K. Evaluation of bauxites from the view point of alumina production, Thesis University for Chemical Industry, Veszprem, 1973.
5.   Gohl M. Chem. Abstr. 63 (1965) 9504 h.
6.   Jonas K, Solymar K and Orban M, Acta Chim. Acad. Scient. Hung Tomas, 81 (4) (1974) 443-453.
7.   Uveges J and Mariassy M, Chem. Abstr. 53 (1959) 10677 e.
8.   Uveges J, Chem. Abstr. 53 (1959) 16486 a.
9.   Dobos G, Nemeez E, Solymar K and Elek S., Chem. Abstr. 66 (1967) 78499 b.
10. Feschenko, Z.1, Skobeey I.K. and Sergeevan V.N., Chem. Abstr. 75 (1971) 154048 g.
11. Ketomo A.G. swizz pat 274, 574, 2 July 1951 Chem. Abstr. 46 (1952) 10674 d.
12. Forges and Camergue Fr. Pat., 978, 108, 10 April 1951 Chem. Abstr. 47 (1953) 6171 e.
13. Armine F, Horst W and George B, Ger. (East) Pat. 19854 20 September 1960 Chem. Abstr. 55 (1961) 19296 h.
14. Riesel W, Peters R and Horst W, Ger. (East) Pat. 62820, 20 July 1968 Chem. Abstr. 70 (1969) 89894 k.
15. Reisner K and Mayer S, Ger. (East) Pat. 67, 107 5 June 1969 Chem. Abstr. 71 (1969) 114764 y.
16. Union Industrial Research Lab., 1021 Kuang-Fu Road Hsinchu, Taiwan, Republic of China. Personal communications.

FLEXIBLE LIQUID CRYSTALLINE MAIN CHAIN THERMOTROPIC COPOLYESTERS

P.K. Kaicker, Sudha Tyagi, Sangeeta Khanna and A.M. Khan

Shriram Institute for Industrial Research
19, University Road, Delhi - 110 007 India

INTRODUCTION

Main chain thermotropic copolyesters of p-hydroxybenzoic acid, terephthalic acid and hydroquinone possess remarkable thermal and mechanical properties. However, these rigid systems show high melting points and are dificult to process[1,2]. Hence, many efforts have been made to structurally modify them in order to reduce their melting points and thus rendering them melt processable. Inclusion of unsymetric substituents on the phenyl rings, modification to include flexible spacers or kinky units and use of crankshaft monomers are the most frequently employed approaches[3]. Jin et al.[4,5] and Krigbaum et al.[6] described how these modifications change the liquid crystalline and other properties of the resulting polyester. In general, relatively expensive reactants like dibenzoic acids, biphenols and naphthalene units have been used.[3,7]

In this study the combined effect of rigid and kinky moieties on liquid crystallinity and melting points of a p-hydroxy based thermotropic copolyesters was investigated. A series of copolyesters consisting of p-hydroxybenzoic acid (POB), hydroquinone (HQ) and isophthalic acid (IA) were synthesised. The last commonomer creats a kinky structure.

EXPERIMENTAL

The commercially available p-hydroxybenzoic acid, isophthalic acid and hydroquinone were characterised by infrared spectroscopy and analyzed for the purity by TLC. Hydroquinone was recrystallized from hot water.

p-Hydroxybenzoic acid (p-HBA), used in the synthesis, does not by itself have sufficient reactivity and undergoes decarboxylation at higher temperature i.e. above 200°C. Two synthetic routes were, therefore, followed depending upon whether p-HBA was taken in the form of (i) acetate i.e. acetoxy route[8] or (ii) phenyl ester i.e. phenylester route. The typical polymerization procedure was as follows:

In a four necked 1 L flask 0.25 mol of HQ, 0.25 mol of p-HBA and 0.25 mol of isophthalic acid were charged along with 2.92 mol of acetic

Frontiers of Polymer Research, Edited by P.N. Prasad and J.K. Nigam, Plenum Press, New York, 1991

anhydride and 1% sodium phosphate and liquid paraffin as heat transfer media. The reaction mixture was refluxed initially at 140°C for 4 hrs. Temperature was increased with slow and controlled heating in steps along with distillation side by side. Maximum temperature attained was 330°C and the total time of various steps (140°C - 200°C, 200°C - 280°, 280°C - 300°C and 300°C - 330°C) was 8 hrs. The copolyester formed was taken out and extracted with acetone to remove unreacted monomers.

CHARACTERIZATION

The thermal properties were determined by Dupont DSC 1090 thermal analyser. Thermogravimetric analysis was carried out on Stanton Redcroft TG - 770 thermobalance. DSC studies were carried out in nitrogen atmosphere and at a heating rate of 20°C/min. Transition temperatures were taken corresponding to maxima of enthalpy peaks. The liquid crystalline texture was observed on a Leitz Laborlux - 12 POL Polarizing microscope equipped with a Mettler FD 82 hot stage. Thin flim of 6 - 10 micron thickness were made between glass slides and coverslips at melting temperature for texture studies.

TABLE 1. Properties of Synthesised Copolyesters

Copo-lyester	Composition in Mol %			Temperature (°C)			Texture	Synthetic Route
	p-HBA	IA	HQ	Tg	Tm	Ti		
A	33.3	33.3	33.3	104	269	300	nematic	Acetoxy
B	33.3	33.3	33.3	107	265	337	nematic	Phenylester
C	36.6	36.6	26.4	117	246	307	nematic	Acetoxy
D	36.6	36.6	26.4		350	384	nematic	Phenylester
E	44.4	44.4	11.1	202	300		nematic	Acetoxy
F	40.0	40.0	20.0		239		nematic	Acetoxy
G	40.0	20.0	40.0	no endotherm			no texture	Acetoxy
H	40.0	20.0	40.0		331		no texture	Phenylester
I	36.6	26.4	36.6	205	no endotherm		no texture	Phenylester
J	36.6	26.4	36.6	no endotherm			no texture	Acetoxy

RESULT AND DISCUSSION

(a) Thermal behaviour : - All copolyesters were investigated by thermogravimetric analysis between room temperature and 600°C and found to be thermally stable upto 320°C, a 50% weight loss being detected between 460 - 500°C. The value of glass transitions and melting transitions are given in table - 1. It is evident from the table that copolyesters containing at least 66 mol% and a maximum of 88% of POB and IA units, taken together are crystalline and exhibited melting endotherms (Tm the melting temperature on DSC thermogram). Copolyesters with less than 66 mol % of POB and IA units, either showed very small endotherm (Tm) or it is completely missing. This clearly indicates that these copolyesters are amorphous in nature. It was also found that IInd endotherm of crystalline copolyesters corresponds to the temperature of

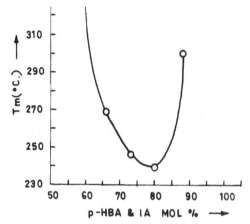

Fig. 1.   Effect of Combined MOL % of p-Hydroxy
          Benzoic Acid(p-HBA) and Isophthalic
          Acid(IA) on Melting Points of
          Copolyesters.

of isotropization (Ti) and the same was also confirmed by texture
studies.   Glass transitiions(Tg), observed in some copolyesters were
above 100°C.

    Fig. 1 shows that as the mol% of mesogenic units increases,
melting point of the copolyester decreases, being minimum at 80 mol %
of POB and IA units. But above this mol %, melting point increases.
It may be due to the increased % of POB unit which is rigid as compare
to IA unit present in the copolyester in same amount.

    (b) Optical behaviour : - Optical texture of copolyester (A & B)
are shown in Fig. 2.   Highly dense particle type texture was
characteristic of crystalline copolyesters.   The mesophase was
identified as nematic. Fig. 2 a shows mesophase of copolyester - B

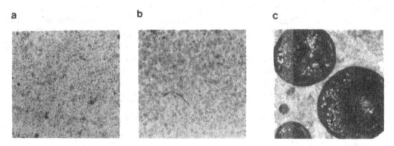

Fig. 2. Optical texture of copolyesters (megnification 100)
        (a) Nematic phase of copolyester B at 266°C (Tm)
        (b) Isotropic phase of copolyester B at 335°C (Ti)
        (c) Mesophase of copolyester A showing rings in nematic
            phase at 290°C.

containing 66 mol % of POB and IA unit at 266°C (Tm) and Fig. - 2b shows isotropic phase of the same copolyester at 335°C (Ti). It is clear that between 266°C (corresponding to Tm) and 335°C (corresponding to Ti), the copolyester is in mesophase i.e. anisotropic phase. At 335°C, it changes into an isotropic liquid which becomes completely developed on further heating. Some of the copolyesters showed rings in their mesophase (Fig-2c). The size of the ring gradually increased with time and they could freely move in the nematic phase. The emergence of these rings, according to Mackley et al.[7] was associated with some form of chemical or physical segregation process due to difference in chemical composition or molecular weight of the copolyester.

It has been observed that copolyesters containing less than 66 mol % of POB and IA unit did not develop well defined texture on heating. It is evident from table No. 1 that they did not show any transition in DSC thermograms. The mol % of nonmesogenic rigid-diol (HQ) was also found to influence the thermal transitions and consequently the optical behaviour of these copolyesters. As the content of HQ increases from 33 mol%, the copolyester becomes highly rigid in nature, similar behaviour of HQ unit was reported by Tsai et al.[10] at a mol % of 80.

CONCLUSION

Among the various copolyesters synthesised, all copolyesters containing more than 66 mol % of POB and IA units taken together are thermotropic. They all show nematic particle type texture on heating which are reproduced on cooling. Jin et al.[11], however, found that compositions with more than 65 mol % of the p - oxybenzoyl unit (POB) exhibit thermotropic behaviour. In the present study, it has also been observed that copolyesters more than 33 mol % of non-mesogenic rigid-diol (HQ) unit, are rigid in structure.

Therefore, it can be drawn that it is the total mol % of mesogenic units present in the copolyester that affect liquid crystallinity.

REFERENCES

1. C.W. Calundann and M. Jaffe, "Anisotropic Polymers, their Synthesis and Properties", Proceed. of the Robert A Walch Conference on Chemical Research XXVI, Synthetic Polymer (1982).
2. W.J. Jackson, Liquid Crysal Polymer XI. Liquid Crystal Aromatic Polyesters: Early History and Future Trends, Mol. Cryst. Liq. Cryst., 169 : 23 (1989).
3. T.S. Chung, The Recent Deavelopment of Thermotropic Liquid Crystalline Polymers, Polym. Engin. Sci., 26 (13) : 901 (1986).
4. J.I. Jin, E.J. Choi and S.C. Ryu, Thermotropic Polyesters with Main Chain Isomeric Naphthalene Trans - 1,4 - Cyclohexylene and 2,5 - Pyridinediyl Units : The Importance of the Order of Ester Linkage in The Mesogenic Unit, J. Polym., Chem. Ed. 25 : 241 (1987).
5. J.I. Jin and E.J. Choi, Properties of Poly (p-Phenylene Terephthalates) Prepared from 2 - Nitro and 2 - Bromoterephthalic Acids and Substituted Hydroquinones, Macromolecules, 20 : 934 (1987).
6. W.R. Krigbaum, H. Hakemi and R. Kotex, Non Mesogenic Polymers Having Rigid Chain 1. Substituted Poly (p-Phenylene terephthalates), Macromolecules 18 : 965 (1985).
7. C.K. Ober and T.L. Bluhm, Thermotropic Liquid Crystalline Polyesters Containing Naphthalenic Mesogenic Groups, Polym. Bull., 15:233 (1986).

8. O.D. Deex, Liquid Crystal Copolyester, U.S. Pat, 444980 (1984).
9. M.R. Mackley, F. Pinaud and G. Siekmann, Observation of Disclinations and Optial Anistoropy in a Mesomorphic Copolyester, Polymer, 22 : 437 (1981).
10. H.B. Tsai, C. Lee, N.S. Chang and A.T. Hu, Thermotropic Copolyesters Modified with Non-Mesogenic Regid Groups, Polym.Bull., 24: 293 (1990).
11. J.Jin, S. Lee, H. Park and J. Kim, Liquid Crystalline Properties of the Copolyesters Prepared from Resorcinol, p-Hydroxybenzoic acid and Terephthalicacid, Polym. J. 21 : 615 (1989).

INDEX

Printed in the United States
by Baker & Taylor Publisher Services